The Molecular Astrophysics
of Stars and Galaxies

International Series on Astronomy and Astrophysics

Series Editors

A. Dalgarno   M. Davis   G. Efstathiou
N. Kaifu   G. Morfill

---

1. E. N. Parker: *Spontaneous current sheets in magnetic fields with applications to stellar X-rays*
2. C. F. Kennel: *Convection and substorms: paradigms of magnetospheric phenomenology*
3. L. F. Burlaga: *Interplanetary magnetohydrodynamics*
4. T. W. Hartquist and D. A. Williams (eds): *The molecular astrophysics of stars and galaxies*

The Molecular Astrophysics of Stars and Galaxies

Thomas W. Hartquist
Max-Planck-Institut für extraterrestrische Physik
and
Department of Physics and Astronomy, University of Leeds

and

David A. Williams
Department of Physics and Astronomy
University College London

CLARENDON PRESS · OXFORD
1998

Oxford University Press, Great Clarendon Street, Oxford OX2 6DP
Oxford New York
Athens Auckland Bangkok Bogota Buenos Aires Calcutta
Cape Town Chennai Dar es Salaam Delhi Florence Hong Kong Istanbul
Karachi Kuala Lumpur Madrid Melbourne Mexico City Mumbai
Nairobi Paris São Paolo Singapore Taipei Tokyo Toronto Warsaw
and associated companies in
Berlin Ibadan

Oxford is a registered trade mark of Oxford University Press

Published in the United States by
Oxford University Press Inc., New York

© *Oxford University Press, 1998*

All rights reserved. No part of this publication may be
reproduced, stored in a retrieval system, or transmitted, in any
form or by any means, without the prior permission in writing of Oxford
University Press. Within the UK, exceptions are allowed in respect of any
fair dealing for the purpose of research or private study, or criticism or
review, as permitted under the Copyright, Designs and Patents Act, 1988, or
in the case of reprographic reproduction in accordance with the terms of
licences issued by the Copyright Licensing Agency. Enquiries concerning
reproduction outside those terms and in other countries should be sent to
the Rights Department, Oxford University Press, at the address above.

This book is sold subject to the condition that it shall not,
by way of trade or otherwise, be lent, re-sold, hired out, or otherwise
circulated without the publisher's prior consent in any form of binding
or cover other than that in which it is published and without a similar
condition including this condition being imposed
on the subsequent purchaser.

A catalogue record for this book is available from the British Library

Library of Congress Cataloging in Publication Data
(Data available)

ISBN 0 19 850158 7

Typeset using LaTeX
Printed by Thomson Press (India) Ltd

This volume is dedicated to
Alexander Dalgarno

Dedication

Some rare individuals of exceptional vision and imagination are able to predict the prospect of new opportunities and to contemplate the arrival of new fields of research. A very few of these far-sighted people have the ability and energy to lay the foundation required and to build much of a new edifice. Alexander Dalgarno is one such energetic and able visionary. His huge contributions to atomic and molecular physics not only constitute a major part of a fundamental science, but they have also provided large parts of the foundations for the exceptional growth of two vast, modern scientific edifices: aeronomy, and molecular astrophysics. In the early 1960s, when only four molecular species were known to exist in space, before the detection of molecular hydrogen or the development of the techniques that would eventually allow the identification of a hundred varieties of interstellar molecules, Alex Dalgarno advised one of the authors of this Dedication to study molecules in astronomy. This piece of advice was gratefully received and closely followed. It certainly wasn't obvious at the time that interstellar space was as beautifully complex in physical and chemical terms as we later learned it to be, but now we all accept that molecules contain the bulk of the non-stellar baryonic matter in the Universe. Molecules are found in all sufficiently dense ($\gtrsim 1$ H nuclei cm^{-3}) objects in the Universe that are as cool or cooler than sunspots.

Alex's achievements are equivalent to the combined outputs of several talented scientists. His recognition of the fundamental importance of atomic and molecular physics led him to establish the Institute for Atomic and Molecular Physics at the Harvard-Smithsonian Center for Astrophysics. The growth and success of this Institute in a difficult financial environment, attracting many to enjoy the opportunity of working with him, has been a source of pride and pleasure to Alex. The Institute has certainly become a powerful—even dominant—influence on the world of atomic and molecular physics during the last decade. Through the Institute Alex has been able to carry forward a huge programme in atomic and molecular physics, a great portion of which is relevant to aeronomy, and to the Jovian atmosphere. His current interests in long range interactions are relevant to the now very exciting areas of very low energy collisions and the creation of Bose–Einstein condensates, but echo his early work in the 1950s. It must be a great pleasure to him to see that the foundations he laid so long ago have once again become of topical interest.

Ten years ago David Bates wrote "There is no greater figure than Alex in the history of atomic physics and its applications" The applications have indeed been phenomenally successful, and—together with their extension into molecular physics—nowhere more so than in astrophysics. Alex, with the school he built at Harvard, showed the way in detailed modelling of the complex physical and chemical systems that are molecular clouds, and honed these techniques so that they could be applied to situations as diverse as supernovae ejecta and the early Universe. He is universally regarded as the father of molecular astrophysics.

The details of Alex's career up to his 60th birthday have been described eloquently in several volumes and Festschrifts to celebrate that happy event. His many achievements, and honours received up to that time, will not be described further here. Honours and responsibilities continue to flow his way.

He has, thankfully, been able to shed some of the more onerous administrative duties. Time for tennis- and squash-playing continues to be a high priority, and the research output never slackens its imperious flow. But time remains for Alex a precious commodity. It has often been remarked how research papers issue perfectly formed from his pen. In these modern times, it is better to remark how Alex's e-mails are exemplary messages: brief, to the point of terseness; yet conveying with instant clarity the essential message, and no more. Would that all e-mails were so phrased! It is said that George Bernard Shaw wrote at the end of a lengthy letter "Excuse this long letter ... I haven't enough energy to write a short one". Alex recognizes the force of that statement but has the energy to ensure that all his thoughts emerge without hindrance of redundancy.

The leading role that Alex occupies in these several branches of science is not a remote one. All who have worked with him are struck with his accessibility, his response to new ideas, and his kindness and consideration to all who try to emulate his standards. Alex cares about the future of the subjects in which he has worked, vis à vis other branches of science. He is concerned about the health of branches of physics and astronomy to which he continues to devote his energy and imagination. He continues to travel widely to promote the sciences that he loves, in a punishing schedule that would drain the energy of a much younger man. He generously adds his inimitable contributions to conferences and workshops worldwide, and is known particularly for his penetrating summings-up—gently deflating overblown claims and pointing out hidden values of new results. Some do indeed find precious new results; if Alex is not himself responsible, he will explore their unexpected implications faster than anyone.

His service to the community is nowhere better exemplified than in his (near) quarter century as editor of *Astrophysical Journal Letters (ApJL)*. This is quite simply, under his leadership, the finest and most prestigious journal in astrophysics. Alex's exacting standards and insistence on deadlines ensure the highest quality and timeliness. Alex is reputed to act sometimes with piratical fervour on behalf of *ApJL*. It has been reported to the writers that a very important preprint, sent out of courtesy to him, was accepted by return post for *ApJL* while the authors were preparing a submission to another journal. His record of public service is daunting. Since his 60th birthday, he has been on various national committees (including the Executive Committee of the APS, and the National Committee of the IAU), and has been a member of numerous visiting committees both in the US and abroad.

The continued very high regard and affection in which he is held is testified by the large number of prestigious invitations to lecture. Not including a number of very prestigious events in the USA, he has since 1988 been Marchon Lecturer, University of Newcastle upon Tyne, Elizabeth Laird Memorial Lecturer, University of Western Ontario, Spiers Memorial Lecturer at the Faraday Division of the Royal Society of Chemistry, Sir Joseph Larmor Lecturer at the Queen's University of Belfast, and Mortimer and Raymond Sachler Lecturer at Tel Aviv University. He has received the Spiers Memorial Medal of the Royal Society of Chemistry (1992) and the John A. Fleming Medal of the American Geophysical Union (1995). He has been elected an Honorary Member of the Royal Irish Academy (1989), a Fellow of UMIST (1992), and a Member of the International Academy of Quantum Molecular Science (1993).

At the meeting on "Molecules and Grains in Space", Mont St Odile, 1994, Alex Dalgarno concluded a masterly summary of the state of the subject (in particular, the PAH hypothesis) with the following words: "A combination of laboratory measurements, astronomical observations and theoretical calculations in a continuing dialogue between chemistry and astronomers ... is needed to answer the question". He has always sought to bring these communities of scientists together. Let us hope that we shall be celebrating the fruits of his labour in these communities in a decade's time.

Preface

In the past decade the field of molecular astrophysics has developed into one that impacts on almost all branches of astrophysics. While important work on interstellar molecular clouds remains to be done, major advances in investigations of the chemistries of whole classes of objects, that a decade ago had been considered only in preliminary ways by molecular astrophysicists, have greatly broadened the field. Now such topics as the chemistry in outflows of young stellar objects, the formation of dust in and its effects on the winds of highly evolved stars, molecules in supernovae and masers in the discs around black holes at the centres of active galactic nuclei are sufficiently well developed that molecular emissions serve as diagnostics of the dynamics of such sources, and the roles of chemistry in governing the dynamics in many of them are subjects of increasingly sophisticated theoretical models and detailed observational study.

The contents of this volume reflect the change in the nature of molecular astrophysics as a field. The interstellar medium of our Galaxy receives attention here, but dominates the subject matter of only a couple of chapters. The few chapters in which galactic interstellar sources figure heavily are of considerable relevance to other varieties of objects as well. The largest group of chapters in the book deals with evolved stars. In a decade's time the major part of a book surveying molecular astrophysics may very well concern starburst galaxies and active galactic nuclei, which are the subjects of several chapters in this work.

In addition to presenting a balanced overview of the field, we wanted to produce a volume that is useful to a wide range of astrophysicists, to chemists who are not actively involved in astronomical research, and to scientists and postgraduate students wishing to specialize in the field. To help to satisfy such a diverse audience, we have included a short chapter containing a basic introduction to the structures and spectra of simple molecules and a chapter on stellar evolution; in addition, a number of authors have made very concerted efforts to define astronomical terms that are possibly unfamiliar to non-specialists.

The first part of the book consists of two chapters dealing with quantum mechanics. Chapter 1 is the introduction to the structures and spectra of molecules. Many of the advances in molecular astrophysics in the last decade have concerned the chemistries of initially dust-free environments, such as young stellar outflows, winds of evolved stars, novae and supernovae. Radiative processes and radiative association in particular are important for the initiation of the chemistries in such environments. Theoretical results for the rates of radiative processes important in dust-free environments have become available in the last couple of years, and Chapter 2 contains a review of them. Many other classes of reactions are introduced and described in Chapter 4, but due to limitations of space and the existence of previous reviews on the theory (including that by D. R. Bates in the 1990 volume honouring Alex Dalgarno edited by T. W. Hartquist) and on laboratory measurements (see that by D. Smith, N. G. Adams and E. E. Ferguson in the same volume) we have not included chapters on theoretical or experimental studies of those reactions. Particularly important advances in recent years in the understanding of neutral–neutral reactions at very low temperatures ($\ll 100$ K) have been described by I. R. Sims and I. W. M. Smith (referenced in Chapter 4). Significant progress in the study of branching ratios in dissociative recombination has also been made as reported by T. L. Williams, N. G. Adams, L. M. Babcock, C. R. Herd and M. Geoghegan (also referenced in Chapter 4). Other recent developments in the studies of reactions important in molecular astrophysics were reviewed by several authors in the proceedings of the 178th International Astronomical Union Symposium edited by E. F. van Dishoeck (also see Chapter 4). A large critically evaluated set of rate coefficients of a wide variety of reactions have been prepared by T. J. Millar, P. R. A. Farquhar and K. Willacy (referenced in Chapter 8), as are papers by A. Sternberg and A. Dalgarno and by W. G. Roberge, D. Jones, S. Lepp and A. Dalgarno who gave rates for a number of photoionization and photodisssociation reactions. Many other sources of specific reaction rate data of particular significance are cited throughout the volume.

The second part of the book concerns chemistry during the formation of galaxies and present-day stars. Chapter 3 covers molecular processes in the Universe before elements more massive than lithium became abundant; the formation of H_2, which acted as a major coolant, affected the types of structures to form as the first galaxies were born. Chapter 4 is the largest in the volume and is a thorough and broad exposition on the chemistry of dusty media with elemental abundances similar to those that obtain throughout much of the Galaxy's interstellar medium; its contents are relevant to many topics addressed by this volume's other authors and must be mastered before many of the other chapters can be fully appreciated. Chapter 5 is a description of theoretical and observational investigations of molecules in star forming regions; a good fraction of that chapter is focused on the means by which chemistry controls the rate of collapse during star formation and how the interpretation of signatures (displayed in molecular emission features) of infall in star forming regions must be refined before observational results can be used to infer the dynamics of stellar birth.

Even during the formation process, young stars affect their environments by emitting radiation and losing mass in powerful winds. The third part of the volume is devoted to the topics of the outflows of young stars and the response of the cloud material to stars. Chapter 6 is a presentation of a theory of winds driven from young stars having masses comparable to the Sun; the mass lost actually comes from a disc whose structure is affected by the rotation of the central star's magnetic field, and at large distances from the star the flow is collimated into a jet. The winds of such stars are sources of molecular emissions from which the mass-loss rates and the early evolution of the stars can be inferred; the chemistry of the winds is the subject of Chapter 7. The winds drive shocks into interstellar material that did not accrete on to the disc or the star; the physical structures of such shocks and the chemistry in heated gas behind them are addressed in Chapter 8. Chapter 9 is an exposition of the physical properties and chemistry in regions affected more by the absorption of ultraviolet radiation from a central source than by its mass loss. Herbig–Haro objects, which constitute the subject of Chapter 10, are optical emission regions near young stars and are formed by the interactions of the collimated stellar outflows with the surrounding interstellar material; efforts to model the Herbig–Haro objects and the H_2 infrared emission from them have been informed by some of the considerations presented in Chapters 8 and 9.

High mass loss rates characterize highly evolved stars as well as young stars, and the chemistry in outflows from highly evolved stars is the theme of the work's fourth part. Most of the chapters in this part of the book are about stars with masses comparable to that of the Sun, but Chapter 17 is relevant to stars that are considerably more massive. Chapter 11 is an introduction to the evolution of stars and provides descriptions of the different types of evolved stars of interest to molecular astrophysicists. The chemistry in the outflows of many evolved stars gives rise to the formation of dust, and as described in Chapter 12 the pressure of stellar radiation on dust formed in a stellar envelope has huge effects on the mass loss rate and the terminal wind speed. That chapter concerns stars with envelopes containing more carbon than oxygen. Chapter 13 concerns stars with envelopes containing more oxygen than carbon and presents a number of new results on the chemistry leading to the production of dust, a topic that is probably less well understood for oxygen-rich envelopes than carbon-rich. In oxygen-rich envelopes, chemistry produces SiO and H_2O shortly before dust forms, and excitation conditions of these molecules result in maser emissions in some lines of these species. As material moves further from a star, radiation in the interstellar medium will dissociate H_2O to form OH, and OH maser features are also seen. Models of maser emissions from such stars is the subject of Chapter 14. Masers also form near some very massive young stars but these sources are not treated in this book; however, masers in starburst galaxies and active galaxies are mentioned in several chapters in the final part of this book. As mentioned above the radiation field in the interstellar medium dissociates molecules in some regions of evolved star's outflows; Chapter 15 concerns the rich chemistry initiated by that dissociation and observed consequences of it. As the central star around which an envelope is outflowing evolves, the star may, depending on its mass, begin to emit enough ionizing radiation to photoionize all of the envelope not contained in condensed clumps; such ionization leads to the births of extended nebulae called planetary nebulae. The chemistry that gives rise to the molecules that trace the evolution of planetary nebulae is addressed in Chapter 16. Chapter 17 is a review of the very new subject of dust formation near stars much hotter than the evolved stars addressed in Chapters 12 and 13. The existence of dust in some of these environments was unexpected, and a surprise when it was first discovered.

More violent mass loss from stars is associated with novae and supernovae, the topics of the penultimate part of the volume. Chapter 18 is concerned with novae, and Chapter 19, supernovae. Theoretical work on chemistry in supernovae has resulted in the inference, from observed molecular emissions, that instabilities occurring during the supernova explosion itself distort the interfaces between regions having different nuclear compositions but do not result in "microscopic" inter-region mixing. It is a remarkable testament to the wide range of astronomical conditions to which chemistry is relevant that molecular emissions can be used to study the processes occurring during an explosive event deep in the interior of a star.

The final part of the work is about molecules and dust in starburst galaxies and active galaxies, which are believed to emit a large fraction of their luminosities due to accretion on to black holes having masses of hundreds of millions times that of the Sun. Chapter 20 contains definitions of many of the different types of galaxies and descriptions of what is revealed by maps of molecular emission features about the properties of such galaxies at distances of roughly a hundred parsecs from their centres. It also introduces the topic of masers much nearer to the central black holes. Chapter 21 is focused on the central few parsecs of active galaxies and how radiation from the central regions affects molecular excitation conditions and, thus, the observed emissions. The radiation field of an active galaxy also affects the physical conditions of the gas and its chemical composition; the active galaxies are X-ray sources as well as ultraviolet sources, and Chapter 22 concerns the response of a dusty region to X-rays, a problem that is somewhat related to the subject of Chapter 9. Chapter 22 is relevant to sources in the Milky Way, as well as extragalactic objects, because molecular material around some compact objects in the Galaxy also gives rise to observable X-ray irradiated sources. As described in Chapter 23 water maser emissions arise within the central parsec of some active galaxies; these emissions allow the inference of the rotation velocities at some points in some discs around the central engines of two active galaxies, allowing estimates of the masses of their black holes to be made. This volume closes with Chapter 24 which has as its topic the influence of an active galaxy's radiation field on the chemistry of dust formation in evolved stars near its centre. It is argued that dust formation can be suppressed in roughly the central ten parsecs of some active galaxies.

We hope that by the time that a reader has looked at least briefly at all of the chapters in this book it will be evident that molecules are extremely relevant to a huge portion of modern astrophysics.

We have been aided immensely by a number of people. Carol Broad typed several versions of most of the chapters of which the editors are coauthors, as well as all the parts of the book that are the editors' sole responsibility; her good humour and warmth often improved a day for us. Deborah Ruffle's assistance with numerous technical problems and devotion of time to the project was far beyond the call of duty. Jonathan Rawlings gave crucial technical help and his scientific advice to the editors is much valued. Vito Graffagnino kindly helped deal with a bothersome technical problem and gave patient and kind instruction, as did Stephen Taylor.

Garching, Germany   T. W. H.
London, England   D. A. W.
November 1997

Contents

Part I Some Relevant Quantum Mechanics

1 The Basics of the Structures and Spectra of Simple Molecules   1
  T.W. Hartquist and S. Viti
2 Molecule Formation in Dust-poor Environments   11
  J.F. Babb and K.P. Kirby

Part II Chemistry at the Births of the Galaxies and Stars

3 Molecules in the Early Universe and Primordial Structure Formation   37
  S. Lepp and P.C. Stancil
4 The Chemistry of Diffuse and Dark Interstellar Clouds   53
  E.F. van Dishoeck
5 The Chemistry of Star Forming Regions   101
  T.W. Hartquist, P. Caselli, J.M.C. Rawlings, D.P. Ruffle and D.A. Williams

Part III Young Stellar Objects and Herbig–Haro Objects

6 The Magnetohydrodynamics of Outflows from Low-mass Young Stellar Objects   141
  T.W. Hartquist and D.P. Ruffle
7 Chemistry in the Winds of Young Stellar Objects   161
  A.E. Glassgold
8 Shock Chemistry   179
  T.W. Hartquist and P. Caselli
9 Photon-dominated Regions   201
  A. Sternberg
10 Molecular Hydrogen Emission from Herbig–Haro Objects   221
  J.C. Raymond

Part IV Evolved Stars

11 Introduction to Stellar Evolution   237
  D. Schönberner and T. Blöcker
12 Dust Formation in Carbon-Rich AGB Stars   265
  I. Cherchneff
13 Dust Formation in M Stars   285
  H.-P. Gail and E. Sedlmayr

14	Models of Circumstellar Masers D. Field	313
15	Molecular Synthesis in the External Envelopes of AGB Stars T.J. Millar	331
16	The Chemistry of Planetary Nebula Formation D.A. Howe and D.A. Williams	347
17	Dust Formation in the Environments of Hot Stars H. Beck and E. Sedlmayr	371

Part V Novae and Supernovae

18	Dust Formation in Novae J.M.C. Rawlings	393
19	Supernova Chemistry W. Liu	415

Part VI Starburst Galaxies and Active Galactic Nuclei

20	Molecular Gas, Starbursts and Active Galactic Nuclei R.S. Booth and S. Aalto	437
21	Excitation and Detectability of Molecules in Active Galactic Nuclei J.H. Black	469
22	X-ray Dominated Regions S. Lepp and S. Tiné	489
23	Water Molecules in the Circumnuclear Regions of Active Galaxies D.A. Neufeld	507
24	The Suppression of Dust Formation in Evolved Stars near Active Galactic Nuclei T.W. Hartquist, F. Bertoldi, R.H. Durisen, J.E. Dyson, R.J.R. Williams, J.M.C. Rawlings and D.A. Williams	517
	Index	525

PART I

Some Relevant Quantum Mechanics

1
The Basics of the Structures and Spectra of Simple Molecules

T. W. Hartquist
Max-Planck-Institut für extraterrestrische Physik

and

S. Viti
Department of Physics and Astronomy, University College London

1 Introduction

Atomic spectroscopy concerns the electronic states of atoms. Molecular spectra are richer than atomic spectra because they depend on the properties of the motions of the nuclei as well as the electronic characteristics of the systems.

Section 2 of this chapter describes the Born–Oppenheimer approximation which is at the heart of the description of molecular states. Section 3 treats diatomic molecules. When considering polyatomic molecules we focus mostly on their rotational properties because, though (particularly in the local interstellar medium) a number of astronomically abundant diatomic species are detectable in transitions that are between states differing by more than their rotational properties, observations of polyatomic astronomical molecules are made primarily in purely rotational transitions at wavelengths usually (but not always) in the range of a few tenths of millimetres to a few centimetres.

Linear polyatomic molecules constitute the subject of Section 4. Symmetric and asymmetric top molecules are the topics of Sections 5 and 6 respectively.

More detailed treatments of aspects of this chapter's subject may be found in numerous texts and standard monographs including those by Herzberg (1945; 1950), Townes and Schawlow (1955), Davydov (1976), Landau and Lifschitz (1977) and Bernath (1995).

2 The Born–Oppenheimer Approximation

The wavefunction, $\psi_{ij}(\mathbf{R}_1, \mathbf{R}_2, \ldots, \mathbf{R}_m, \mathbf{r}_1, \mathbf{r}_2, \ldots, \mathbf{r}_n)$, of the i,jth state of a molecule depends on the positions of its nuclei, $(\mathbf{R}_1, \mathbf{R}_2, \ldots, \mathbf{R}_m)$ and of its electrons, $(\mathbf{r}_1, \mathbf{r}_2, \ldots, \mathbf{r}_n)$, where m and n are the numbers of nuclei and electrons respectively. The Schrödinger equation for a molecule is

$$\left[\sum_{l=1}^{m}\left(\frac{-\hbar^2}{2M_l}\nabla_{\mathbf{R}_l}^2\right) + \sum_{l=1}^{n}\left(\frac{-\hbar^2}{2m_e}\nabla_{\mathbf{r}_l}^2\right) + V_N + V_e + V_{eN} - E_{ij}\right]\psi_{ij} = 0 \quad (2.1)$$

where $\hbar$, M_l and $\nabla_{\mathbf{R}_l}^2$ are Planck's constant divided by 2π, the mass of the lth nucleus and the Laplacian operator for the lth nucleus. m_e and $\nabla_{\mathbf{r}_l}^2$ are the electron mass and the Laplacian operator for the lth electron. V_N, V_e and V_{eN} are potentials that depend on the nuclear coordinates only, the electron coordinates only and the electron and the nuclear coordinates respectively. E_{ij} is the eigenenergy.

The adiabatic or Born–Oppenheimer approximation is based on the assumption that to a zeroth approximation nuclei are at rest and that their motions may be taken into account at higher approximations. The adiabatic or Born–Oppenheimer approximation is that

$$\psi_{ij} \approx \Phi_{ij}(\mathbf{R}_1, \mathbf{R}_2, \ldots \mathbf{R}_m)\varphi_j(\mathbf{R}_1, \mathbf{R}_2, \ldots, \mathbf{R}_m, \mathbf{r}_1, \mathbf{r}_2, \ldots, \mathbf{r}_n) \quad (2.2)$$

where φ_j is a solution of

$$\left[\sum_{l=1}^{n}\left(\frac{-\hbar^2}{2m_e}\nabla_{\mathbf{r}_l}^2\right) + V_e + V_{eN} - \epsilon_j\right]\varphi_j = 0 \quad (2.3)$$

and ϵ_j is an eigenvalue that depends on the nuclear coordinates only. Φ_{ij} is a solution of

$$\left[\sum_{l=1}^{m}\left(\frac{-\hbar^2}{2M_l}\nabla_{\mathbf{R}_l}^2\right) + V_N + \epsilon_j - E'_{ij}\right]\Phi_{ij} = 0 \quad (2.4)$$

where E'_{ij} is an eigenvalue of eqn (2.4), and if the Born–Oppenheimer approximation is a good one $E'_{ij} = E_{ij}$. In short, eqn (2.3) is solved for one set of nuclear coordinates at a time in order to obtain a function that can be used to approximate the electrons' probability distributions in space when the nuclei are at the assumed coordinates. The solution to eqn (2.3) gives no direct information about the state of the nuclei; however, the dependence of the energy associated with the electronic state (ϵ_j) on the nuclear coordinates does influence the nuclear wavefunction as seen in (2.4). $V_N + \epsilon_j$ is called a potential curve when the molecule is a diatomic and a potential surface when it is polyatomic.

Writing

$$\psi_{ij} = \sum_{kl} a_{kl}\Phi_{kl}\varphi_l \quad (2.5)$$

(with the a_{kl} being constants determined by normalization and the solution of (2.1)) and substituting into (2.1) and using (2.3) and (2.4), one finds that in the adiabatic approximation terms involving derivatives of the φ_l with respect to nuclear coordinates are neglected. In many interesting cases the neglect of these terms is justified. However, in each of many other interesting cases these terms cause a transition between adiabatic electronic states (i.e. a reaction). We will consider cases in which the terms may be neglected.

3 Diatomic Molecules

Figure 1 shows potential curves for several different electronic states of H_2. If a potential curve has a deep enough minimum at a finite separation between two nuclei, the electronic state supports a number of discrete rovibrational levels, the energies of which can be approximated from the solution of (2.4), and a rovibrational continuum. If no minimum exists in the potential curve only a continuum of rovibrational levels is associated with the electronic state. The electronic states are labelled by the multiplicity corresponding to the total electronic spin and by the magnitude, Λ, of the component of the electronic angular momentum along the internuclear axis. (Throughout this chapter all angular momenta are given in units of $\hbar$.) $\Lambda = 0, 1$ and 2 states are called Σ, Π and Δ states respectively. Homonuclear molecules are also either symmetric (g for *gerade*) or antisymmetric (u for *ungerade*) with respect to interchange of the two nuclei. Superscripts + and − indicate respectively that the electronic state does not change sign and does change under reflection at the origin. The ground electronic state of H_2 is a $^1\Sigma_g^+$ state; it has a total electronic spin of 0 and $\Lambda = 0$ and supports discrete rovibrational levels as well as a rovibrational continuum. The first excited electronic state of H_2 is a $^3\Sigma_u^+$; it has a total electronic spin of $\hbar$ (and, hence, is a triplet which accounts for the superscript 3) and $\Lambda = 0$ and has no discrete rovibrational levels.

The nuclear motion in a potential may be separated into radial (vibrational) and angular (rotational) components. The discrete vibrational levels denoted by $v = 0, 1, 2, \ldots$ are separated by about 0.2 eV (2321K) in many diatomics but by about 0.5 eV in the H_2 ground electronic state. The discrete rotational levels labelled by $N = 0, 1, 2, \ldots$ are separated by about $2BN$ (if N is taken to be the rotational quantum number of the higher rotational level) and B is a constant. For H_2 and hydrides $2B \approx 170\,\text{K}$ and for other cosmically abundant diatomic molecules such as CO, $2B \approx 5\text{K}$.

Because it is a homonuclear molecule, H_2 possesses a symmetry which does not characterize heteronuclear diatomics and that requires the entire wavefunction to retain the same sign if all nuclear and electronic coordinates change sign (i.e. a mirror reflection takes place). As a consequence in the ground electronic state, when N is even the sum of the nuclear spins must be 0 and the state is said to be a para-state while when N is odd the sum of the nuclear spins must be 1 and the state is said to be an ortho-state. The statistical weight of a para-$^1\Sigma$ state is $(2N + 1)$ while the statistical weight of an ortho-$^1\Sigma$ state is $3(2N+1)$. Radiative transitions between an ortho-state and a para-state are highly forbidden, as are transitions between an ortho-state and a para-state induced by nonreactive collisions. The proton exchange reaction $H^+ + H_2 \rightarrow H_2 + H^+$ does cause ortho to para transitions (Dalgarno, Black and Weisheit

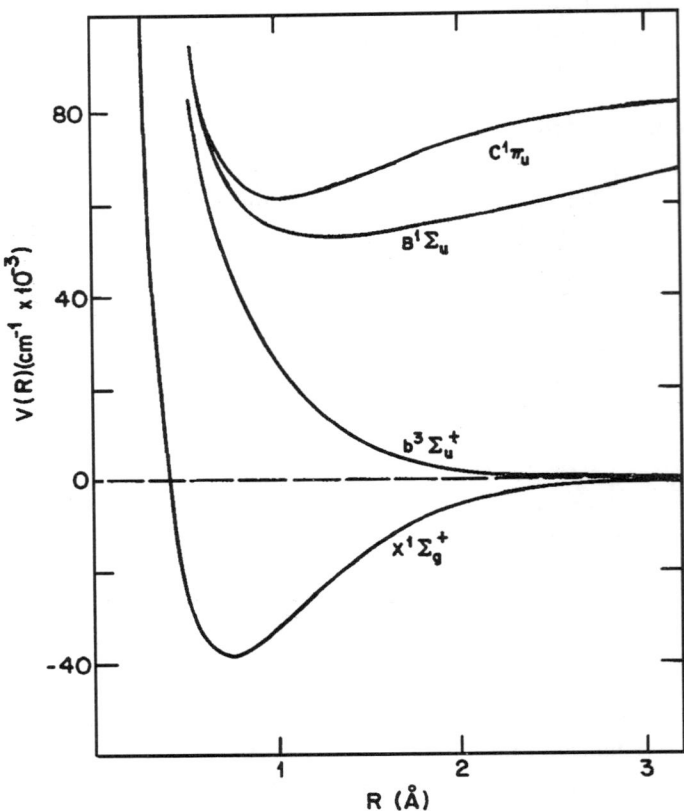

FIG. 1. Potential curves for some low-lying states of H_2.

1973), but in regions of low H^+ number density H_2 often behaves as though it is two separate gases.

In general, the interactions between the rotation of a diatomic molecule and its electronic spin and its electronic angular momentum must be taken into account. Just as electronic spin–electronic orbital angular momentum coupling removes degeneracies in atoms, these rotation–spin–orbital angular momentum interactions do so in molecules. The ground electronic states of CO and CH, like that of H_2, are $^1\Sigma$ states. For a $^1\Sigma$ state the sum, J, of the rotational angular momentum, the electronic angular momentum and the electronic spin is just the rotational angular momentum, and $J = N$ is used to specify the rotational level of the molecule. For example, one refers to the $CO(J = 1 \to 0)$ transition. However, the ground electronic states of some other astronomically important molecules are not $^1\Sigma$ states and interactions between spins and different angular momenta are important. For instance, the ground electronic states of OH and of CH are $^2\Pi$ states. We will consider OH specifically. (For some other molecules it is appropriate to treat the angular momentum coupling differently.) Figure 2 is an energy level diagram of OH in the lowest vibrational level of the ground electronic state. Ω is the magnitude of the sum of the projections of the electronic spin and the electronic orbital angular momentum along the internuclear axis. The ground electronic state of OH has two separate rotational ladders with different values of Ω. One is the $^2\Pi_{3/2}$ ladder, and the other is the $^2\Pi_{1/2}$ ladder, where the subscript is Ω. The lowest rotational level in the $^2\Pi_{3/2}$ ladder has $J = 3/2$, while the lowest rotational level in the $^2\Pi_{1/2}$ ladder has $J = 1/2$. Each J level in the $^2\Pi_{3/2}$ and $^2\Pi_{1/2}$ ladders is split into four sub-levels.

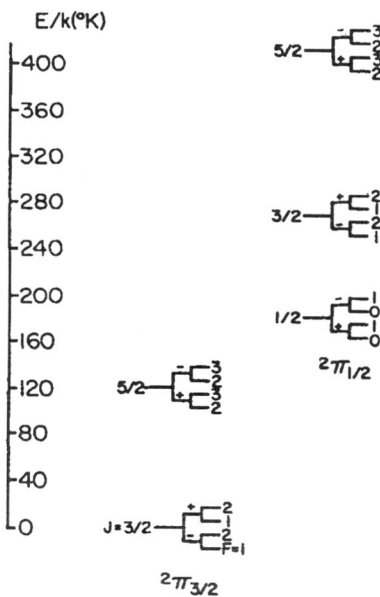

FIG. 2. Rotational levels of OH

One cause of the splitting is the so-called Λ-doubling effect which arises due to the interaction of the electronic orbital angular momentum and spin with the rotation; the energy of the state consequently depends not only on Λ but also on whether the projection of the total electronic angular momentum onto the internuclear axis is positive or negative. Further splitting called hyperfine splitting is due to the interaction of $\mathbf{J}$ and $\mathbf{I}$, the total spin of the nuclei; this causes the energy to depend on $F \equiv |\mathbf{F}|$ with $\mathbf{F} \equiv \mathbf{J} + \mathbf{I}$. Radiation at four different wavelengths near 18 cm is emitted and absorbed in the four transitions between the two hyperfine sub-levels of the upper Λ-doublet level and the two hyperfine sub-levels of the lower Λ-doublet level of the lowest rotational level.

Radiative transitions obey selection rules. For electric dipole transitions $\Lambda \to \Lambda$, $\Lambda \pm 1$ and $g \leftrightarrow u$. Also $+ \leftrightarrow -$. For light molecules changes in the magnitude of the electronic spin are forbidden.

There are no absolute selection rules for changes in vibrational quantum numbers.

For transitions in which the electronic state changes $\Delta J = 0, \pm 1$ unless both the initial and final electronic states are Σ states in which case $\Delta J = \pm 1$; also $J = 0 \to J = 0$ transitions are not allowed.

For transitions in a heteronuclear diatomic molecule in which no change of electronic state takes place $\Delta J = 0, \pm 1$ (except when the electronic state is a Σ state in which case $\Delta J = \pm 1$ or when the initial state has $J = 0$ in which case $\Delta J = +1$). For transitions in a homonuclear molecule in which no change of electronic state occurs $\Delta J = 0, \pm 2$ except when initially $J = 0$ in which case $\Delta J = +2$ or when initially $J = 1$ in which case $\Delta J = 0, +2$. A transition in which the J of the more energetic state is 2 greater, 1 greater, the same as, 1 less or 2 less than the J of the less energetic level is called an S, R, Q, P or O branch transition.

4 Linear Polyatomic Molecules

In the ground electronic states of many linear polyatomic molecules the projection of the electronic orbital angular momentum along the internuclear axis is zero and the electrons are paired. Thus, we will consider first those linear molecules for which the total angular momentum minus the nuclear spins, **J**, is equal to **N**, the angular momentum associated with the rotation of the molecule. Then the energy levels of the lowest vibrational level of the ground electronic state are given approximately simply by $BJ(J+1)$ where B is a rotation constant. In nonsymmetric molecules, such as HCO^+ or HCN (which are linear when in their ground electronic states but have nonlinear configurations in some excited electronic states), the radiative transitions within the lowest vibrational level of the ground electron state lead to $\Delta J = \pm 1$. Symmetric linear molecules, such as $O = C = O$, have no dipole allowed transitions between rotational levels in this vibrational level.

C_2S is a linear molecule but has a $^3\Sigma$ ground electronic state (Saito et al. 1987) which causes its rotational levels to be labelled differently than those of the more usual $^1\Sigma$ ground electronic states of linear molecules. The notation J_N indicates the magnitude of the total angular momentum given by the sum of the electronic spin and the molecular rotational angular momentum **N**. $J = N \pm 1$ except for $N = 0$ in which case $J = 1$.

In a linear molecule vibration in a "stretching mode" is along the length of the molecule while vibration in a "bending mode" is perpendicular to it. Two quantum numbers specify a bending mode; one is associated with the radial coordinate in a cylindrical coordinate system while the other is associated with the angular coordinate.

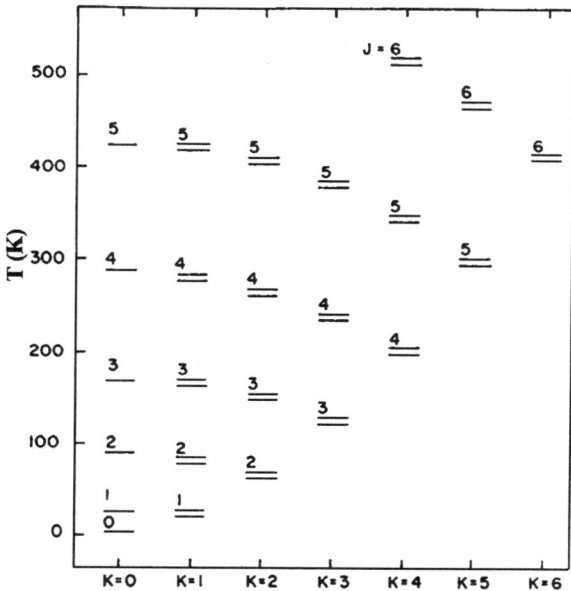

FIG. 3. Partial energy level diagram of rotation states for NH$_3$. J is the total angular momentum quantum number. K is the projected angular momentum along the molecular axis. The splitting of inversion levels is shown.

5 Symmetric Tops

The rotational level energies of a nonlinear polyatomic molecule in its lowest vibrational level of its ground electronic state can usually be fitted well by the energies of rigidly rotating tops. A top has three principal moments of inertia I_A, I_B and I_C.

Spherical top molecules (e.g. CH$_4$) are those for which $I_A = I_B = I_C$, but they do not have dipole moments, so that transitions between rotational levels in the ground electronic states are weak.

For symmetric tops $I_A = I_B$; for oblate symmetric tops $I_A < I_C$ and for prolate symmetric tops $I_A > I_C$. We assume the electronic orbital angular momentum and electronic spin to be zero. Then the energies of a symmetric top are given by

$$E(J,K) = \frac{\hbar^2}{2I_A}J(J+1) + \frac{\hbar^2}{2}\left(\frac{1}{I_C} - \frac{1}{I_A}\right)K^2 \qquad (5.1)$$

with K taking integer values from $-J$ to $+J$.

K is the projection of $\mathbf{J}$ onto the axis about which I_C is the moment of inertia. For a given J, levels with $K = |K|$ and $K = -|K|$ are degenerate.

The inversion properties of a symmetric top molecule are often important. Inversion or reflection about the origin means that the sign of all of the coordinates

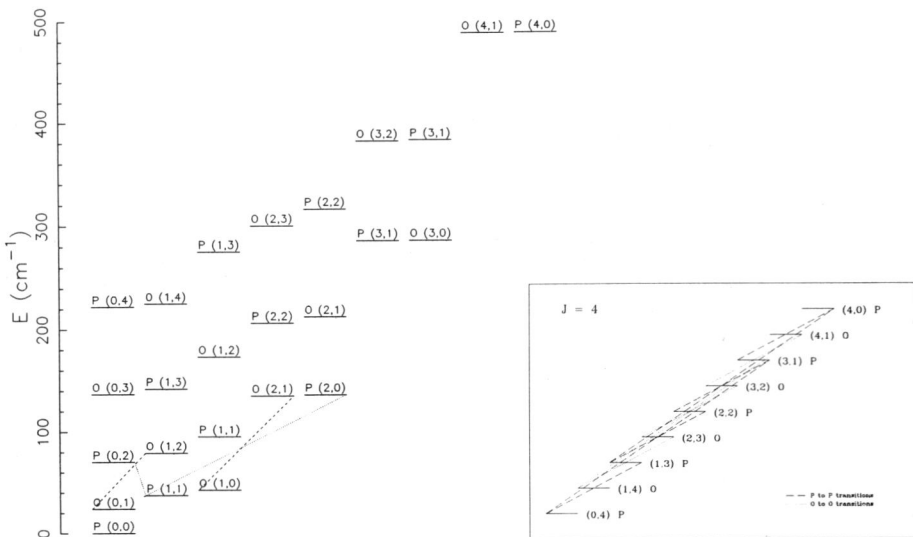

FIG. 4. Partial energy level diagram for H_2O from $J = 0$ to $J = 4$. The (n,m) notation is used to give K_A, K_C. P and O labels identify para-states and ortho-states respectively. The allowed transitions from $J = 1$ to $J = 2$ are indicated. In the right hand box, the allowed $\Delta J = 0$ transitions for $J = 4$ are also shown.

are changed. As stated in Section 3, a state for which a wavefunction maintains the same sign under inversion is a + state, and one associated with a change of sign is a − state. NH_3 is an example of a symmetric top molecule for which the inversion properties of states are important for astrophysical spectroscopy. Along the symmetry axis, a potential maximum obtains when the N atom is located in the plane defined by the H atoms. While (because of the existence of this maximum) in a classical description of a low energy state of NH_3 the N atom would be on one side of the H atom plane or the other, in a quantum mechanical description the N atom has an equal probability of being on either side of the plane. Then, the NH_3 wavefunction must be either of the + or − type. Because the + wavefunction need not give a zero probability of the N atom being in the plane of the H atoms and the − wavefunction must, the − state is of lower energy than the + state for a finite midplane potential maximum. Considerations of symmetry and nuclear statistics show that $K = 0$ levels are not split into inversion sub-levels, but all other (K, J) levels are. Each of the inversion levels of NH_3 is further split into hyperfine sub-levels by the interaction of nuclear moments with **J** and with each other (e.g. Ho and Townes 1983). Figure 3 is a partial energy level diagram of NH_3. Observed NH_3 inversion transition features occur at wavelengths around 1.3 cm.

In a symmetric top molecule dipole transitions connect + states to − states but do not connect + states to + states or − states to − states, and $\Delta K = 0$ and $\Delta J = 0, \pm 1$ unless initially $J = 0$ in which case $\Delta J = +1$.

Table 1 Dipole transition rules for an asymmetric molecule. The $\Delta J = 0$ transitions correspond to the Q branch, while the $\Delta J = \pm 1$ transitions produce the R and P branches. If the dipole moment is parallel to the A axis (signified by $\|A_{axis}$), the strongest transitions are the ones with $\Delta K_A = 0$ and $\Delta K_C = \pm 1$. If the dipole moment is parallel to the B axis, the strongest transitions are the ones with $\Delta K_A = \pm 1$ and $\Delta K_C = \pm 1$. Finally if the dipole moment is parallel to the C axis, the strongest transitions correspond to the $\Delta K_A = \pm 1$ and $\Delta K_C = 0$ ones.

$\Delta J = 0, \pm 1$		
Dipole		
$\|A_{axis}$	$\|B_{axis}$	$\|C_{axis}$
$\Delta K_A = 0, \pm 2, \pm 4, \ldots$	$\Delta K_A = \pm 1, \pm 3, \pm 5, \ldots$	$\Delta K_A = \pm 1, \pm 3, \pm 5, \ldots$
$\Delta K_C = \pm 1, \pm 3, \pm 5, \ldots$	$\Delta K_C = \pm 1, \pm 3, \pm 5, \ldots$	$\Delta K_C = 0, \pm 2, \pm 4, \ldots$

6 Asymmetric Tops

Asymmetric top molecules (e.g. H_2O and H_2CO) are nonlinear rotors with $I_A \neq I_B \neq I_C$. The Schrödinger equation for the asymmetric top has no general analytic solution. Energy levels and wavefunctions are computed numerically for a suitable potential surface.

Levels are specified with the quantum numbers $J_{K_A K_C}$ where K_A and K_C are the projections of **J** on the A and C principal axes respectively. The dipole transition rules are summarized in Table 1. Figure 4 shows a partial energy level diagram of H_2O which is an asymmetric top molecule with its dipole moment parallel to the B axis. Like H_2, H_2O possesses ortho and para modifications as a consequence of the symmetry arising from the identity of the two H nuclei; radiative transitions between an ortho-state and a para-state are forbidden.

Bibliography

1. Bernath, P. F. (1995). *Spectra of atoms and molecules*. Oxford University Press, Oxford.
2. Dalgarno, A., Black, J. H. and Weisheit, J. C. (1973). *Ap Lett*, **14**, 77.
3. Davydov, A. S. (1976). *Quantum mechanics*. Pergamon Press, Oxford.
4. Herzberg, G. (1945). *Molecular spectra and molecular structure – II. Infrared and Raman spectra of polyatomic molecules*. Van Nostrand, New York.
5. Herzberg, G. (1950). *Molecular spectra and molecular structure – I. Spectra of diatomic molecules*. Van Nostrand, New York.
6. Ho, P. T. P. and Townes, C. H. (1983). *Ann Rev Astron Ap*, **21**, 239.

7. Landau, L. D and Lifschitz, E. M. (1977). *Quantum mechanics*. Oxford, Pergamon Press.
8. Saito, S., Kawaguchi, K., Yamamoto, S., Ohishi, M., Suzuki, H. and Kaifu, H. (1987). *Ap J*, **317**, L115.
9. Townes, C. H. and Schawlow, A. L. (1955). *Microwave spectroscopy*. McGraw-Hill, New York.

2
Molecule Formation in Dust-poor Environments

J. F. Babb and K. P. Kirby
Harvard-Smithsonian Center for Astrophysics

1 Introduction

Molecules have been found to exist in a number of astrophysical environments that at first glance would appear hostile to their formation or survival. Regions which are observed to have negligible amounts of dust and/or are subjected to strong radiation fields such as the ejecta from novae (cf. Chapter 18), supernovae (cf. Chapter 19), and certain stellar winds (cf. Chapters 7, 12, 13 and 17) contain molecules. No detections of primordial molecules have been made as yet. However, theoretical studies of the early Universe to examine the evolution of primordial structure leading to the first astronomical objects show the important role that molecules may have played in initiating and sustaining gravitational collapse (cf. Chapter 3). In this chapter the main processes involved in making molecules under the unusual conditions that obtain in these sources will be described and their importance assessed.

Molecule formation in dust-poor environments often takes place through two-body association reactions, as densities are usually too low to allow for the more common ternary association reactions. In order to conserve momentum in the formation of a molecule from two colliding species, either a photon or an electron must also be given off. The primary process considered here is radiative association in which two species, A and B, approach each other and in the course of the collision emit a photon, thereby stabilizing the collision complex:

$$A + B \rightarrow AB^* \rightarrow AB + h\nu. \tag{1.1}$$

The photon may be emitted spontaneously or its emission may be stimulated by the ambient radiation field. So far, in the several cases discussed to date, the stimulated process does not contribute significantly to the overall molecule formation rate (Stancil and Dalgarno 1997a, 1997b). In the simplest case A and B are atoms, but they can also be neutral molecules. In certain cases A or B may be ionized, and then the process of radiative association may compete with radiative or non-radiative charge transfer. Usually A and B are considered to be in their respective ground electronic states. However, if one collision partner is in an excited electronic state due to some sort of optical or collisional pumping mechanism, the rate for radiative association may be considerably enhanced. All of these particular cases will be considered in greater detail and examples of each will be given later in this chapter.

Radiative association is currently thought to play a significant role in the formation of large molecules in dense interstellar clouds (cf. Herbst 1985; Bates 1987; Herbst and Dunbar 1991). This is because the rate generally increases with reactant complexity. In diffuse clouds radiative association of C^+ and H may be a source of CH^+ and hydrocarbon chemistry. The chemistry of the interstellar medium, however, will not be discussed further in this chapter.

Another two-body association mechanism, which plays an important role in molecular hydrogen formation in situations where there are no grains present, is associative detachment:

$$A^- + B \to AB + e^-. \tag{1.2}$$

In order for this process to be important, particular conditions which allow for the formation and stability of the negative ion must exist. Of the first-row elements in the periodic table, H, Li, B, C, F and O have positive electron affinities, indicative of the stability of the associated negative ions. This mechanism appears to be very limited in its astrochemical impact, with the exception of H_2 formation in dustless environments.

Finally, there is the two-body process, associative ionization, which usually must involve an electronically excited collision partner:

$$A^* + B \to AB^+ + e^-. \tag{1.3}$$

This process can in some cases clearly compete with radiative association. Detailed numerical calculations are often necessary to assess the relative importance of these processes, and to obtain branching ratios as a function of collision energy.

In this chapter, the theoretical foundations of the radiative association process are introduced in Section 2. The natures of associative detachment and associative ionization processes are briefly described in Section 3. Section 4 discusses the role of these molecule formation mechanisms in the chemistry of the early Universe and supernovae.

Very few rate coefficients or cross sections for the two-body association reactions listed above have been measured reliably in the laboratory. At normal laboratory densities, measurements of these processes are usually swamped by three-body association reactions. Most of the information on rates of formation of individual molecules comes from detailed theoretical calculations, an area in which Professor Alex Dalgarno has been a leader. The papers of Dalgarno, his students and collaborators dominate the literature, exploring the radiative association process and its roles in molecule formation in SN 1987A, the early Universe, and planetary atmospheres. He is a continuing source of inspiration for all of us who enjoy collaborating with him.

2 Theory of Radiative Association

For the radiative association process there is both a *direct* and an *indirect* mechanism, the latter sometimes known also as inverse predissociation. These

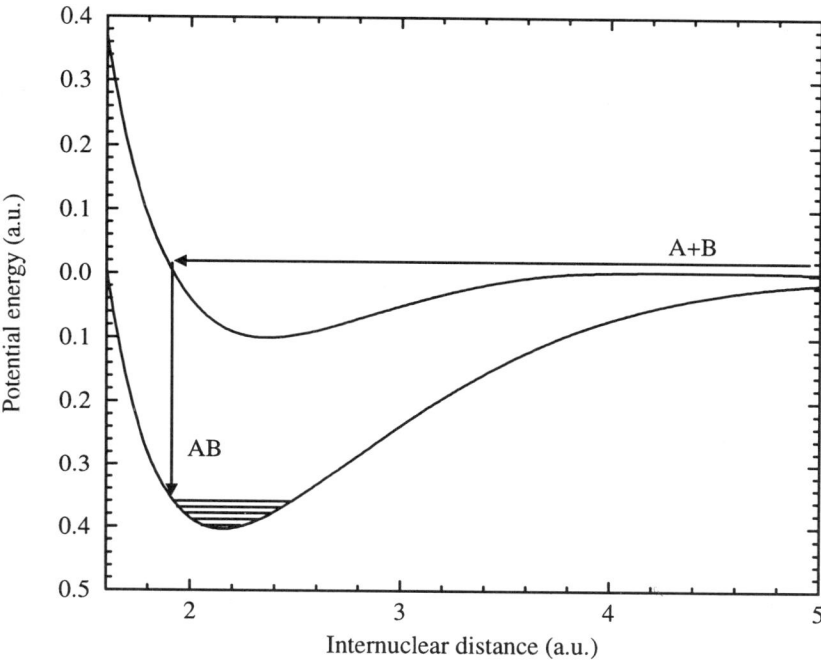

FIG. 1. Illustration of the direct radiative association process. The two atoms A and B approach in the vibrational continuum of an excited electronic state and emit a photon, thereby forming a molecule in a vibrational–rotational state. The data plotted are the CO $A^1\Pi$ (excited) and $X^1\Sigma^+$ (ground) electronic states.

mechanisms will be discussed separately in this section because their theoretical descriptions are quite different. Each of these mechanisms has a characteristic emission spectrum that in principle may be used as evidence that the process is occurring. The direct process always occurs for every molecule, albeit often with negligible cross sections. Because the indirect process involves a "crossing" or coupling of the incoming potential energy curve with an intermediate bound state which radiates to a lower lying state, in many cases the indirect process may not occur at all.

2.1 Direct Process

In direct radiative association, the largest cross section results when A and B approach along an excited state potential energy curve from which a strong dipole transition can be made to the ground state. Transitions occur from the vibrational continuum of the upper state to the discrete vibrational levels of the lower state; see Fig. 1. The usual dipole selection rules must be obeyed, (cf. Herzberg 1950). The process is usually quite inefficient, as spontaneous emission must take place during the collision and the radiative lifetime (usually 10^{-9} to 10^{-7} s for allowed transitions) is several orders of magnitude larger than the collision time (of the order of a molecular vibration, 10^{-12} to 10^{-13} s).

If atoms A and B with electron orbital angular momentum L_A and L_B and electron spin angular momentum S_A and S_B approach each other, a total of $(2L_A + 1)(2L_B + 1)(2S_A + 1)(2S_B + 1) = g_{AB}$ molecular states are possible. The probability of approach along any particular molecular potential energy curve, $V_{\Lambda'S}$, with electronic orbital angular momentum projection on the internuclear axis Λ' and spin state S, is given by

$$p_{\Lambda'S} = g_{\Lambda'S}/g_{AB} \qquad (2.1)$$

where

$$g_{\Lambda'S} = (2S+1)(2-\delta_{0,\Lambda'}). \qquad (2.2)$$

If E is the initial energy of relative motion in the collision, the cross section for radiative association is given by

$$\sigma(E) = \sum_{\Lambda',S} p_{\Lambda'S} \sigma_{\Lambda'S}(E) \qquad (2.3)$$

in which the cross sections $\sigma_{\Lambda'S}(E)$ can be calculated either semiclassically or quantum mechanically as given below. The summation is over all the different initial states $\Lambda'S$ formed in the approach of A and B that can radiate to a lower-lying bound state—usually the ground state of the molecule AB. The rate coefficient for the process at a temperature T is obtained from an average over a Maxwellian distribution in the usual manner:

$$\alpha(T) = \left(\frac{8}{\mu\pi}\right)^{1/2} \left(\frac{1}{k_b T}\right)^{3/2} \int_0^\infty E\sigma(E) \exp(-E/k_b T)\, dE, \qquad (2.4)$$

where k_b is the Boltzmann constant and μ is the reduced mass of the molecule.

2.1.1 Semiclassical Treatment

A semiclassical description of the nuclear motion may be completely adequate for collisions of heavy species. This approach was first formulated correctly by Bates (1951). Here we follow the notation as given by Dalgarno, Du and You (1990). In this case, the cross section $\sigma_{\Lambda'S}(E)$ is obtained from the integration of the radiative transition probability $A(R)$, expressed as a function of internuclear separation R, over all the classical trajectories. The radiative transition probability from an initial state $\Lambda'S$ to a lower-lying electronic state $\Lambda''S$ is

$$A(R) = 2.03 \times 10^{-6} s^{-1} \frac{(2S+1)}{g_{\Lambda'S}} (2 - \delta_{0,\Lambda'+\Lambda''}) \nu(R)^3 |D(R)|^2 \qquad (2.5)$$

in which $D(R)$ (in atomic units) is the dipole transition moment connecting states Λ' and Λ'', and $\nu(R)$ is the photon energy emitted (in cm^{-1}), which is the sum of the initial energy of relative motion, E, and the energy difference between the molecular potential energy curves $V_{\Lambda'S}$ and $V_{\Lambda''S}$ associated, respectively, with states Λ' and Λ''. Thus the expression for the cross section is

$$\sigma_{\Lambda'S}(E) = 4\pi \left(\frac{\mu}{2E}\right)^{1/2} \int_0^\infty b\,db \int_{R_c}^\infty \frac{A(R)}{[1 - V_{\Lambda'S}(R)/E - b^2/R^2]^{1/2}}\,dR \quad (2.6)$$

in which R_c is the classical distance of closest approach and b is the impact parameter.

2.1.2 Quantum Mechanical Treatment

In the quantum mechanical treatment, cross sections for transitions from the vibrational continuum of the upper state to particular bound vibrational and rotational levels of the lower state can be calculated and then summed appropriately. At a collision energy $E = \hbar^2 k^2/2\mu$ and for the initial state partial wave characterized by the rotational N' and total J' angular momentum quantum numbers, the partial cross section for a transition to a bound level, with vibrational, rotational and total angular momentum quantum numbers v'', N'' and J'' and binding energy $E_{v''N''}$, is given by

$$\sigma_{v''N''N'}(E, J', J'') = \frac{4\pi^2 \alpha^3}{3\mu E}(E + E_{v''N''})^3 S_{J''J'} |M_{N''N'}(E)|^2. \quad (2.7)$$

[In this section we have followed the common practice in atomic physics of using atomic units, where $\hbar$, e, and m, the electron mass, are all set to unity. Accordingly the cross sections appear in units of a_0^2 and can be readily converted to cm using $a_0 = 5.29177 \times 10^{-9}$ cm.] In the expression for the partial cross section, eqn (2.7), we have suppressed Λ' and S, α is the fine structure constant, $S_{J''J'}$ is the appropriate line-strength factor (or Hönl–London factor), and

$$M_{N''N'}(E) = \int_0^\infty \chi_{v''N''}(R)\,D(R)\,\chi_{kN'}(R)\,dR. \quad (2.8)$$

The bound and continuum vibrational wavefunctions, $\chi_{v''N''}(R)$ and $\chi_{kN'}(R)$ respectively, are usually obtained by numerical integration of the appropriate radial Schrödinger equations:

$$\left\{\frac{d^2}{dR^2} - \frac{N'(N'+1) - \Lambda'^2}{R^2} - \frac{2\mu}{\hbar^2}[V_{\Lambda'S}(R) - V_{\Lambda'S}(\infty)] + k^2\right\}\chi_{kN'}(R) = 0 \quad (2.9)$$

and

$$\left\{\frac{d^2}{dR^2} - \frac{2\mu}{\hbar^2}[V_{\Lambda''S}(R) - V_{\Lambda''S}(\infty)] - \frac{N''(N''+1) - \Lambda''^2}{R^2} - \frac{2\mu}{\hbar^2}E_{v''N''}\right\}\chi_{v''N''}(R) = 0. \quad (2.10)$$

The continuum function is normalized according to the asymptotic form

$$\chi_{kN'}(R) \sim (2\mu/\pi k)^{1/2} \sin(kR - \tfrac{1}{2}N'\pi + \delta_{N'}) \tag{2.11}$$

where $\delta_{N'}$ is the phase shift. The relation between the total angular momentum quantum number J and the rotational and spin quantum numbers is:

$$J' = |N' + S| \quad \text{and} \quad J'' = |N'' + S|. \tag{2.12}$$

The quantum number S is used for upper and lower states, as it does not change during the transition, and in each case it is understood that N' (and N'') take integer values $\geq \Lambda'$ (and $\geq \Lambda''$).

Dipole selection rules dictate that $\Delta J = 0, \pm 1$. For given N' and S, J' values are determined, which dictate through the selection rules the J'' and thus N'' terms which enter the cross section. Summing over the J' and J'' quantum numbers, one obtains

$$\sigma_{v''N''N'}(E) = \sum_{J'}\sum_{J''} \sigma_{v''N''N'}(E, J'J''). \tag{2.13}$$

Except in the case of very low temperatures, the approximation $J \approx N$ can be made and the cross section can be written for $N'' = N' \pm 1$ and $N'' = N'$:

$$\begin{aligned}\sigma_{v''N'}(E) = \frac{4\pi^2\alpha^3}{3\mu E}[&(E + E_{v''N'+1})^3 S_{N'+1\,N'}|M_{N'+1\,N'}(E)|^2 \\ &+ (E + E_{v''N'})^3 S_{N'\,N'}|M_{N'\,N'}(E)|^2 \\ &+ (E + E_{v''N'-1})^3 S_{N'-1\,N'}|M_{N'-1\,N'}(E)|^2]. \end{aligned} \tag{2.14}$$

Various simplifications of this formula can be made with the assumption that the dipole matrix elements do not vary greatly with N', $|M_{N'-1\,N'}| \approx |M_{N'\,N'}| \approx |M_{N'+1\,N'}|$, and through the neglect of the small energy differences between the bound vibration-rotation levels, $E_{v''N'-1} \approx E_{v''N'+1} \approx E_{v''N'}$. Summing over all initial partial waves N' and all the bound vibrational levels of the lower state, one obtains

$$\sigma_{\Lambda'S}(E) = \sum_{N'}\sum_{v''} \sigma_{v''N'}(E). \tag{2.15}$$

For a collision energy E, creating molecule AB in the $v''N''$ vibrational-rotational state, a photon is emitted with energy $h\nu = E + E_{v''N''}$. For gas temperature T characterized by a Maxwellian distribution, the spectral signature is an emission continuum. The total cross section for radiative association is obtained by summing the cross sections $\sigma_{\Lambda'S}(E)$ over all initial electronic states $\Lambda'S$ and it is given by eqn (2.3) with the probability of approach $p_{\Lambda'S}$ given by eqn (2.1).

Table 1 Rate coefficients $\alpha(T)$ for the formation of H_2^+, He_2^+, HD, and LiH by radiative association.

Molecule	T(K)	$\alpha(T)$(cm^3 s^{-1})	Ref.
H_2^+	10	1.55(−20)	a,b
	100	7.85(−20)	a,b
	1 000	5.3(−18)	a,b
	4 000	6.2(−17)	a,b
He_2^+	100	7.18(−21)	b
	1 000	5.82(−19)	b
	10 000	2.62(−17)	b
HD	100	0.83(−26) / 1.35(−26)*	c
	1 000	1.01(−26) / 2.2(−26)*	c
	10 000	0.40(−26) / 0.62(−26)*	c
LiH	100	3.17(−20) / 4.33(−20)**	d,e / f
	1 000	2.08(−20) / 2.71(−20)**	d,e / f
	5 000	4.0(−21) / 5.0(−21)**	d,e / f

*) Includes stimulated association in a blackbody radiation field of 10 000 K.
**) Includes stimulated association in a blackbody radiation field of 1000 K.
a) Ramaker and Peek (1976); b) Stancil, Babb, and Dalgarno (1993); c) Stancil and Dalgarno (1997b); d) Dalgarno, Kirby, and Stancil (1996), tabulated here; e) Gianturco and Gori Giorgi (1997); f) Stancil and Dalgarno (1997a).

2.1.3 Discussion and Examples

In order to understand the large variation in the magnitude of the radiative association rate coefficients appearing in Table 1, it is useful to list a number of salient features of the radiative association process. First, although the process has been portrayed as occurring in transitions between two electronic states, radiative association can occur within a single electronic state. This latter mechanism, however, is very slow. Second, the cross section scales as the third power of the emitted photon energy, so energetic transitions are highly favoured. In collisions of ground state atoms, cross sections are larger for formation of molecules with large binding energies. Radiative association involving electronically excited atoms may be particularly favoured, but a significant excitation mechanism must exist in order for this route to be important. Third, spontaneous emission of a photon in the association process can be enhanced by stimulated emission if there is a strong background radiation field present. Lastly, there are competing processes which can occur for the more energetic radiative association reactions. In the case of molecular ions, the radiative association process competes with radiative and non-radiative charge transfer (see, for example, Kimura et al. 1993). In collisions involving an excited atom, the processes of associative ionization or of Penning ionization may completely overwhelm the formation of neutral molecules by radiative association.

The case of LiH^+ formation, as described by Dalgarno, Kirby and Stancil (1996), illustrates a number of the above points. There are two channels for formation: $Li^+ + H$ and $Li + H^+$. As can be seen from Table 2, the rate coefficient for formation of LiH^+ from $Li^+ + H$ is approximately eight orders of magnitude smaller than that starting from $Li + H^+$. Radiative association of $Li^+ + H$ takes place along the ground state potential energy curve of LiH^+. Transitions take place from the vibrational continuum of this state to the bound vibrational levels, governed by the dipole moment of the $X^2\Sigma^+$ state. The bound and continuum vibrational wavefunctions of a molecular electronic state are mutually orthogonal. If the dipole moment function were constant as a function of internuclear separation, the matrix elements of the continuum-bound vibrational transitions would be identically zero. The dipole moment function varies with R, however, and can be expanded around the equilibrium internuclear distance R_e,

$$D(R) = D_o(R_e) + D'(R_e)(R - R_e) + D''(R_e)(R - R_e)^2 + \cdots, \qquad (2.16)$$

where D' and D'' are first and second derivatives of the dipole moment, respectively.

Because of the orthogonality of the wavefunctions, the matrix elements $M_{v''N''kN'}$ do not depend on the magnitude of the dipole moment, $D_o(R_e)$, but only on the first and higher order derivatives. The matrix elements are usually very small. In addition, transitions taking place from continuum to bound vibrational levels within a single electronic state emit very low energy photons, so the ν^3 factor in the cross section is quite small compared to transitions between two electronic states which are separated asymptotically by a large energy (e.g. transitions from the $Li + H^+$ asymptote). Stancil, Lepp and Dalgarno (1996) have examined the effect of these radiative reactions on the formation of LiH^+ in the early Universe, as discussed in Section 4.

The importance of a particular channel in overall molecule formation depends not only on the rate constant but also on the densities of the relevant atoms and ions. Thus in many cases the higher energy channel with significantly greater

Table 2 Rate coefficients $\alpha(T)$ for the formation of HeH^+ and LiH^+ by radiative association.

Molecule	T(K)	$\alpha(T)$(cm^3 s^{-1})	Ref.
HeH^+	100	9.8(−16)*	a,b,c
(He^+ + H channel)	1 000	4.1(−16)*	a,b,c
	10 000	2.6(−16)*	a,b,c
HeH^+	6 000	2.7(−20)	d
(H^+ + He channel)	10 000	1.3(−20)	d
	40 000	1.5(−21)	d
LiH^+	100	1.33(−22)	e
(Li^+ + H channel)	1 000	2.39(−23)	e
	5 000	2.9(−24)	e
LiH^+	100	5.53(−15)	e
(Li + H^+ channel)	1 000	1.81(−15)	e
	5 000	8.1(−16)	e

∗) Corrected by multiplying by a factor of $\frac{1}{4}$ as indicated in Stancil and Zygelman (1996).
a) Zygelman and Dalgarno (1990), tabulated here; b) Stancil and Zygelman (1996); c) Kraemer, Špirko and Jurek (1995); d) Kimura et al. (1993); e) Dalgarno, Kirby and Stancil (1996).

cross section may not be important because the appropriate collisional species are underabundant.

Stancil and Zygelman (1996) have carried out quantum mechanical calculations of the radiative charge transfer process

$$Li + H^+ \to Li^+ + H + h\nu \qquad (2.17)$$

for temperatures from 10 to 40 000 K. This process occurs when in the course of the Li + H^+ collision a transition to the vibrational continuum of the ground state of LiH^+ occurs with emission of a photon. The radiative charge transfer rate coefficient ($\sim 1.2 \times 10^{-13}$ cm^3 s^{-1} at $T = 1000$ K) varies extremely slowly with T, and is approximately 100 times larger than the radiative association rate for $T > 1000$ K. For decreasing temperatures, the radiative association rate increases steadily but is always less than the charge transfer rate by more than a factor of 20. The recently revised rate coefficients for the combined process of radiative charge transfer and association by Kimura, Dutta and Shimakura (1995) have been shown by Stancil and Zygelman (1996) to be in fair agreement with rate coefficients calculated using a semiclassical method.

Stancil and Dalgarno (1997a) have explored the mechanism of stimulated radiative association of Li + H in the presence of the cosmic background radiation field:

$$\text{Li} + \text{H} + h\nu_b \to \text{LiH} + h\nu + h\nu_b \tag{2.18}$$

in which $h\nu_b$ is the photon energy in a blackbody radiation field at temperature T_b. They have derived an expression for the total (stimulated plus spontaneous) cross section:

$$\sigma_{v''N'}(E) = \sigma^{sp}_{v''N'}(E)[1 - \exp(-h\nu/k_b T_b)]^{-1}, \tag{2.19}$$

where k_b is the Boltzmann constant and $h\nu$ is the radiated photon energy. The rate coefficient for the combined process, stimulated plus spontaneous, is obtained by taking a Maxwellian average as in eqn (2.4) for a velocity distribution characterized by a matter temperature T_m. As can be seen from eqn (2.19), significant enhancements over the spontaneous radiative association rate can occur for small ν and large T_b. In the case of LiH, Stancil and Dalgarno (1997a) showed that for a range of T_m, from 10 to 10 000 K, the enhancement for $T_b = $ 500, 1000, and 5000 K is approximately 10%, 30%, and 300%, respectively.

In obtaining radiative association cross sections for two open-shell colliding species which give rise to a large manifold of molecular states, one finds that the cross section is dominated by the one or two channels which have the largest oscillator strength and the largest ν^3 factors. For example, the separated atoms $C(^3P) + O(^3P)$ give rise to 18 molecular states, six each of singlet, triplet and quintet multiplicities: $\Sigma^+(2)$, Σ^-, $\Pi(2)$, Δ. Transitions to the lowest-lying electronic state within each spin manifold will optimize the ν^3 factor, and only those transitions that are driven by a large dipole transition moment need be considered. In many cases, such as for the CO triplet system, the vital molecular data are lacking. The singlet transitions appear to be dominated by the strong $A^1\Pi - X^1\Sigma^+$ transition. Dalgarno, Du and You (1990) determined the rate coefficient assuming that approach occurs along the $A^1\Pi$ potential curve with transitions to the ground $X^1\Sigma^+$ state (as in Fig. 1). A number of channels may be ineffective for radiative association at low temperatures due to repulsive barriers at long range. In the case of ion–atom collisions, such as $C^+ + O$, the long-range interaction is attractive and more excited states may contribute. However, data on most of these transitions are not available. Dalgarno, Du and You (1990) used the $A^2\Pi - X^2\Sigma^+$ transition in a calculation of the formation of CO^+ by radiative association. The process of inverse predissociation, discussed in Section 2.2, below, may contribute to CO^+ formation.

The rate coefficients for radiative association of an excited state atom with a ground state atom may be significantly larger than for two ground-state atoms. This enhancement is due to an increased ν^3 factor and, in many cases, to a strong

Table 3 Rate coefficients $\alpha(T)$ for the formation of H_2 by radiative association of $H(n = 1)$ and $H(n = 2)$.

Molecule	T(K)	$\alpha(T)(\mathrm{cm}^3\,\mathrm{s}^{-1})$	Ref.
H_2	100	2.2(−14)	a
$H(2s) + H$	1 000	1.2(−14)	a
	10 000	1.1(−14)	a
H_2	100	1.2(−14)	a
$H(2p) + H$	1 000	3.4(−14)	a
	10 000	4.2(−14)	a

a) Latter and Black (1991).

dipole transition moment. Latter and Black (1991) have carried out semiclassical calculations for the excited atom process:

$$H(n = 2) + H(n = 1) \rightarrow H_2 + h\nu. \quad (2.20)$$

They obtained rates for the radiative association as functions of temperature for both $H(2p)$ and $H(2s)$. Photon-pumping, such as with trapped Lyman α radiation, populates only $H(2p)$ but collisions can populate $H(2s)$. The combined rate coefficient over a broad temperature range is on the order of 5×10^{-14} cm^3 s^{-1} (see Table 3). In most astrophysical environments, densities are high enough such that the collision time is much shorter than the radiative lifetime, so that once excited atoms are formed, they will usually collide before radiating. Thus, only in environments with a large amount of Lyα photon trapping is this process significant. As the colliding gas temperature increases, the process of associative ionization will start to compete with the radiative association. This additional two-body process is described in Section 3.

2.2 Inverse Predissociation

In inverse predissociation the approaching atoms make a radiationless transition into a predissociated electronic state where spontaneous emission takes place:

$$A + B \rightarrow AB^* \rightarrow AB + h\nu, \quad (2.21)$$

as illustrated in Fig. 2. The spectrum is a band emission spectrum, in contrast to the continuous emission occurring in direct radiative association. Following Herzberg (1950), various classifications of inverse predissociation mechanisms, depending on the nature of the predissociating and predissociated states, have been given (Carrington 1972; Golde and Thrush 1973; Julienne and Krauss 1973). The coupling between the states may be a spin–orbit coupling or an electronic–rotational coupling (Lefebvre-Brion and Field 1986).

Table 4 Rate coefficients $\alpha(T)$ for the formation of CO, CS, SiO and O_2 by direct radiative association and of O_2 by inverse predissociation.

Molecule	T(K)	$\alpha(T)$(cm^3 s^{-1})	Ref.
CO	300	0.21(−18)	a
	700	3.31(−18)	a
	1 900	15.3(−18)	a
	9 500	41.6(−18)	a
CS	300	1.24(−18)	b
	700	1.75(−18)	b
	1 500	2.26(−18)	b
	5 100	3.05(−18)	b
SiO	300	2.87(−17)	b
	700	4.27(−17)	b
	1 500	5.63(−17)	b
	5 100	7.83(−17)	b
O_2	100	1.3(−26)	c
	500	1.9(−24)	c
	1 000	{ 2.9(−23) 9.2(−24)*	c
	3 000	{ 2.5(−22) 7.8(−20)*	c
	5 000	{ 6.1(−22) 3.9(−19)*	c
	10 000	{ 1.7(−20) 9.0(−19)*	c

*) Inverse predissociation.
a) Dalgarno, Du and You (1990); b) Andreazza, Singh and Sanzovo (1995); c) Babb and Dalgarno (1995).

2.2.1 Theoretical Formulation

The cross section for inverse predissociation can be written using the Breit–Wigner theory for isolated resonances (Carrington 1972; Du and Dalgarno 1990).

$$\sigma(E) = \frac{g'}{g}\left(\frac{\pi\hbar^2}{2\mu E}\right)\sum_{v'}\sum_{N'}\frac{(2N'+1)\Gamma_r(v'N')\Gamma_d(v'N')}{[E-E(v'N')]^2 + \frac{1}{4}[\Gamma_r(v'N')+\Gamma_d(v'N')]^2}, \quad (2.22)$$

where the sum is over all levels $v'N'$ above the crossing point of the electronic states, E is the relative kinetic energy of the approaching atoms, g is the statistical weight of the atom–atom pair, g' is the statistical weight of the predissociated

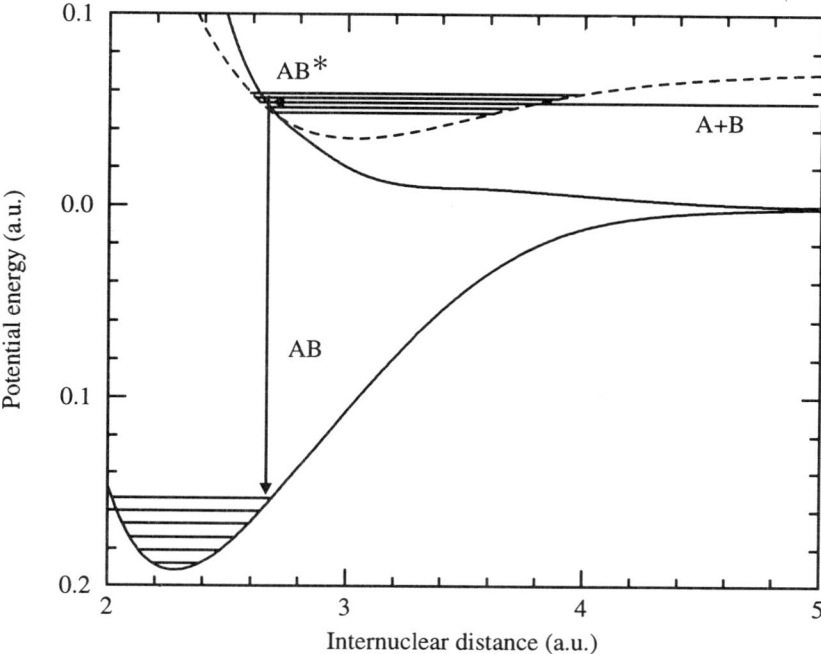

FIG. 2. Illustration of inverse predissociation (2.21). In this example, the two atoms A and B approach in the vibrational continuum of a repulsive excited electronic state, make a radiationless transition into a vibrational–rotational state of an excited molecular electronic state (dotted lines) forming AB*, and then emit a photon thereby forming a molecule in a vibrational–rotational state of electronic state AB. The data plotted are the O_2 $1^3\Pi_u$ (repulsive), $B^3\Sigma_u^-$ (excited or AB*), and $X^3\Sigma_g^-$ (ground) electronic states.

electronic state, $E(v'N')$ is the energy level of the predissociated state, $\Gamma_r(v'N')$ and $\Gamma_d(v'N')$ are, respectively, the radiative and predissociative widths of level v', N', with

$$\Gamma_r(v', N') = \hbar A(v'N') \equiv \hbar \sum_{v''} \sum_{N''} A(v'N'v''N''), \qquad (2.23)$$

where $A(v'N'v''N'')$ is the Einstein A-coefficient defined according to Schadee (1978). The width $\Gamma_d(v'N')$ of each predissociated vibrational–rotational level is related to the lifetime, τ_d, of the level by $\Gamma_d = \hbar/\tau_d$ and it may be obtained from experimental measurements of line broadening or it may be calculated using a Golden Rule approximation from the expression

$$\Gamma_d = 2\pi|\langle\Psi_{v'N'}|H'|\Psi_{EN}\rangle|^2, \qquad (2.24)$$

where H' is the perturbation causing the predissociation. The wavefunctions Ψ_{EN} and $\Psi_{v'N'}$ appearing in (2.24) are each Born–Oppenheimer product wavefunctions of an electronic wavefunction and a vibrational wavefunction. The perturbation H' arises from various physical mechanisms: see, for example, Lefebvre-Brion and Field (1986). Averaging over a Maxwellian distribution, as in (2.4), with the assumption that the radiative lifetime is much larger than the predissociation lifetime, so that $\Gamma_r(vN) \ll \Gamma_d(vN)$, one finds that

$$\alpha(T) \approx \frac{g'}{g} \hbar^2 \left(\frac{2\pi}{\mu kT}\right)^{3/2} \sum_v \sum_N (2N+1) \Gamma_r(v, N) \exp(-E_{vN}/kT). \quad (2.25)$$

The radiative width can be determined either from theoretical calculations if potential energy curves and transition dipole moments are available or from experimental data.

2.2.2 Discussion and examples

Calculations of the rate coefficients for the formation of CH, OH, NO and O_2 by inverse predissociation have been carried out and experimental studies of inverse predissociation have been performed for NO and O_2. The formation by inverse predissociation of OH (Julienne and Krauss 1973; Smith and Zweibel 1976) and of CH (Julienne and Krauss 1973; Brzozowski et al. 1976) has been studied, but these species are not directly relevant to the astronomical environments discussed in the present volume.

Estimates of the rate coefficients for formation of CO, C_2 and CN by inverse predissociation were given by Julienne and Krauss (1973). Dalgarno, Du and You (1990) argued that formation of CO by inverse predissociation is negligible, because the highest lying singlet, triplet and quintet states are repulsive, a conclusion differing from that of Julienne and Krauss, who roughly estimated a rate coefficient of order 10^{-17} cm^3 s^{-1}. Mechanisms for inverse predissociation in N_2 were reviewed by Golde and Thrush (1973). Du and Dalgarno (1990) and Sun and Dalgarno (1992) investigated the formation of NO via inverse predissociation. The process is important for planetary atmosphere modelling.

Babb and Dalgarno (1995) studied the formation of oxygen molecules through the inverse predissociation of ground state oxygen atoms approaching along the $1^3\Pi_u$ state that crosses the $B^3\Sigma_u^-$ state; see Fig. 2. The process is insignificant for low temperatures, but for $T > 1000$ K the rate coefficient rapidly becomes larger than the rate coefficient for direct radiative association along the $1^3\Pi_u$ state. The formation of O_2 by this mechanism can provide an additional source of CO or SiO in SN 1987A, besides direct radiative association of C or Si with O, via (Dalgarno and Fox 1994; Liu and Dalgarno 1994, 1996)

$$\left\{ \begin{array}{c} C \\ Si \end{array} \right\} + O_2 \rightarrow \left\{ \begin{array}{c} CO \\ SiO \end{array} \right\} + O. \quad (2.26)$$

Table 5 Rate coefficients $\alpha(T)$ for the formation of CO^+, CS^+, O_2^+, and SiO^+ by radiative association.

Molecule	$T(K)$	$\alpha(T)(\text{cm}^3\,\text{s}^{-1})$	Ref.
CO^+	300	2.5(−18)	a
	700	2.5(−18)	a
	1 900	2.3(−18)	a
	9 500	1.8(−18)	a
CS^+	300	0.54(−18)	b
	700	0.78(−18)	b
	1 500	1.01(−18)	b
	5 100	1.18(−18)	b
O_2^+	100	2.6(−19)	c
	500	2.7(−18)	c
	1 000	4.2(−18)	c
	5 000	5.6(−18)	c
SiO^+	300	6.89(−18)	b
	700	6.24(−18)	b
	1 500	5.49(−18)	b
	5 100	4.45(−18)	b

a) Dalgarno, Du and You (1990); b) Andreazza, Singh and Sanzovo (1995); c) Babb, Fan and Dalgarno (1994).

3 Other Two-Body Molecule Formation Processes

The processes of associative detachment and associative ionization may contribute to molecule formation under certain conditions. The theoretical description of these processes is considerably more complex than for radiative association because they involve both an electronic and a nuclear continuum, as evident in eqns (1.2) and (1.3). In general, although these processes have been included in the astrophysical models, they have not been found to play a major role in molecule formation. There is one major exception—H_2 formation in the early Universe occurring at later times, which appears to proceed via associative detachment (Lepp and Shull 1984; Dalgarno and Fox 1994; Chapter 3). A brief discussion of both associative ionization (AI) and associative detachment (AD) is included here, however, for completeness. As pointed out by Rawlings, Drew and Barlow (1993), these processes are unlikely to compete with each other, because very different conditions are required for AI and AD to become significant. In order for associative ionization to occur, there must be a way of electronically exciting one of the collision partners. This usually involves an optical pumping mechanism, requiring a significant radiation field. For the AD process to occur, an environment must exist which is conducive to negative ion formation and stability. This involves cool, neutral gas, far away from strong radiation sources.

Associative ionization has not been studied as extensively as its inverse process, dissociative recombination:

$$AB^+ + e^- \to A + B. \tag{3.1}$$

Dissociative recombination is a major destruction process for molecular ions in a variety of astrophysical environments. An extensive summary of theoretical methods, experiments and applications relevant to dissociative recombination (DR) work through 1993 can be found in a book edited by Rowe, Mitchell and Canosa (1993). Measurements of the overall DR cross section can now be carried out very efficiently in storage rings (Larsson 1995; Larsson et al. 1993), but it is still difficult to obtain information over a range of initial vibrational states of the molecular ion and on the final excitation states of the neutral atoms produced. Therefore, an associative ionization cross section is not easily obtained by applying the principle of microscopic reversibility.

Urbain et al. (1991), using a merged-beam apparatus, measured an absolute cross section for

$$H(2s) + H(1s) \to H_2^+ + e^- \tag{3.2}$$

for relative energies from 1 to 10 eV. The $H(2p)$ atoms decay too rapidly to enter into the reaction in their apparatus. This work also includes calculations of the associative ionization reaction which combine a multichannel quantum-defect theory (MQDT) description of the ionization with a close coupling treatment of the colliding atoms. The calculations are in harmony with experiment. Rawlings, Drew and Barlow (1993) used the cross section results, including an inferred cross section for the $H(2p)$ channel, to obtain a rate coefficient for each channel, separately and in combination, over a temperature range from 3000 K to 15 000 K (see Table 6). At $\sim$3000 K, the total associative ionization rate is $\sim 4.2 \times 10^{-13}$ cm^3 s^{-1}, approximately an order of magnitude greater than the radiative association rate obtained by Latter and Black (1991) for the process (2.20). At 10 000 K the AI rate coefficient is 300 times greater than the radiative association rate coefficient.

Crosas and Weisheit (1993) have explored the contribution of collisions between excited and ground-state atomic hydrogen to the H_2 molecule formation via both associative ionization and radiative association in quasar broad-line region clouds. They found that these processes are not efficient in comparison to the associative detachment process which will now be discussed in some detail.

The associative detachment process, eqn (1.2), may be viewed formally as a collision process and as such its theoretical description has been called "perhaps the most difficult problem in atomic collision theory" (Haywood and Delos 1990). The difficulties, as discussed above, stem essentially from the presence of both

Table 6 Selected rate coefficients $\alpha(T)$ for the formation of H_2^+ by associative ionization and of H_2 by associative detachment.

Formation reaction	T(K)	$\alpha(T)$(cm^3 s^{-1})	Ref.
$H(2s) + H(1s) \rightarrow H_2^+ + e^-$	3 000	3.1(−13)*	a
	5 000	2.1(−12)*	a
	10 000	1.11(−11)*	a
$H(2p) + H(1s) \rightarrow H_2^+ + e^-$	3 000	4.6(−13)*	a
	5 000	3.7(−12)*	a
	10 000	1.78(−11)*	a
$H^- + H \rightarrow H_2 + e^-$	300	1.89(−9)	b
	1 000	1.3(−9)	c
	2 500	1.1(−9)	c

*) A net rate coefficient for $H(2s, 2p)$ can be obtained from a weighted average of the rate coefficients for $H(2s)$ and for $H(2p)$: $\alpha(T; H(2s, 2p)) = \frac{1}{4}[\alpha(T; H(2s)) + 3\alpha(T; H(2p))]$; a) Rawlings, Drew and Barlow (1993); b) Bieniek and Dalgarno (1979); c) Launay, Le Dourneuf and Zeippen (1991), potential $V^{(2)}$.

an electronic continuum and a nuclear continuum and also, if there is binding, from the description of the temporary negative molecular ion complex.

A simplification is achieved if the process is viewed as the formation of a negative molecular ion AB^-, followed by autodetachment leading to a free electron and molecule AB (Bates and Massey 1954). In the Born–Oppenheimer approximation, a local complex potential can be used to describe the scattering, where the imaginary part of the potential describes the possibility of making a transition to the system AB and a free electron (Bieniek and Dalgarno 1979).

Formation of H_2 via associative detachment

$$H + H^- \rightarrow H_2(v''J'') + e^- \tag{3.3}$$

appears to be important in dust-poor astrophysical environments. It has been proposed that associative detachment could be detected by the emission spectra (Bieniek and Dalgarno 1979; Black, Porter and Dalgarno, 1981) from the subsequent cascading quadrupole radiation (Turner, Kirby-Docken and Dalgarno, 1977) of the populated $H_2(v''J'')$ molecules. Since the vibrational–rotational level population of molecules from the AD reaction is different from that formed by other processes (Black, Porter and Dalgarno, 1981), it might be possible at least in principle to detect environments in which associative detachment is operative.

The local complex potential can be written

$$W(R) = V(R) - \tfrac{i}{2}\Gamma(R), \tag{3.4}$$

where $\Gamma(R)$ is the width of the intermediate molecular state H_2^- of energy $V(R)$ and lifetime $\tau = \hbar/\Gamma(R)$. Specific quantum-mechanical calculations for the process (3.3) using a local complex potential have been performed by Bieniek and Dalgarno (1979), Bieniek (1980) and Launay, Le Dourneuf and Zeippen (1991). The vibrational wavefunction $F_{kJ'}(R)$ for the approaching atom and negative ion is now complex and can be obtained from the equation

$$\left[\frac{d^2}{dR^2} - \frac{J'(J'+1)}{R^2} - \frac{2\mu}{\hbar^2}[W(R) - E]\right] F_{kJ'}(R) = 0, \qquad (3.5)$$

where $E = \hbar^2 k^2/2\mu$ is the initial energy of relative motion and the wavefunction is normalized according to

$$F_{kJ'}(R) \sim (2\mu/\pi k)^{1/2} \sin(kR - \tfrac{1}{2}J'\pi + \eta_{J'}), \quad R \sim \infty, \qquad (3.6)$$

and the complex phase shift is

$$\eta_{J'} = \delta_{J'} + i\gamma_{J'}. \qquad (3.7)$$

The wavefunction of the final vibrational–rotational state $v''J''$ of H_2 is obtained from the solution of eqn (2.10) with $V_{\Lambda''S}(R)$ the $X^1\Sigma_g^+$ potential of H_2. The coupling between the initial and final states can be treated in perturbation theory, but data on the values of the coupling matrix elements are scarce. Bieniek and Dalgarno (1979) used a golden rule formulation to estimate the coupling $V(\epsilon l, R)$, where ϵ and l are, respectively, the energy and angular momentum of the detached electron. The width $\Gamma(R)$ in the local complex potential $W(R)$ was constructed semiempirically. They calculated the cross sections for formation of $\mathrm{H}_2(v''J'')$ at relative energies of 0.0129 and 0.129 eV and the rate coefficient at a temperature of 300 K, which is given in Table 6 and is in harmony with the value measured by Schmeltekopf, Fehsenfeld and Ferguson (1967). The validity of the local complex potential description of the associative detachment process (3.3) was investigated by Bieniek (1980).

Launay, Le Dourneuf and Zeippen (1991) constructed a local complex potential that gave results for rate coefficients consistent with experimental data for both associative detachment and dissociative attachment. Their values are generally within a factor of two of the results of Bieniek and Dalgarno (1979) and demonstrate the sensitivity of the calculations to values used for the H+H$^-$ potential $V(R)$ appearing in eqn (3.4). Their results for the AD reaction at 1000 K and 2500 K are given in Table 6.

4 Applications to the Early Universe and SN Ejecta

There is considerable interest in exploring the mechanisms for making molecules in the early Universe (Chapter 3; Puy *et al.* 1993; Dalgarno and Fox 1994; Stancil, Lepp and Dalgarno 1996). It has been recognized that molecules may have played a critical role in cooling the primordial gas clouds, thus enabling gravitational collapse leading to formation of the first stars. From the atomic nuclei created in the primordial nucleosynthesis—primarily H and ^{4}He, with trace amounts of D, ^{3}He and ^{7}Li—a very limited variety of molecules can be made. Molecule formation could only occur after the Universe had cooled sufficiently that a significant fraction of ambient neutral species was present.

The first element to become neutral was He, with an ionization potential of 24.6 eV. Thus the first molecules to be formed were He_2^+, HeH^+ and $LiHe^+$ from the radiative association of He^+, H^+ and Li^+ with neutral He, respectively. Rate coefficients for the first two molecules are found in Tables 1 and 2, but the radiative association rate coefficient for $LiHe^+$ has not been calculated. Its formation would proceed through the single electronic state, $X^1\Sigma^+$, associated with the Li^+ + He asymptotic limit. Based on comparisons with the rate coefficients for radiative association of H^+ + He and of Li^+ + H, a radiative association rate coefficient of 1×10^{-21} cm^3 s^{-1} is estimated. These molecular ions would have been rapidly destroyed by dissociative recombination.

By $T \approx 3000$ K, the recombination of protons and electrons was accomplished and the Universe had become almost completely neutral. With the formation of neutral hydrogen, the H_2 molecule could also be formed, but *not* through the process of radiative association of $H(1s) + H(1s)$. Only two states correlate to the H+H ground state separated atom limit: $b^3\Sigma_u^+$ and $X^1\Sigma_g^+$. Formation along the $b^3\Sigma_u^+$ potential energy curve does not occur, as it would necessitate a highly forbidden triplet–singlet radiative transition. Formation along the $X^1\Sigma_g^+$ curve is not possible because the dipole moment of H_2 is zero.

Two main mechanisms for H_2 formation in the early Universe (Chapter 3) as well as in the winds of young stellar objects (Chapter 7) have been generally accepted. The first, occurring at earliest times, involves formation of H_2^+ by either radiative association or proton transfer from collisions of HeH^+ with H. Radiative association of H + H^+ takes place through transitions from the $^2\Sigma_u^+(2p\sigma_u)$ repulsive electronic state to the $^2\Sigma_g^+(1s\sigma_g)$ bound state of H_2^+. Ramaker and Peek (1976) explored this process both semiclassically and quantum mechanically. More recent work of Stancil, Babb and Dalgarno (1993) obtained rate coefficients in excellent agreement with the Ramaker and Peek (1976) values. The formation of H_2 is achieved by a final charge transfer of the H_2^+ with H.

The second mechanism for H_2 formation involves the associative detachment process (3.3). This process can only occur when the radiation field has cooled sufficiently that the H^- ions are not instantly photodetached.

Latter and Black (1991) explored whether radiative association of $H(n = 2)$ + H could contribute significantly to H_2 formation. Despite the sizable rate coefficient, they found that this process only contributes at the onset of the recombination epoch. This is due to the strong coupling of the radiation field and matter, leading to an enhancement of the $H(n = 2)$ population due to resonant scattering of Lyα photons. As was noted in Section 3, rate coefficients for associative ionization are generally much larger than those for radiative association. This route to H_2^+ formation also should contribute to H_2 formation at the earliest times, as long as the H_2^+ can charge transfer with a hydrogen atom before it is destroyed by photodissociation or dissociative recombination.

Considering both spontaneous and stimulated radiative association Stancil and Dalgarno (1997b) have calculated rate coefficients for the radiative association of H and D to form HD. Association takes place through the $X^1\Sigma^+$ state, driven only by the very small dipole moment of HD. The spontaneous radiative association rate at a collision temperature of 1000 K is 1×10^{-26} cm^3 s^{-1} and this is increased by 33% if the stimulated process is included at a blackbody radiation temperature of 3000 K. The combined rate, however, is too slow to compete with the two other mechanisms for making HD which are exactly analogous to the H_2 formation mechanisms.

The rate coefficients for formation of LiH$^+$ and LiH by radiative association have been calculated by Dalgarno, Kirby and Stancil (1996) and by Gianturco and Gori Giorgi (1997) with quantum mechanical methods. The two channels for formation of LiH$^+$, with rates differing by five orders of magnitude as shown in Table 2, are operative under very different conditions. At high redshifts formation is mainly due to Li$^+$ + H, whereas at later times, the Li + H$^+$ channel also contributes substantially, even though the abundance of H$^+$ is approximately $10^{-4}n_H$. Formation of LiH occurs through the $X^1\Sigma^+$ ground state. Close agreement between the two groups of investigators for all the various rate coefficients leading to LiH$^+$ and LiH was found, as shown explicitly in tables by Gianturco and Gori Giorgi (1997). The correct quantum mechanical rate coefficient for LiH formation is three orders of magnitude smaller than that originally estimated by Lepp and Shull (1984).

Stancil and Dalgarno (1997b) also examined the contribution of the stimulated radiative association process to the formation of LiH. They found only ~30% enhancement at a blackbody radiation field temperature of 1000 K. It appears that stimulated radiative association does not contribute substantially to the overall rate for blackbody temperatures which pertain to currently used early Universe models.

Stancil, Lepp and Dalgarno (1996) have also considered the formation of LiH due to associative detachment of Li + H$^-$ and Li$^-$ + H. They assumed a rate coefficient for both processes of approximately 4.0×10^{-10} cm^3 s^{-1}. These formation channels are significant only at later times (redshifts < 80).

In the ejecta of the supernova SN 1987A the infrared emission features of several molecules have been observed. Positive identifications have been made for CO vibrational (first overtone) transitions (5-3, 4-2, 3-1, 2-0) around 2.3 µm (Spyromilio et al. 1988) and for an SiO fundamental transition ($\Delta v = 1$) around 8–10 µm (Aitken et al., 1988). Other less certain identifications have been suggested: CO$^+$, with a feature at 2.26 µm (Petuchowski et al. 1989) and H_3^+ with features at 3.41 µm and 3.53 µm (Miller et al. 1992). All of these molecular emissions were found to appear approximately 100 days after the supernova explosion and to disappear 400–500 days later. At the time that the molecular bands were first detected, the temperature of the ejecta was about 5000 K, and the density was $\sim 10^{11}$ cm^{-3}, too low for three-body processes to be effective in forming molecules (Dalgarno, Liu and Lepp 1991). There was no evidence of dust formation in the ejecta until approximately 500 days post explosion. Just recently, another type II supernova, SN 1995ad, was observed to exhibit strong infrared emission consistent with that of first overtone transitions in CO (Spyromilio and Leibundgut 1996).

All of the mechanisms for making the above molecules in the supernova ejecta involve radiative association, associative ionization and associative detachment (Lepp, Dalgarno and McCray 1990; Petuchowski et al. 1989; Miller et al. 1992; Liu and Dalgarno 1994). Because the ejecta are layered, with successive shells consisting of Fe, Si, O, C, He and H, molecules will preferentially be made from constituents within a shell or in adjacent shells. Clearly CO and SiO are obvious candidates.

There are several possible formation mechanisms for CO. Petuchowski et al. (1989) proposed the radiative association of C$^+$ and O to form CO$^+$, followed by charge transfer with atomic oxygen to yield CO. Calculations of the rate coefficients for formation of CO and CO$^+$ by Dalgarno, Du and You (1990) showed that the formation of CO was much more rapid than the formation of CO$^+$ at all but the very lowest temperatures (see Table 5). At the temperatures pertaining to the supernova ejecta during molecule formation, the rate for CO formation is an order of magnitude greater than for CO$^+$ formation.

Radiative association rate coefficients for the formation of SiO and SiO$^+$ were calculated by Andreazza, Singh and Sanzovo (1995). In analogy with CO and CO$^+$, the molecule-forming transitions were assumed to occur via $A^1\Pi$–$X^1\Sigma^+$ and $A^2\Pi$–$X^2\Sigma^+$, respectively (see Section 2.1.2). Their calculations for SiO also included formation through the $E^1\Pi$–$X^1\Sigma^+$ transitions, even though a barrier of ~ 0.5 eV exists in this channel. As can be seen from Table 4, at temperatures between 1500 K and 5000 K, the radiative association rate coefficient for SiO is greater by more than a factor of 10 than that for SiO$^+$.

In a paper identifying several infrared features in SN1987A as due to H$_3^+$, Miller et al. (1992) postulated significant amounts of H$_2^+$ made by associative ionization involving H($n = 2$) or H($n = 3$) with H, in addition to that which is formed by radiative association of H + H$^+$. An additional source of H$_2^+$ comes also from the radiative association of He$^+$ + H to form HeH$^+$ which can exchange a proton with H to form H$_2^+$. As described in the previous section, H$_2^+$ charge exchanges with neutral hydrogen atoms to yield H$_2$. The molecular ion H$_3^+$ is then formed (Miller et al. 1992) by

$$H_2 + H_2^+ \rightarrow H_3^+ + H. \tag{4.1}$$

Chemical models of molecular formation in SN 1987A reported by Lepp, Dalgarno and McCray (1990) also include the process of associative detachment involving O$^-$ colliding with O, C and H to form O$_2$, CO, and OH, as well as H$^-$ colliding with O and C to produce OH and CH. Rate coefficients for these reactions are not well known. Besides predicting the observed species, the

Lepp, Dalgarno and McCray (1990) models give significant abundances of other molecules: CS, O_2, H_2, HeH^+ and OH.

Rate coefficients for CS and CS^+ radiative association have been computed by Andreazza, Singh and Sanzovo (1995) and appear to be very similar: $1-2\times10^{-18}$ cm^3 s^{-1} over a temperature range of 4000 K. Recently, Andreazza and Singh (1997) have explored the formation of Si_2, C_2, C_2^+ and N_2^+ by radiative association over a temperature range from 300 to 14 700 K.

Liu (1997) has studied the formation of molecules in Type Ia supernovae. He has developed a chemical model for molecule formation based on two-body association processes. However, the high ionization fraction as well as the significantly lower densities in type Ia supernovae, as compared to type II, lead to predictions of very small masses of molecules formed in the ejecta.

Bibliography

1. Aitken, D. K., Smith, C. H., James, S. D., Roche, P. F., Hyland, A. R. and McGregor, P. J. (1988). *Mon. Not. R. Astron. Soc.*, **231**, Short Comm., 7P.
2. Andreazza, C. M., Singh, P. D. and Sanzovo, G. C. (1995). *Astrophys. J.*, **451**, 889.
3. Andreazza, C. M. and Singh, P. D. (1997). *Mon. Not. R. Astron. Soc.*, **287**, 287.
4. Babb, J. F. and Dalgarno, A. (1995). *Phys. Rev. A.* **51**, 3021.
5. Babb, J. F., Fan, Z. and Dalgarno, A. (1994). *J. Quant. Spect. Rad. Transfer*, **52**, 161.
6. Bates, D. R. (1951). *Mon. Not. R. Astron. Soc.*, **111**, 303.
7. Bates, D. R. (1987). *Astrophys. J.*, **312**, 363–366.
8. Bates, D. R. and Massey, H. S. W. (1954). *Phil Mag.*, **45**, 111.
9. Bieniek, R. J. (1980). *J. Phys. B*, **13**, 4405.
10. Bieniek, R. J. and Dalgarno, A. (1979). *Astrophys. J.*, **228**. 635.
11. Black, J. H., Porter, A. and Dalgarno, A. (1981). *Astrophys. J.*, **249**, 138.
12. Brzozowski, J., Bunker, P., Elander, N. and Erman, P. (1976). *Astrophys. J.*, **207**, 414.
13. Carrington, T. (1972). *J. Chem. Phys*, **57**, 2033.
14. Crosas, M. and Weisheit, J. C. (1993). *Mon. Not. R. Astron. Soc.*, **262**, 359.
15. Dalgarno, A., Babb, J. F. and Sun, Y. (1992). *Planet. Space. Sci.*, **40**, 243.
16. Dalgarno, A., Du, M. L. and You, J. H. (1990). *Astrophys. J.* **349**, 675.
17. Dalgarno, A. and Fox, J. L. (1994). In *Unimolecular and bimolecular reaction dynamics*, eds. C.-Y. Ng, T. Baer and I. Powis. Wiley, NY, p.1.
18. Dalgarno, A., Liu, W. and Lepp, S. (1991). In *Chemistry and spectroscopy of interstellar molecules*, eds. D. K. Bohme, E. Herbst, N. Kaifu, and S. Saito. University of Tokyo Press, Tokyo, p.221.
19. Dalgarno, A., Kirby, K. and Stancil, P. C. (1996). *Astrophys. J.*, **458**, 397.

20. Du, M. L. and Dalgarno, A. (1990). *J. Geophys. Res.*, **95**, A8, 12265.
21. Gianturco, F. A. and Gori Giorgi, P. (1997) *Astrophys. J.*, **479**, 560.
22. Golde, M. F. and Thrush, B. A. (1973). *Rep. Prog. Phys.*, **36**, 1285.
23. Haywood, S. E. and Delos, J. B. (1990). *Chem. Phys.*, **145**, 253.
24. Herbst, E. (1985). In *Molecular astrophysics*, eds. G. H. F. Diercksen et al. D. Reidel, Dordrecht, p.237.
25. Herbst, E. and Dunbar, R. C. (1991). *Mon. Not. R. Astr. Soc.*, **253**, 341.
26. Herzberg, G. (1950). *Spectra of diatomic molecules*. Van Nostrand, NY.
27. Julienne, P. S. and Krauss M. (1973). In *Molecules in the galactic environment*, eds. M. A. Gordon and L. E. Snyder, Wiley, NY, p.354.
28. Kimura, M., Dutta, C. M. and Shimakura, N. (1995). *Astrophys. J.*, **454**, 545.
29. Kimura, M., Lane, N. F., Dalgarno, A. and Dixson, R. G. (1993). *Astrophys. J.*, **405**, 801.
30. Kraemer, W. P., Špirko, V. and Juřek, M. (1995). *Chem. Phys. Lett.*, **236**, 177.
31. Larsson, M. (1995). *Physica Scripta* **T59**, 270.
32. Larsson, M., Danared, H., Mowat, J. R., Sigray, P., Sundstrom, G., Bostrom, L., Filevich, A., Kallberg, A., Mannervik, S., Rensfelt, K. G. and Datz, S., (1993). *Phys. Rev. Lett.*, **70**, 430.
33. Latter, W. B. and Black, J. H. (1991). *Astrophys. J.* **372**, 161.
34. Launay, J. M., Le Dourneuf, M. and Zeippen, C. J. (1991). *Astron. Astrophys.*, **252**, 842.
35. Lefebvre-Brion, H. and Field, R. W. (1986). *Perturbations in the spectra of diatomic molecules*. Academic Press, Orlando, FL.
36. Lepp, S., Dalgarno, A. and McCray, R. (1990). *Astrophys. J.*, **358**, 262.
37. Lepp, S. and Shull, J. M. (1984) *Astrophys. J.*, **280**, 465.
38. Liu, W. (1997). *Astrophys. J.*, **479**, 907.
39. Liu, W. and Dalgarno, A. (1994). *Astrophys. J.*, **428**, 769.
40. Liu, W. and Dalgarno, A. (1996). *Astrophys. J.*, **471**, 480.
41. Miller, S., Tennyson, J., Lepp, S. and Dalgarno, A. (1992). *Nature*, **355**, 420.
42. Petuchowski, S. J., Dwek, E., Allen, J. E. Jr. and Nuth, J. A. III, (1989). *Astrophys. J.*, **342**, 406.
43. Puy, D., Alecian, G., LeBourlot, J., Léorat, J. and Pineau des Forêts, G. (1993). *Astron and Astrophys.*, **267**, 337.
44. Ramaker, D. E. and Peek, J. M. (1976). *Phys. Rev. A.*, **13**, 58
45. Rawlings, J. M. C., Drew, J. E. and Barlow, M. J. (1993). *Mon. Not. R. Astron. Soc.*, **265**, 968.
46. Rowe, B. R., Mitchell, J. B. A. and Canosa, A. (eds.) (1993). *Dissociative*

recombination: Theory, experiment and applications. Plenum Press, NY.
47. Schadee, A. (1978). *J. Quant. Spect. Rad. Transfer*, **19**, 451.

48. Schmeltekopf, A. L., Fehsenfeld, F. C. and Ferguson, E. E. (1967). *Astrophys. J.*, **148**, L155.
49. Smith, W. H. and Zweibel, E. G. (1976). *Astrophys. J.*, **207**, 758.
50. Spyromilio, J. and Leibundgut, B. (1996). *Mon. Not. R. Astron. Soc.*, **283**, L89.
51. Spyromilio, J., Meikle, W. P. S., Learner, R. C. M. and Allen, D. A. (1988). *Nature*, **334**, 327.
52. Stancil, P. C, Babb, J. F. and Dalgarno, A. (1993). *Astrophys. J.*, **414**, 672.
53. Stancil, P. C, Lepp, S. and Dalgarno, A. (1996). *Astrophys. J.*, **458**, 401.
54. Stancil, P. C. and Dalgarno, A. (1997a). *Astrophys. J.*, **479**, 543.
55. Stancil, P. C. and Dalgarno, A. (1997b). *Astrophys. J.*, **490** (to be published).
56. Stancil, P. C. and Zygelman, B. (1996). *Astrophys. J.* **472**, 102.
57. Sun, Y. and Dalgarno, A. (1992) *J. Geophys. Res.*, **97**, A5, 6537.
58. Turner, J., Kirby-Docken, K. and Dalgarno, A. (1977). *Astrophys. J. Suppl.*, **35**, 281.
59. Urbain, X., Cornet, A., Brouillard, F. and Guisti-Suzor, A. (1991). *Phys. Rev. Lett.*, **66**, 1685.
60. Zygelman, B. and Dalgarno, A. (1990). *Astrophys. J.* **365**, 239.

PART II

Chemistry at the Births of the Galaxies and Stars

3
Molecules in the Early Universe and Primordial Structure Formation

Stephen Lepp and Phillip C. Stancil
Physics Department, University of Nevada, Las Vegas

1 Introduction

Molecular astrophysics began in the recombination era. As soon as electrons and nuclei combined to form atoms these atoms began combining to form molecular ions and molecules. The appearance of the first molecules marked the dawn of chemistry and set the stage for the subsequent evolution of the Universe.

The early Universe is a unique area of astrophysical study as the amount of observational data is currently severely limited. To predict the behaviour of the primordial gas, we must rely upon our knowledge of nonequilibrium chemistry, physical cosmology, hydrodynamics, and nucleosynthesis with constraints provided only by the nature of the Universe we see today and the few objects from the observable distant past, such as quasars and Lyman α absorption systems. Advances in parallel computing, numerical hydrodynamics, and chemical reaction rate data have made it possible to improve our description of this era. Emerging technologies and proposed satellite missions, such as the Next Generation Space Telescope, offer the hope of observing the first generation of cosmological objects.

In this chapter, we review recent progress in the understanding of the chemical evolution of the early Universe, particularly how molecules influenced the formation of the first stars and galaxies. We discuss the formation of the first molecules, how molecules helped determine the scale of the first objects, and how they might have helped amplify galaxy formation and lead to large-scale structure. We present an overview of the current status, remaining controversies, and anticipated direction of the field. In the interests of pedagogy we provide a minimum of references, which should be seen as starting places for further reading. Reviews of the subject have been given by Dalgarno and Lepp (1987), Black (1990), Shapiro (1992), and Dalgarno and Fox (1994). A popularized discussion can be found in Hartquist and Williams (1996). For an understanding or review of basic cosmology we recommend Peebles (1993).

We begin in Section 2 with the recombination era, the importance of chemistry, and the role molecules may have played. Section 3 discusses the advances made in understanding the gravitational collapse of the first objects in free-fall and with numerical hydrodynamics, the role of molecular cooling, and the effect of reheating by ultraviolet (UV) radiation from the first stars. Section 4 addresses structure formation behind radiative shocks and in protogalaxies, while high redshift Lyman α absorption systems are discussed in Section 5. Finally, our work in this area, and in others, has in large part been in collaboration with or motivated by Alex Dalgarno. Alex has been a conscientious and patient mentor and we are grateful for his generous encouragement and friendship.

2 The Recombination Era and the First Molecules

The recombination era began at a redshift of about $z = 2500$ when the temperature of the Universe became cool enough for nuclei and electrons to combine to form atoms. Before this time, the primordial gas was hot, dense, and completely ionized as a result of efficient photoionization due to the cosmic background radiation (CBR) field produced by the big bang. The matter and radiation were strongly coupled. As the Universe expanded the total particle density fell as $n \propto (1+z)^3$ and the temperature T_r of the CBR decreased as $T_r = 2.726(1+z)$ K. Today the redshift $z = 0$ and $T_r = 2.726$ K, the temperature of the currently observed microwave background. A timeline of important events in the early Universe is presented in Table 1.

The chemistry of this era was relatively simple since dust grains, cosmic rays, the interstellar radiation field, and stellar UV radiation had yet to appear. The only nuclei that were present were H, ^{4}He, D, ^{3}He, and ^{7}Li, with trace amounts of Be and B, as predicted by the Standard Big Bang Nucleosynthesis (SBBN) model (cf. Smith et al. 1993). The heavier elements, lumped under the designation "metals", were produced later in the much hotter and denser interiors of stars. As such, the metallicity Z of the recombination era was practically zero.

As the temperature of the gas fell, electrons and nuclei combined and formed atomic ions and neutral atoms through radiative recombination:

$$He^{++} + e^- \rightarrow He^+ + \nu, \quad (2.1)$$

$$He^+ + e^- \rightarrow He + \nu, \quad (2.2)$$

$$H^+ + e^- \rightarrow H + \nu. \quad (2.3)$$

"Recombination era" is a bit of a misnomer, as in fact at this time electrons and nuclei combined for the first time. Since the hydrogen nuclei account for $\sim$76% of the mass of the baryonic-gas, its formation at $z \sim 1300$ resulted in a phase transition to a mostly neutral Universe and a decoupling of matter from radiation.

The appearance of neutral atoms, through reactions (2.2) and (2.3), allowed for the formation of molecules. The molecular ions, He_2^+, HeH^+, and HeD^+, formed first as they can be made directly by radiative association, but they never reached significant abundances. H_2^+, HD^+, H_2, and HD were then produced with H_2 obtaining the largest fractional abundance $n(H_2)/n_H \sim 10^{-6}$. H_2^+ is formed

Table 1 Early Universe formation timeline as a function of redshift z. Partially adopted from Peebles (1993).

		redshift	time[a] (yrs)
Big bang		$z = \infty$	0
Recombination era:	He, He_2^+, and HeH^+	$z \sim 2500$	1×10^5
	H	$z \sim 1300$	3×10^5
	Li	$z \sim 450$	1×10^6
	H_2 and LiH	$z \sim 300$	3×10^6
Structure formation era:	Onset of linear collapse	$z < 100$	1×10^7
	Spheroids of galaxies	$z \sim 20$	1×10^8
	Pop III stars	$z \sim 12$–20	3–1×10^8
	First $\sim 0.02\%$ of metals	$z \sim 14$	2×10^8
	Reheating of IGM	$z \sim 10$–20	4–1×10^8
	AGN and QSO engines	$z \sim 10$	4×10^8
	Lyman limit clouds	$z \sim 10$	4×10^8
	Pop II stars	$z < 10$	$> 4 \times 10^8$
	Reionization of IGM	$z \sim 7$	6×10^8
	Lyman α forest clouds	$z \sim 5$	9×10^8
	Protogalaxies	$z \sim 5$	9×10^8
	Damped Lyman α clouds	$z \sim 2$–4	3–1×10^9
	First $\sim 10\%$ of metals	$z \sim 3$	2×10^9
	Galaxy clusters	$z \sim 2$	3×10^9
	Milky way disc	$z \sim 1$	5×10^9
	Superclusters, walls, and voids	$z \sim 1$	5×10^9
Today		$z = 0$	1.3×10^{10}

[a] For an assumed flat ($\Omega_0 = 1$) Einstein–de Sitter Universe with Hubble constant $H_0 = 50$ km s^{-1} Mpc^{-1}.

by the radiative association reaction

$$H^+ + H \rightarrow H_2^+ + \nu, \qquad (2.4)$$

but H_2, since it lacks a dipole moment, cannot be produced by similar processes. Instead, it is formed via the charge transfer reaction

$$H_2^+ + H \rightarrow H_2 + H^+, \qquad (2.5)$$

following (2.4) or

$$H^+ + He \rightarrow HeH^+ + \nu, \qquad (2.6)$$

followed by

$$HeH^+ + H \rightarrow H_2^+ + He. \qquad (2.7)$$

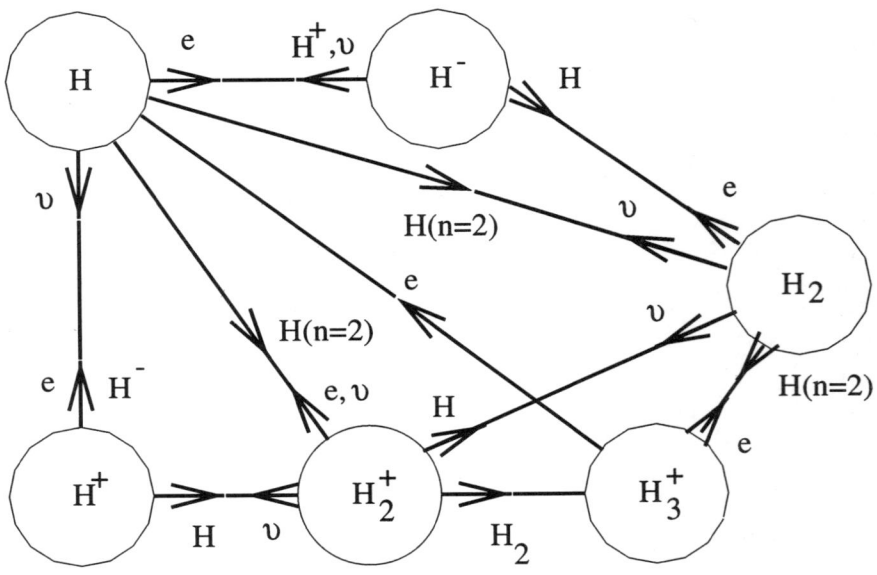

FIG. 1. A schematic of the important reactions in producing molecular hydrogen in the early Universe. Adapted from Dalgarno and Lepp (1996).

Reactions (2.6) and (2.7) were important for H_2 formation for $z > 500$, while process (2.5) dominated for $100 < z < 500$. For $z < 100$, H_2 was primarily produced through the so-called H^- sequence,

$$H + e^- \rightarrow H^- + \nu, \qquad (2.8)$$

$$H^- + H \rightarrow H_2 + e^-, \qquad (2.9)$$

as photodetachment of H^- became inefficient due to the decline of the CBR. Figure 1 displays the full reaction network for the hydrogen molecules. Other minor molecules like LiH, LiH$^+$, H_3^+, and H_2D^+ were also formed. Further details of the molecule production processes including compilations of the most recent reaction rate coefficients can be found in Abel et al. (1997b), Stancil et al. (1997), and Stancil et al. (1996) for the hydrogen, helium and deuterium, and lithium chemistries, respectively.

Figure 2 presents the fractional abundances of the ions, atoms, and molecules in the recombination era as a function of z. The abundances of most species settled to a constant value as $z \rightarrow 0$ since their timescales for formation became larger than the age of the Universe due to the decrease in density. The gas would have remained in this state unless some type of perturbation was introduced. Perturbations may have resulted from the gravitational collapse of clouds which were overdense compared to the background gas (see Section 3) or reheating due to the UV radiation of the first stars (see Section 3) and radiative shocks (see Section 4). Collapse is expected to have begun near $z \sim 100$–20 while reheating is believed to have occurred for $z < 20$. Therefore the results of Fig. 2 are only applicable down to these redshifts, but are of interest as the abundances of atoms and molecules produced in the recombination era provided the initial conditions for the next stage in the evolution of the Universe—the epoch of structure formation.

It is expected that recombination era primordial gas cannot be directly observed. Puy *et al.* (1993) calculated the flux from vibrational–rotational transitions of H_2, HD, and LiH integrated over redshifts $z = 1000$ to $z = 0$. They found the flux to be 10^{-8} times smaller than the measured microwave background precluding the possibility of detection. However, Dubrovich (1993) and Maoli *et al.* (1994) have suggested that Thomson scattering of CBR photons by primordial molecules with large dipole moments, such as H_2D^+ and LiH, may have attenuated spatial CBR anisotropies, remnants of density fluctuations in the Universe at the time of recombination, thereby leaving their imprint on the CBR spectrum. This CBR imprint may be observable if sufficient molecular abundances were produced. Palla *et al.* (1995) and Stancil *et al.* (1996, 1997) have found that LiH and H_2D^+ had abundances many orders of magnitude too small to have a significant effect, casting further doubt on the possibility of observing primordial molecules.

The predicted LiH abundance could be increased through the inclusion of the esoteric, and previously uninvestigated, process of stimulated radiative association,

$$\text{Li} + \text{H} + \nu_{\text{CBR}} \rightarrow \text{LiH} + \nu + \nu_{\text{CBR}}. \tag{2.10}$$

The LiH formation is enhanced by the presence of the CBR field since the frequency range of the CBR, with frequency ν_{CBR}, is comparable to the frequency ν of the radiative association transitions. Stancil and Dalgarno (1997) found that stimulated association could only increase the LiH abundance by $\sim 25\%$ in the recombination era, giving an abundance still too small to be observable.

The atomic and molecular abundances obtained in the nonequilibrium chemistry calculations are dependent on the assumed elemental and isotopic abundances, typically adopted from the SBBN model. The validity of this model is constrained by the elemental abundances observed in the Solar System and the interstellar medium. This material is not pure primordial gas as it has been processed in stars and galaxies. Estimates of the primordial elemental and isotopic abundances are then inferred from the application of theories of stellar and galactic evolution. Ideally, observations of pure primordial matter would be preferred, but as discussed above, the prospects appear dim. However, observations of material at relatively later epochs, at which primordial clouds began to collapse for example, may provide some useful constraints.

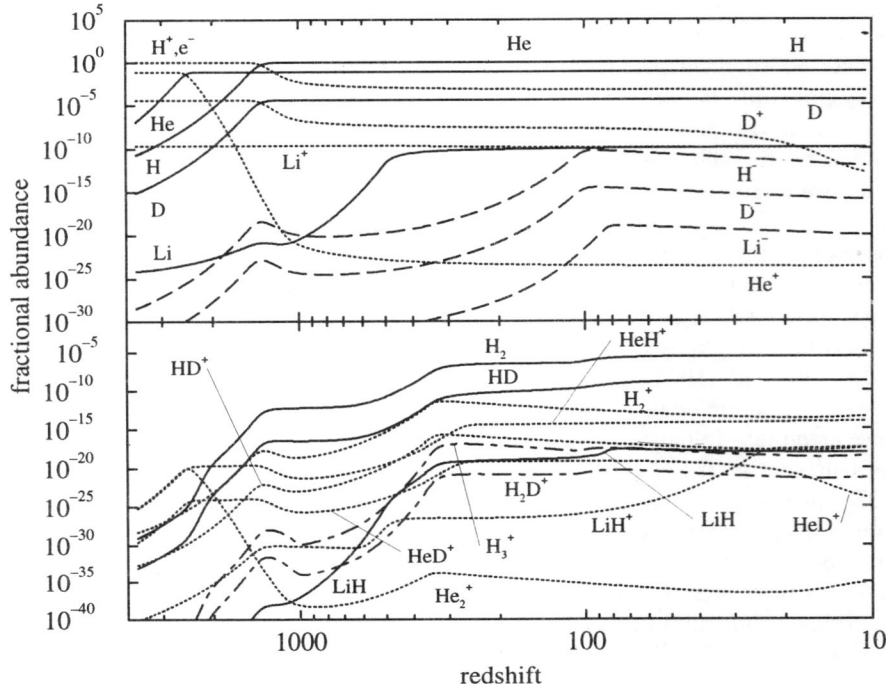

FIG. 2. The fractional abundances (relative to the number density of hydrogen nuclei) in the recombination era for an SBBN model (closure parameter $\Omega_0 = 1$, baryonic-matter fraction $\Omega_b = 0.0367$, Hubble constant $H_0 = 67$ km s^{-1} Mpc^{-1}, and primordial fractional abundances $n_{^4He}/n_H = 8.04 \times 10^{-2}$, $n_D/n_H = 4.0 \times 10^{-5}$, and $n_{Li}/n_H = 2.3 \times 10^{-10}$). Neutral species (solid lines), positively charged ions (dotted lines), negatively charged ions (dashed lines), and triatomics (dot–dashed lines). Adapted from Stancil et al. (1997).

3 Collapse of the First Objects

The formation of the first generation of stars (Population III) and galaxies has long been a fundamental problem in astronomy. The initial mass function (IMF) of Pop III stars and their epoch of formation are uncertain. We do know that the Universe was reionized by $z \sim 5$ due to the lack of a Gunn–Peterson trough in QSO (quasi-stellar object) spectra, placing the appearance of luminous objects prior to that time.

Due to the homogeneity of Milky Way globular clusters (GCs) and GCs observed in other galaxies, Peebles and Dicke (1968) suggested that GCs were the first objects to form. In addition, GCs have baryonic masses $M_b \sim 10^5$ M$_\odot$, comparable to the Jeans mass M_J of the recombination era, where

$$M_J \approx 2 \times 10^5 \text{ M}_\odot \left(\frac{T}{100}\right)^{3/2} n^{-1/2}, \qquad (3.1)$$

T is the matter temperature in K, n is in cm^{-3}, and M_J is the baryonic mass above which a self-gravitating cloud is unstable to collapse. As a cloud contracts, it will be heated adiabatically increasing its Jeans mass. If M_J becomes greater than M_b, the collapse will terminate as pressure forces become comparable to gravitational forces. Collapse can only continue if the gravitational energy can be removed. In contemporary Pop I star formation, energy removal is provided by radiative cooling from metal atoms, molecules, and ions (cf. Hartquist *et al.* 1993; Stahler 1994). Conversely, cooling in primordial gas is primarily provided by H$_2$ rotational and vibrational line emission since the metallicity Z, the sums of the fractional abundances of the elements heavier than helium, was practically zero. H$_2$ cooling is efficient for $T \sim 10^2 - 10^4$ K, while for $T > 10^4$ K, H line cooling will dominate. With H$_2$ cooling, the Jeans mass of a protoglobular cluster (PGC) is reduced, allowing it to collapse and possibly fragment into objects destined to become stars.

A preliminary investigation of the role of H$_2$ cooling in primordial star formation was conducted by Palla *et al.* (1983). Assuming a free-fall collapse in pressure equilibrium and nonequilibrium hydrogen chemistry, they demonstrated for an isolated spherical cloud that M_J would continuously fall from $\sim 10^5$ M$_\odot$ for $n = 10^2$ cm^{-3} to ~ 0.1 M$_\odot$ for $n = 10^{22}$ cm^{-3} as a consequence of the ability of H$_2$ cooling to slow the rise in temperature. All the H was converted to H$_2$ for $n > 10^{12}$ cm^{-3} due to three-body reactions, such as,

$$\text{H} + \text{H} + \text{H} \rightarrow \text{H}_2 + \text{H}, \qquad (3.2)$$

further enhancing the cooling. Palla *et al.* showed that primordial gas overdensities are able to fragment into low-mass stars, even given the lack of cooling from metals, but that M_J cannot provide a useful prediction of the Pop III IMF.

Due to the density and temperature dependence of the H$_2$ cooling function, any initial pressure gradients are likely to increase and the assumption of free-fall collapse might break down. Hence, to determine the Jeans mass that can be reached during the collapse requires detailed hydrodynamical modelling. Such hydrodynamical calculations have only been possible recently due to advances in computation. Haiman *et al.* (1996b) have followed the gas dynamics in addition to the nonequilibrium hydrogen chemistry of a 1D isolated object of spherical shells and found that H$_2$ cooling would have allowed the collapse to continue once the object had already virialized. The simulations produce bound objects with $M_b \sim 10^2 - 10^3$ M$_\odot$, depending on the dark matter overdensity amplitude; such objects could have been the progenitors of massive Pop III stars. Preliminary 3D simulations have been performed (Abel *et al.* 1997a), but are limited in mass resolution to $> 10^3$ M$_\odot$ and baryonic densities $n < 10^4$ cm^{-3}.

While 3D calculations are computationally expensive, their further development is considered to be a Grand Challenge problem as the first structures to collapse are believed to be sheet- or pancake-like as opposed to spherically symmetric.

Even 1D simulations are time-consuming and prohibitive for extensive parameter studies. To investigate the $M_{tot}-z$ distribution of the first bound objects (where M_{tot} is the sum of dark and baryonic mass), Tegmark et al. (1997) used a simple top-hat model for the density evolution. The density model gives good agreement with 3D simulations. For a cloud of mass M_{tot}, they followed the initial collapse and H_2 production until virialization. Whether a cloud continues to collapse depends on its ability to cool. If it can produce a sufficient amount of H_2, runaway cooling and collapse will follow; otherwise its cooling timescale will exceed a Hubble time, the cloud will remain pressure supported and fail to produce a luminous object. Lyman α clouds may be the descendants of these failed events. The minimum baryonic mass needed to collapse was shown to have a strong dependence on redshift requiring $\sim 10^3$ $M_\odot$ at $z_{vir} \sim 100$ and rising to $\sim 10^6$ $M_\odot$ at $z_{vir} \sim 10$ for a standard cold dark matter (CDM) model, where z_{vir} is the redshift of virialization. At z_{vir} the object is overdense by a factor $18\pi^2$ compared to the background gas or intergalactic medium (IGM) and its density remains constant at this value thereafter.[1] The actual magnitude of the $M_{tot}-z_{vir}$ curve depends sensitively on the adopted density-dependent H_2 cooling function, as shown in Fig. 3. Improvements to the H_2 cooling function are in progress.

Lepp and Shull (1984) investigated the effect of cooling from other primordial molecules, including HD and LiH, in a free-fall collapse model. HD and LiH might be expected to be potentially important for cooling at temperatures below which H_2 is an effective coolant because the energies of their $J = 1$ levels are much lower than the $J = 2$ level of H_2, which because H_2 lacks a permanent dipole moment (cf. Chapter 1) is the lowest level that can contribute to radiative cooling. While HD and LiH are also expected to survive longer than H_2 since they have slower collisional dissociation rates, their abundances remain too small to be significant coolants (cf. Fig. 2) for low to medium densities. However, during the later stages of Pop III star formation, LiH may dominate the cooling as H_2 emission becomes optically thick for $n > 10^{13}$ cm^{-3} trapping its cooling radiation. Improved studies incorporating the D and Li chemistries with modern reaction rates and cooling functions are in progress, but hydrodynamical simulations approaching stellar densities remain on the horizon.

The production of Pop III stars may have had profound effects on subsequent structure evolution due to the generation of UV radiation and metals. Ostriker and Gnedin (1996) have recently modelled the era of Pop III star formation including reheating, reionization, early metal production, and gas clumping. The simulation involves the evolution of a cube of space ~ 3 Mpc on a side with a

---

[1] The $18\pi^2$ factor is an analytical result from the theory of spherical nonlinear collapse (Padmanabhan 1993).

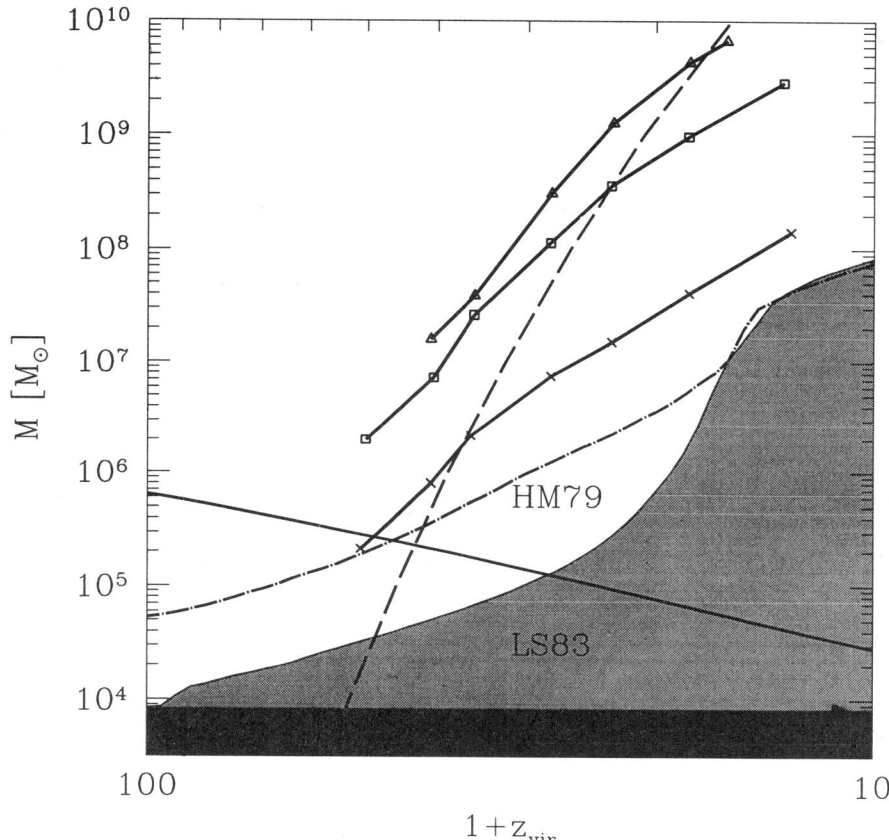

FIG. 3. Collapsed total mass M_{tot} vs. redshift of virialization z_{vir} for a CDM Universe with $\Omega_{\text{b}} = 0.06$. The dark shaded region depicts the mass scale for which the virial temperature equals the CBR temperature. Only above the light shaded area labelled with LS83 are structures believed to be able to collapse (Tegmark *et al.* 1997). The dotted lines are results from the 3D hydrodynamic simulations of Abel *et al.* (1997a) for different cell grid sizes; the crosses representing the finest. The dashed upward sloping line depicts a CDM spectrum scaled appropriately for 4σ peaks. The solid downward sloping line depicts the Jeans mass for $18\pi^2$ times the background density. The dot–dashed line depicts the delimiting line computed by Tegmark *et al.* (1997) who adopted a modified form of the Hollenbach and McKee (1979) H_2 cooling function. LS83 was determined with the H_2 cooling function of Lepp and Shull (1983). Adapted from Abel *et al.* (1997a).

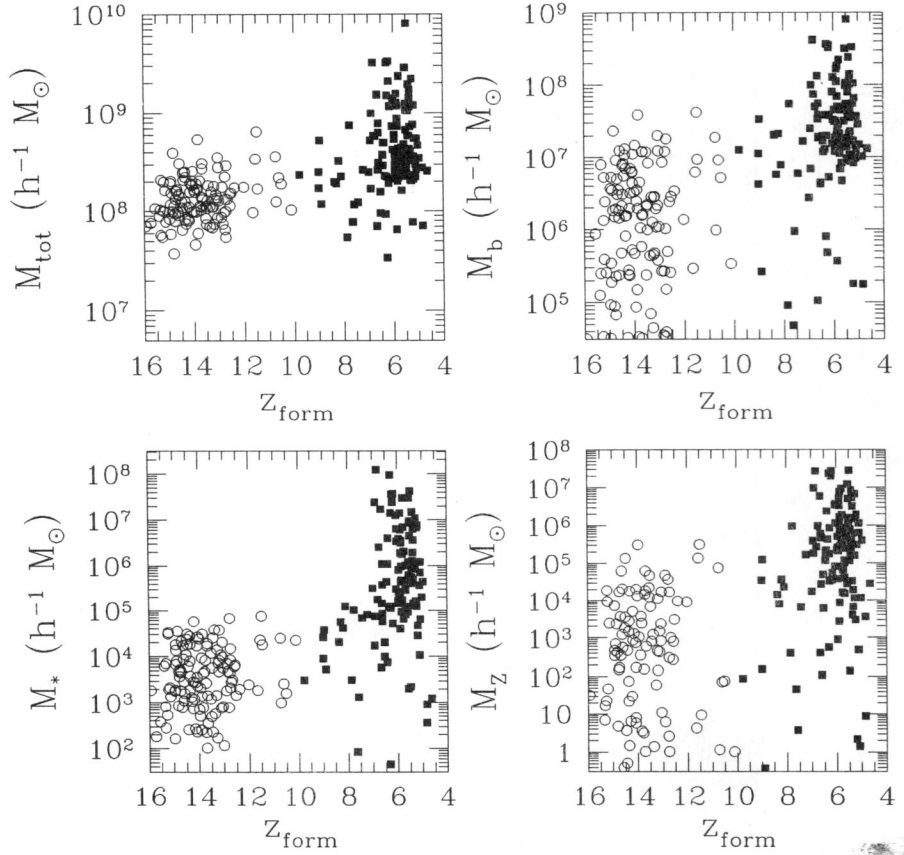

FIG. 4. Total mass M_{tot}, baryonic mass M_b, stellar mass M_*, and mass in metals M_Z for all bound objects as a function of average redshift of star formation at $z = 4.3$. Open circles and filled squares show distinct stellar populations corresponding to Pop III and Pop II, respectively. Adapted from Ostriker and Gnedin (1996).

resolution of ~1 kpc using a 3D hydrodynamics code including H_2 nonequilibrium chemistry, but with star formation treated phenomenologically. The simulation suggests, as displayed in Fig. 4, that a burst of star formation (Pop III) began at $z \sim 20$ due to H_2 cooling producing stars of mass $\sim 10^2$–10^5 $M_\odot$. These stars would have generated UV radiation which would have begun to reheat the IGM, raising M_J, and eventually suppressing star formation by $z \sim 12$. Once the temperature reached $\sim 10^4$ K, H line cooling became efficient and a second burst of star formation (Pop II) began at $z < 10$. The UV radiation from the Pop III and the earliest Pop II stars reionized the surrounding gas creating the

IGM observed today. The reionization would have allowed for the suppression of spatial anisotropies in the CBR, the task which LiH was too underabundant to fulfil (see Section 2). Some Pop III stars evolved to Type II supernovae and expelled metals into the primordial gas. Ostriker and Gnedin (1996) found that by $z \sim 14$ the metallicity $Z \sim 10^{-3.7} Z_\odot$. This small metallicity could have contributed to the Pop II star formation by providing cooling in addition to that of H line emission.

Objects, such as PGCs, originally containing Pop III stars accreted mass and produced Pop II stars when T reached 10^4 K. Sequential Pop III and Pop II supernovae enhanced Z until eventually Pop I stars were made evolving a PGC into a Pop I GC with relic Pop II stars, similar to the objects seen in the halo of the Milky Way. An interesting point is that for the same redshift, objects dominated by Pop II stars were much older than objects dominated by Pop III stars. The objects dominated by Pop II stars were all objects dominated by Pop III at an earlier time which continued to evolve and form Pop II stars (Ostriker and Gnedin 1996).

While there is a wealth of direct observational data for $z < 5$ from QSO emission lines and Lyman α absorption systems (see Section 5), data at larger redshifts from the era of structure formation is nonexistent. The only possible exception is a measurement by de Bernardis et al. (1993) of LiH vibrational–rotational lines in a collapsing protocloud at $z \sim 180$. The observation has not been confirmed. Additionally, Maoli et al. (1996) have proposed a search for LiH rotational lines. In modelling the collapse of massive protoclouds, $M_b = 10^9$–10^{12} $M_\odot$, they predicted that the rotational line intensities would reach a maximum of $\sim 10^{-3}$ of the CBR intensity just after the onset of the nonlinear collapse regime for $z \sim 3$–12. While they found very narrow line widths $\Delta\nu/\nu < 10^{-4}$, their assumed constant fractional LiH abundance, $n(\mathrm{LiH})/n_H \sim 10^{-9}$, seems overoptimistic in light of the small $\sim 10^2$ density enhancement during the collapse and the recombination era fractional abundance of $\sim 10^{-17}$ calculated by Stancil et al. (1996) (see Fig. 2). Puy and Signore (1996) have followed the collapse of protoclouds with similar properties, but found only a factor of ~ 2 increase in the LiH fractional abundance for a density enhancement of ~ 5. It appears difficult to achieve the $\sim 10^8$ increase in $n(\mathrm{LiH})/n_H$ necessary to obtain the rotational intensities predicted by Maoli et al. (1996); however as the densities approached the protostellar range, three-body reactions, for example

$$\mathrm{Li + H + H \rightarrow LiH + H}, \qquad (3.3)$$

may have significantly increased the LiH fractional abundance (Stancil et al. 1996). Rate coefficients for most lithium reactions are, however, unknown.

The best hope for observing high redshift primordial material may lie in the path to protostellar formation. Lepp and Shull (1984) have estimated that $\sim 10^{47}$ ergs $M_\odot^{-1}$ of gravitational energy must be radiated away for a protocloud to collapse to a protostar from $z = 50$. This energy can only leave the cloud in the form of molecular cooling line emission. The evolutionary path of a protostar on a Hertzsprung–Russell diagram is nearly vertical since H_2 cooling keeps the temperature fairly constant over a large density range. The luminosity from a collapsing 5 $M_\odot$ primordial protostar approaches 10^4 $L_\odot$ just prior to its becoming a pre-main-sequence star, and unlike in Pop I star formation, the optical radiation will not be degraded to the infrared since the protocloud lacks dust. See Stahler (1986) for a review of primordial star formation.

4 Shocks and Galaxy Formation

Shocks played an important role in the coalescing of clouds to form the first galaxies and stars. Shocks may have occurred during collapse of the first objects, in cloud–cloud collisions, in blast waves driven by the nuclear burning of earlier objects or in expanding voids. Shocks compress and heat the gas. Increasing the density of the gas increases the rates of chemical reactions and the rate at which the gas cools. This additional radiation from shock-heated gas is crucial in allowing continued collapse of protogalaxies. Shocks are also important in determining how a protogalaxy fragments into stars. In a bottom-up scenario of galaxy formation, numerous PGCs assemble into a protogalaxy spheroid. Collisions between the PGCs would induce shocks which could trigger fragmentation and subsequent star formation.

In many of the biased galaxy formation scenerios, shocks played a crucial role in the intergalactic medium. Galaxy formation was stimulated either by galactic explosions (Ostriker and Cowie 1981) or by the expansion of a void. Both these led to a self-similar expansion in which intergalactic material was swept up in a shell. A shock formed in the shell and if there was sufficient cooling in the dense post-shock gas, the gas collapsed to stimulate more galaxy formation.

The cooling in the postshock region was critical in determining the fragmentation size, or even if the postshock region would collapse. In the absence of heavy elements, cooling much below 10 000 K must come from molecules. In order for the shells to have become Jeans unstable they must have cooled to low temperatures; without molecular cooling the shells would not have collapsed to form molecules (Vishniac et al. 1985). Without this cooling they may have still formed Lyman α systems.

Models of molecular formation in the cooling shells in the intergalactic medium have been constructed by Wandel (1985), who used an equilibrium chemistry and by MacLow and Shull (1986), Shapiro (1986), Shapiro and Kang (1987) and Kang and Shapiro (1992) with increasing degrees of sophistication for nonequilibrium chemistry. The main result has remained much the same: for a large range of parameters the gas turns partially molecular with an H_2 fractional abundance of order 10^{-3}. This allows the gas to cool to temperatures of order 100 K.

The gas behind a shock was initially highly ionized and at a temperature of over 10 000 K. The gas cooled by inverse Compton scattering and then Lyman α emission. The cooling time was shorter than the recombination time and so the gas was still highly ionized. Lyman α cooling is effective only above about

7000 K and below this molecules must cool the gas. Molecular hydrogen is dissociated for temperatures higher than about 3000 K (cf. Chapter 8), and so cooling initially must have been due to molecules that formed but were then dissociated behind the shock. The molecules formed in a similar manner to that of the early Universe, through radiative association leading to H_2^+ (reaction 2.4) and radiative attachment to H^- (reaction 2.8). The gas also recombined and a competition between recombination and molecular formation meant that at most a fractional abundance of approximately 10^{-3} of H_2 was produced.

A higher formation rate would have resulted if a somewhat more rapid ionization was induced. However, a too high ionization rate would have prevented gas from cooling below 10 000 K and molecules from forming. Kang and Shapiro (1992), in a model that included both secondary ionization made in the shock itself and an estimate of the ionization due to quasars, found that many shocks never cooled to below 10 000 K within the age of the Universe. Thus, galaxy formation by this process must have turned off at some redshift. However, for $z \sim 2$–4, Haiman et al. (1996a) have suggested that quasar UV radiation could actually enhance gravitational collapse due to increased ionization and, hence, increased H_2 production.

Shocks also play a role in GC formation. For a nonequilibrium model where gravitational collapse is driven by H_2 cooling, Vietri and Pesce (1995) found that a PGC could fragment into stellar-sized masses after a shock. This fragmentation occurred if the radius of the PGC was large enough that the fragments could cool prior to the return of the shock after reflecting from the core. Otherwise the reflected shock would have returned the PGC material to the IGM. The predicted masses agree well with observed Galactic GCs.

5 The Intergalactic Medium and Lyman α Clouds

Sometime around a redshift of $z = 5$ the Universe was reionized. This is required by the results of the Gunn–Peterson test which puts a limit on the neutral hydrogen abundance in the intergalactic medium based on absorption in Lyman α lines. QSO emission is used as a background source of radiation to search for resonant Lyman α absorption from diffuse neutral hydrogen. The limits placed are much smaller than the total baryon density. This failure to detect neutral hydrogen suggests either that much of the gas has already collapsed to form compact objects or that the gas has been ionized. The standard model is that the reionization occurs when QSOs and/or Pop III stars form and turn on to produce large fluxes of UV and X-ray ionizing radiation.

Shapiro et al. (1994) have shown that in a Universe dominated by cold dark matter, QSOs would not have been able to ionize the intergalactic medium, but that it may be ionized by radiation emitted from collapsed baryonic virialized structures such as young star-forming protogalaxies. This ionizing radiation can eventually inhibit collapse altogether. The authors argued that this feedback mechanism can explain the distribution of quasars in redshift with the earliest quasars appearing at about $z = 5$ and the last ones at around $z = 2$.

Lines of sight toward QSOs also show a forest of absorption lines. The absorption lines come mostly from low column density $[N(\text{H}) \sim 10^{13}$–$10^{15} \text{ cm}^{-2}]$ objects called Lyman α forest clouds. If the column density of a cloud exceeds that needed to be opaque at the ionizing threshold $[N(\text{H}) > 3 \times 10^{17} \text{ cm}^{-2}]$, then the cloud is called a Lyman limit cloud. The abundance of Lyman α forest and Lyman limit clouds decreases with increasing column density. The total hydrogen mass in the Lyman α clouds is small compared to that in galaxies and they are dissipating.

Lyman α clouds, because they are objects containing at most small abundances of heavy metals synthesized in stars, are the best sources at present in which to measure the primordial abundances. Because of their low metallicity it is unlikely that the material has had any significant amount of reprocessing. Recent measurements (Songaila et al. 1994; Rugers and Hogan 1996) with the Keck telescope suggest that the deuterium to hydrogen ratio, D/H, in a Lyman α cloud at $z \sim 2.8$ is about 1.9×10^{-4}, nearly 10 times the ratio inferred for the local interstellar medium. The measurement is made in observations of absorption in deuterium's equivalent of the H Lyman α line. The deuterium line, since D is twice as massive, should have a thermal width of $\frac{1}{\sqrt{2}}$ of the hydrogen line. The detection is not only at the correct wavelength; it also has the expected width for a deuterium line, making the certainty of detection very high. On another line of sight, also observed with the Keck telescope, Tytler et al. (1996) detected a deuterium line of the correct wavelength and width relative to a Lyman α line at $z \sim 3.57$. Their D/H ratio, though, is a factor of 10 smaller, D/H = 1.6×10^{-5}. Clearly more data are needed, but eventually the Lyman α clouds should provide the best measure of the primordial D/H ratio.

The largest QSO absorption systems are called damped Lyman α clouds because they display a damped Lyman α profile. They have column densities $N(\text{H}) > 10^{20} \text{ cm}^{-2}$ which are comparable to interstellar gas in present era spiral galaxies. Because the column densities are so large, it has been suggested that detectable amounts of molecules may be present. Searches for CO have been unsuccessful, but Foltz et al. (1988) have detected lines from the Lyman and Werner bands of molecular hydrogen in one of these damped Lyman α systems at $z = 2.811$. The detection, confirmed by Cowie and Songaila (1995), also allows a measurement of the electron to proton mass ratio at this redshift because the wavelength differences between the molecular lines depend on this mass ratio. Cowie and Songaila found that the variation in m_e/m_p is less then 7×10^{-4} over this time period, $\sim 10^{10}$ years.

6 Summary

In this chapter we have reviewed the role of atoms and molecules in the early Universe. The influence of atomic and molecular processes was very large, ultimately controlling the formation of the first objects, from large-scale structure to the first galaxies to the first stars. Without the cooling provided by primordial atoms and molecules, these objects would never have formed. Details in the atomic and molecular processes can have a large influence on the sizes and structures of the final objects.

Atoms and molecules, because they emit in lines rather than continua, can also provide the best probe into these regions. The easiest objects to detect are the Lyman α clouds. These provide an additional probe to the mass distribution and structure formation in the Universe. Ultimately, one would like to detect emission from the collapsing objects themselves and from the shocks which may be triggering these events.

Acknowledgements

We thank Drs T. Abel and N. Gnedin for kindly providing some of the figures and Drs T. Hartquist and T. Abel for comments on the manuscript. This work was supported by NSF grant OSR-9353227.

Bibliography

1. Abel, T., Anninos, P., Norman, M. L. and Zhang, Y. (1997a). *New Astron.*, **2**, 209.
2. Abel, T., Anninos, P., Zhang, Y. and Norman, M. L. (1997b). *New Astron.*, **2**, 181.
3. Black, J. H. (1990). In *Molecular Astrophysics*, ed. T. W. Hartquist. Cambridge Univ. Press, Cambridge, p.473.
4. Cowie, L. L. and Songaila, A. (1995). *ApJ*, **453**, 596.
5. Dalgarno, A. and Fox J. L. (1994). In *Unimolecular and Bimolecular Reaction Dynamics*, eds. C. Y. Ng, T. Baer, and I. Powis. Wiley, Chichester, 1.
6. Dalgarno, A. and Lepp, S. (1987). In *Astrochemistry*, ed. S. P. Tarafdar and M. P. Varshni. Reidel, Dordrecht, p.109.
7. Dalgarno, A. and Lepp, S. (1996). In *Handbook of Atomic and Molecular Physics* ed. G. Drake, AIP Press, New York.
8. de Bernardis, P. *et al.* (1993). *A&A*, **269**, 1.
9. Dubrovich, V. K. (1993). *Astron. Lett.*, **19**, 132.
10. Foltz, C. B., Chaffee, F. H. and Black, J. H. (1988). *ApJ*, **324**, 267.
11. Haiman, Z., Rees, M. J. and Loeb, A. (1996a). *ApJ*, **467**, 522.
12. Haiman, Z., Thoul, A. A. and Loeb, A. (1996b). *ApJ*, **464**, 523.
13. Hartquist, T. W. and Williams, D. A. (1996). *The Chemically Controlled Cosmos*, Cambridge Univ. Press, Cambridge.
14. Hartquist, T. W., Rawlings, J. M. C., Williams, D. A. and Dalgarno, A., (1993). *QJRAS*, **34**, 213.
15. Hollenbach, D. and McKee, C. F. (1979). *ApJS*, **342**, 555.

16. Kang, H. and Shapiro, P. R. (1992). *ApJ*, **386**, 432.
17. Lepp, S. and Shull, J. M. (1983). *ApJ*, **270**, 578.
18. Lepp, S. and Shull, J. M. (1984). *ApJ*, **280**, 465.
19. MacLow, M.-M. and Shull, J. M. (1986). *ApJ*, **302**, 585.
20. Maoli, R., Melchiorri, F. and Tosti, D. (1994). *ApJ*, **425**, 372.
21. Maoli, R., Ferrucci, V., Melchiorri, F., Signore, M. and Tosti, D. (1996). *ApJ*, **457**, 1.
22. Ostriker, J. P. and Cowie, L. L. (1981). *ApJ*, **234**, L127.
23. Ostriker, J. P. and Gnedin, N. Y. (1996). *ApJ*, **472**, L63.
24. Padmanabhan, T. (1993). *Structure Formation in the Universe*. Cambridge Univ. Press, Cambridge.
25. Palla, F., Salpeter, E. E. and and Stahler, S. W. (1983). *ApJ*, **271**, 632.
26. Palla, F., Galli, D. and Silk, J. (1995). *ApJ*, **451**, 44.
27. Peebles, P. J. E. (1993). *Principles of Physical Cosmology*. Princeton Univ. Press, Princeton.
28. Peebles, P. J. E. and Dicke, R. H. (1968). *ApJ*, **154**, 891.
29. Puy, D., Alecian, G., Le Bourlet, J., Léorat, J. and Pineau des Forêts, G. (1993). *A&A*, **267**, 337.
30. Puy, D. and Signore, M. (1996). *A&A*, **305**, 371.
31. Rugers, M. and Hogan, C. J. (1996). *ApJ*, **459**, L1.
32. Shapiro, P. R. (1992). In *Astrochemistry of Cosmic Phenomena*, ed. P. D. Singh. Inter. Astron. Union, Kluwer, Dordrecht, p.73.
33. Shapiro, P. R. and Kang, H. (1987). *ApJ*, **318**, 32.
34. Shapiro, P. R., Giroux, M. L. and Babul, A. (1994). *ApJ*, **427**, 25.
35. Smith, M. S., Kawano, L. H. and Malaney, R. A. (1993). *ApJS*, **85**, 219.
36. Songaila, A., Cowie, L. L., Hogan, C. J. and Rugers, M. (1994). *Nature*, **368**, 599.
37. Stahler, S. W. (1986). *PASP*, **98**, 1081.
38. Stahler, S. W. (1994). *PASP*, **106**, 337.
39. Stancil, P. C. and Dalgarno, A. (1997). *ApJ*, **479**, 543.
40. Stancil, P. C., Lepp, S. and Dalgarno, A. (1996). *ApJ*, **458**, 401.
41. Stancil, P. C., Lepp, S. and Dalgarno, A. (1997). Submitted.
42. Tegmark, M., Silk, J., Rees, M. J., Blanchard, A., Abel, T. and Palla, F. (1997). *ApJ*, **474**, 1.
43. Tytler, D., Fan, X.-M. and Burles, S. (1996). *Nature*, **381**, 207.
44. Vietri, M. and Pesce, E. (1995). *ApJ*, **442**, 618.
45. Vishniac, E. T., Ostriker, J. P. and Bertschinger, E. (1985). *ApJ*, **291**, 399.
46. Wandel, A. (1985). *ApJ*, **294**, 385.

4
The Chemistry of Diffuse and Dark Interstellar Clouds

Ewine F. van Dishoeck
Leiden Observatory

1 Introduction

Ever since the realization early this century that the space between the stars is not empty, the study of the physics and chemistry of the interstellar gas has played a prominent role in astrophysics (Eddington 1926). Most of the early work focused on absorption line observations of diffuse interstellar clouds, which do not completely obscure the light from bright background stars. The electronic transitions of atoms and molecules can then be seen as sharp lines superposed on the stellar spectra at visible and ultraviolet wavelengths. After the development of millimetre telescopes in the early 1970s, most of the interest shifted to dense, dark molecular clouds where a rich chemistry is revealed through the rotational emission lines of molecules (see Table 1). Diffuse and dark clouds form the essential link in the lifecycle between the death and birth of stars: late-type stars return heavy elements and dust grains back to the diffuse interstellar medium, whereas new stars are born deep inside dense molecular clouds. A good understanding of these processes is therefore a prerequisite for interpreting observations of the interstellar medium and star formation in the local and distant Universe.

The first interstellar molecules CH, CH^+ and CN were identified between 1937 and 1941. Hints that much more complex species are present go back at least 15 years earlier, when Heger (1922) noted the diffuse interstellar bands. Sixty years later, both the abundance of CH^+ and the carrier of the diffuse bands are still subject to intense debate. Their presence may well hold important clues to the physical processes occurring in interstellar clouds.

Over the last 60 years, many models have been developed to interpret the observed abundances of interstellar molecules (see van Dishoeck 1990a for a short historical account). Two basic schemes have been considered for the formation of molecular bonds in interstellar space: the first occurs primarily in the gas phase by ion–molecule reactions (Bates and Spitzer 1951; Herbst and Klemperer 1973; Black and Dalgarno 1973), whereas the second involves mostly reactions on the surfaces of interstellar grains (Hollenbach and Salpeter 1971; Tielens and Hagen 1982). In recent years, observations of an increasing number of molecules

Table 1 Identified interstellar and circumstellar molecules

Simple hydrides, oxides, sulphides, halogens and related molecules

H_2 (IR)	CO	NH_3	CS	$NaCl^*$
HCl	SiO	SiH_4^* (IR)	SiS	$AlCl^*$
H_2O	SO_2	C_2 (IR)	H_2S	KCl^*
N_2O	OCS	CH_4 (IR)	PN	AlF^*

Nitriles and acetylene derivatives

C_3 (IR,UV)	HCN	CH_3CN	HNC	$C_2H_4^*$ (IR)
C_5^* (IR)	HC_3N	CH_3C_3N	HNCO	C_2H_2 (IR)
C_3O	HC_5N	CH_3C_5N ?	HNCS	
C_3S	HC_7N	CH_3C_2H	HNCCC	
C_4Si^*	HC_9N	CH_3C_4H	CH_3NC	
	HC_2CHO	CH_3CH_2CN	HCCNC	
		CH_2CHCN		

Aldehydes, alcohols, ethers, ketones, amides and related molecules

H_2CO	CH_3OH	HCOOH	CH_2NH	CH_2CC
H_2CS	CH_3CH_2OH	$HCOOCH_3$	CH_3NH_2	CH_2CCC
CH_3CHO	CH_3SH	$(CH_3)_2O$	NH_2CN	
NH_2CHO	$(CH_3)_2CO$	H_2CCO	CH_3COOH	

Cyclic molecules

C_3H_2	SiC_2	c-C_3H	CH_2OCH_2	

Molecular ions

CH^+ (VIS)	HCO^+	$HCNH^+$	H_3O^+	HN_2^+
HCS^+	$HOCO^+$	HC_3NH^+	HOC^+	H_3^+ (IR)
CO^+	H_2COH^+	SO^+		

Radicals

OH	C_2H	CN	C_2O	C_2S
CH	C_3H	C_3N	NO	NS
CH_2	C_4H	$HCCN^*$	SO	SiC^*
NH (UV)	C_5H	CH_2CN	HCO	SiN^*
NH_2	C_6H	CH_2N	MgNC	CP^*
HNO	C_7H	NaCN	MgCN	
	C_8H			

From Ohishi (1997) with detections as of May 1997 added; Species denoted with * have only been detected in the circumstellar envelope of carbon-rich stars. Most molecules have been detected at radio and millimetre wavelengths, unless otherwise indicated (IR, VIS or UV).

indicate that both schemes must be taken into account, with gas-phase processes dominating the formation of some molecules and grain surface chemistry that of others.

The interstellar medium has a complex, inhomogeneous structure, in which many different types of clouds can be distinguished on the basis of their physical characteristics. In this chapter, the chemistry of both the diffuse, low density molecular clouds with $A_V \approx 1$ mag and dark, high density clouds with $A_V > 5$ mag are considered. Here A_V is the visual extinction in magnitudes, which equals 1.086 times the optical depth at 5550 Å due to absorption and scattering by dust. In the diffuse clouds, ultraviolet radiation dissociates molecules, ionizes species with ionization potentials less than that of atomic hydrogen, and heats the gas. In cold dark clouds the molecules are shielded from external radiation to a sufficient extent that cosmic rays provide the major source of ionization. The translucent clouds with $A_V \approx$ 1–5 mag form the bridge between these two types of clouds. These regions are probably the best testing ground for the basic physical structures and chemical networks. A good knowledge of the chemical state of quiescent clouds prior to star formation is needed to determine quantitatively the effects that young stars have on their surroundings.

This chapter starts with a review of the basic molecular processes in space (§2) and the principal chemical routes in networks of diffuse and dark clouds (§3). The main characteristics of depth- and time-dependent models are presented in §4. The observational techniques for studying diffuse, translucent and dark clouds are summarized in §5, and the available diagnostic tools are reviewed. In §6–8, recent results on these clouds are illustrated with a few examples. Finally, some outstanding problems are mentioned in §9. Because of space limitations, it is not possible to mention all recent developments in this vast area, but references to key reviews are given throughout.

The excitation and abundances of molecules contain key information on the physical structure and evolution of interstellar clouds. Alex Dalgarno's basic research and excellent training of several generations of molecular astrophysicists have been essential for the development of this field.

2 Basic Chemical Reactions in Space

Modern networks developed to describe the chemistry in diffuse and dark clouds contain up to 4000 different reactions between several hundred species (Millar et al. 1997; Lee et al. 1996a). However, only a few different types of reactions occur, which are summarized in Table 2. They have been described in detail in reviews by Dalgarno (1987) and van Dishoeck (1988). At the low densities in the interstellar gas, three-body processes are unimportant, so that only two-body reactions need to be considered. The rate of a reaction between species X and Y is given by $\alpha n(X) n(Y)$ in cm^{-3} s^{-1}, where α is the reaction rate coefficient in cm^3 s^{-1} and n is the concentration in cm^{-3}.

2.1 Formation Processes

There are two basic processes by which molecular bonds can be formed. The first one is radiative association of atoms or molecules, in which a new molecule is stabilized by the emission of a photon (see Chapter 2). The second process involves formation on the surfaces of grains, in which the grain carries off the released energy. A third process, called associative detachment, plays a minor

Table 2 Types of molecular processes

Bond Formation Processes	
Radiative association	$X + Y \rightarrow XY + h\nu$
Grain surface formation	$X + Y{:}g \rightarrow XY + g$
Associative detachment	$X^- + Y \rightarrow XY + e$
Bond Destruction Processes	
Photodissociation	$XY + h\nu \rightarrow X + Y$
Dissociative recombination	$XY^+ + e \rightarrow X + Y$
Collisional dissociation	$XY + M \rightarrow X + Y + M$
Bond Rearrangement Processes	
Ion–molecule exchange	$X^+ + YZ \rightarrow XY^+ + Z$
Charge–transfer	$X^+ + YZ \rightarrow X + YZ^+$
Neutral–neutral	$X + YZ \rightarrow XY + Z$

role in dense cloud chemistry but is important, for example, in the chemistry of the early Universe (see Chapter 3).

Many of the rate coefficients for these processes have considerable uncertainties of up to an order of magnitude. The radiative association process is so slow that it is very difficult to measure properly in a laboratory on Earth, where other processes dominate (Gerlich and Horning 1992). Theoretical calculations can give accurate results for the formation of diatomic molecules, but rapidly become more cumbersome for large species with many degrees of freedom. The effective rate coefficients for reactions on surfaces depend on the probability that the atoms or molecules stick to the grains upon collision, their mobility on the surface, the probability that molecule formation occurs, and finally the probability that the molecule is released back into the gas phase (Tielens and Allamandola 1987; see §3.8). Light species such as H, H_2, C, N and O can hop over the surface to find a reaction partner, whereas heavier species are immobile at low dust temperatures $T_D \approx 10$ K.

2.2 Destruction Processes

Diffuse and translucent clouds are permeated by intense ultraviolet radiation, which can destroy molecular bonds through the process of photodissociation. The average interstellar radiation field in the solar neighbourhood has been estimated by different techniques and is characteristic of a diluted early B star, peaking in the far ultraviolet between 1200 and 1600 Å, with a sharp cut-off at the Lyman limit of 912 Å (13.6 eV) (see van Dishoeck 1994 for an overview). For some molecules such as H_2, CO and CN, the photodissociation can take place only at very short wavelengths between 912 and 1100 Å, whereas for other species such as CH and OH photodissociation can occur by radiation out to 3000 Å. The ultraviolet photons can also ionize atoms, thereby increasing the electron abundance in the cloud. In the unshielded radiation field, typical lifetimes against photodissociation or photoionization are only 10^2–10^3 yr.

Inside a cloud, the ultraviolet radiation is reduced because of absorption and scattering by grains (e.g. Roberge et al. 1991). Deep inside dark clouds, little of the ambient radiation penetrates, but a weak ultraviolet field can be maintained by cosmic-ray induced photons, resulting from the excitation of H_2 by secondary electrons following cosmic ray ionization of H_2 (Gredel et al. 1989). For a typical cosmic ray ionization rate (see §3.1), the resulting photodestruction rates are several orders of magnitude lower than those in the unshielded interstellar field, indicating that the two become comparable only deep into the cloud.

Molecular ions are efficiently destroyed by the process of dissociative recombination, which is very fast at low temperatures. The dissociative recombination of H_3^+—a key species in the chemistry—has been subject to considerable discussion in the last decade (Dalgarno 1994), but recent experiments appear to converge on a relatively rapid value of $\sim 10^{-7}$ cm^3 s^{-1} at low temperatures (Sundström et al. 1994; Smith and Španel 1993). Rate coefficients for the dissociative recombination of other abundant molecular ions in 10–30 K gas are typically 10^{-7}–10^{-6} cm^3 s^{-1}. A major uncertainty in the models is the branching ratio to the various products. Theoretical determinations have produced differing results, and complete experimental data are only just becoming available. Recent results indicate that three-body product channels (e.g. $H_3O^+ + e \rightarrow OH + H + H$ or $O + H_2 + H$) have a much larger probability than thought previously (Williams et al. 1996; Vejby-Christensen et al. 1997).

In dense clouds, destruction of neutral molecules can also occur through chemical reactions. The He^+ ion, formed by the cosmic-ray ionization of He at a rate of $\sim 10^{-18}$ s^{-1}, is particularly effective in breaking bonds (e.g. $He^+ + N_2 \rightarrow He + N^+ + N$). Collisional dissociation of molecules is only important in regions of very high temperature (>3000 K) and density such as shocks in the vicinity of young stars (cf. Chapter 8).

2.3 Rearrangement Processes

Once molecular bonds have been formed, they can be rearranged by chemical reactions leading to more complex species. Over the last 20 years, ion–molecule reactions have been considered, primarily because the vast majority are very rapid down to temperatures of 10 K. If the reaction is exothermic, the simple Langevin theory states that the rate coefficient is independent of temperature, and depends only on the polarizability of the neutral molecule and the reduced mass of the system, leading to typical values of $\sim 10^{-9}$ cm^{-3} s^{-1} (e.g. $C^+ + O_2$, see Fig. 1). It was subsequently realized that reactions between ions and molecules with a permanent dipole (e.g. $C^+ + H_2O$) may be factors of 10–100 larger at low temperatures, because of the enhanced long-range attraction (e.g. Clary 1985).

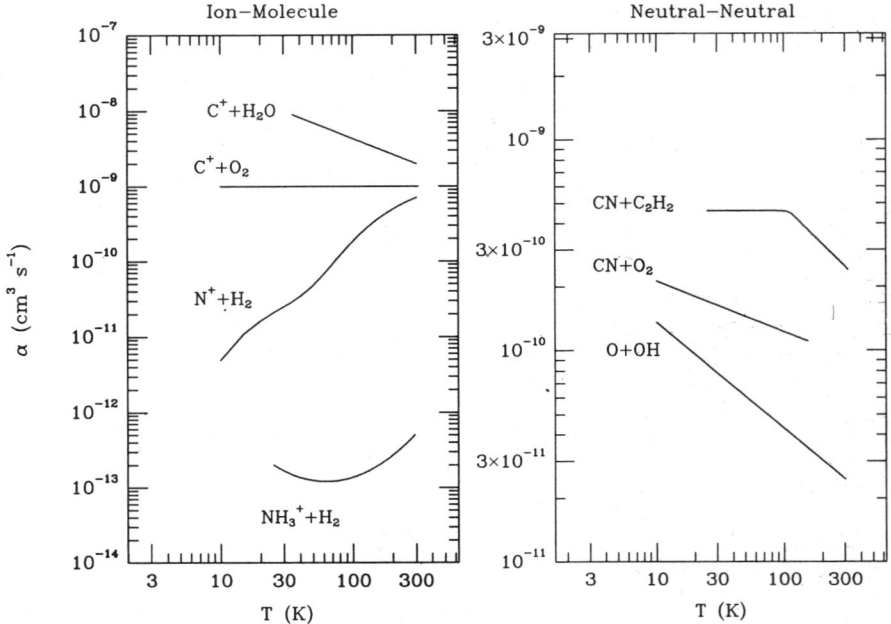

FIG. 1. Measured rate coefficients for several important interstellar ion–molecule and neutral–neutral reactions, illustrating the different temperature behaviours that can occur (based on M. Smith 1993 and I. Smith 1997).

Until recently, neutral–neutral reactions were thought to be very slow compared with ion–molecule reactions at low temperatures. However, experimental work has demonstrated that radical–radical (e.g. CN + O_2) and radical–unsaturated molecule (e.g. CN + C_2H_2) reactions have rate coefficients that are only a factor of ∼5 lower than the ion–molecule reactions (see Fig. 1), and even some radical-saturated molecule reactions can occur in cold clouds (e.g. CN + C_2H_6). Also, reactions between radicals and atoms with a non-zero angular momentum (e.g. $O(^3P_2)$ + OH) are fast at low temperatures (Sims and Smith 1995; Smith 1997). These examples illustrate that basic chemical physics studies remain essential to progress in interstellar chemistry.

3 Chemistry Networks

The molecular processes described above are incorporated into chemical models of diffuse and dense interstellar clouds. The chemical networks have been described extensively in the literature, e.g. by Dalgarno and Black (1976), Watson (1978), Prasad et al. (1987), Herbst and Winnewisser (1987), Dalgarno (1987), Turner (1989), van Dishoeck (1988, 1990a), Millar (1990) and Herbst (1995). Only some basic concepts will be reiterated here.

Table 3 Solar elemental abundances

Element	Abundance	Element	Abundance
H	1.00	Si	3.5(−5)
He	0.10	S	1.6(−5)
O	7.4(−4)	P	2.8(−7)
C	4.0(−4)	Cl	1.1(−7)
N	9.3(−5)	K	1.3(−7)
Na	2.1(−6)	Ca	2.3(−6)
Mg	3.8(−5)	Fe	3.2(−5)

Abundances by number based on Grevesse and Noëls (1993). See text and Meyer (1997) for interstellar abundances. The notation $a(-b)$ indicates $a \times 10^{-b}$.

In deciding which reactions are most important in the formation of a certain species, one should keep a few simple facts in mind. First, the abundances of the elements plays an important role. Table 3 lists the current best estimates of the solar abundances relative to hydrogen. It has recently become clear that these abundances are not representative of the interstellar medium, where they appear to be ∼30% lower at least for O and C (Meyer 1997). Significant fractions of heavier elements with high condensation temperatures such as Si, Mg and Fe are incorporated into grains (Jenkins 1987). Because hydrogen is so much more abundant than any other element, reactions with H and H_2 dominate the networks if they are exothermic. This is only the case for small ions. Most reactions of neutrals and large ions with H or H_2 have substantial energy barriers, so that they do not proceed at low temperatures. Reactions with minor species such as C^+ then become more important.

The abundances of the various ions which drive the ion–molecule chemistry are a second point for consideration. The ionization potentials of O and N are larger than 13.6 eV, so these elements are mostly neutral in interstellar clouds. In diffuse clouds and at the edges of dense clouds, C, S, Na, Mg, Cl, ... are singly ionized, because their ionization thresholds are lower than that of hydrogen. C^+ is particularly important, because it participates actively in the ion–molecule chemistry leading to more complex hydrocarbon molecules.

In the following, the networks for molecules containing hydrogen, carbon, oxygen, nitrogen, sulphur and chlorine will be discussed in more detail.

3.1 Hydrogen and Deuterium Chemistry

Because H_2 has no permanent dipole moment, radiative association of H(1s) + H(1s) is exceedingly slow. Other gas phase routes through $H^+ + H \rightarrow H_2^+$ or $H^- + H \rightarrow H_2$ (cf. Chapter 3) are also not sufficient to explain the observed abundances of H_2 in diffuse clouds, because of the low concentrations of H^+ and H^-. The only plausible formation route of H_2 is therefore on the surfaces of grains (Hollenbach and Salpeter 1971).

H$_2$ is destroyed mostly by photons through the two step process of spontaneous radiative dissociation. Because the destruction is initiated by line absorptions, the lines can become optically thick, thereby shielding molecules deeper in the cloud from the dissociating radiation. Inside dark clouds, chemical reactions also become effective in destroying H$_2$. The excitation and chemistry of molecular hydrogen is discussed in more detail in Chapter 9.

H$_2$ is ionized by cosmic rays at a rate ζ_o of the order of 10^{-17}–10^{-16} s^{-1}; 97% of the ionizations produce H$_2^+$. The well-known reaction

$$H_2^+ + H_2 \rightarrow H_3^+ + H \tag{3.1}$$

rapidly leads to the stable H$_3^+$ ion, which plays a pivotal role in the subsequent ion–molecule chemistry through proton transfer. Its recent detection in interstellar space by Geballe and Oka (1996) after years of searching provides strong observational support for the gas-phase chemical networks (see Fig. 2).

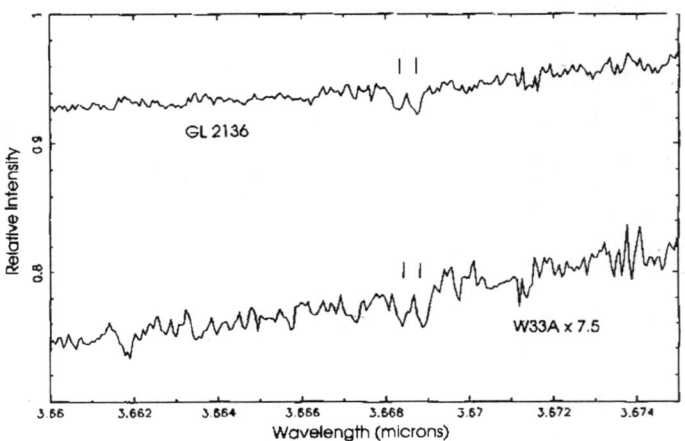

FIG. 2. Detection of interstellar H$_3^+$ through infrared absorption line observations toward young stars embedded in dense molecular clouds (Geballe and Oka 1996).

In contrast with H$_2$, HD can be readily produced by gas-phase reactions, once H$_2$ has been formed:

$$H^+ + D \rightleftarrows H + D^+ \tag{3.2}$$

$$D^+ + H_2 \rightleftarrows H^+ + HD. \tag{3.3}$$

HD is destroyed much more rapidly by photodissociation than H$_2$, because the HD lines do not become self-shielding owing to its much lower abundance. The

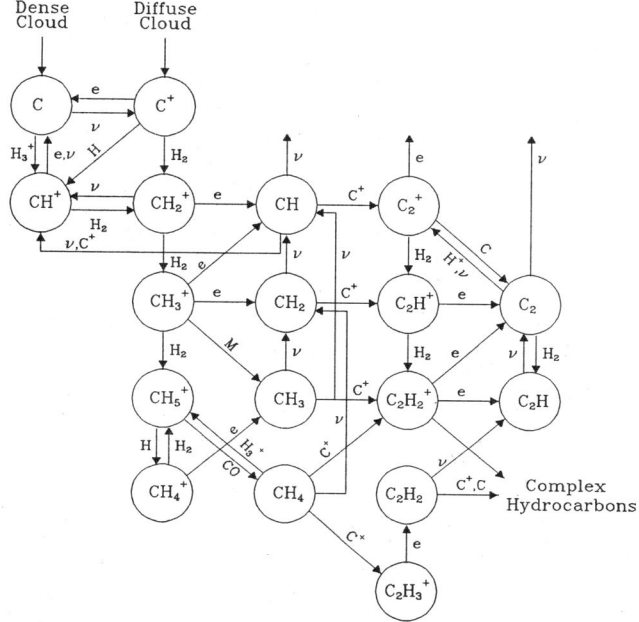

FIG. 3. Initial steps in the carbon chemistry in diffuse and dark clouds.

net balance of the more rapid formation and destruction of HD in diffuse clouds is an HD/H_2 abundance ratio which is somewhat lower than the elemental [D]/[H] ratio of $\sim 1.6 \times 10^{-5}$ (Linsky et al. 1993). The H^+ ions involved in reaction (3.1) are produced by cosmic ray ionization of H. Thus, in diffuse clouds where some of the deuterium is not locked up in HD, the observed abundance of HD can be used as a measure of the cosmic ray ionization rate.

In dense clouds, most of the deuterium is thought to be contained in HD, although a small amount of atomic deuterium of order 10% can be maintained through the dissociative recombination of $DCO^+ + e \rightarrow D + CO$ (Dalgarno and Lepp 1984).

3.2 Carbon Chemistry

In diffuse clouds, atomic carbon is mostly in the form of C^+, because its ionization potential is less than 13.6 eV. Formation of CH^+ through the reaction

$$C^+ + H_2 \rightarrow CH^+ + H - 4640 \text{ K} \tag{3.4}$$

does not proceed at low temperatures, however, because the reaction is endothermic by about 0.4 eV. Black and Dalgarno (1973) therefore suggested that the carbon chemistry is initiated by the slow radiative association reaction

$$C^+ + H_2 \rightarrow CH_2^+ + h\nu. \tag{3.5}$$

Recent theoretical and laboratory results seem to converge on a rate coefficient of $\alpha_5 \approx 1 \times 10^{-15}$ cm^3 s^{-1} (Gerlich and Horning 1992; Smith 1989). CH_2^+ reacts rapidly with H_2 to form CH_3^+, which is followed by dissociative recombination with electrons to produce neutral hydrides like CH and CH_2.

The presence of abundant C^+ in diffuse and translucent clouds leads to rapid formation of other hydrocarbons through ion–molecule reactions of C^+ with neutral molecules such as CH and C_2H (see Fig. 3). The build-up of polyatomic hydrocarbons is limited, however, by rapid photodissociation, which dominates the removal of most neutral small molecules up to $A_V \approx 2$ mag. At larger depths, reactions of CH, CH_2 and CH_3 with atomic O leading to species such as CO and H_2CO become the dominant destruction routes. Once the carbon is locked up in the very stable CO molecule, formation of the more complex hydrocarbons ceases.

In dense clouds shielded from ultraviolet radiation, the amount of C^+ is very low. The reaction of C with H_3^+ now initiates the carbon chemistry through the reactions $C + H_3^+ \rightarrow CH^+ \rightarrow CH_2^+ \rightarrow CH_3^+$. Because the reaction of CH_3^+ with H_2 to form CH_4^+ is endothermic, only slow radiative association leading to CH_5^+ can occur. The rate coefficient for this reaction is uncertain (Gerlich and Horning 1992), but is orders of magnitude slower than that of typical exothermic ion–molecule reactions. Thus, reactions of CH_3^+ with other abundant species such as O, CO and HCN become competitive, leading to ions which are the precursors of CO, CH_2CO (ketene) and CH_3CN (methylcyanide) respectively (see Fig. 4).

The production of complex hydrocarbons in dense clouds occurs via three types of pathways (Herbst 1995): (i) carbon insertion reactions (e.g. $C^+ + CH_4 \rightarrow C_2H_2^+$ or $C + C_2H_2 \rightarrow C_3H + H$); (ii) condensation reactions (e.g. $CH_3^+ + CH_4 \rightarrow C_2H_5^+ + H_2$ or $C_2H + C_2H_2 \rightarrow C_4H_2 + H$); and (iii) radiative association reactions (e.g. $C^+ + C_n \rightarrow C_{n+1}^+ + h\nu$). In general, carbon insertion with C^+ is thought to be the dominant route. Since this leads to loss of one hydrogen and since the larger ions $C_nH_m^+$ do not react rapidly with H_2, it is not surprising that gas-phase chemistry produces strongly unsaturated hydrocarbons, in agreement with observations (see Table 1). The necessary C^+ ions are produced in small amounts through destruction of CO by He^+

$$He^+ + CO \rightarrow C^+ + O + He. \qquad (3.6)$$

where the He^+ production is induced by cosmic rays. Reactions with C may be competitive if they are as rapid as suggested by recent laboratory experiments.

3.3 Oxygen Chemistry

Because oxygen is mostly neutral even in diffuse clouds, the gas-phase oxygen chemistry is initiated by cosmic ray ionization of H and H_2 to form H^+ and H_2^+, with the latter leading rapidly to H_3^+ by reaction (1). OH^+ then forms through the reactions

$$O + H^+ \rightleftharpoons O^+ + H - 227 \text{ K} \tag{3.7}$$

$$O^+ + H_2 \to OH^+ + H \tag{3.8}$$

$$O + H_3^+ \to OH^+ + H \text{ or } H_2 \tag{3.9}$$

Once OH^+ is formed, rapid hydrogen abstraction reactions and dissociative recombination lead to neutral oxygen-bearing molecules like OH and H_2O. Reaction (7) is a near-resonant charge-transfer reaction which requires temperatures of more than 100 K to proceed. Thus, the relative importance of the two routes to OH^+ in diffuse clouds depends sensitively on the temperature structure, as well as the H^+/H_3^+ ratio.

Most neutral molecules are destroyed by photodissociation and reaction with C^+. Specifically, reaction of OH with C^+ leads to CO^+, which reacts with H_2 to yield HCO^+; HCO^+ then dissociatively recombines with electrons, completing one of the major production routes of CO. CO can be destroyed only by photons with wavelengths shorter than 118 Å. Like H_2, the photodissociation of CO occurs through discrete lines, which become optically thick deeper into the cloud. In addition, shielding by dust and coincident lines of H and H_2 plays a role (van Dishoeck and Black 1988b). The isotopic species ^{13}CO and $C^{18}O$ can be partly shielded by ^{12}CO.

In cold dense clouds, the reaction

$$OH + O \to O_2 + H \tag{3.10}$$

rapidly transforms a significant part of the oxygen into O_2. Reactions of O and OH with H_2 to transform oxygen into H_2O are endothermic by nearly 2000 K, and therefore only take place in shocks and warm photon-dominated regions (PDRs) (see Chapters 8 and 9), and possibly in regions heated by the dissipation of turbulence (cf. §9.1).

3.4 Nitrogen Chemistry

In contrast with the carbon and oxygen chemistries, the reactions that initiate the nitrogen chemistry are still not well understood. In diffuse clouds, most of the nitrogen is in neutral atomic form. The reaction of N with H_3^+ to form NH^+ is endothermic and the exothermic reaction

$$N + H_3^+ \to NH_2^+ + H \tag{3.11}$$

is unusual because the transfer of two protons is involved. Theoretical calculations indicate large barriers along the reaction path which may prevent reaction at low temperatures. The alternative formation route is through

$$N^+ + H_2 \to NH^+ + H, \tag{3.12}$$

which is endothermic by $\sim$100 K for ground-state H_2 $J = 0$ (Le Bourlot 1991). Once NH^+ or NH_2^+ are formed, hydrogen abstraction reactions lead to NH_4^+, and

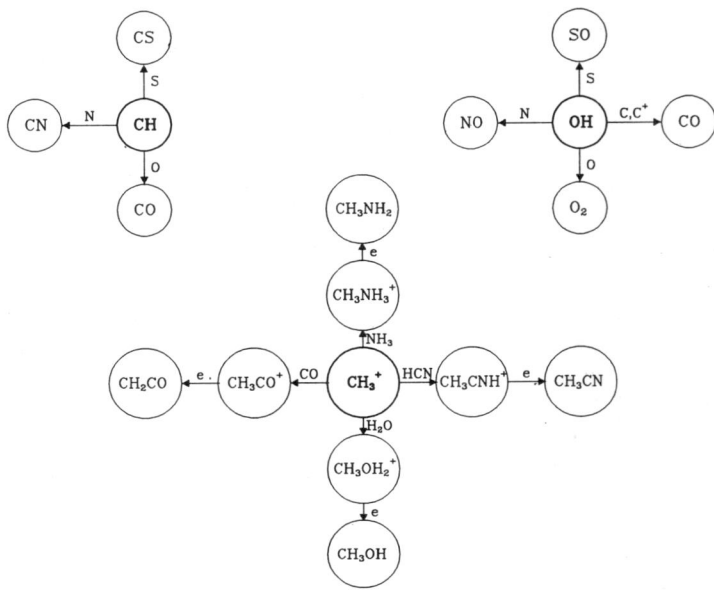

FIG. 4. Some important connections between the carbon, oxygen, nitrogen and sulphur networks in diffuse and dense clouds.

dissociative recombination to NH, NH_2, and NH_3. The N^+ ions are produced by cosmic ray ionization. In dense clouds, additional N^+ comes from the reaction of He^+ with N_2, which breaks the strong triple N–N bond. Some of this N^+ is formed with sufficient energy to overcome the barrier of reaction (12).

Neutral–neutral reactions play an important role in the formation of other nitrogen-containing species, both in diffuse and dense clouds. Reactions of atomic N with CH, OH and NO lead to CN, NO and N_2, respectively. Reactions of CN with C_2H_2 are probably a major source of the cyanopolyynes ($HC_{2n}CN$), which are prominently observed in some dark clouds (see §8 and Chapter 4)).

3.5 Sulphur Chemistry

The sulphur chemistry in cold clouds constitutes a puzzle, since most of the formation routes involve at least one endothermic reaction. Sulphur exists mostly as S^+ in diffuse clouds, but the reaction of S^+ with H_2 to form SH^+ is endothermic, whereas the radiative association reaction is predicted to be slow. S can react with H_3^+ to form SH^+, but the subsequent reaction of SH^+ with H_2 is again endothermic. Thus, sulphur hydrides are not expected to be abundant in cold clouds.

In contrast, other sulphur-containing species can be readily produced through ion–molecule and neutral–neutral reactions. For example, CS results from reactions of S^+ and S with CH and C_2, whereas SO stems from reactions with OH. The subsequent reaction of SO with OH is expected to produce copious amounts of SO_2. The detection of H_2S in translucent clouds and the lack of abundant SO_2 in dark clouds indicates an incomplete understanding of either the sulphur chemistry or the physical structure (see §7.3).

3.6 Chlorine Chemistry

The chlorine chemistry in interstellar clouds is particularly simple. Reaction of Cl^+ with H_2 leads to HCl^+, which subsequently reacts to H_2Cl^+ and dissociatively recombines to HCl. The first step in this chain is slow, but the other steps are rapid. The detections of HCl in both diffuse and dense clouds form a test of these models and can be used to place limits on the depletion of Cl (Federman et al. 1995; Schilke et al. 1995). Similarly, the recent detection of HF gives constraints on the gas-phase fluorine abundance (Neufeld et al. 1997).

3.7 Isotopic Fractionation

At the low temperatures in interstellar clouds, significant enhancement of various isotopes can occur as a result of a process called "fractionation". An extreme example is the deuterium fractionation initiated by isotope exchange reactions (Wootten 1987; Millar et al. 1989)

$$H_3^+ + HD \rightleftarrows H_2D^+ + H_2 + 227 \text{ K} \tag{3.13}$$

$$CH_3^+ + HD \rightleftarrows CH_2D^+ + H_2 + 370 \text{ K}, \tag{3.14}$$

which occur preferentially in the forward direction at low temperatures. These reactions are followed by deuterium transfer reactions such as

$$H_2D^+ + CO \rightarrow DCO^+ + H_2 \tag{3.15}$$

$$CH_2D^+ + N \rightarrow DCNH^+ + H_2 \tag{3.16}$$

which lead to DCO^+/HCO^+ and DCN/HCN abundance ratios which are up to a factor of 1000 larger than the overall [D]/[H] ratio.

^{13}C can be incorporated preferentially in various molecules through exchange reactions such as

$$^{13}C^+ + {}^{12}CO \rightleftarrows {}^{12}C^+ + {}^{13}CO + 36 \text{ K}, \tag{3.17}$$

which are most effective in diffuse and translucent clouds where not all carbon is incorporated in CO. Isotope selective photodissociation of ^{13}CO tends to counteract the enhancement at low temperature through reaction (17).

3.8 Large Molecules and Metals

Large molecules such as polycyclic aromatic hydrocarbons (PAHs) can affect the chemistry in both diffuse and dense clouds through charge-transfer and mutual neutralization reactions of the type (Omont 1986; Lepp et al. 1988)

$$\text{PAH} + \text{X}^+ \rightarrow \text{PAH}^+ + \text{X} \qquad (3.18)$$

$$\text{PAH}^- + \text{X}^+ \rightarrow \text{PAH} + \text{X}. \qquad (3.19)$$

Thus, the main effect of these species is to neutralize both atomic and molecular ions. Large molecules and small grains are also important in the thermal balance of the cloud (d'Hendecourt and Léger 1987; Lepp and Dalgarno 1988), and their presence affects the extinction curve at the shortest wavelengths where CO and CN photodissociate (Cardelli 1988; van Dishoeck and Black 1989).

Metals also play an important role in neutralizing ions in dense clouds through charge-transfer reactions:

$$\text{M} + \text{X}^+ \rightarrow \text{M}^+ + \text{X}. \qquad (3.20)$$

3.9 Grain Surface Chemistry

The chemistry on the surfaces of interstellar grains has received ample discussion in the literature (e.g. Tielens and Hagen 1982; d'Hendecourt et al. 1985; Tielens and Allamandola 1987; Herbst 1993). Because hydrogen is so abundant and mobile, grain surface chemistry is expected to lead primarily to hydrogenated species, such as H_2O, NH_3 and CH_4. Hydrogenation of CO can lead to H_2CO and CH_3OH. At higher densities, the amount of atomic hydrogen in the gas phase decreases, and reactions with atomic oxygen become important, e.g. CO + O → CO_2. Tunnelling reactions with H_2 also occur competitively once $H_2 \gg H$. Within the icy mantles, photochemical reactions can be triggered either by external ultraviolet photons or by internal photons produced by cosmic rays. Photodissociation can lead to radicals (e.g. $H_2O \rightarrow OH + H$ or $O + H_2$), which can subsequently react to form other molecules (e.g. CO + OH → CO_2). Explosive exothermic reactions between stored radicals at higher temperatures produce more complex organics (d'Hendecourt et al. 1982).

4 Models

4.1 Gas-phase models

The chemistry networks described in §3 can be solved provided that a physical model of the cloud is specified. In general, two different classes of models are considered: (i) Steady-state, depth-dependent models, in which the abundances of the molecules do not change with time, but are functions of depth into the cloud. Models of diffuse and translucent clouds and dense photon-dominated regions (see Chapter 9) fall in this category. In these clouds, the time scales for photoprocesses are sufficiently short—a few hundred years in the unshielded radiation field—that the assumption of steady state is justified. The only exception is formed by the chemistry of H_2. (ii) Time-dependent, depth-independent models, in which the concentrations are computed as functions of time at a single position deep into the cloud. Models of dark and dense clouds fall into this category. The time scale for reaching chemical equilibrium is comparable to the time required for the number of cosmic-ray-induced ionizations of hydrogen per unit volume to be equal to the number density of all elements more massive than helium, and is typically $\sim 10^6$–10^7 yr. Some combined time- and depth-dependent models have been made for diffuse clouds (e.g. Wagenblast and Hartquist 1989), dense PDRs (e.g. Goldshmidt and Sternberg 1995) and dark clouds (e.g. Lee et al. 1996b) which are useful to explore the effects of the assumptions.

Several parameters enter the models. In both cases (1) the elemental abundances of C, O, N, S, metals...; and (2) the cosmic ray ionization rate ζ_o in s^{-1} need to be specified. In steady-state, depth-dependent models, additional parameters are (3) the geometry (e.g. plane-parallel, spherical, ...); (4) the density $n_H = n(H) + 2n(H_2)$ as a function of position; (5) the incident radiation field, specified by a factor I_{UV} times the standard interstellar radiation field as given by, e.g., Draine (1978); and (6) the grain parameters, i.e. the extinction curve, albedo and scattering function. The temperatures of the gas and dust as functions of position in the cloud can be obtained self-consistently from the balance of heating and cooling. Alternatively, they can be constrained from observations and provided as an additional input parameter.

In time-dependent models, the additional parameters besides (1) and (2) are (3) the density, usually taken to be constant with time (so-called pseudo time-dependent models); and (4) the visual extinction A_V at the position in the cloud, usually taken to be so large that photo-processes can be neglected. The temperature can be obtained from the thermal balance, but is almost always set at 10 K for both the gas and dust, typical of a dark cloud shielded from ultraviolet radiation and heated by cosmic rays only.

4.2 Gas-grain Models

Gas-grain chemistry has been incorporated in chemical models in two different approaches (Tielens and Whittet 1997). In the "accretion limited" regime, the time scale for a mobile species to scan the surface is much less than the accretion time of the co-reactant. The chemistry is then limited by the accretion rate of new species. In the "reaction limited" regime, the opposite holds so that a species trapped in a site can react only with migrating species that visit that site. Most of the chemical models which incorporate gas-grain interactions have been formulated in the "reaction limited" regime through the use of rate equations for computational convenience (e.g., Hasegawa and Herbst 1993; Shalabiea and Greenberg 1994; Bergin et al. 1995). However, under dark cloud conditions the surface chemistry is likely to be in the accretion-limited regime and can only be properly treated by a Monte–Carlo method (e.g. Tielens and Hagen 1982). Recently, attempts have been made to modify the rate equations to take into account the shortcomings of the reaction-limited approach (Caselli et al. 1997).

Another important parameter in the gas-grain models is the mechanism for returning the molecules to the gas phase. If no desorption mechanism is included, the gas-phase molecules accrete on the grains on a time scale of $\sim 2 \times 10^9 / n_H y_S$ yr, where the sticking coefficient y_S is thought to lie between 0.1 and 1.0 (Williams 1993). Thus, for typical dark cloud densities of 10^4 cm^{-3} all molecules disappear from the gas phase in less than 10^6 yr, inconsistent with the observed widespread molecular emission. The possible desorption mechanisms have been summarized by Schutte and Greenberg (1991) and Schutte (1996). Thermal evaporation is effective only at higher temperatures, $T_D > 20$ K. The energy liberated by the formation of molecules can heat the grains, but this mechanism is only efficient locally or for very small (< 100 Å) grains. Similarly, photodesorption by ultraviolet radiation is ineffective compared with other processes, except perhaps in PDRs. A related mechanism has been suggested by Williams et al. (1992) and Dzegilenko and Herbst (1995), in which absorption of infrared photons causes vibrational excitation of surface species, which can lead to desorption of (adjacent) molecules. The most effective desorption mechanisms in cold gas are likely to be cosmic-ray spot heating and explosive desorption. The latter can be either cosmic-ray induced or triggered by grain–grain collisions at velocities greater than $\sim$0.1 km s^{-1}. More basic laboratory experiments are needed to determine whether these mechanisms can indeed return most species back to the gas phase on a time scale comparable with the accretion time scale. In particular H$_2$O ice, which contains strong hydrogen bonds, is difficult to remove in cold clouds.

4.3 Column Densities and Abundances

Steady-state, depth-dependent models provide column densities integrated over cloud structure, which can be compared directly with observations. For a simple plane-parallel geometry, the column density of species X in cm^{-2} integrated over depth z is

$$N_X(z) = \int_0^z n_X(z')dz'. \tag{4.1}$$

In contrast, the time-dependent models give local concentrations $n_X(t)$ in cm^{-3} and abundances $x_X(t) = n_X(t)/n_H(t)$ as functions of time. The concentration ratios are then assumed to be equal to the column density ratios N_X/N_H in comparison with observations. While this procedure may be valid for most molecules in dark clouds with $A_V > 5$ mag and little internal structure, it can lead to erroneous conclusions for other situations. Note that for standard diffuse cloud conditions, N_H and A_V are related through $N_H/A_V \approx 1.6 \times 10^{21}$ cm^{-2} mag^{-1}.

This point is illustrated in Fig. 5, which shows on the left hand side three typical distributions of abundances as functions of depth into a cloud. The first case involves a species such as C$^+$ whose abundance peaks at the edge of the cloud, but drops rapidly for $A_V > 1$ mag. The second case is for a molecule such as CH or CN whose abundance peaks in the outer zone of the cloud ($A_V \approx 1$–2 mag), but decreases in the inner region. The third case is for a molecule like CO or HCN, which reaches its maximum abundance only deep inside the cloud where it is shielded from radiation. The right-hand side

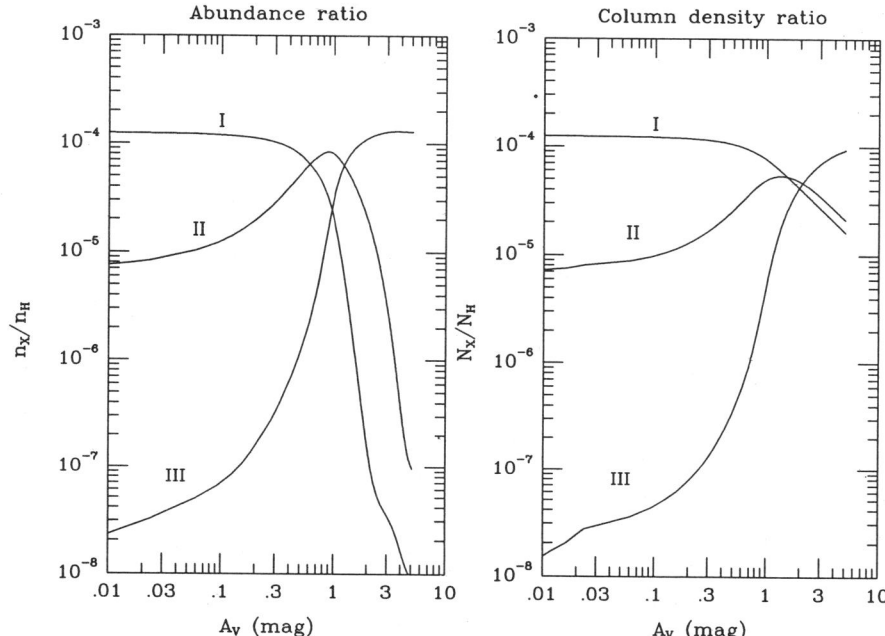

FIG. 5. Abundances (left) and column densities (right) as functions of depth for three typical cases in a cloud with $n_H = 1500$ cm^{-3} and $I_{UV} = 1$. I: Abundance peaks at edge of cloud (e.g. C^+); II: Abundance peaks at intermediate depth in cloud (e.g. C, CH, CN, C_2H); III: Abundance peaks in shielded interior of cloud (e.g. CO, HCN).

of Fig. 5 shows the column densities in the three cases integrated up to depth z, expressed here in A_V, divided by the total hydrogen column density up to that depth, $N_X(z)/N_H(z)$. It is clear that for large A_V (dark clouds), only case III gives column density ratios which are similar to the local abundance ratios. For species which belong to cases I and II, a significant part of the column density originates in a different, outer part of the cloud. These considerations are particularly important when interpreting observations of translucent clouds with A_V of a few mag.

4.4 The $C^+ \to C \to CO$ Transition

The principal chemical characteristics of both the depth- and time-dependent models are governed by the transition of carbon from atomic to molecular form (see Fig. 5). In the depth-dependent case, C^+ recombines to C around $A_V \approx 1$ mag, followed by the conversion to CO around $A_V \approx 2$ mag. The CO photodissociation rate as a function of depth is crucial in this transition, and its calculation requires a careful treatment of the complicated radiative transfer

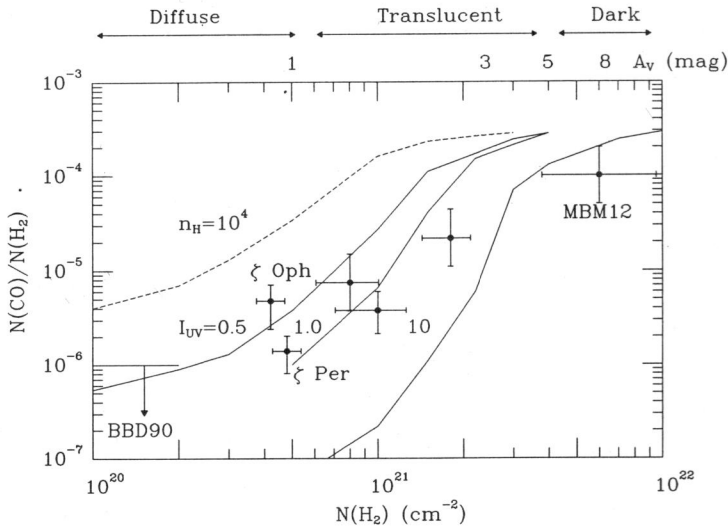

FIG. 6. Computed and observed $^{12}\text{CO}/\text{H}_2$ column density ratios as functions of total H_2 column density. The full lines from left to right are model results for $I_{UV} = 0.5$, 1.0 and 10, respectively, and $n_H \approx 500\text{--}2000$ cm^{-3}. The dashed line is for models with $I_{UV} = 1$ and $n_H = 10^4$ cm^{-3}. Observed values for a number of diffuse, translucent and dark clouds discussed in this chapter are included. The unlabelled points refer to HD 210121, PKS0528+134 and HD 154368 from left to right. Adapted from van Dishoeck and Black (1988b) and van Dishoeck (1992).

(van Dishoeck and Black 1988b; Viala et al. 1988; Warin et al. 1996; Lee et al. 1996b). Thanks to the recent laboratory work of Eidelsberg et al. (1991, 1992), Stark et al. (1993), Ubachs et al. (1994) and Yoshino et al. (1995), the basic molecular parameters are now better understood, reducing the uncertainties in the models.

The precise location of the $C^+ \to C \to CO$ transition depends on the physical parameters, in particular the density and incident radiation field, as illustrated in Fig. 6. This figure provides a convenient framework in which to characterize various clouds, with the diffuse clouds occupying the low CO/H_2 part of the figure and the dark clouds the high CO/H_2 regime. The various translucent and high-latitude clouds can be found in the intermediate region. Figure 6 shows that small variations in physical parameters by a factor of ~ 2 can result in an order of magnitude difference in CO column densities in the translucent regime. Since CO maps of such clouds usually have a rather small dynamic range of a factor of ~ 5, much of the observed structure may be due to $\sim 30\%$ fluctuations in density, total column density or radiation field.

The $C^+ \to C \to CO$ transition also affects the chemistry of other species. At the edge of the cloud, only the simplest diatomic molecules are found. Around $A_V = 1$–2 mag, their abundances increase, but both C and C^+ are still abundant. This results in an increase in the abundance of hydrocarbon molecules such as CN and C_2H. Deeper inside the cloud, atomic carbon is no longer available, and destruction by O becomes more important than photodissociation. Stable species such as CH_4, C_2H_2 and HCN become dominant.

Many of the same features are observed in the pseudo time-dependent models, if depth is replaced by time. These models usually are based on the assumption that dark clouds originate from diffuse gas, so that all species except H_2 are initially in atomic form, with carbon present as C^+. On a time scale of $\sim 10^3$ yr, C^+ recombines to C, which subsequently transforms to CO after $\sim 10^5$ yr (see Fig. 7). Since the presence of atomic carbon is essential to build up more complex organic species such as HC_3N, they are abundant only at early times, but not at steady state.

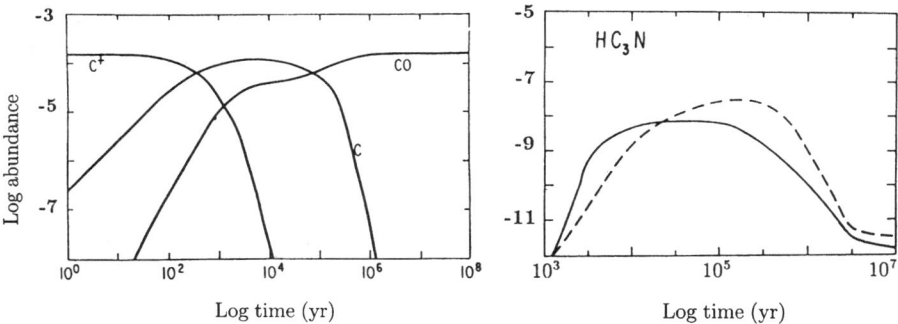

FIG. 7. Time dependence of the abundances of C^+, C and CO (left) and HC_3N (right) for a cloud with $n_H = 2 \times 10^4$ cm^{-3}. The latter behaviour is typical of all complex carbon-bearing molecules. The full and dashed lines refer to two different assumptions about the rate coefficients for reactions between ions and polar molecules (Herbst and Leung 1986).

4.5 Beyond the Standard Models

Virtually all models to date are highly simplified because they are based on a homogeneous distribution of gas and dust. There is ample evidence from CO maps, IRAS observations, radio-interferometric observations, and optical observations toward resolvable binaries that the interstellar medium has a very inhomogeneous structure on scales ranging from 100 AU to several pc (see Falgarone 1997 for an overview). It has proved difficult to quantify this structure in such a way that it can be readily incorporated into models. Nevertheless, inhomogeneous models, in which the continuum (Boissé 1990) and line (Spaans 1996) radiative transfer can be fully treated for an arbitrary density distribution and geometry, have been constructed. Spaans (1996) presents several models representative of inhomogeneous translucent clouds which include a simple chemical network. It is found that in the regime $A_V \approx 1$–5 mag, most of the C^+ is in the lower density, "interclump" gas, whereas CO and ^{13}CO are primarily in the denser "clumps" Neutral atomic carbon C appears to trace the surfaces of the clumps.

Dynamical models which go beyond the pseudo time-dependent model have been developed by various groups. Tarafdar et al. (1985) and Prasad et al. (1991) consider the case in which the density increases from diffuse cloud to dense cloud values. More recently, such models have been extended to describe the collapse phase of dark clouds to form stars (e.g. Rawlings et al. 1992), and are discussed in more detail in Chapter 5. Another class is formed by the dynamical mixing models, in which parcels of gas cycle between the diffuse and dense medium due to turbulence (Boland and de Jong 1982; Chièze et al. 1991; Xie et al. 1995). This process increases the amount of atomic carbon and related molecules inside the clouds. However, Rawlings and Hartquist (1997) have suggested that the microscopic mixing leading to reactions between material from initially different fluid elements is likely to be restricted to the turbulent boundary layers at the edges of clouds in the presence of a magnetic field. De Boisanger and Chièze (1992) and De Boisanger et al. (1992) have made initial attempts to couple hydrodynamical simulations with chemical models. Specifically, the effects of variation in the cloud illumination and the resulting formation of transient clumps have been studied.

4.6 Bistability

It has recently been found that the time-dependent models have at least two different steady-state solutions in certain regions of parameter space (Pineau des Forêts et al. 1992; Le Bourlot et al. 1993, 1995). Since the rate equations are non-linear differential equations, multiple solutions are indeed possible mathematically, but had not been encountered in practice. Two phases can be distinguished: (a) low ionization phase, which is the "normal" solution found in dark cloud chemistry. Here proton-transfer reactions with H_3^+ dominate the formation routes of molecules and the C/CO abundance ratio is low; (b) high ionization phase, where the H_3^+ abundance is low so that charge-transfer reactions with H^+ are dominant, which destroy large molecules and lead to a large C/CO ratio (Smith et al. 1994). Depending on the initial conditions, either of the two solutions can be found. It is not clear, however, what the astrophysical consequences of this bistability are. For example, Shalabiea and Greenberg (1995) show that if different elemental abundances are used and gas-grain chemistry is included, the bistability disappears or shifts to densities below 10^3 cm^{-3} which are no longer characteristic of dark clouds. Inclusion of photoprocesses in the models, which are important at the edges of the clouds, removes the bistability. It will therefore be very difficult to find unambiguous observational evidence for this phenomenon, since only column densities integrated over depth are observed. Gerin et al. (1997) suggest that a high CS/SO abundance ratio can be used to establish the presence of the high ionization phase along the line of sight. However, since the abundance of CS peaks more at the edge than that of SO (case II vs. case III; Fig. 5), such variations could also be due to variations in column density or incident radiation field, or even changes in the gas-phase elemental $[C]_g/[O]_g$ ratio (see §8.3).

5 Observations

Diffuse, translucent and dark clouds can be observed from ultraviolet to millimetre wavelengths with a large variety of techniques. The data provide diagnostics of the physical conditions and constraints on the molecular abundances. It should be emphasized that clouds studied with different observational methods are not necessarily of a different nature. For example, the translucent clouds studied with optical absorption, CO emission or IRAS 100 μm emission often have similar physical characteristics. As explained in §4.4, many of the perceived differences are likely due to relatively small changes in the parameters. Thus, although the diffuse, translucent and dark clouds are discussed separately in the following, it is clear that they are physically related and that multiple components are likely to be present along any line of sight.

5.1 Diffuse Clouds

Table 4 summarizes the various diagnostics that are available to constrain the parameters entering the chemical models of diffuse clouds, such as density, temperature and radiation field. Most of them are based on optical absorption lines. The ultraviolet spectra such as those obtained with the *Copernicus* satellite are particularly powerful. The analysis of these data has been discussed extensively in the literature (e.g. van Dishoeck and Black 1988a; van Dishoeck 1990a, 1992; Wagenblast and Williams 1993). Briefly, the density and temperature structure can be constrained from the H/H_2 ratio, the atomic fine-structure excitation (e.g. C, C^+ and O), and the C_2 and CO rotational excitation. The high-J levels of H_2 are thought to be populated by ultraviolet pumping, by the H_2 formation process on grains and possibly by high-temperature collisional processes. If the contributions from these latter processes are small, the observations can be used to constrain the incident radiation field. Ultraviolet observations are also the only way to accurately determine the elemental abundances, and constrain grain properties such as the extinction curve.

In terms of the chemistry, only a limited set of (mostly) diatomic molecules has been detected in diffuse clouds. Although few in number, their column densities and abundances are known with high accuracy, because H and H_2 can be measured directly at ultraviolet wavelengths. Millimetre emission from molecules other than CO is very weak, because of the low densities.

5.2 Translucent Clouds

Translucent clouds with $A_V \approx 1$–5 mag have the advantage that they can be studied not only with optical absorption line techniques provided a suitable background star is available, but also with millimetre emission lines. Unfortunately, the absorption observations are mostly limited to visible wavelengths; virtually no ultraviolet data have yet been taken, except for the line of sight toward HD 154368 (Snow et al. 1996). As Table 4 shows, this means that fewer diagnostics are available; specifically, constraints on I_{UV} and elemental abundances are lacking, and H_2 cannot be observed directly. On the other hand, the ability to map the cloud in CO emission lines provides more information on cloud structure, and the thicker clouds show detectable millimetre emission from other molecules (Turner 1996 and references cited).

Recently, a third, very powerful technique for studying diffuse and translucent clouds has been developed; it involves the use of millimetre interferometers to measure molecular absorption lines against the continuum of bright, distant radio sources (Lucas and Liszt 1994, 1996, 1997; Hogerheijde et al. 1995; see Fig. 9). The excitation of high-dipole moment molecules in these low density clouds is controlled by the cosmic background radiation, with a slight contribution from collisions with electrons and neutrals. Because the optical depth and excitation of the line is known, this method allows an accurate determination of the column density, even from a single line. The observations of triatomic and polyatomic molecules are particularly useful to test the chemistry.

5.3 Dark Clouds

Dark clouds with $A_V > 5$ mag are studied primarily with millimetre emission lines. The ratios of the strengths of emission lines from species such as CO, CS, HCO^+, NH_3 and HC_3N are used to constrain the temperature and density in the clouds (e.g. Walmsley 1987; Genzel 1992). More than 50 different species, ranging from simple diatomic molecules to complex organics can be detected in dark clouds (see Table 1) so that the chemistry can be studied in much more detail than in diffuse and translucent clouds. However, the derived column densities and abundances have substantial uncertainties, and are averaged over a beam of $20''$–$60''$ which is orders of magnitude larger than the pencil beam line of sight of $<0.01''$ appropriate for absorption line observations.

Dark clouds can also be studied with infrared absorption lines (Whittet 1993). The spectra toward massive young stars embedded in dense clouds show strong lines of gas-phase molecules like CO, and solid-state species such as H_2O, CO, CO_2 and CH_3OH (Whittet et al. 1996; van Dishoeck et al. 1996; see Fig. 8). For these lines of sight the chemistry has probably been affected by the star-formation process (see Chapter 5). A few infrared objects have been found, however, which are thought to be true background sources, such as Elias 16 behind the Taurus dark cloud. Such data give important constraints on the grain chemistry and gas-grain interactions in dark clouds. The *Infrared Space Observatory* (ISO) will provide a wealth of new information on the ice composition in dense clouds with and without star formation in the next few years.

Table 4 Available diagnostics for parameters entering chemical models

Species	Phys. Par.	Method	Diffuse	Translucent			Dark
				Opt^a	Mm abs^a	Mm em^a	
H_2 low J	T	UV abs	+	–	–	–	–
H_2 high J	I_{UV}, form.	UV abs	+	–	–	–	–
H_2 $v > 0$	I_{UV}	UV abs	+	–	–	–	–
C_2 low J	T	VIS abs	+	+	–	–	–
C_2 high J	I_R/n_H	VIS abs	+	+	–	–	–
C, C^+, O(J)	$n_H T$	UV abs	+	–^{b}	–	–	–
CN low J	$n(e), n_H$	VIS abs	+	+	–	–	–
	$n(e), n_H$	mm em	–	+	–	+	+
CO low J	$n_H T$	UV abs	+	–^{b}	–	–	–
		mm em	+	+	+	+	+
		mm abs	–	–	+	–	–
HCO^+, CS.. low J	$n(e), n_H$	mm abs	–	–	+	–	–
HCO^+, CS.. low J	n_H	mm em	–	+	–	+	+
Ca^+/Ca, ..	$n(e)$	VIS abs	+	+	–	–	–
C^+, O, ..	El. abund.	UV abs	+	–^{b}	–	–	–
UV cont.	Ext. curve	UV abs	+	+	–	–	–
Examples			ζ Oph	HD169454	PKS0528	MBM16	TMC-1
			ζ Per	HD210121	NRAO530	CB228	L134N

^{a} Principal technique used to identify translucent cloud
^{b} Technique starting to become feasible with HST

5.4 Determination of $N(H_2)$

A problem in determining abundances in translucent and dark clouds is that in general no direct observations of H_2 are available. Several methods have been developed to infer $N(H_2)$ indirectly (see van Dishoeck and Black 1987; van Dishoeck et al. 1992 for reviews). First, one can use observations of other molecules whose abundances are thought to be well known. For diffuse and translucent clouds, CH may be suitable (Magnani and Onello 1995), whereas for dark clouds CO is most commonly used. The latter can be calibrated by the direct determination of $CO/H_2 = 2.7 \times 10^{-4}$ in at least one (warm) dense cloud by Lacy et al. (1994), which is a factor of 3 higher than the value of 8×10^{-5} commonly used for dark clouds. HCO^+ may also be a good tracer, once its abundance is better understood (see §7.3). Because of their different behaviour with depth (see Fig. 5), care should be taken in the use of either CH, CO or HCO^+ in translucent clouds. Second, the "standard" conversion factor $X = N(H_2)/\int T_A(CO)dV$ can be adopted, where $\int T_A dV$ is the observed integrated antenna temperature of CO. This method may be valid for large GMC ensembles, but not necessarily for diffuse and translucent clouds, where the derived values of X vary by more than an order of magnitude (Magnani and Onello 1995). Third, the visual extinctions obtained from optical absorption line observations can be converted into a total hydrogen column density, on the assumption that the N_H/A_V relation derived for diffuse clouds also holds for translucent clouds. Fourth, IRAS 100 μm observations can be used with

$I_{100\,mm}/N_H$ calibrated in a global three-dimensional correlation fit between H I. ^{12}CO and 100 µm maps (Bloemen, Deul and Thaddeus 1990). Finally, X–ray shadowing observations can be employed to determine N_H directly but such data are available for only a few clouds (Mebold et al. 1994).

In the following, the recent observations and models of these regions will be discussed in more detail and illustrated by a few specific cases. The uncertainties in the derived abundances should be kept in mind.

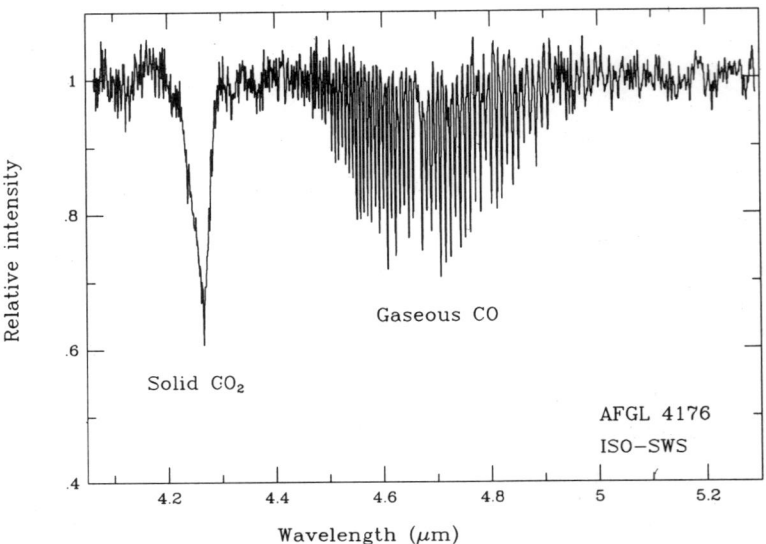

FIG. 8. Normalized spectrum obtained with the *Short Wavelength Spectrometer* on board the *Infrared Space Observatory* toward the massive young stellar object AFGL 4176 embedded in a dense molecular cloud. The strong, broad absorption at 4.27 µm is due to solid CO_2, whereas the characteristic ro–vibrational P- and R-branch structure at 4.4–4.9 µm indicates the presence of warm, gaseous CO along the line of sight (van Dishoeck et al. 1996).

6 Diffuse Molecular Clouds

Diffuse neutral gas is seen throughout the Galaxy with various techniques, e.g. HI emission, IRAS 100 µm emission and NaI absorption lines. Here only the subset of diffuse molecular clouds, in which the fraction of hydrogen in molecular form is larger than ~50%, is considered. These clouds typically have number densities n_H of a few hundred cm^{-3} and total hydrogen column densities up to 10^{21} cm^{-2}. Only those clouds which happen to lie in front of bright stars have been modelled in any detail, because of the wealth of available diagnostics (see Table 4). General discussions can be found e.g. in van Dishoeck (1990a,b) and

Wagenblast and Williams (1993). As an example, the well-studied line of sight toward ζ Oph will be discussed here.

6.1 The ζ Oph Cloud

6.1.1 Physical Structure

Several new diagnostics of diffuse clouds such as that toward ζ Oph have become available in recent years. Very weak absorption lines of H_2 out of vibrational levels $v = 3$ and 4 have been detected with the *Goddard High-Resolution Spectrograph* (GHRS) on board the *Hubble Space Telescope* (Federman et al. 1995, 1997b). The populations of these levels should be a better diagnostic of I_{UV} than those of the $v = 0$, J levels, because no contribution from collisions is expected (van Dishoeck and Black 1986). For the ζ Oph cloud, the inferred enhancement of the radiation field from the $v > 0$ populations is only $I_{UV} \approx 1$–2, about a factor of 2 smaller than thought before.

The Mulliken D $^1\Sigma_u^+$–X $^1\Sigma_g^+$ system of C_2 near 2300 Å has also been detected with HST (Lambert et al. 1995), thereby complementing observations of the Phillips A $^1\Pi_u$ – X $^1\Sigma_g^+$ system at 8750 Å from the ground. Lines can be observed originating from rotational levels up to $J = 24$, providing even tighter constraints on the density. New ultraviolet observations of the fine-structure excitation of neutral atomic carbon and the rotational excitation of CO have also led to improved diagnostics of the product $n_H T$ (Lambert et al. 1994).

Other important new constraints from HST data concern the elemental abundances, since the earlier determinations using *Copernicus* data had large uncertainties owing to the high optical depths of the lines. With the HST-GHRS, the weak, optically thin intersystem transitions of C^+ at 2325 Å and O at 1356 Å have been observed in a few clouds. The inferred column densities with respect to hydrogen imply gas-phase abundances $[O]_g/[H]=(3.16 \pm 0.09)\times 10^{-4}$ and $[C]_g/[H]=(1.40 \pm 0.2)\times 10^{-4}$, with $[C]_g/[O]_g = 0.44$. The abundances of many other important elements such as N, Mg, Si, S, Fe, ... have been determined as well, and can be used as input to the models (see Meyer 1997 for a summary; §3).

Observations at ultra-high spectral resolving power ($\lambda/\Delta\lambda \approx 10^6$) of atomic and molecular lines provide insight into the velocity structure of a cloud and the question whether or not the species co-exist (Lambert et al. 1990; Crane et al. 1995; Crawford et al. 1994; Crawford 1995; Crawford and Barlow 1996; Sembach et al. 1996; see also Jenkins et al. 1989). In some cases, the line profile is seen to be composed of several sub-clouds, which makes the chemical analysis based on unresolved profiles less reliable without prior knowledge on how each species is distributed over the various velocity components. Thus, caution should be exercised in trying to model all lines with a single component structure.

With the increased sensitivity of SIS receivers, it is possible to observe millimetre emission lines in diffuse clouds. Such observations have the advantage of very high velocity resolution of better than 0.1 km s^{-1} and provide independent information on the physical parameters. ^{12}CO $J = 1$–0 and 2–1 lines are readily detected (e.g. Le Bourlot et al. 1989; Liszt 1993; Kopp et al. 1996), but observations of species with high dipole moments such as CN (Crane et al. 1989), CS (Drdla et al. 1989) or HCO$^+$ 1–0 (Liszt and Lucas 1994) prove very difficult under diffuse cloud conditions. Only collisions with electrons can provide a slight population excess in the excited levels over the cosmic background (Drdla et al. 1989).

In summary, the analysis of the new data on the ζ Oph cloud indicates that the inferred density is not much changed compared with that in earlier models: $n_H \approx$ 300–500 cm^{-3} at the centre. The lack of CO $J = 3$–2 emission limits the density of the bulk of the gas to less than 800 cm^{-3} (van Dishoeck et al. 1991). Most of the gas has $T \sim$ 20–30 K, but a small fraction of warm gas with $T \sim$ 200 K is present as well, presumably at the edge. The radiation field is significantly lowered from $I_{UV} = $ 3–6 to 1–2 based on the results of vibrationally excited H$_2$. This means that the steady-state ultraviolet pumping models can no longer reproduce the observed H$_2$ $v = 0$, $J = $ 5–7 populations toward ζ Oph—even with a modified H$_2$ formation model—suggesting either that time-dependent effects must be invoked (Wagenblast 1992) or that some physical component to the cloud is still missing which populates the $v = 0$, J levels through collisions. The elemental gas-phase abundances are generally lower than adopted in the earlier models, although not by much.

6.2 Chemistry

The comparison between models and observations for diffuse clouds such as ζ Oph has been reviewed extensively by van Dishoeck (1990a,b). For the carbon chemistry, the combination of the lower elemental abundance and the lower radiation field implies few changes for species such as CH, C$_2$ and C$_2$H. The branching ratio to CH$_2$ + H in the dissociative recombination of CH$_3^+$ has recently been measured to be 0.4 (Vejby-Christensen et al. 1997), resulting in a factor of two lower CH$_2$ column density. The case of CH$^+$ is discussed separately in §9.1, since it traces a different, "energetic" chemistry in the cloud.

In the oxygen chemistry, the largest change stems from the enhanced H$_3^+$ dissociative recombination rate. The resulting drop in the H$_3^+$ abundance implies that the O + H$^+$ reaction (7) rather than the O + H$_3^+$ channel (9) dominates the initial steps. Since the rate of reaction (7) depends sensitively on the temperature structure, the predicted OH and H$_2$O column densities are more uncertain. Even with a reduced radiation field and reduced removal by C$^+$, the cosmic ray ionization rate still needs to be fairly high, $\zeta_o \approx (7 \pm 3) \times 10^{-17}$ s^{-1} to reproduce the observed OH column densities in steady-state models. The observed narrow OH radio line widths of 1–2 km s^{-1} toward ζ Oph (Crutcher 1979) and translucent clouds in general (Magnani and Siskind 1990) suggests little contribution from energetic processes.

The observed HD column densities can also be used to constrain the cosmic ray ionization rate, now that the deuterium abundance in the local interstellar medium has been well determined by HST to $(1.65^{+0.07}_{-0.18}) \times 10^{-5}$ (Linsky et al. 1993). Similar results for ζ_0 are found as from the OH data.

The observed CO column density toward ζ Oph is still difficult to reproduce in homogeneous models. Although the models are compatible with $N(\text{CO}) \leq$ a few $\times 10^{14}$ cm^{-2} found in other diffuse clouds, they fall short by nearly an order of magnitude for clouds such as ζ Oph, even with a reduced radiation field. Because the CO abundance is very sensitive to density and radiation field in the regime where the ultraviolet lines just become self-shielding (see Fig. 6), an inhomogeneous medium with a small fraction of the gas at somewhat higher density is needed to approach the observed value (Spaans 1996).

The only new molecule detected by absorption lines toward ζ Oph in the last decade is HCl (Federman *et al.* 1995). The observed abundance agrees well with the model predictions of van Dishoeck and Black (1986) and updated reaction rates. NH has been detected in two other diffuse clouds (Meyer and Roth 1991) but not yet toward ζ Oph, and is further discussed in §9.2. Its abundance can be used to constrain the importance of grain-surface chemistry. Finally, HCO$^+$ is seen by millimetre emission toward ζ Oph (Liszt and Lucas 1994). Its inferred abundance is uncertain but is likely to be several orders of magnitude larger than predicted by models. This is partly due to the fact that the model CO abundances are lower than observed, thus resulting in a reduced rate of formation of HCO$^+$ by the CO + H$_3^+$ reaction (see also §7.3 and §9.1).

In summary, the ion–molecule and neutral–neutral gas-phase chemistry discussed in §3 can produce the observed abundances of most diatomic molecules in diffuse clouds, given plausible values for the other parameters entering the networks. However, the observations of species such as NH signify the importance of other chemical processes which need to be taken into account, whereas the CH$^+$, HCO$^+$ and CO abundances indicate that the physical structure may not yet be fully understood.

7 Translucent Clouds

Many different types of translucent clouds with $A_V \approx$ 1–5 mag occur in the interstellar medium. They can be either small, isolated clouds, clumps within giant molecular clouds (GMCs) detected in CO surveys, the outer edges of dark molecular clouds or (high–latitude) cirrus clouds. They share the characteristic that photoprocesses play an important role in the chemistry throughout the cloud. No signatures of pre-main sequence stars have yet been found in any isolated low extinction (<5 mag) translucent cloud (Magnani *et al.* 1996a) and their narrow line widths indicate quiescent gas conditions without shocks. Thus, these clouds are very well suited to test our understanding of the next step in the chemistry networks, that leading to the formation of simple polyatomic molecules.

7.1 Clouds Selected by Optical Absorption

Optical absorption line observations of more than 50 stars lying behind translucent and high-latitude clouds with $A_V = $ 1–5 mag have been performed in the last decade (Crutcher 1985; van Dishoeck and Black 1989; Cardelli *et al.* 1990;

Gredel et al. 1991, 1992, 1993; Crawford 1989, 1990; Penprase 1993; Federman et al. 1994; Gredel 1997). Most studies were focused on CH, CH$^+$, CN and/or C$_2$. It is found that the column densities of all four species continue to increase with extinction and are up to an order of magnitude larger than those found in diffuse clouds. The CN column density increases strongly with density, consistent with the gas-phase formation routes. The data have been used to test the model predictions of different chemical zones and the nature of the transitions between them, in particular the C$^+ \to$ CO transition and the transition from destruction by ultraviolet radiation to destruction by chemical processes (Fig. 5).

Complementary millimetre ^{12}CO and ^{13}CO 1–0, 2–1, and 3–2 emission data have been obtained for several clouds (e.g. van Dishoeck et al. 1991; Gredel et al. 1994; Magnani et al. 1996b). The corresponding CO column densities range from 10^{15} to a few times 10^{16} cm^{-2}, up to two orders of magnitude larger than found in diffuse clouds. The physical parameters inferred from the C$_2$ and CO data indicate densities $n_H \approx$300–2000 cm^{-3}, with temperatures varying from $\sim$10 to 40 K. Some of these clouds (e.g. Cyg OB2 no. 12) appear similar to (a combination of) diffuse clouds, whereas others (e.g. HD 29647) resemble darker clouds. [C I] emission at 492 GHz has been observed from a number of high latitude clouds (Stark and van Dishoeck 1994; Ingalls et al. 1994; Stark et al. 1996), providing further constraints on the models and the thermal balance.

In order to obtain more information on the chemistry, Crutcher (1985) and Gredel et al. (1994) searched for millimetre emissions of simple diatomic and polyatomic molecules such as CS, CN, C$_2$H, HCO$^+$, and HCN in these optically selected translucent clouds. The success of these searches has been limited, presumably because of the low densities. Surprisingly, C$_2$H is not detected in any of these clouds in spite of the large CH and C$_2$ abundances. The observed column densities and upper limits are consistent with the extension of the diffuse cloud models into the translucent regime, provided that the gas-phase carbon abundances are reduced by factors of 2–4. Observational constraints on the gas-phase carbon and oxygen elemental abundances in translucent clouds are badly needed. C$_3$ has been tentatively detected by Haffner and Meyer (1995) in one translucent cloud, at a level slightly below the earlier predictions.

7.2 Clouds Selected by Millimetre Absorption

The lack of strong emission makes the millimetre absorption line technique much better suited for the study of the chemistry in translucent clouds. In earlier work by, e.g., Nyman and Millar (1989), Tieftrunk et al. (1994) and Greaves and Nyman (1996), absorption features were observed against strong galactic continuum sources such as Cas A, SgrB2 and W49, but in these cases the absorption is difficult to disentangle from the emission. Lucas and Liszt (1994, 1996, see 1997 for an overview) surveyed several simple molecules—CO, HCO$^+$, CN, HCN, HNC, CS, SO, H$_2$S, C$_2$H, C$_3$H$_2$ and H$_2$CO—in a set of $\sim$15 diffuse and translucent clouds toward extragalactic sources and found CO column densities ranging from 10^{14} to 10^{16} cm^{-2}. The inferred temperatures and densities of a

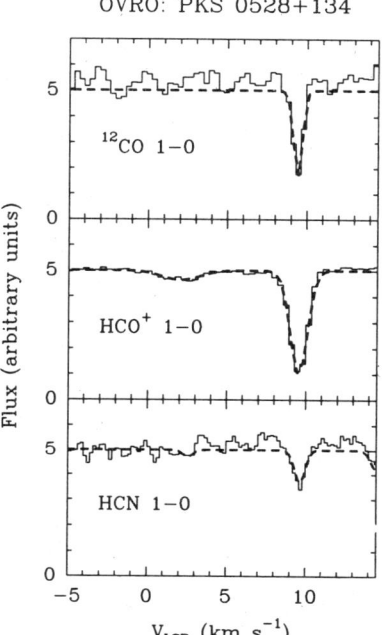

FIG. 9. Millimetre absorption spectra toward the quasar PKS 0528+134, showing the presence of triatomic molecules in local diffuse and translucent clouds (Hogerheijde et al. 1995).

few hundred to a few thousand cm^{-3} are similar to those found from the optical absorption line data of other clouds discussed above. Some of these lines of sight have also been observed at radio wavelengths for species such as OH (Crutcher 1980; Liszt and Lucas 1996), H_2CO (Colgan et al. 1986; Liszt and Lucas 1995) and C_3H_2 (Cox et al. 1988).

Several remarkable conclusions follow from this unique data set. First, HCO$^+$ is readily detected in all clouds, even those that do not show ^{12}CO absorption (see Fig. 9). Second, an abrupt increase in the CO column density occurs around $N(CO)=10^{15}$ cm^{-2}, which is the range where the CO ultraviolet lines become self-shielding and carbon is transformed from atomic to molecular form (see Fig. 6). The $^{12}CO/^{13}CO$ ratio varies between 15 and 100, but is in no case as large as the value of 167±15 found toward ζ Oph by Lambert et al. (1994). Third, a tight correlation is found between the column densities of HCO$^+$, OH, HCN, HNC, CN, CS, SO and H_2S over several orders of magnitude. Surprisingly, the abundance ratios are very similar to those derived for dark clouds, in spite of the presence of ultraviolet radiation which dissociates the molecules. The expected order-of-magnitude variations in molecular abundances between the diffuse and dark cloud regime are not observed for most of these species. Only C_2H and

Table 5 Abundances $N(X)/N(H_2)$ of simple molecules in translucent clouds

Species	HD169454[a]	Mm abs[b]	Turner[c] $A_V^o \approx 1.5$	Turner[c] $A_V^o \approx 2.0$	TMC-1[d] S	L134N[d]
CH	3(-8)	–	–	–	2(-8)	1(-8)
CN	3(-8)	1(-8)	–	–	3(-8)	<3(-9)
HCO$^+$	2(-9)	1(-9)[b]	7(-10)	7(-10)	6(-9)	8(-9)
C$_2$H	<2(-8)	1(-8)	–	–	8(-8)	<5(-8)
HCN	<1(-8)	3(-9)	2(-9)	6(-9)	1(-8)	4(-9)
CS	2(-8)	1(-9)	1(-9)	5(-9)	1(-8)	1(-9)
SO	–	6(-10)	6(-9)	3(-8)	5(-9)	2(-8)
SO$_2$	–	–	2(-9)	6(-9)	<1(-9)	4(-9)

[a] Gredel et al. (1994); [b] Lucas and Liszt (1997) assuming $x(HCO^+)=1(-9)$; [c] Turner (1994, 1995, 1996 and ref. cited); [d] Ohishi et al. (1992)

C$_3$H$_2$ are more abundant relative to HCO$^+$ in translucent clouds, suggesting that these molecules occur in the most outer regions of the clouds (cf. Fig. 5). The resulting abundances are summarized in Table 5.

7.3 Cirrus Cores and Clemens–Barvainis Objects

A complementary set of observations designed to test chemical models of simple polyatomic molecules in translucent clouds has been performed by Turner in a series of papers (Turner 1996 and references cited). Turner searched for millimetre emission lines of more than a dozen molecules in 9 cirrus high–latitude cores and 21 translucent clouds from the Clemens–Barvainis (CB) (1988) catalogue, selected on the basis of optical extinction. The CB objects are found to have typical central densities $n_H \approx 600$–2000 cm^{-3} and temperatures around 15 K, whereas estimates of the incident radiation field range from $I_{UV} = 0.3$–2.0. The total extinctions A_V are in the range 2–5 mag (central extinctions $A_V^o \approx 1$–2.5), suggesting that these clouds represent the lower I_{UV}, higher density and total column density end of the translucent clouds studied by Gredel et al. and Lucas and Liszt. The clouds appear to be in hydrostatic equilibrium, although models with constant density or density inversely proportional to radius fit equally well. Their lifetimes are likely to be long enough that steady-state chemistry has been achieved.

With deep integrations, many of the same molecules observed by Lucas and Liszt have been detected in emission in these clouds. In addition, species such as N$_2$H$^+$, HCNH$^+$, SO$^+$, SO$_2$, OCS, H$_2$CS, and HCS$^+$ have been observed, providing more detailed tests of the nitrogen and sulphur networks. Using physical models designed to reproduce the observed ^{13}CO and C^{18}O distributions, Turner has derived abundance profiles for the various species. Contrary to Lucas and Liszt, he concludes that most abundances increase significantly with extinction, as expected for the transition from diffuse to dense cloud chemistry. The principal exception is HCO^+. Table 5 includes his derived abundances.

Comparison with translucent cloud models containing simple gas-phase chemical networks indicates that good agreement can be obtained for all species except H_2CO, NH_3 and H_2S. Grain-surface chemistry is suspected to be the dominant formation mechanism for these hydrides. The models allow tests of the major chemical routes, discussed in §3. Specifically, the observed constancy of the HCO^+ abundance in the translucent cloud regime can be explained by formation through the $C^+ + OH \rightarrow CO^+ + H$ route in the low A_V case where CO is not yet self-shielding, and the $H_3^+ + CO \rightarrow HCO^+$ reaction at high A_V. This also explains in part why the $HCO^+/^{12}CO$ ratio is a factor of 10 higher in the diffuse clouds. In contrast, N_2H^+ is detected in only a few translucent clouds, consistent with the fact that this ion cannot be formed efficiently in diffuse clouds, but only in dense clouds through the $N_2 + H_3^+ \rightarrow N_2H^+$ reaction when most of the nitrogen has been converted to molecular form.

The observed abundances of SO and SO_2 can be reproduced with the chemistry described in §3.5, but only if the OH abundance in the models is high, of order 10^{-5}. Such high OH abundances are not implausible, but appear significantly larger than those observed in diffuse clouds (Liszt and Lucas 1996; Felenbok and Roueff 1996). The observed SO/SO_2 abundance ratio varies from 1 to 15 in the clouds, and is very different from the ratio of ~ 0.1 predicted for the centre of dark clouds, where most of the sulphur is transformed into SO_2. It is also much lower than the ratio of ~ 1000 predicted for shocks. At low A_V, the S–O and S–C chemistries appear decoupled, but at higher A_V they are connected through the $C^+ + SO$ reaction.

Table 5 shows that for some species, the derived abundances for the various clouds are consistent within the observational errors, whereas for others, differences of factors of a few are found. These variations can be traced back partly to observational uncertainties (especially in the determination of $N(H_2)$), and partly to the fact that the molecules behave differently with increasing A_V (case II vs case III, Fig. 5). Nevertheless, the similarity in the abundances in translucent and dark clouds even for species such as SO_2 suggests that perhaps the interiors of dark clouds are not as shielded from ultraviolet radiation as thought previously, but have an inhomogenous structure in which most of the molecules "see" an extinction more representative of that in translucent clouds.

8 Dark Clouds

Cold, dark molecular clouds were discovered originally as dark patches on the sky but have subsequently been traced in molecular line emission. Maps of two dark clouds, B5 and TMC–1, in various species are given in Chapter 5. Dark clouds are often part of large complexes such as those in Taurus or Perseus, which contain up to 10^4 $M_\odot$ of material. Low-mass (< 2 $M_\odot$) stars form throughout these clouds (see Fig. 5 in Chapter 5). Thus, in the study of the chemistry of quiescent dark clouds, care has to be taken that the gas has not yet been affected by the star-formation process, especially since young stars have powerful outflows which may extend over several pc.

Typical densities in dark cores derived from ratios of millimetre emission lines are $\sim 10^4$–10^5 cm^{-3}, whereas temperatures are low, $T = 10$–15 K. Infrared observations give information on the presence of ice mantles along a line of sight. The threshold for water ice mantle formation occurs at $A_V \sim 3$ mag in Taurus, increasing to ~ 12 mag in Ophiuchus (Whittet 1993). Solid water is likely to be the second most abundant heavy molecule in these clouds after gas-phase CO. For other species, information on the amount of depletion is very difficult to infer from millimetre observations (Mundy and McMullin 1997).

Most observational and theoretical studies of dark cloud chemistry have focused on TMC-1 and L134N, two regions which show particularly rich chemistries (see Ohishi et al. 1992 for an overview). As discussed in §4.4, pseudo time-dependent models can fit the observed high abundances of carbon chain molecules in TMC-1 only at early times, $t \approx 10^5$ yr. Is this characteristic for all dark clouds? Suzuki et al. (1992) performed a systematic study of a few molecules (C_2S, HC_3N, HC_5N and NH_3) in a set of dark cores identified by Myers and Benson (1983). Cores with and without stars were selected. The abundances of the various carbon-chain molecules were found to correlate well with each other, but not with NH_3, which was found to be more abundant in older cores where stars have already formed. The observed C_2S/NH_3 abundances were reproduced quantitatively in models which start from diffuse gas and form dense cores over a period of 10^5 to 2×10^6 yr. Thus, this ratio may be a particularly useful tracer of cloud evolution. Further discussion of observations and models of specific dark clouds is contained in Chapter 5.

Other systematic surveys have been performed by Bergin et al. (1997a) and Ungerechts et al. (1997), who mapped several GMC cores (Orion, M17 and Cepheus A) over a $5' \times 5'$ region in various molecules. They identified several positions away from the regions of star formation, where the chemical composition was analysed. The derived abundances are remarkably uniform (within a factor of a few), both within a single GMC and between different GMCs, in contrast with the situation for some dark cores. Bergin et al. (1997b) ran several gas-phase and gas-grain models for comparison. Only pure gas-phase models appear to be able to reproduce the observed abundances at early times ($\sim 10^5$ yr) if a fairly high elemental abundance ratio $[C]_g/[O]_g \approx 0.8$ is adopted. They argue that the observed CS/SO ratio may be used to constrain the $[C]_g/[O]_g$ ratio, as suggested previously for the L134N dark core (Swade 1989).

Further systematic studies on the variations of the abundances of key molecules from cloud to cloud will be very valuable, especially in comparison with the new information available on translucent clouds from millimetre absorption (Lucas and Liszt 1997) and emission (Turner 1996).

9 Some Outstanding Problems

In the previous sections, an overview of observations and models of a variety of clouds has been given. Although for many molecules the low-temperature gas-phase chemistry described in §3 provides a satisfactory explanation of the observed abundances, there are some species whose abundances point to different processes.

9.1 The CH^+ Mystery

The high abundance and ubiquity of CH^+ in diffuse clouds present a long-standing puzzle to astrochemists (see Dalgarno 1976; Black 1988; Williams 1992 for reviews). The problem stems from the fact that formation of CH^+ in cold diffuse molecular clouds is very slow because of the endothermicity of reaction (4), whereas CH^+ is rapidly destroyed by reactions with H, H_2, and electrons, as well as photodissociation. Quiescent gas-phase models fall short by 1–2 orders of magnitude of the observed values. Some energetic process is therefore needed to stir up the gas and drive reaction (4), but at the same time this should not affect significantly the abundances of other species such as CH and OH which are already reproduced by the standard chemistry and whose line profiles indicate that they are not co-existent with CH^+. Proposed mechanisms include shocks (Elitzur and Watson 1980), reactions of C^+ with vibrationally excited H_2 (Stecher and Williams 1974), and heating of the gas by strong radiation fields such as found in PDRs (White 1984, see Chapter 9). It can readily be shown that the last two suggestions fall short quantitatively in low density diffuse clouds exposed to the normal radiation field.

The shock hypothesis has remained popular for more than a decade, especially after magnetohydrodynamic (MHD) shock models were developed in which the C^+ ions are accelerated with respect to neutrals like H_2 ahead of the shock (Draine and Katz 1986; Pineau des Forêts et al. 1986; Hartquist et al. 1990; Flower 1994). In this case, part of the energy needed to overcome the reaction barrier is supplied by the differential streaming of the ions and neutrals. Although single MHD shock models can approach the observed CH^+ column densities in diffuse clouds, they predict other features which are not observed, in particular a velocity shift between the CH and CH^+ lines. The very high spectral resolution observations by Lambert et al. (1990), Crane et al. (1995) and Crawford (1995) confirm earlier indications that such shifts are virtually absent (<1 km s^{-1}), even though the CH^+ lines are broader.

Additional clues to the CH^+ origin stem from recent systematic searches for CH^+ in more highly reddened lines of sight, which indicate that the CH^+ column density continues to increase with A_V (Gredel et al. 1993, Penprase 1993; Gredel 1997). For randomly oriented clouds, this is not necessarily in contradiction with the shock models, since the more reddened stars are also often located further away so that the lines of sight probably intercept more than one shock. However, Gredel (1997) has demonstrated that the correlation with A_V also holds for different lines of sight with $A_V = 0.5$–4.5 mag through the same cloud. Such a correlation is difficult to reconcile with scenarios where the CH^+ formation is limited to the surface of clouds. Thus, although the observed widths of the CH^+ lines and the increase in CH^+ column density with that of rotationally excited H_2 $J = 3$–5 (Lambert and Danks 1986) indicate an energetic origin of CH^+, it must be well mixed with the more quiescent gas.

In the last few years, models in which random turbulence rather than shocks heat material have been examined. Duley et al. (1992) proposed that the boundary layers between cold molecular clouds and the warm intercloud medium may be heated by dissipation of turbulence leading to CH^+ formation. In their model, the thickness of the turbulent boundary layer increases with cloud size, which would be compatible with the observed correlation of CH^+ with A_V.

Falgarone et al. (1995) and Falgarone and Phillips (1996) have found strong observational evidence for the presence of turbulence, which undergoes intermittent bursts of dissipation to create hot, mostly atomic pockets of gas inside the clouds. These small regions have high temperatures (>1000 K) in which CH^+ can be temporarily formed, and relax back to the cold conditions after $\sim 10^3$ yr. Since the heating and cooling rates used by Falgarone et al. are rather crude, it is not yet clear whether this mechanism can be maintained. Finally, Spaans (1995) uses a model in which the turbulent cascade of kinetic energy creates a non-Maxwellian velocity distribution, in which the fraction of C^+ and H_2 moving at different speeds is sufficiently large to reproduce the observed amounts of CH^+. A better understanding of the physics underlying interstellar turbulence is needed before firmer conclusions can be made.

Regardless of the precise formation mechanism, the gas which contains enhanced CH^+ abundances may affect those of other species. The most obvious case is that of CH, since the reaction sequence $CH^+ + H_2 \to CH_2^+ \to CH$ leads directly to its production. Observationally, some broad CH absorption is seen, but is constrained to be less than $\sim 30\%$ of that of quiescent CH in the case of ζ Oph (Lambert et al. 1990). Federman et al. (1997a) have recently estimated $N(CH)/N(CH^+) \leq 0.4$ for two very low density lines of sight ($n_H \approx 50$ cm^{-3}). Hogerheijde et al. (1995) point out that in gas containing large amounts of CH^+, the reactions $CH^+ + O \to CO^+ \to HCO^+$ and $CH^+ + H_2 \to CH_2^+ \to HCN^+ \to H_2CN^+ \to HCN$ lead to increased abundances of HCO^+ and HCN, which may explain in part the high abundances of these species observed in diffuse and translucent clouds (see §7).

9.2 Large Molecules in Diffuse Clouds

There is growing observational evidence that in addition to simple molecules, large polyatomic molecules are present in the diffuse medium. Recent maps obtained with ISO show that the near-infrared emission features between 3 and 13 μm ascribed to PAHs are widespread, even in clouds exposed to the normal radiation field. The rotational substructure observed in some diffuse interstellar bands (Sarre et al. 1995; Ehrenfreund and Foing 1996) also strongly favours an identification with large gas-phase molecules. In addition to (ionized) PAHs, long carbon-chain molecules of the type C_{2n+1}, HC_{2n+1}^+ and $C_{2n}N^+$ with $n = 7$–18 have been proposed (see Maier 1997 for an overview), and evidence for the presence of C_{60}^+ has been found (Foing and Ehrenfreund 1994, 1997).

Bettens et al. (1995, 1996) have included pathways in their models to produce molecules containing as many as 64 carbon atoms. Unsaturated linear chains, single and triple rings and fullerenes are produced in reasonable amounts in the models, but not PAHs. It is perhaps more plausible that these polyatomic molecules arise in a chemistry quite distinct from the normal ion–molecule chemistry, for example from the erosion of carbonaceous grains in shocks (e.g. Jones et al. 1996) or warm interfaces (Duley et al. 1992). The hydrocarbon fragments can subsequently react with other species in the conventional chemistry after injection into the gas, producing other complex molecules (Thaddeus 1994). Hall and Williams (1995) have followed the normal chemistry after injection of small hydrocarbon molecules in diffuse clouds. Various complex species are formed, but unless they are very large in size, their abundances are maintained only for the duration of the injection period, and decay rapidly on a time scale of photodissociation (~1000 yr). Possible destruction mechanisms of larger PAHs and fullerenes have been discussed by Millar (1992).

Duley et al. (1992) have suggested that the erosion of carbonaceous grains can also lead to CH^+ in diffuse atomic clouds. The widespread appearance of both CH^+ and the diffuse interstellar bands would then imply that this mechanism operates with high frequency in the interstellar medium.

9.3 Importance of Grain-surface Chemistry

Hydrides are thought to be the natural product of grain surface chemistry, but what is the importance of this process in diffuse and dark clouds besides the formation of H_2? Although CH and OH have been observed for decades, NH has been detected in diffuse clouds only recently by Meyer and Roth (1991). The inferred NH column densities for two lines of sight are nearly a factor of 30 smaller than those of CH and OH. This fact is one of the strongest arguments that grain-surface chemistry does not dominate the formation of carbon- and oxygen-bearing species in diffuse clouds (Crutcher and Watson 1976). However, it may contribute at the level of a few % to CH and OH, and dominate the production of NH. Indeed, the gas-phase formation of NH through reactions (11) and/or (12) is so slow that pure gas-phase models fall short of the observed values by an order of magnitude. Mann and Williams (1984) and Wagenblast et al. (1993) have shown that the observed NH column densities can be reproduced if grain-surface chemistry is included. Further searches for NH and the related NH_2 molecule can help constrain the nitrogen-hydride chemistry.

NH_2 submillimetre absorption lines have been detected in the dense cloud(s) along the line of sight toward SgrB2 (van Dishoeck et al. 1993). Its abundance in the quiescent gas is consistent with ion–molecule chemistry. The ratios of unsaturated to saturated hydrides may provide clues to the relative importance of gas-phase and grain-surface chemistry (van Dishoeck 1995).

As discussed in §7.3, gas-phase networks can produce abundances of most simple species observed by Turner, with the exception of NH_3, H_2CO and H_2S. The problem of the large abundance of H_2CO in diffuse and translucent clouds has been discussed for some time in the literature (e.g. Federman and Allen 1991), whereas that of H_2S has also been noted for translucent clouds and denser PDRs by Tieftrunk et al. (1994) and Jansen et al. (1995). For all three molecules, grain-surface formation is the most likely explanation for the discrepancy although detailed quantitative models for the translucent conditions still have to be developed.

Finally, the direct infrared observations of ices in dark clouds prove unambiguously the importance of grain-surface chemistry, since the large amount of solid H_2O cannot result simply from accretion from the gas. In star-forming regions and "hot cores", gas-grain interactions actively modify the gas phase chemistry by evaporation and photochemistry (see Chapter 5). To what extent this is also the case for quiescent dark clouds still remains to be determined.

9.4 The Oxygen Budget

Most models of the chemistry of quiescent dense clouds predict that O_2 and O are the major oxygen-bearing species in the gas (Lee et al. 1996a; Millar et al. 1997). Ground-based searches for O_2 are limited to the $^{16}O^{18}O$ isotopic variety, for which the 2_1–0_1 line at 234 GHz can be observed through the atmosphere. Pagani et al. (1993) have presented a tentative detection of the line in the dark cloud L183, but these observations have not yet been confirmed. The corresponding upper limit is $O_2/H_2 < 10^{-5}$ or $O_2/CO < 0.07$. More sensitive limits of $O_2/CO < 0.012$ have been obtained for external galaxies with sufficiently high velocities to shift the 118 GHz line out of the atmospheric line, but these limits refer to 10 kpc scales (Combes et al. 1991; Liszt 1992).

An independent, elegant method to search for interstellar O_2 is provided by absorption line observations toward bright background quasars. A dense molecular cloud at $z = 0.69$ has been found toward B0218+357 which shows strong $C^{18}O$ absorption. However, a deep search for several O_2 lines has resulted only in upper limits, corresponding to an abundance $O_2/CO < 0.01$ (Combes and Wiklind 1995). For a typical CO/H_2 abundance of 2×10^{-4}, the corresponding limit on O_2 is $< 2 \times 10^{-6}$. Such low O_2 abundances form a serious challenge for chemical models, not only with respect to the oxygen chemistry but also that of other related carbon- and nitrogen-bearing species.

If the O_2 abundances are indeed as low as indicated by these observations, where is all the oxygen? In warm clouds, a significant fraction is driven into H_2O, as discussed in §3.3. In cold clouds, the gas-phase H_2O abundance is low, $\sim 10^{-8}$–10^{-7} (Zmuidzinas et al. 1997; Tauber et al. 1996). H_2O ice is a better candidate, but direct observations of the 3 µm ice band together with the 9.7 µm silicate feature indicate that only $\sim 10\%$ of the oxygen budget is locked up in water ice (Whittet 1993). Solid O_2 is another possibility, but is very difficult to observe owing to its extremely weak infrared bands (Ehrenfreund et al. 1992).

This leaves atomic oxygen as the most likely alternative.

Although it is well known from ultraviolet absorption lines that atomic O is the dominant form of gas-phase oxygen in diffuse and translucent clouds, observations of emission lines in cold dark clouds are hampered by the fact that the lowest $^3P_1$–$^3P_2$ transition at 63 µm lies at 227 K. Thus, a search for this line in absorption is the best method. Poglitsch et al. (1996) reported possible absorption of the [O I] 63 µm line superposed on the broad, shocked emission line in DR 21. Recently, Baluteau et al. (1997) observed the [O I] 63 µm line in absorption toward SgrB2. In both cases, the data imply that more than 40% of the available oxygen is in atomic form. The large abundance of atomic O is consistent with an inhomogeneous structure in which some ultraviolet radiation penetrates and dissociates the O_2.

9.5 Ionization Structure

An important parameter for the dynamical structure of a cloud is the ionization fraction, $x(e) = n(e)/n(H_2)$, since the charged species govern the coupling of the gas to magnetic fields and the ability of a cloud to collapse and form stars (see Chapter 5). In the model calculations, the electron abundance is high ($\sim 10^{-4}$) in diffuse clouds and equal to the gas-phase carbon abundance, its main supplier. At $A_V \approx 0.5$–2 mag (depending on the density and strength of the external radiation field), C^+ recombines to C and the ionization fraction drops. Photoionization of second-row species such as S now control the electron abundance. Finally, deep inside the clouds, cosmic rays are the primary ionization source and the electron fraction drops to a few $\times 10^{-9}$–10^{-8} in the normal "low ionization" solution (see Fig. 2 of Chapter 5 and Fig. 7 of Chapter 9). H_3^+, HCO^+ and H_3O^+ are expected to be the principal molecular ions, in addition to metal ions such as Mg^+, Fe^+ (see §3.8). The ionization balance can be modified by the presence of PAHs, which may contain a significant part of the negative charge. The predicted electron abundance scales with $(\zeta/n_H)^{1/2}$ if the metal abundance is low, and $(\zeta/n_H)^{1/3}$ if metals are not taken to be significantly depleted (Oppenheimer and Dalgarno 1974; Millar 1990).

How do these predictions compare with observations? In diffuse and translucent clouds, the atomic ionization balance (e.g. Ca^+/Ca) and the rotational excitation of species such as CN can be used to determine the electron concentration directly (e.g. Black and van Dishoeck 1991; Gredel et al. 1993) (see Table 4). Combined with independent estimates of the density from diagnostics such as C_2 and CO, whose excitation is not determined by collisions with electrons, $x(e)$ can be determined. Typical values are a few $\times 10^{-5}$–10^{-4}, consistent with models. In the denser translucent clouds such as studied by Turner (1993), the excitation of H_2CO has been used to constrain $x(e)$. Inside dark clouds, the situation is more complex. The traditional method involved observations of the DCO^+/HCO^+ ratio as a measure of the H_2D^+/H_3^+ ratio, which together with a model for the deuterium fractionation and H_3^+ recombination gives a limit on $x(e)$ (Langer 1985). However, this method depends strongly on the adopted H_3^+ dissociative recombination rate. Recently, de Boisanger et al. (1996) used observations of several protonated molecular ions—including detections of the principal ions HCO^+ and H_3O^+ and limits on H_3^+—together with a simple model to constrain $x(e)$ and ζ in two dense clouds with $n_H \approx 10^6$ cm^{-3}. A lower limit of $\sim 3 \times 10^{-9}$ is provided by the observed abundances of the positive ions, whereas typical derived values of $x(e)$ are $\sim 10^{-8}$ with a small uncertainty factor. Detections of H_3^+ along the same lines of sight could be used to provide an independent constraint on ζ. Schilke et al. (1991) inferred $6 \times 10^{-9} < x(e) < 5 \times 10^{-8}$ from detailed modelling of the sulphur chemistry in TMC–1, where the range depends on the number of metals and PAHs included.

10 Concluding Remarks

The gas-phase chemistry models which have been developed over the last 25 years provide results in remarkable agreement with many aspects of interstellar chemistry (Herbst 1997). These include: the abundances of simple hydrocarbons in diffuse and translucent clouds; the presence of H_3^+ and the related high abundances of protonated ions such as HCO^+ and N_2H^+ in dense clouds; the high abundances of unsaturated and metastable molecules in dark clouds; and the large isotopic fractionation such as found for DCO^+/HCO^+ and DCN/HCN. On the other hand, gas-grain interactions cannot be neglected, as evidenced by the abundances of H_2 and NH in diffuse clouds, and of H_2CO, NH_3 and H_2S in translucent and dark clouds. Also, the direct observations of H_2O- and CO-ice signify the importance of freeze-out and grain-surface chemistry in dark clouds. It is gratifying to note that through a series of systematic studies, the chemical and physical relation between the diffuse, translucent and dark clouds is now starting to be better defined.

In spite of these successes, several gaps in our understanding remain. First, basic studies of the chemical processes under interstellar conditions remain crucially important for progress in the field, as illustrated by the recent experiments on neutral–neutral reactions. Second, there are molecules for which the failure of the standard models is not due to uncertainties in the chemical processes, but caused by a lack of understanding of the (inhomogeneous) structure of the cloud and the physical processes which can occur. Examples are provided by the abundances of CH^+ and CO in diffuse and translucent clouds, the excitation of the high rotational levels of H_2 in diffuse clouds, the presence of large polyatomic molecules in diffuse clouds, and the lack of abundant SO_2 and O_2 in dark clouds. It will be a challenge to provide a realistic physical description of processes such as turbulence which can be used to quantitatively test the chemistry. Nevertheless, the mere presence or absence of such species can be used as diagnostics of these physical processes and cloud structure.

On the observational side, exciting new information on the chemistry of diffuse and translucent clouds is becoming available from millimetre absorption line observations. In the near future, the *Space Telescope Imaging Spectrometer* aboard HST may provide much needed complementary information on the atomic and molecular abundances in more highly reddened lines of sight. In addition, direct observations of H_2 will be possible in these clouds with *FUSE*. In the infrared, high resolution echelle spectrographs are now sensitive enough to study gas-phase absorption lines of crucial species such as CO and H_3^+. The *Infrared Space Observatory* will provide important new constraints on the physical structure by observations of the H_2 pure rotational emission lines. Further data on the composition of ices in dark clouds prior to star formation will be important as well. Finally, the development of a new generation of millimetre arrays will allow the chemistry in dark clouds to be studied at much higher angular resolution, which will be important for tracing the origin of small-scale abundance variations. Using these facilities, we may soon be in a better position to address important questions regarding the physical structure and chemical composition of a dark cloud prior to star formation and its ability to collapse and form stars.

11 Acknowledgements

This chapter has benefited from discussions with many colleagues. I am particularly grateful to E. Herbst, R. Lucas and B. Turner for sending preprints in advance of publication, and to T. W. Hartquist for a critical reading. This work is supported by the Netherlands Organization for Scientific Research (NWO/NFRA).

Bibliography

1. Baluteau, J.-P. *et al.* (1997). *A&A*, **322**, L33.
2. Bates, D. R. and Spitzer, L. (1951). *ApJ*, **113**, 441.
3. Bergin, E. A., Langer, W. D. and Goldsmith, P. F. (1995). *ApJ*, **441**, 222.
4. Bergin, E. A., Ungerechts, H., Goldsmith, P. F., Snell, R. L., Irvine, W. M. and Schloerb, F. P. (1997a). *ApJ*, **482**, 267.
5. Bergin, E. A., Goldsmith, P. F., Snell, R. L. and Langer, W. D. (1997b). *ApJ*, **482**, 285.
6. Bettens, R. P. A. and Herbst, E. (1996). *ApJ*, **468**, 686.
7. Bettens, R. P. A., Lee, H. H. and Herbst, E. (1995). *ApJ*, **443**, 664.
8. Black, J. H. (1988). *Adv. Atom. Mol. Phys.*, **25**, 477.
9. Black, J. H. and Dalgarno, A. (1973). *Astrophys. Lett.*, **15**, 79.
10. Black, J. H. and van Dishoeck, E. F. (1991). *ApJ* **369**, L9.
11. Blitz, L., Bazell, D. and Désert, F. X. (1990). *ApJ*, **352**, L13 (BDD).
12. Bloemen, H., Deul, E. and Thaddeus, P. (1990). *A&A*, **238**, 437.
13. Boissé, P. (1990). *A&A*, **228**, 483.
14. Boland, W. and de Jong, T. (1982). *ApJ*, **261**, 110.
15. Cardelli, J. A. (1988). *ApJ*, **335**, 177.
16. Cardelli, J. A., Suntzeff, N. B., Edgar, R. J. and Savage, B. D. (1990). *ApJ*, **362**, 551.

17. Caselli, P., Hasegawa, T. I. and Herbst, E. (1997). *ApJ*, in press.
18. Chièze, J. P., Pineau des Forêts, G. and Herbst, E. (1991). *ApJ*, **373**, 110.
19. Clary, D. C. (1985). *Mol. Phys.*, **54**, 605.
20. Clemens, D. P. and Barvainis, R. E. (1988). *ApJS*, **68**, 257.
21. Colgan, S. W. J., Salpeter, E. E. and Terzian, Y. (1986). *AJ*, **91**, 107.
22. Combes, F. and Wiklind, T. (1995). *A&A*, **303**, L61.
23. Combes, F., Casoli, F., Encrenaz, P., Gerin, M. and Laurent, C. (1991). *A&A* **275**, 558.
24. Cox, P., Güsten, R. and Henkel, C. (1988). *A&A*, **206**, 108.
25. Crane, P., Hegyi, D. J., Kutner, M. L., and Mandolesi, N. (1989). *ApJ*, **346**, 146.
26. Crane, P., Lambert, D. L. and Sheffer, Y. (1995). *ApJS*, **99**, 107.
27. Crawford, I. A. (1989). *MNRAS*, **241**, 575.
28. Crawford, I. A. (1990). *MNRAS*, **244**, 646.
29. Crawford, I. A. (1995). *MNRAS*, **277**, 458.
30. Crawford, I. A. and Barlow, M. J. (1996). *MNRAS*, **280**, 863.
31. Crawford, I. A., Barlow, M. J., Diego, F. and Spyromilio, J. (1994). *MNRAS*, **266**, 903.
32. Crutcher, R. M. (1979). *ApJ*, **231**, L151.
33. Crutcher, R. M. (1980). *ApJ*, **236**, 549.
34. Crutcher, R. M. (1985). *ApJ*, **288**, 604.
35. Crutcher, R. M. and Watson, W. D. (1976). *ApJ*, **209**, 778.
36. Dalgarno, A. (1976). In *Atomic Processes and Applications*, eds. P. G. Burke and B. L. Moiseiwitsch, North-Holland, p.110.
37. Dalgarno, A. (1987). In *Physical Processes in Interstellar Clouds*, eds. G. Morfill and M. S. Scholer, D. Reidel, Dordrecht, p.219.
38. Dalgarno, A. (1994). *Adv. At. Mol. & Opt. Phys.*, **32**, 57.
39. Dalgarno, A. and Black, J. H. (1976). *Rep. Prog. Phys.*, **39**, 573.
40. Dalgarno, A. and Lepp, S. (1984). *ApJ*, **287**, L47.
41. de Boisanger, C. and Chièze, J. P. (1991). *A&A*, **241**, 581.
42. de Boisanger, C., Chièze, J. P. and Meltz, B. (1992). *ApJ*, **401**, 182.
43. de Boisanger, C., Helmich, F. P. and van Dishoeck, E. F. (1996). *A&A*, **310**, 315.
44. d'Hendecourt, L. B. and Léger, A. (1987). *A&A*, **180**, L9.
45. d'Hendecourt, L. B., Allamandola, L. J., Baas, F., and Greenberg, J. M. (1982). *A&A*, **109**, L12.
46. d'Hendecourt, L. B., Allamandola, L. J. and Greenberg, J. M. (1985). *A&A*, **152**, 130.
47. Draine, B. T. (1978). *ApJS*, **36**, 595.
48. Draine, B. T. and Katz, N. S. (1986). *ApJ*, **306**, 655; **310**, 392.

49. Drdla, K., Knapp, G. R. and van Dishoeck, E. F. (1989). *ApJ*, **345**, 815.
50. Duley, W. W., Hartquist, T. W., Sternberg, A., Wagenblast, R. and Williams, D. A. (1992). *MNRAS*, **255**, 463.
51. Dzegilenko, F. and Herbst, E. (1995). *ApJ*, **443**, L81.
52. Eddington, A. S. (1926). *Proc. Roy. Soc. A*, **111**, 424.
53. Ehrenfreund, P. and Foing, B. H. (1996). *A&A*, **307**, L25.
54. Ehrenfreund, P., Breukers, R., d'Hendecourt, L. and Greenberg, J. M. (1992). *A&A*, **260**, 431.
55. Eidelsberg, M., Benayoun, J. J., Viala, Y. and Rostas, F. (1991). *A&AS*, **90**, 231.
56. Eidelsberg, M. et al. (1992). *A&A*, **265**, 839.
57. Elitzur, M. and Watson, W. D. (1980). *ApJ*, **236**, 172.
58. Falgarone, E. (1997). In *CO: Twenty-five Years of Millimeter Spectroscopy*, IAU Symposium 170, eds. W. Latter et al., Kluwer, Dordrecht, p.119.
59. Falgarone, E. and Phillips, T. G. (1996). *ApJ*, **472**, 191.
60. Falgarone, E., Pineau des Forêts, G. and Roueff, E. (1995). *A&A*, **300**, 870.
61. Federman, S. R. and Allen, M. (1991). *ApJ*, **375**, 157.
62. Federman, S. R., Strom, C. J., Lambert, D. L. et al. (1994). *ApJ*, **424**, 772.
63. Federman, S. R., Cardelli, J. A., van Dishoeck, E. F., Lambert, D. L. and Black, J. H. (1995). *ApJ*, **445**, 325.
64. Federman, S. R., Welty, D. E. and Cardelli, J. A. (1997a). *ApJ*, **481**, 795.
65. Federman, S. R., Lambert, D. L., Sheffer, Y., Cardelli, J. A., Andersson, B-G, van Dishoeck, E. F. and Zsargó, J. (1997b). *ApJ*, in preparation.
66. Felenbok, P. and Roueff, E. (1996). *ApJ*, **465**, L57
67. Flower, D. R. (1994). In *The First Symposium on the Infrared Cirrus and Diffuse Interstellar Clouds*, ASP Conference Series no. 58, eds. R. M. Cutri and W. B. Latter, ASP, San Francisco, p.332.
68. Foing, B. H. and Ehrenfreund, P. (1994). *Nature*, **369**, 296.
69. Foing, B. H. and Ehrenfreund, P. (1997). *A&A*, **317**, L59.
70. Geballe, T. R. and Oka, T. (1996). *Nature*, **384**, 334.
71. Genzel, R. (1992). In *The Galactic Interstellar Medium*, eds. D. Pfenniger and P. Bartholdi, Saas Fee Advanced Course 21, Springer, Berlin, p.275.
72. Gerin, M., Falgarone, E., Joulain, K., Kopp, M., Le Bourlot, J., Pineau des Forêts, G., Roueff, E. and Schilke, P. (1997). *A&A*, **318**, 579.
73. Gerlich, D. and Horning, S. (1992). *Chem. Rev.*, **92**, 1509.
74. Goldshmidt, O. and Sternberg, A. (1995). *ApJ*, **439**, 256.
75. Greaves, J. S. and Nyman, L. A. (1996). *A&A*, **305**, 950.
76. Gredel, R. (1997). *A&A*, **320**, 929.
77. Gredel, R., Lepp, S., Dalgarno, A. and Herbst, E. (1989). *ApJ*, **347**, 289.

78. Gredel, R., van Dishoeck, E. F. and Black, J. H. (1994). A&A, **285**, 300.
79. Gredel, R., van Dishoeck, E. F., de Vries, C. P. and Black, J. H. (1992). A&A, **257**, 245.
80. Gredel, R., van Dishoeck, E. F. and Black, J. H. (1991). A&A, **251**, 625.
81. Gredel, R., van Dishoeck, E. F. and Black, J. H. (1993). A&A, **269**, 477.
82. Grevesse, N. and Noëls, A. (1993). In *Origins and Evolution of the Elements*, eds. N. Pratzos et al., Cambridge University Press, p.15.
83. Haffner, L. M. and Meyer, D. M. (1995). ApJ, **453**, 450.
84. Hall, P. and Williams, D. A. (1995). *Astrophys. Space Sci.*, **229**, 49.
85. Hartquist, T. W., Flower, D. R. and Pineau des Forêts, G. (1990). In *Molecular Astrophysics—A volume honoring Alexander Dalgarno*, ed. T. W. Hartquist, Cambridge University Press, p.99.
86. Hasegawa, T. I. and Herbst, E. (1993). MNRAS, **261**, 83; **263**, 589.
87. Heger, M. L. (1922). *Lick. Obs. Bull.*, **10**, 146.
88. Herbst, E. (1993). In *Dust and Chemistry in Astronomy*, eds. T. J. Millar and D. A. Williams, IOP, Bristol, p.183.
89. Herbst, E. (1995). *Ann. Rev. Phys. Chem.*, **46**, 27.
90. Herbst, E. (1997). In *CO: Twenty-five Years of Millimeter Spectroscopy*, IAU Symposium 170, eds. W. Latter et al., Kluwer, Dordrecht, p.71.
91. Herbst, E. and Klemperer, W. (1973). ApJ, **185**, 505.
92. Herbst, E. and Leung, C. M. (1986). ApJ, **310**, 378.
93. Herbst, E. and Leung, C. M. (1989). ApJS, **69**, 271.
94. Herbst, E. and Winnewisser, G. (1987). *Topics in Current Chemistry*, **139**, 121.
95. Hogerheijde, M. R., de Geus, E. J., Spaans, M., van Langevelde, H. J. and van Dishoeck, E. F. (1995). ApJ, **441**, L93.
96. Hollenbach, D. J. and Salpeter, E. E. (1971). ApJ, **163**, 155.
97. Howe, D. A. and Millar, T. J. (1993). MNRAS, **262**, 868.
98. Ingalls, J. G., Bania, T. M. and Jackson, J. M. (1994). ApJ, **431**, L139.
99. Jansen, D. J., van Dishoeck, E. F., Black, J. H., Spaans, M. and Sosin, C. (1995). A&A, **302**, 223.
100. Jenkins, E. B. (1987). In *Interstellar Processes*, eds. D. J. Hollenbach and H. A. Thronson Jr., D. Reidel, Dordrecht, p.533.
101. Jenkins, E. B., Lees, J. F., van Dishoeck, E. F. and Wilcots, E. M. (1989). ApJ, **343**, 785.
102. Jones, A. P., Tielens, A. G. G. M. and Hollenbach, D. J. (1996). ApJ, **469**, 740.
103. Kopp, M., Gerin, M., Roueff, E. and LeBourlot, J. (1996). A&A, **305**, 558.
104. Lacy, J. H., Knacke, R., Geballe, T. R. and Tokunaga, A. T. (1994). ApJ, **428**, L69.
105. Lambert, D. L. and Danks, A. (1986). ApJ, **303**, 401.

106. Lambert, D. L., Sheffer, Y. and Crane, P. (1990). *ApJ*, **359**, L19.
107. Lambert, D. L., Sheffer, Y., Gilliland, R. L. and Federman, S. R. (1994). *ApJ*, **420**, 756.
108. Lambert, D. L., Sheffer, Y. and Federman, S. R. (1995). *ApJ*, **438**, 740.
109. Langer, W. D. (1985). In *Protostars and Planets II*, eds. D. C. Black and M. S. Matthews, Univ. Arizona, Tucson, p.650.
110. Le Bourlot, J. (1991). *A&A*, **242**, 235.
111. Le Bourlot, J., Gerin, M. and Perault, M. (1989). *A&A*, **219**, 279.
112. Le Bourlot, J., Pineau des Forêts, G., Roueff, E. and Schilke, P. (1993). *ApJ*, **416**, L87.
113. Le Bourlot, J., Pineau des Forêts, G. and Roueff, E. (1995). *A&A*, **297**, 251.
114. Lee, H.-H., Bettens, R. P. A. and Herbst, E. (1996a). *A&AS*, **119**, 111.
115. Lee, H.-H., Herbst, E., Pineau des Forêts, G., Roueff, E. and Le Bourlot, J. (1996b). *A&A*, **311**, 690.
116. Lepp, S. and Dalgarno, A. (1988). *ApJ*, **335**, 769.
117. Lepp, S., Dalgarno, A., van Dishoeck, E. F. and Black, J. H. (1988). *ApJ*, **329**, 418.
118. Linsky, J. L. et al. (1993). *ApJ*, **402**, 694.
119. Liszt, H. S. (1992). *ApJ*, **386**, 139.
120. Liszt, H. S. (1993). *ApJ*, **414**, 242; **424**, 510.
121. Liszt, H. S. and Lucas, R. (1994). *ApJ*, **431**, L131.
122. Liszt, H. S. and Lucas, R. (1995). *A&A*, **299**, 847.
123. Liszt, H. S. and Lucas, R. (1996). *A&A*, **314**, 917.
124. Lucas, R. and Liszt, H. (1994). *A&A*, **282**, L5.
125. Lucas, R. and Liszt, H. (1996). *A&A*, **307**, 237.
126. Lucas, R. and Liszt, H. (1997). In *Molecules in Astrophysics: Probes and Processes*, IAU Symposium 178, ed. E. F. van Dishoeck, Kluwer, Dordrecht, p.421.
127. Magnani, L. and Onello, J. S. (1995). *ApJ*, **443**, 169.
128. Magnani, L. and Siskind, L. (1990). *ApJ*, **359**, 355.
129. Magnani, L. et al. (1996a). *ApJ*, **465**, 825.
130. Magnani, L., Hartmann, D. and Speck, B. G. (1996b). *ApJS*, **106**, 447.
131. Maier, J. P. (1997). In *Molecules in Astrophysics: Probes and Processes*, IAU Symposium 178, ed. E. F. van Dishoeck, Kluwer, Dordrecht, p.287.
132. Mann, A. P. C. and Williams, D. A. (1984). *MNRAS*, **209**, 33.
133. Mebold, U., Kerp, J., Moritz, P., Engelmann, J. and Herbstmeier, U. (1994). In *The First Symposium on the Infrared Cirrus and Diffuse Interstellar Clouds*, ASP Conference Series no. 58, eds. R. M. Cutri and W. B. Latter, ASP, San Francisco, p.45.

134. Meyer, D. M. (1997). In *Molecules in Astrophysics: Probes and Processes*, IAU Symposium 178, ed. E. F. van Dishoeck, Kluwer, Dordrecht, p.407.
135. Meyer, D. M. and Roth, K. (1991). *ApJ*, **376**, L49.
136. Millar, T. J. (1990). In *Molecular Astrophysics—A volume honoring Alexander Dalgarno*, ed. T. W. Hartquist, Cambridge University Press, Cambridge, p.115.
137. Millar, T. J. (1992). *MNRAS*, **259**, P35.
138. Millar, T. J., Bennett, A. and Herbst, E. (1989). *ApJ*, **340**, 906.
139. Millar, T. J., Farquhar, P. R. A. and Willacy, K. (1997). *A&AS*, **121**, 139.
140. Mundy, L. and McMullin, J. P. (1997). In *Molecules in Astrophysics: Probes and Processes*, IAU Symposium 178, ed. E. F. van Dishoeck, Kluwer, Dordrecht, p.183.
141. Myers, P. C. and Benson, P. J. (1983). *ApJ*, **266**, 309.
142. Neufeld, D. A., Zmuidzinas, J., Schilke, P., Phillips, T. G. (1997). *ApJ Letters*, **488**, L141.
143. Nyman, L. A. and Millar, T. J. (1989). *A&A*, **222**, 231.
144. Ohishi, M. (1997). In *Molecules in Astrophysics: Probes and Processes*, IAU Symposium 178, ed. E. F. van Dishoeck, Kluwer, Dordrecht, p.61.
145. Ohishi, M., Irvine, W. M. and Kaifu, N. (1992). In *Astrochemistry of Cosmic Phenomena*, IAU Symposium 150, ed. P. D. Singh, Dordrecht, Kluwer, p.171.
146. Olano, C. A., Walmsley, C. M. and Wilson, T. L. (1988). *A&A*, **196**, 194.
147. Omont, A. (1986). *A&A*, **164**, 159.
148. Oppenheimer, M. and Dalgarno, A. (1974). *ApJ*, **192**, 29.
149. Pagani, L., Langer, W. D. and Castets, A. (1993). *A&A*, **274**, L13.
150. Penprase, B. E. (1993). *ApJS*, **88**, 433.
151. Pineau des Forêts, G., Flower, D. R., Hartquist, T. W. and Dalgarno, A. (1986). *MNRAS*, **220**, 801.
152. Pineau des Forêts, G., Roueff, E. and Flower, D. R. (1992). *MNRAS*, **258**, 45.
153. Poglitsch, A. *et al.* (1996). *ApJ*, **462**, L43.
154. Prasad, S. S., Tarafdar, S. P., Villere, K. R. and Huntress, W. J., Jr. (1987). In *Interstellar Processes*, eds. D. Hollenbach and H. A. Thronson, D. Reidel, Dordrecht, p.631.
155. Prasad, S. S., Heere, K. R. and Tarafdar, S. P. (1991). *ApJ*, **373**, 123.
156. Rawlings, J. M. C. and Hartquist, T. W. (1997). *ApJ*, **487**, 672.
157. Rawlings, J. M. C., Hartquist, T. W., Menten, K. M. and Williams, D. A. (1992). *MNRAS*, **255**, 471.
158. Roberge, W. G., Jones, D., Lepp, S. and Dalgarno, A. (1991). *ApJS*, **77**, 287.

159. Sarre, P. J. et al. (1995). *MNRAS*, **277**, L41.
160. Schilke, P., Walmsley, C. M., Millar, T. J. and Henkel, C. (1991). *A&A*, **247**, 487.
161. Schilke, P., Phillips, T. G. and Wang, N. (1995). *ApJ*, **441**, 334.
162. Schutte, W. A. (1996). In *Cosmic Dust Connection*, ed. J. M. Greenberg, Kluwer, Dordrecht.
163. Schutte, W. A. and Greenberg, J. M. (1991). *A&A*, **244**, 190.
164. Sembach, K. R., Danks, A. C. and Lambert, D. L. (1996). *ApJ*, **460**, L61.
165. Shalabiea, O. and Greenberg, J. M. (1994). *A&A*, **290**, 266.
166. Shalabiea, O. and Greenberg, J. M. (1995). *A&A*, **296**, 779.
167. Sims, I. and Smith, I. W. M. (1995). *Ann. Rev. Phys. Chem.*, **46**, 109.
168. Smith, D. and Španel, P. (1993). *Int. J. Mass. Spectrom. Ion Proc.*, **129**, 163.
169. Smith, D., Španel, P. and Millar, T. J. (1994). *MNRAS*, **266**, 31.
170. Smith, I. W. M. (1989). *ApJ*, **347**, 282.
171. Smith, I. W. M. (1997). In *Molecules in Astrophysics: Probes and Processes*, IAU Symposium 178, ed. E. F. van Dishoeck, Kluwer, Dordrecht, p.253.
172. Smith, M. A. (1993). *J. Chem. Soc. Far. Trans.*, **89**, 2216.
173. Snow, T. P. et al. (1996). *ApJ*, **465**, 245.
174. Spaans, M. (1995). PhD thesis, University of Leiden.
175. Spaans, M. (1996). *A&A*, **307**, 271.
176. Stark, G. et al. (1993). *ApJ*, **410**, 837.
177. Stark, R. and van Dishoeck, E. F. (1994). *A&A*, **286**, L43.
178. Stark, R., Wesselius, P. R., van Dishoeck, E. F. and Laureijs, R. J. (1996). *A&A*, **311**, 282.
179. Stecher, T. P. and Williams, D. A. (1974). *MNRAS*, **168**, 51P.
180. Sundström, G. et al. (1994). *Science*, **263**, 785.
181. Suzuki, H. et al., (1992). *ApJ*, **392**, 551.
182. Swade, D. A. (1989). *ApJ*, **345**, 828.
183. Tarafdar, S. P., Prasad, S. S., Huntress, W. T., Villere, K. R. and Black, D. C. (1985). *ApJ*, **289**, 220.
184. Tauber, J., Olofsson, G., Pilbratt, G., Nordh, L. and Frisk, U. (1996). *A&A*, **308**, 913.
185. Thaddeus, P. (1994). in *Molecules and Grains in Space*, Am. Inst. Phys. Conf. Proc., **312**, ed. I. Nenner and L. Trojanowski, Am. Inst. Phys., New York, p.711.
186. Tieftrunk, A., Pineau des Forêts, G., Schilke, P. and Walmsley, C. M. (1994). *A&A*, **289**, 579.
187. Tielens, A. G. G. M. and Allamandola, L. J. (1987). In *Interstellar Processes*, eds. D. Hollenbach and H. A. Thronson, D. Reidel, Dordrecht, p.379.
188. Tielens, A. G. G. M. and Hagen, W. (1982). *A&A*, **114**, 245.

189. Tielens, A. G. G. M. and Whittet, D. C. B. (1997). In *Molecules in Astrophysics: Probes and Processes*, IAU Symposium 178, ed. E. F. van Dishoeck, Kluwer, Dordrecht, p.45.
190. Turner, B. E. (1989). Space *Science* Rev., **51**, 235.
191. Turner, B. E. (1993). *ApJ*, **410**, 140.
192. Turner, B. E. (1994). *ApJ*, **420**, 661.
193. Turner, B. E. (1995). *ApJ*, **449**, 635.
194. Turner, B. E. (1996). *ApJ*, **468**, 694.
195. Ubachs, W. *et al.* (1994). *ApJ*, **427**, L55.
196. Ungerechts, H., Bergin, E. A., Goldsmith, P. F., Irvine, W. M., Schloerb, F. P. and Snell, R. L. (1997). *ApJ*, **482**, 245.
197. van Dishoeck, E. F. (1988). In *Millimetre and Submillimetre Astronomy*, ed. R. D. Wolstencroft and W. B. Burton, Kluwer, Dordrecht, p.117.
198. van Dishoeck, E. F. (1990a). In *Molecular Astrophysics—A volume honoring Alexander Dalgarno*, ed. T. W. Hartquist, Cambridge University Press, Cambridge, p.55.
199. van Dishoeck, E. F. (1990b). In *The Evolution of the Interstellar Medium*, ASP Conference Series no. 12, ed. L. Blitz, ASP, San Francisco, p.207.
200. van Dishoeck, E. F. (1992). In *Astrochemistry of Cosmic Phenomena*, IAU Symposium 150, ed. P. D. Singh, Kluwer, Dordrecht, p.143.
201. van Dishoeck, E. F. (1994). In *The First Symposium on the Infrared Cirrus and Diffuse Interstellar Clouds*, ASP Conference Series no. 58, eds. R. M. Cutri and W. B. Latter, ASP, San Francisco, p.319.
202. van Dishoeck, E. F. (1995). In *Physics and Chemistry of Interstellar Molecular Clouds*, eds. G. Winnewisser and G. H. Pelz, Springer, Berlin, p.225.
203. van Dishoeck, E. F. and Black, J. H. (1986). *ApJS*, **62**, p.109.
204. van Dishoeck, E. F. and Black, J. H. (1987). In *Physical Processes in Interstellar Clouds*, eds. G. Morfill and M. S. Scholer, Reidel, Dordrecht, p.241.
205. van Dishoeck, E. F. and Black, J. H. (1988a). In *Rate Coefficients in Astrochemistry*, eds. T. J. Millar and D. A. Williams, Kluwer, Dordrecht, p.209.
206. van Dishoeck, E. F. and Black, J. H. (1988b). *ApJ*, **334**, 771.
207. van Dishoeck, E. F. and Black, J. H. (1989). *ApJ*, **340**, 273.
208. van Dishoeck, E. F., Black, J. H., Phillips, T. G. and Gredel, R. (1991). *ApJ*, **366**, 141.
209. van Dishoeck, E. F., Glassgold, A. E., Guélin, M. *et al.* (1992). In *Astrochemistry of Cosmic Phenomena*, IAU Symposium 150, ed. P. D. Singh, Kluwer, Dordrecht, p.285.
210. van Dishoeck, E. F., Jansen, D. J., Schilke, P. and Phillips, T. G. (1993). *ApJ*, **416**, L83.
211. van Dishoeck, E. F. *et al.* (1996). *A&A*, **315**, L349.

212. Vejby-Christensen, L., Andersen, L. H., Heber, O., Kella, D., Pedersen, H. B., Schmidt, H. and Zajfman, D. (1997). *ApJ*, **483**, 531.
213. Viala, Y. P., Letzelter, C., Eidelsberg, M. and Rostas, F. (1988). *A&A*, **193**, 265.
214. Wagenblast, R. (1992). *MNRAS*, **259**, 155.
215. Wagenblast, R. and Hartquist, T. W. (1989). *MNRAS*, **237**, 1019.
216. Wagenblast, R. and Williams, D. A. (1993). In *Dust and Chemistry in Astronomy*, eds. T. J. Millar and D. A. Williams, IOP Publishing, Bristol, p.171.
217. Wagenblast, R., Williams, D. A., Millar, T. J. and Nejad, L. A. M. (1993). *MNRAS*, **260**, 420.
218. Walmsley, C. M. (1987). In *Physical Processes in Interstellar Clouds*, eds. G. Morfill and M. S. Scholer, Reidel, Dordrecht, p.161.
219. Warin, S., Benayoun, J. J. and Viala, Y. P. (1996). *A&A*, **308**, 535.
220. Watson, W. D. (1978). *ARAA*, **16**, 585.
221. White, R. E. (1984). *ApJ*, **284**, 695.
222. Whittet, D. C. B. (1993). In *Dust and Chemistry in Astronomy*, eds. T. J. Millar and D. A. Williams, IOP Publishing, Bristol, p.9.
223. Whittet, D. C. B. *et al.* (1996). *A&A*, **315**, L357.
224. Williams, D. A. (1992). *Planet. Spa. Sci.*, **40**, 1683.
225. Williams, D. A. (1993). In *Dust and Chemistry in Astronomy*, eds. T. J. Millar and D. A. Williams, IOP Publishing, Bristol, p.143.
226. Williams, D. A., Hartquist, T. W., Whittet, D. C. B. (1992). *MNRAS*, **258**, 599.
227. Williams, T. L., Adams, N. G., Babcock, L. M., Herd, C. R. and Geoghegan, M. (1996). *MNRAS*, **282**, 413.
228. Wootten, A. (1987). In *Astrochemistry*, eds. M. S. Vardya and S. P. Tarafdar, Kluwer, Dordrecht, p.311.
229. Xie, T. L., Allen, M. and Langer, W. D. (1995). *ApJ*, **440**, 674.
230. Yoshino, K. *et al.* (1995). *ApJ*, **438**, 1013.
231. Zmuidzinas, J., Blake, G. A., Carlstrom, J., Keene, J., Miller, D., Schilke, P. and Ugras, N. G. (1997). *ApJ*, submitted.

5
The Chemistry of Star Forming Regions

T. W. Hartquist
Max-Planck-Institut für extraterrestrische Physik

P. Caselli
Osservatario Astrofisico di Arcetri

J. M. C. Rawlings, D. P. Ruffle and D. A. Williams
Department of Physics and Astronomy, University College London

1 Introduction

The first generation of stars formed in gas that was almost entirely hydrogen and helium, and the chemistry that was important during their births, is treated in Chapter 3. In this chapter we focus on star forming regions in which elements more massive than helium are important in the chemistry that controls star formation and creates molecules which are observed to diagnose the process of stellar birth.

Low-mass stars, those having masses less than 4 $M_\odot$, are generally thought to form as a consequence of a gradual weakening of the magnetic fields in the initially magnetically supported objects that are their progenitors (e.g. Mouschovias 1987; Shu, Adams and Lizano 1987). We will give a more detailed description of the relevant processes later in this chapter.

In contrast, the ratio of the magnetic flux to the mass of a clump in which a high-mass star (one having a mass of at least 8 $M_\odot$) forms is thought to be too small for the magnetic field to prevent collapse if the pressure external to the clump increases above a critical value (e.g. Mouschovias 1987). Hence, the formation of high-mass stars can be triggered by the increase of the pressure in the medium surrounding a clump (with a sufficiently small magnetic flux to mass ratio) induced by the winds and supernovae of other high-mass stars (e.g. Elmegreen and Lada 1977; Mouschovias 1987). Sections 3–7 of this chapter are concerned solely with regions of low-mass star formation, whereas Section 8 is focused on regions of high-mass star formation.

Section 2 gives an overview of the structures of giant molecular cloud complexes which contain many of the Galaxy's star forming regions; the complexes contain clumps in which the presence of waves plays a significant role in their support, and we describe how the ionization structure is important in determining the lifetimes of the waves and, thus, affects the gravitational stability of the clumps. The eventual collapse of one of these clumps leads to the formation of objects known as dense cores, which can be identified as the progenitors of stars. Dense cores *sometimes* are found to exist together in what we will call a "core cluster", a collection of several cores near enough to one another that the birth of even a low-mass star in one core may regulate the evolution of the other cores in the cluster. Possible scenarios for the development of a core cluster (such as B5) in which low-mass stars form and atomic and molecular observations that might help in the establishment of the relevance of one of these scenarios are the topics of Section 3. Section 4 concerns a particular set, known as TMC-1, of dense cores which lie in a ridge; TMC-1 is one of the closest and most thoroughly observed sets of dense cores, and variations along the ridge in the chemical composition provide information about the dynamical properties of this source which lies in the vicinity of a recently formed low-mass star. Because young low-mass stars are found near the dense cores, ablation (which is mentioned in Sections 3 and 4) of the cores by the winds of young stars occurs; ablation is a process of great importance in a wide variety of diffuse astrophysical plasmas (e.g. Hartquist and Dyson 1993; Hartquist and Dyson 1996) from the interstellar medium to the envelopes of evolved stars as well as line forming regions of active galactic nuclei, and Section 5 concerns the diagnosis of interfaces between dense cores and the winds of young low-mass stars. Sections 6 and 7 treat the observational and theoretical studies of the collapse of dense cores to form low-mass stars. Section 8 closes the chapter and contains descriptions of observations of collapse in regions in which high-mass stars are forming and the chemistry of hot cores, dense cores heated to temperatures of 80 K or more due to the proximity of recently born nearby high-mass stars. The role of evaporation of material from grain surfaces in such gas is mentioned as is the presence of 2000 K gas in such star forming regions.

2 Clumpy Giant Molecular Cloud Complexes

Much of the molecular material in the Milky Way exists in Giant Molecular Cloud (GMC) complexes having masses typically in the range of 10^4 $M_\odot$ to 10^6 $M_\odot$ and linear extents of 30 pc to 100 pc. A detailed analysis of ^{12}CO($J = 1$–0) and ^{13}CO($J = 1$–0) emission maps of the Rosette Molecular Cloud (RMC), at a distance of 1600 pc, has been made (Williams, Blitz and Stark 1995) and provided insight into the structure of a GMC. The RMC has a projected cross section of about 2200 pc^2, mass of about 1–2×10^5 $M_\odot$, a mean H$_2$ column density of 4×10^{21} cm^{-2} and a mean H$_2$ number density of 30 cm^{-3}. Most of the molecular material exists in about 70 clumps having masses between about 30 $M_\odot$ and 2500 $M_\odot$; the number of clumps with masses between M and $M + \mathrm{d}M$ scales as $M^{-1.27} \mathrm{d}M$. Less massive clumps may also exist, but only a few were detected at the sensitivity of the observations. Figure 1 shows the integrated ^{12}CO($J = 1$–0) emission line map of the RMC.

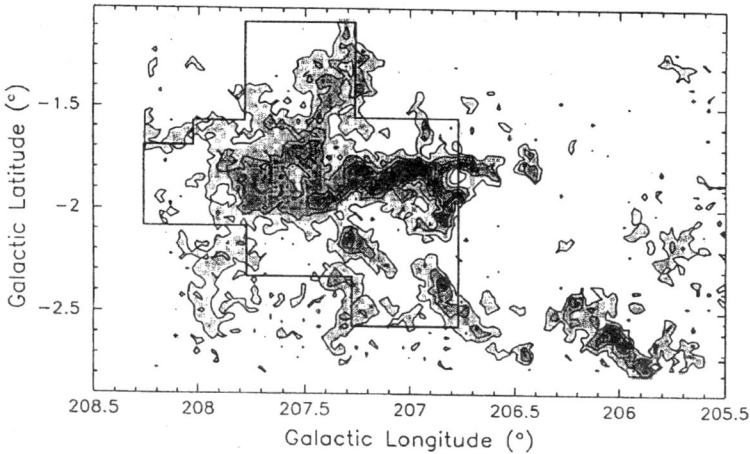

FIG. 1. Map of the Rosette Molecular Cloud in ^{12}CO($J = 1\text{--}0$) emission. $1° \approx$ 28 pc. From Williams, Blitz and Stark (1995).

The identified clumps fill only about 8% of the volume of the RMC, and in most clumps $n(H_2) \approx 200$ cm^{-3}; a variation of about a factor of 4 in this value is found, but the measured variation may be more limited than the real variation in density because (due to the small dipole moment of CO) the CO rotational level population distribution becomes thermalized at $n(H_2)$ of several hundred cm^{-3}, making CO a poor tracer of denser gas. The peak ^{13}CO column density through a clump scales roughly as $M^{0.63}$ and is about 10^{16} cm^{-2} for $M = 10^3$ $M_\odot$; the H_2 column density is assumed to be a factor of 5×10^5 larger. Embedded stars exist in the three most massive and the fifth, seventh, eleventh, and eighteenth most massive clumps. The clumps that contain stars are all amongst the 16 clumps having the highest ^{13}CO column densities of about 10^{16} cm^{-2} and more; as we argue below the collapse of a clump may begin if a critical column density is exceeded.

The line of sight full width at half maximum of a feature formed in an individual clump lies in the range of about 0.9 km s^{-1} to 3.3 km s^{-1}; for comparison, the value for ^{13}CO that is thermally broadened only and is at 30 K (roughly the maximum temperature obtaining in the clumps) is only 0.22 km s^{-1}. Comparison of the velocity dispersion with the escape velocity estimated from the mass and radius of each clump shows that about half of the clumps are not bound by their own gravitational fields, a result that seems to apply also to clumps in other GMCs (Bertoldi and McKee 1992). The clumps that are not bound by their own gravity must be bound by the pressure of an interclump medium, which may consist primarily of gas at about 10^4 K.

The magnetic fields in the clumps are important in supporting them (e.g. Mouschovias 1987). Along the magnetic field lines the turbulent pressure associated with the broad lines is important for the support. The turbulence almost certainly consists of a superposition of Alfvén waves (Arons and Max 1975; Mouschovias and Psaltis 1995). The damping rate of an Alfvén wave in a weakly ionized medium due to ion–neutral friction (Kulsrud and Pearce 1969) is

$$\Gamma_D \approx \omega^2/2\nu_{ni} \quad \text{for } \omega \ll \nu_{in} \tag{2.1}$$

$$\Gamma_D > \omega \quad \text{for } \nu_{ni} < \omega < \nu_{in} \tag{2.2}$$

$$\Gamma_D \approx \nu_{in}/2 \quad \text{for } \omega \gg \nu_{in} \tag{2.3}$$

where ω is the angular frequency of the wave, ν_{in} is the inverse stopping time of an ion by the neutrals, and ν_{ni} is the inverse stopping time of a neutral by the ions. For ions of mass m_i colliding with H$_2$ (Osterbrock 1961)

$$\nu_{ni} \approx 1.8 \times 10^{-9} \text{ s}^{-1} \left(\frac{m_i}{m_i + m_{H_2}}\right)^{1/2} \left(\frac{n_i}{\text{cm}^{-3}}\right) \tag{2.4}$$

and

$$\nu_{in} = \nu_{ni} \left(\frac{m_{H_2}}{m_i}\right) \left(\frac{n(H_2)}{n_i}\right) \tag{2.5}$$

where m_{H_2} is the mass of an H$_2$ molecule and n_i is the number density of ions.

From equations (2.1) and (2.4) one sees clearly that the damping rate of the waves comprising the turbulence that supports a clump against gravitationally induced collapse primarily along the magnetic field lines depends on the ionization structure in the cloud. As discussed in Chapter 4, the fractional ionization in the outer regions of a molecular cloud is generally at least 10^{-4} because the gas phase carbon remains primarily in C$^+$ due to the efficiency of the photodissociation and photoionization processes. As also mentioned in that chapter, in dark regions, in which the ionization is due almost entirely to cosmic rays the fractional ionization is several orders of magnitude lower than this.

We have seen earlier in this section that a maximum ^{13}CO column density of 10^{16} cm^{-2} appears to be critical for the onset of star formation. The visual extinction in magnitudes (where five magnitudes corresponds to a decrease in intensity at 5550 Å by a factor of 100) for a clump having this maximum ^{13}CO column density is 2.5 magnitudes between the centre of the clump and the clump edge. Here, we have assumed the standard relationship of $A_V = 5 \times 10^{-22}$ mag (H nuclei cm^{-2})$^{-1}$ between visual extinction and H nuclei column density (Savage and Mathis 1979), together with a ratio of ^{13}CO to hydrogen nuclei, $n(^{13}\text{CO})/n_H = 1 \times 10^{-6}$, where $n(^{13}\text{CO})$ and n_H are the ^{13}CO and hydrogen nuclei number densities. A visual extinction, A_V, of 2.5 magnitudes is below that of about 4 or 5 magnitudes at which cosmic rays begin to dominate the ionization.

Figure 2 contains the results of a calculation of the fractional ionization in each of two clumps in which $n_H = 1.5 \times 10^3$ cm^{-3} and 1.5×10^5 cm^{-3}, as a

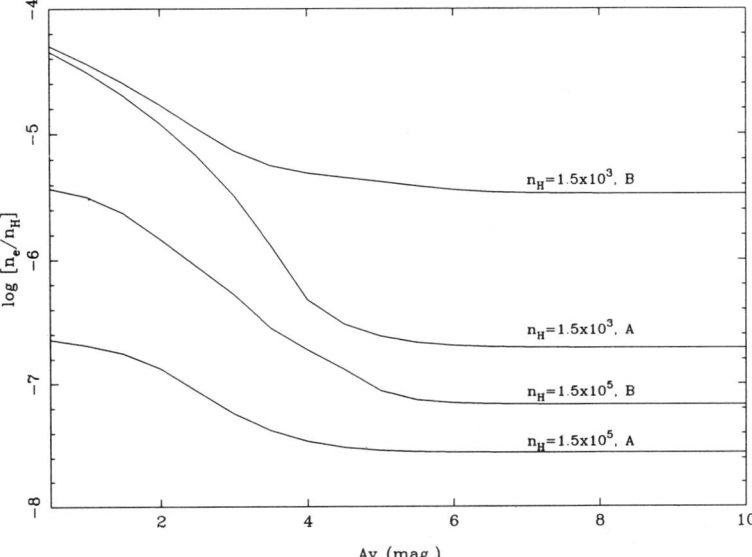

FIG. 2. Fractional ionization as a function of visual extinction. Results are given for two densities and for two sets of depletions (see text).

function of the visual extinction, A_V, to the nearest edge of the clump. A standard interstellar background radiation field is assumed to be incident on one side of each semi-infinite plane parallel clump, and a steady state chemical model has been assumed in which species that collide with grains undergo simple chemical processing on the grains (resulting in the production of saturated species such as H_2O, H_2S, CH_4 and NH_3 from many unsaturated species) followed by immediate return to the gas phase. Results for two sets of elemental fractional abundances with respect to hydrogen nuclei are given; for set A the fractional abundances of He, C, N, O, S, Si and more massive metals are 0.07, 1×10^{-4}, 2×10^{-5}, 2×10^{-4}, 2×10^{-8}, 7×10^{-9} and 2×10^{-7} while for set B they are 0.07, 1×10^{-4}, 2×10^{-5}, 2×10^{-4}, 3×10^{-6}, 1×10^{-6} and 1×10^{-6}. (Set A and set B may represent those in dark and diffuse clouds, respectively; see below.) Clearly the depletions of S, Si and more massive metals greatly affect the ionization structure (cf. Shalabiea and Greenberg 1995); for $n_H = 1.5 \times 10^3$ cm^{-3} the magnitude of the logarithmic derivative of the fractional ionization with respect to A_V is much larger for case A (higher depletions of low ionization potential species) than for case B, and from (2.1) and (2.4) a large magnitude of the logarithmic derivative is what is required for the turbulent damping to be a sensitive function of A_V. Furthermore, extreme values of the logarithmic derivative obtain for values of A_V in the range of the minimum value of A_V of RMC clumps containing stars (note that the A_V in the plots is only one half of the A_V through the entire clump and should be doubled when comparison is made with the minimum value of $A_V (\approx 5)$ of RMC clumps containing stars); this agreement between the minimum A_V for the existence of stars and the A_V range of extreme logarithmic derivative of the fractional ionization is in harmony with the suggestion (Hartquist et al. 1993) that the minimum A_V for the existence of stars is associated with a maximum A_V at which turbulent support of clumps is effective in preventing collapse along magnetic field lines. Unfortunately, the depletions in $n_H \approx 10^3$ cm^{-3} clumps like those in the RMC are unknown; while the set B depletions are closer to those measured in absorption line studies of diffuse interstellar clouds, the set A depletions are typical of those required for dense core chemical models to reproduce measured fractional abundances of species. A key outstanding set of questions in the chemistry and dynamics of star formation is that which concerns the nature of depletion and desorption processes of elemental S, Si, Fe, Na and Mg in translucent and dark clouds; it is remarkable that those elements are so much more depleted in many regions than O, C and N but still have high enough gas phase abundances in those regions for their presence in the gas phase to have significant chemical consequences.

We therefore present in Fig. 3 model results for the fractional abundances of species as functions of A_V to the nearest cloud edge for the $n_H = 1.5 \times 10^3$ cm^{-3}, case A depletion model. It is disappointing that CH and OH seem to be the only molecules other than CO that are potentially detectable in emission from RMC-type clumps (with $A_V \lesssim 5$), and model results show that their abundances are insensitive to the depletions of S, Si and more massive elements, making the inference of the values of those depletions in such objects a very difficult goal. The dominant ion throughout an RMC-type clump is C$^+$; it is so abundant because $n(C^0)/n(CO)$ is maintained at a high value by the dissociation of CO by photons emitted in the interaction of fast electrons (produced by cosmic-ray-induced ionization of H$_2$) with H$_2$ (cf. Chapter 4); C^0 is ionized by the absorption of photons of external origin.

We do not know what mechanism induces a clump to reach a higher visual extinction. Perhaps, the occurrences of supernovae or clump–clump collisions produce increased visual extinctions by compression. If the increase in wave damping allows collapse driven either by the pressure of the surrounding interclump medium or by gravity to occur, collapse in a clump in which the magnetic field contributes significantly to the support will be primarily along the magnetic field lines. As an illustration we give in Fig. 4 results for the time-dependent chemistry in gas collapsing at a constant A_V of 3.0 from the nearest edge and infinity to the furthest edge until n_H reaches 2×10^4 cm^{-3} due to a one-dimensional gravitational free-fall from an initial steady state equilibrium at $n_H = 1500$ cm^{-3}. The chemical processes included are just the same as those included in the generation of the results displayed in Fig. 2 and Fig. 3. After n_H reaches 2×10^4 cm^{-3} collapse is taken to occur in such a manner that

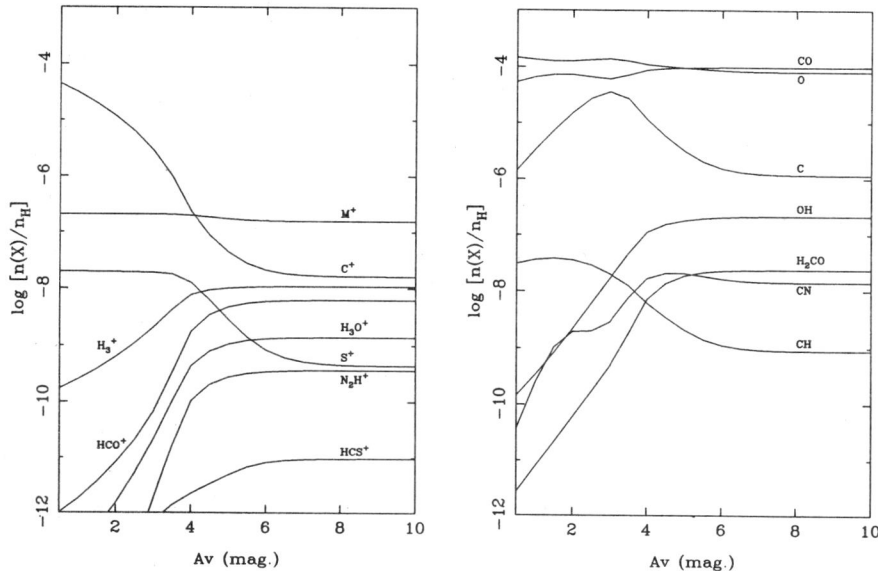

FIG. 3. Fractional abundances as functions of visual extinction in an $n_H = 1500$ cm^{-3} clump with Case A depletions.

$$A_V = 0.5 + 2.5 \left(\frac{n_H}{2 \times 10^4 \text{ cm}^{-3}}\right)^{2/3} \quad (2.6)$$

with the constant of 0.5 being attributed to an interclump medium. The collapse was assumed to occur thereafter on $6.95(n_i/10^{-3} \text{ cm}^{-3})(n_H/2 \times 10^4 \text{ cm}^{-3})^{-1/2}$ times the spherical free fall timescale as is consistent with the object being magnetically subcritical (i.e. supported by the magnetic field) and collapse taking place due to ambipolar diffusion (Mouschovias 1987), which is the relative drift of charged particles with respect to neutrals arising because charged particles are directly subjected to electromagnetic forces, while neutrals are not and are affected by those forces only by the frictional coupling between charged and neutral fluids. During the collapse the chemistry proves to be very sensitive to A_V with ratios such as $n(C^0)/n(CO)$ and fractional abundances such as $x(C^+)$ showing marked changes as soon as equation (2.6) is assumed to apply; the sensitivity to A_V shown in the collapse chemistry is fully consistent with the sensitivity of the chemistry to A_V at $A_V \approx 3$ displayed in Fig. 3 for the $n_H = 1.5 \times 10^3$ cm^{-3}, case A depletion results.

In some regions the emission from young low-mass stars may have a significant effect on the ionization structures of dense cores (McKee 1989) even though those stars are not strong sources of far ultraviolet (FUV) radiation. For some

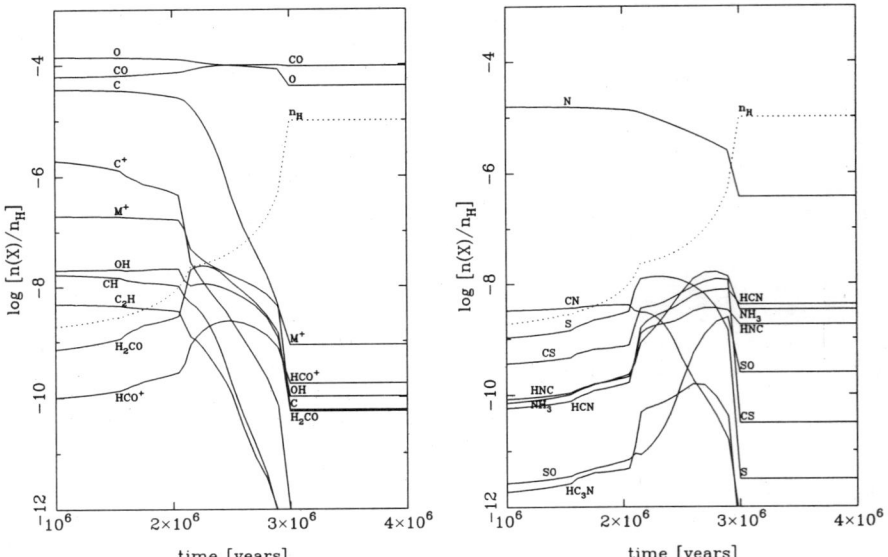

FIG. 4. Fractional abundances of species during collapse from $n_H = 1500$ cm^{-3} for Case A depletions. The time evolution of $(n_H/10^{12}$ cm$^{-3})$ is plotted on a dashed curve.

gas phase abundances of elements such as magnesium and sodium, photons from nearby low-mass stars may maintain a much higher fraction of those metals in ionized states than would be expected on the basis of the sorts of calculations that we have performed. Non-FUV radiation may be responsible for the fractional ionizations that have been inferred for some dense core material from observational data to be sufficiently large that the ambipolar diffusion timescales are at least 10^7 yr (Myers and Khersonsky 1995). It should be remembered that the results in Fig. 3 are for gas that is collapsing directly from an initially much lower density object in a relatively starless region; the results are of most relevance to the chemistry in the cores in which the first stars to form locally are born.

3 Low-mass Star Formation in a Cluster of Dense Cores

The collapse of a clump like those found with CO observations of the RMC leads to fragmentation and the formation of objects known as dense cores (e.g. Myers 1990) many of which have been mapped in ammonia emission. Dense cores typically have $n(H_2) \approx 10^4$–10^5 cm^{-3}, $T \approx 10$–30 K, and masses of one to several tens of solar masses each. However, dense cores with $n(H_2)$ up to roughly 10^7 cm^{-3} and masses of more than one hundred solar masses are also found, particularly in regions in which massive stars form. Figure 5 shows the

The Chemistry of Star Forming Regions

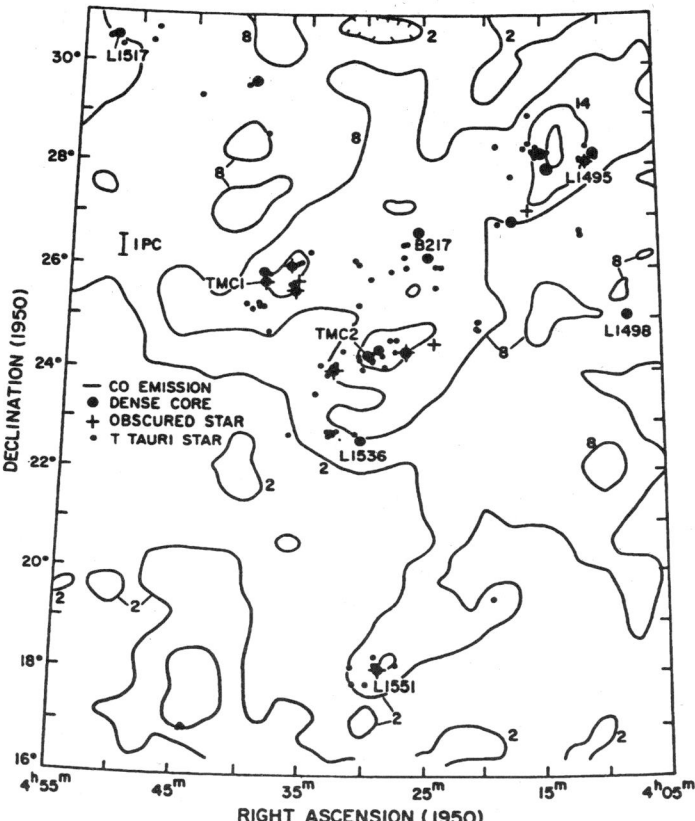

FIG. 5. The distribution of dark cores and low-mass stars in the Taurus–Auriga complex. From Myers (1986).

distribution of dense cores in the Taurus–Auriga complex. About half of all dense cores have young low-mass stars associated with them and the dense cores are considered to be the immediate progenitors of protostars.

Many (but not all) of the dense cores are near other dense cores. Figure 6 shows a $C^{18}O$ emission line map of Barnard 5 (B5), an object containing a cluster of five dense cores and four young low-mass stars. Gravitationally induced collapse of a core no doubt is an important step in the formation of a star, but once young stars have formed in a region their winds may affect the collapse of the neighbouring cores. In a scenario illustrated in Fig. 7 and which is a modified version of one put forward by Norman and Silk (1980) the ablation of a core by the supersonic wind of a young low-mass star (cf. Chapters 6 and 7) creates a stellar wind-ablated material mixture which moves supersonically until

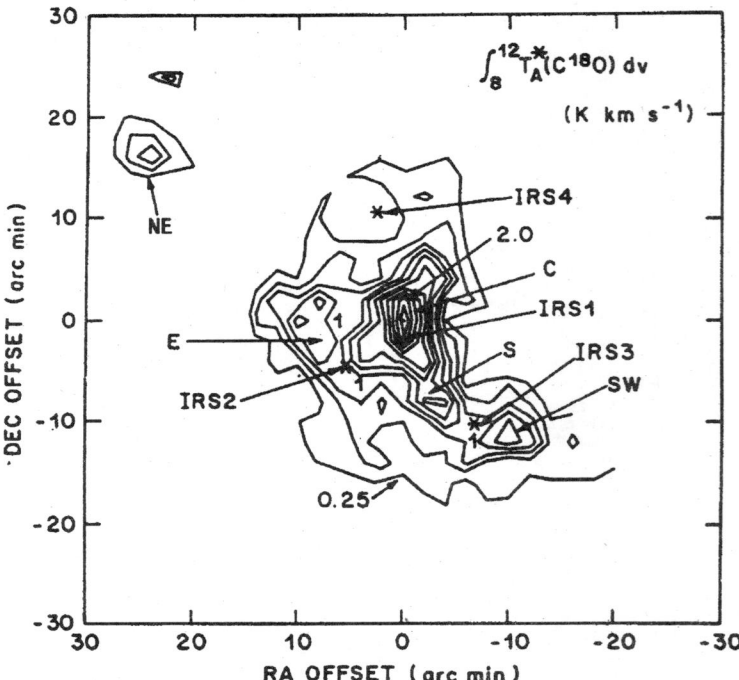

FIG. 6. $C^{18}O$ contour map of B5. The positions of four infrared sources associated with young stars are shown. From Goldsmith, Langer and Wilson (1986).

it collides with the similarly mass-loaded winds of the other nearby stars. The stellar wind-ablated material mixture passes through a shock as it is decelerated in a collision with another wind. Radiative cooling occurs in the shocked gas and leads to the formation of irregular shells separating the winds of different stars. Thus, an intercore medium of regions of supersonic wind-ablated material mixtures and shells of decelerated mixtures comes into existence. The shells probably fragment, *perhaps* leading to the formation of subsequent generations of cores. Alternatively, the ablation of cores may be so inefficient that nearly all cores simply collapse to form stars before being significantly eroded and whatever material that was in a core but does not go into a star may be blown by the stellar winds to a large enough distance from the cluster of cores that it cannot be considered to be associated with the cluster.

We would like to use molecular and atomic compositions of cores in a cluster and of the intercore gas to determine whether cores in clusters form in successive generations (Williams, Hartquist and Caselli 1996) or whether cores form only during the initial collapse of the sorts of clumps that constitute the main topic of the previous section. This goal proves to be an extremely difficult one. If a

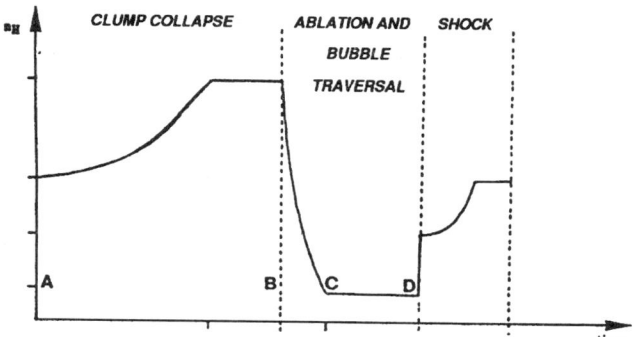

FIG. 7. The evolution of n_H for a parcel of gas in a cyclic model. At point A collapse of the interclump medium begins; the collapse continues until a more slowly evolving core is formed. At point B the parcel is ablated from the dense core by a stellar wind. H^+ and He^+ are assumed to mix from the wind into the ablated gas as it is accelerated. At point C the ablated gas is fully incorporated into the wind. The ablated gas–stellar wind mixture expands freely until at point D it passes through a termination shock inside the interface of the mixture and the ambient intercore medium. The density of the decelerated mixture increases as it cools; after cooling is complete the mixture becomes part of the cool phase of the ambient intercore medium. The parcel of gas may then pass through a qualitatively similar cycle again. Adapted from Charnley et al. (1988).

parcel of gas retains its identity as it passes through the sort of dynamical cycle described in the previous paragraph it goes through a low density ($n_H \approx 10^3$ cm^{-3}), low visual extinction ($A_V \approx 1$ or 2) intercore phase and remains in it long enough that its chemistry reaches a state that is similar to that of the chemistry throughout much of an RMC-type clump; the chemistry in the formation of subsequent generations of cores then will not deviate substantially from that of an RMC-type core collapsing. On the other hand, a parcel of gas may not retain its identity; as it is ablated from a core by a stellar wind, mixing of ablated core material with wind material may occur. The mixing of H^+ and He^+ with molecular gas will lead to the formation of neutral atomic species and atomic ions including C^+ and C^0; hence, mixing of wind with ablated material may significantly affect the chemical structure. However, the effects of such mixing are very similar to those that occur due to the cosmic-ray-generated photons and photons of external origin when a parcel of gas retaining its identity simply cycles through the low density–low visual extinction intercore phase. Consequently, it is very hard to see how the chemical conditions just prior to the onset of core formation from an RMC-type clump are likely to differ much from those that obtain just prior to the onset of core formation from an intercore medium consisting of stellar wind and material ablated from a previous generation of cores. The conclusion presented in the immediately preceding sentence contrasts markedly with what was believed (Williams, Hartquist and Caselli 1996) before the realization that the cosmic-ray-generated photons could be so effective in maintaining high values of $n(C^0)/n(CO)$ in RMC-type clumps.

CI emission has been observed towards B5 (Langer 1987) and towards a ridge of dark cores called TMC-1 (Schilke et al. 1995). Such emission is not likely to arise in an intercore medium, because its density would probably be too low. However, the results shown in Fig. 3 and Fig. 4 suggest that the CI emission arises in material at an $A_V \leq 3$ from the nearest edge of the core unless the matter at $A_V > 3$ collapsed much more rapidly than the timescale on which ambipolar diffusion occurs.

4 TMC-1

Figures 8 and 9 show maps of the NH_3 $(J, K) = (1, 1)$ inversion transition and HC_3N $J = 5-4$ emission arising in TMC-1 (Hirahara et al. 1992). The location of a young star is also shown in Fig. 8. While core B is the site of the peak in the NH_3 emission, the cyanopolyyne emission peak is at core D. At a distance of 140 pc and at a location indicated in Fig. 5, TMC-1 is one of the most thoroughly observed dense core sources; Hirahara et al. (1992) have estimated the densities of the various cores from single-dish observations of $C^{34}S$ $J = 1-0$ and 2-1 and C_2S $J_N = 4_3-3_2$ and 2_1-1_0 emissions and found $n(H_2) \approx 4 \times 10^4$ cm^{-3}, 2.4×10^5 cm^{-3}, and 4×10^5 cm^{-3} in cores D, C and B respectively. From interferometric studies of C_2S emissions of core D, Langer et al. (1995) found core D to be composed of fragments with $n(H_2)$ ranging from roughly 3 to 8×10^4 cm^{-3} in the largest ones to 10^6 cm^{-3} in the smallest. An understanding of the chemical variations along TMC-1 would lead to insight into how TMC-1 has reached its present physical state and, more generally, into dense core formation and evolution during the process of stellar birth. We describe several different ways in which the variations of the HC_3N fractional abundance along TMC-1 might have arisen.

The first model for the HC_3N variation that we consider concerns "early time" chemistry. Cyanopolyynes, including HC_3N and HC_5N, have been identified as "early time" molecules (e.g. Millar 1990), those that have abundance maxima before chemical equilibrium is reached in cold, dark gas in which the heavy elements in the gas phase were initially primarily in atomic form. (The timescale for equilibrium to be approached, for gas phase fractional abundances of elements more massive than helium of about 10^{-4} with respect to hydrogen nuclei and a cosmic-ray-induced ionization rate of H_2 of about 10^{-17} s^{-1}, is around 10^6 years.) Figure 10 from Ruffle et al. (1997) shows the time evolution of the fractional abundances, relative to hydrogen nuclei, of several species for a model of a dark region in which $n_H = 2 \times 10^4$ cm^{-3} and all elements more massive than helium were initially in neutral atomic or ionic atomic (e.g. carbon was in

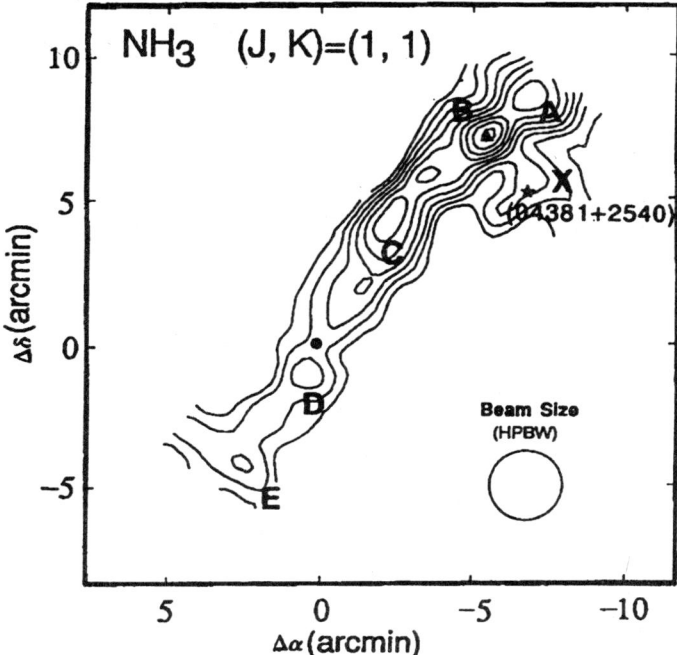

FIG. 8. Contour map of the NH_3 emission from TMC-1. The location of a star is indicated. From Hirahara et al. (1992).

C^+) form but almost all hydrogen was in H_2; neutral species more massive than helium were assumed to freeze out on to dust grains at a rate of 7.0×10^{-14} s^{-1} $(m_x/\text{amu})^{-1/2}$ where m_x is the mass of a particle of species x, while ionized species (except for heavy metals like Na^+) were assumed to freeze out 18 times more rapidly due to each grain carrying one negative charge (e.g. Draine and Sutin 1987), and the cosmic ray ionization rate was set equal to 1.3×10^{-17} s^{-1}. The results in Fig. 10 show that the HC_3N abundance peaks at 1.6×10^6 yr.

The peaks at early times of the HC_3N and C_2H abundances in particular can be understood from an inspection of Fig. 11 which shows the dominant routes to HC_3N and C_2H formation. Clearly, reasonably large abundances of C^+ and CH_3 or CH_4 must obtain simultaneously for these routes to be effective. A substantial fraction of a chemical timescale must have passed for CH_3 and CH_4 to become abundant, but after the passage of many chemical timescales in undepleted dark gas C^+, CH_3 and CH_4 become rarer as the CO abundance increases.

In contrast, the gas phase NH_3 abundance does not decrease significantly with time as long as depletion onto dust grains is either inoperative or counterbalanced by desorption. Thus, one possible explanation for the cyanopolyyne

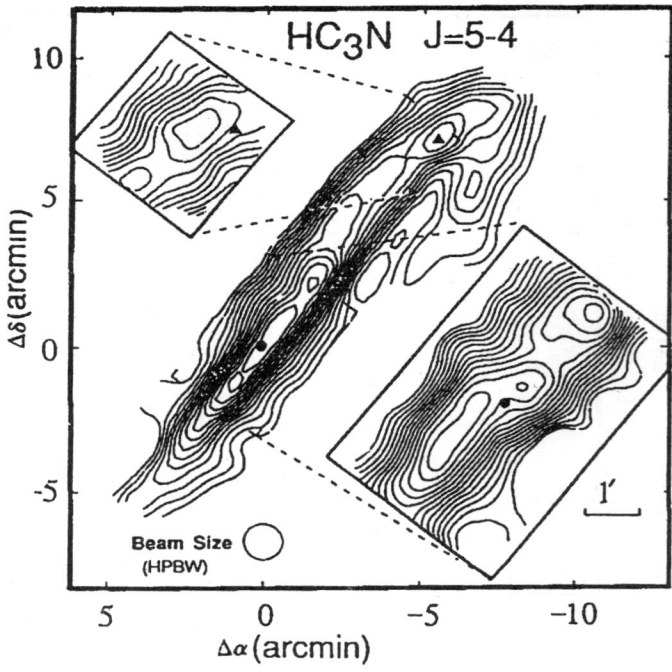

FIG. 9. Contour map of HC$_3$N emission from TMC-1. From Hirahara et al. (1992).

and ammonia variations along TMC-1 is that the more southerly cores collapsed from states in which the gas phase species containing elements more massive than helium were primarily atomic more recently than the more northerly cores and sufficiently recently that chemical equilibrium has not been reached (Hirahara et al. 1992; Howe, Taylor and Williams 1996). The greater average density and the existence of a protostar near the northern end of TMC-1 has been considered to be evidence that the collapse of TMC-1 at that end is more advanced than in more southerly parts.

In the description given so far of "early time" chemistry, the role of grains in the chemistry has been largely ignored. We now consider modifications that grains may introduce into "early time" chemistry. Freeze-out of gas phase species like C, CO and H$_2$O onto dust grains occurs on a timescale of about 10^{10} yr $(n_H/\text{cm}^{-3})^{-1}$ if the temperature is 10 K, the sticking coefficient is unity, and grains responsible for most of the standard visual extinction (cf. the fifth paragraph of Section 2) have radii of about 10^{-5} cm. At TMC-1 densities, freeze-out at this rate is somewhat more than an order of magnitude faster than the chemical timescale implying (as noted by Taylor et al. 1996) that

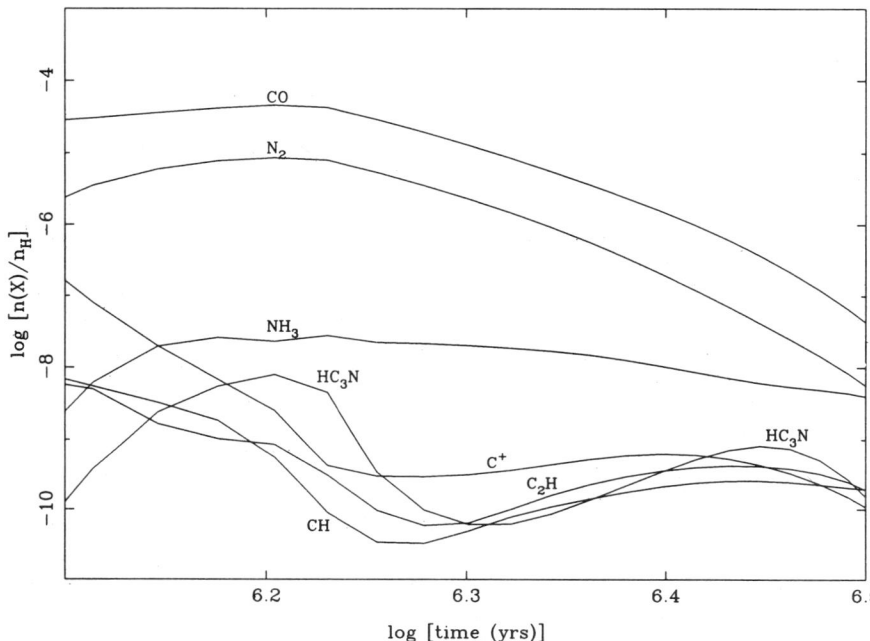

FIG. 10. Evolution of fractional abundances as depletion occurs. From Ruffle et al. (1997).

freeze-out in TMC-1 must be occurring at a rate that is at least an order of magnitude below the "standard rate" if core B has evolved dynamically on a timescale longer than the depletion timescale and "early time" chemistry is to account for the high NH_3 abundance in core B. A reduced fractional abundance of grains and a reduced freeze-out rate can be caused by grains moving out of the dense gas as the magnetic flux in the core decreases due to ambipolar diffusion (Ciolek and Mouschovias 1996). Alternatively, possibly the freeze-out dilemma in the early-time chemistry scenario is averted by only a small fraction ($\lesssim 0.1$) of the collisions between species more massive than helium and grains leading to sticking or by the action of a mechanism that desorbs ices from grain surfaces; much of the desorbed carbon not in CO would be returned to the gas phase as CH_4 (cf. the final section of this chapter) which might cause HC_3N and C_2H formation to be rapid (cf. Fig. 11) before sufficient time had passed for the gas phase chemistry to cause most of the carbon to be contained in CO. Another possibility is that the freeze-out dilemma in TMC-1 was circumvented by that source undergoing recent rapid compression driven by the ram pressure of the nearby star's wind, as suggested by Hartquist, Williams and Caselli (1996) and mentioned more fully later in this chapter.

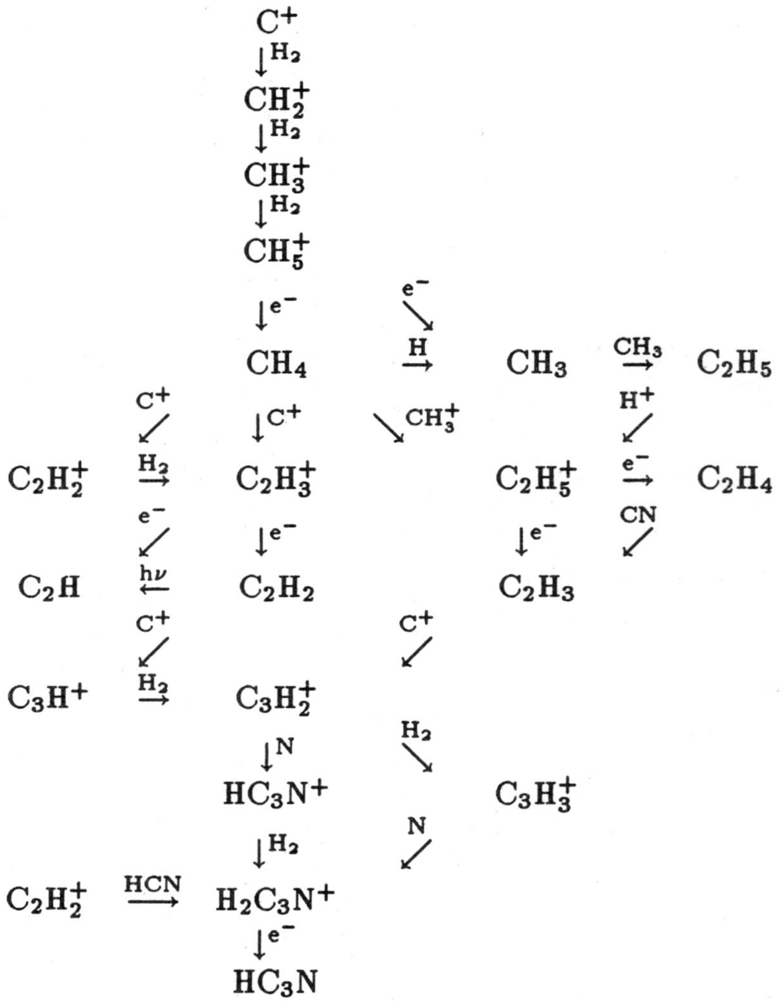

$$C^+$$
$$\downarrow H_2$$
$$CH_2^+$$
$$\downarrow H_2$$
$$CH_3^+$$
$$\downarrow H_2$$
$$CH_5^+$$

$$CH_5^+ \xrightarrow{e^-} CH_4 \qquad \xrightarrow{H} CH_3 \xrightarrow{CH_3} C_2H_5$$

$$C_2H_2^+ \xrightarrow{H_2} C_2H_3^+ \qquad \searrow CH_3^+ \qquad C_2H_5^+ \xrightarrow{e^-} C_2H_4$$

with downward C^+, e^-, $h\nu$, CN arrows leading to:

$$C_2H \qquad C_2H_2 \qquad C_2H_3$$

$$C_3H^+ \xrightarrow{H_2} C_3H_2^+ \qquad \qquad C_3H_3^+$$

$$\downarrow N$$
$$HC_3N^+$$
$$\downarrow H_2$$

$$C_2H_2^+ \xrightarrow{HCN} H_2C_3N^+$$
$$\downarrow e^-$$
$$HC_3N$$

FIG. 11. Routes to HC$_3$N formation.

Hartquist et al. (1996) have argued that the "early time" interpretation given by Hirahara et al. (1992) and Howe et al. (1996) is not a unique explanation of the TMC-1 chemical variations. Hartquist et al. (1996) have offered several alternative conjectures, which constitute the subject of most of the remainder of this section.

In the second type of TMC-1 model that we consider, we propose that the high HC$_3$N to NH$_3$ abundance ratio in core D indicates that core D has been collapsed a longer rather than a shorter time than core B. Hartquist and Williams

(1989) addressed the question of how to observe dark molecular regions in which depletion has substantially reduced the gas phase abundances of elements more massive than helium; they noted that as CO freezes on to dust grains the production rate per unit volume of C^+ by the reaction $He^+ + CO \rightarrow C^+ + O$ does not decrease as long as that reaction remains the main mechanism for removing He^+ (which is produced at a constant rate per helium atom by the cosmic-ray-induced ionization of He). As depletion causes the reduction of the gas phase abundances of oxygen-bearing species including H_2O, the C^+ is increasingly more likely to be removed by $C^+ + H_2 \rightarrow CH_2^+ + h\nu$ rather than by reactions, such as $C^+ + H_2O \rightarrow HCO^+ + H$ (which is followed by $HCO^+ + e \rightarrow CO + H$), that lead rather directly to the re-formation of CO. Consequently, the formation rate per unit volume of CH actually increases until depletion becomes so substantial that He^+ is removed primarily by reactions with H_2 rather than by CO. In addition, as the abundance of elemental oxygen in the gas phase decreases the rate at which CH is removed by reactions including $CH + O \rightarrow HCO^+ + e$ decreases. Clearly, as the production rate per unit volume of C^+ remains constant and its removal rate per unit volume decreases due to depletion removing species that react with C^+, the abundance of C^+ also increases. As Hartquist et al. (1996) noted and is evident from Fig. 11, situations in which the gas phase abundances of C^+ and CH (and along with CH, CH_3 and CH_4) are high are favourable for the production of HC_3N. Hartquist et al. (1996) suggested that a significant increase in the abundance ratio of HC_3N to NH_3 should take place as depletion occurs. The validity of this suggestion has been confirmed with detailed chemical modelling (Ruffle et al. 1997), some results of which are shown in Fig. 10. Hence, the relatively high HC_3N to NH_3 abundance ratio in core D may simply indicate that heavy species in the gas are more depleted in core D than in core B.

The third explanation that we offer for the variation of the HC_3N abundance in TMC-1, like the second one, relies on the result that the HC_3N abundance increases with increasing depletion but, if correct, implies that we can learn nothing about the relative ages of core B and core D. In this third picture, the higher depletion in core D is attributed to desorption from grain surfaces being more efficient in core B than in core D, a possible consequence of the proximity of core B to the protostar. (This proposed explanation is not in direct contradiction with the discussion given earlier of the modifications that grains make to the "early time" chemistry, as the explanation currently being described is based on the assumption that the chemistry has reached an equilibrium in each core.) Williams, Hartquist and Whittet (1992) have maintained that the large cloud-to-cloud variation of the minimum visual extinction at which H_2O-ice features (due to vibrational transitions) at a wavelength of 3.1 µm are observable in absorption against embedded young stars implies that desorption must be driven by radiation that is attenuated significantly in material through which the magnitude of the visual extinction is a few to ten. Infrared radiation at a wavelength of 3.1 µm experiences such attenuation. Hartquist et al. (1996) pointed out that the infrared radiation field arising from the nearby star's interaction with diffuse material around it may limit depletion in core B but have little effect on depletion in core D.

A fourth possible origin of the higher HC_3N abundance in core D is formation in the boundary layer between the dense core and the stellar wind. (The next section describes boundary layers in much more detail.) Williams and Hartquist (1991) and Hartquist et al. (1996) have suggested that mixing of stellar wind ions into dense core material in the interface between the stellar wind material and the dense core material in TMC-1 creates conditions favourable for HC_3N production. Though they explored mixing within a somewhat different physical context, the model results of Chièze, Pineau des Forêts and Herbst (1991) are consistent with this suggestion. In the boundary layer mixing picture the peak in the HC_3N is at core D rather than core B because the thickness of a boundary layer and, hence, the degree of mixing (if mixing occurs at all) increases along the boundary layer.

A noticeable characteristic of the HC_3N emission from TMC-1 is its sharp gradient downwind from the cores. This sharp boundary may be at the location of a wind termination shock positioned just upstream of a contact discontinuity between the stellar wind (mass-loaded with the material that it has ablated and with which it has mixed from the cores) and an ambient intercore or circumcore medium into which the wind-blown bubble is propagating.

The existence of a plethora of explanations for chemical variations in TMC-1 requires that more species be mapped because their abundance variations will no doubt differ between models. Also, considerably more theoretical exploration of the models must be undertaken.

We close this section with some comments on the possible dynamics that established the current physical state of TMC-1. These comments are of particular relevance to the "early-time" and boundary-layer models of the production of HC_3N in core D but they bear on other chemical models as well.

The decrease of core density with distance from the star, as found by Hirahara et al. (1992), is compatible with the pressure in each core being equal to the local ram pressure of the stellar wind. Cores in a variety of stages of structural and chemical evolution may have been present before the stellar wind turned on and then been compressed by shocks driven into them by the stellar wind. A shock speed of only 1 km s^{-1} gives rise to a density increase of a factor of about 10 if it propagates into 10 K gas in which the magnetic pressure is negligible, and it takes only about 10^5 yr to implode a core having a radius of 0.1 pc. This implosion timescale is shorter than the gas phase chemical timescale, which implies that the chemical composition of a pre-implosion state would survive as depletion becomes significant. For standard assumptions depletion takes place on a timescale comparable to 10^5 yr at densities around those given by Hirahara et al. (1992) for core D. Thus, a less dense pre-implosion core with a chemical age (in the early-time chemical picture) of several times 10^5 yr could have been the site of the production of the HC_3N to be observable in core D for the next 10^5 yr (Hartquist et al. 1996).

Another possibility is that the entire TMC-1 ridge was formed by the stellar wind interacting with an initially more extended clump that fragmentated into separate cores due to instabilities arising during the interaction. In this dynamical picture the birth of a star within a single core near a more extended clump triggers the formation of other cores.

5 Turbulent Boundary Layers Between Dense Cores and Stellar Winds

Stellar wind–dense core interactions have already been mentioned in Sections 3 and 4 of this chapter. The natures of the boundary layers arising in these interactions can have important consequences for the chemical compositions and molecular line emissions of dense cores, particularly if microscopic mixing (i.e. mixing that is so thorough that chemical reactions can proceed between material that was previously in the stellar wind and material that was previously in the dark core) occurs. However, even the gross properties of a wind–dark core boundary layer remain unknown and are not subject to rigorous first principles theoretical investigation. We cannot even be certain that microscopic mixing operates. Semi-empirical diagnostic studies provide the only means of learning these properties.

To perform such studies one must first make as many informed guesses as possible about what the boundary layer properties might be and then explore models for the guessed properties and others in a range about those guessed. The boundary layers are magnetized as magnetic fields are associated with the winds and cores. However, in obtaining some guessed characteristics we will assume that the magnetic field is unimportant and use arguments based on ordinary hydrodynamics rather than magnetohydrodynamics.

We first address the issue of boundary thickness (Hartquist and Dyson 1988). The thickness of a boundary layer is ordinarily roughly the length of the obstacle (i.e. the core) divided by the square root of the Reynolds number of the upstream flow, as calculated for the flow velocity measured in the frame of the obstacle. Typically, if the molecular viscosity is assumed in its evaluation, the Reynolds number of a stellar wind moving through the interstellar medium greatly exceeds 10^3 to 10^4, which is the range of minimum Reynolds numbers associated with the onset of turbulence in many situations. In such low molecular viscosity environments, turbulence may develop in such a way that it produces a turbulent viscosity that when used in the evaluation of the Reynolds number gives a value in this minimum range for turbulent onset; thus, turbulence may regulate itself in such a way that the effective turbulent Reynolds number is just below a certain value so that further development of the turbulence is suppressed. If we assume the effective turbulent Reynolds number to be in the range of 10^3 to 10^4, the corresponding thickness of the boundary layer is about a few to one per cent; given that the line of sight passes through the surface of a totally convex core twice, we might expect a maximum of the order of ten percent of a core's column density to be in the turbulent boundary layer. The existence of much broader turbulent boundary layers than this would probably create wings in a number of molecular emissions that would be stronger and broader than consistent with observations (e.g. Myers and Benson 1983).

If the entire energy dissipated in the interaction of a wind with a core were converted to heat in a boundary layer of thickness Δ around a core the heating rate per particle in the boundary layer would be of order

$$\Gamma_{H,max} \approx \frac{k_B T_{core} V_w}{\Delta}$$

$$\approx 10^{-23} \text{erg s}^{-1} \left(\frac{T_{core}}{10\,\text{K}}\right) \left(\frac{V_w}{400\,\text{km s}^{-1}}\right) \left(\frac{\Delta}{0.003\,\text{pc}}\right)^{-1} \quad (5.1)$$

where the thermal pressure in the boundary layer has been taken to be comparable to the ram pressure of the upstream supersonic wind, V_w is the upstream speed of the wind, k_B is Boltzmann's constant, and T_{core} is the temperature on the coreward side of the boundary layer. $\Gamma_{H,max}$ is an approximate upper-bound to the value that the heating rate per particle, Γ, can obtain near the coreward edge of the boundary layer. Of course, the heating rate per particle may increase in the hotter, more tenuous, windward part of the boundary layer.

If all of the wind material striking a core were to mix directly as fully ionized material into the boundary layer, the rate per neutral particle at which H^+ would be injected into the neutral material near the coreward edge of the boundary layer would be of order

$$\zeta_{max} \approx \frac{k_B T_{core}}{\mu_w V_w \Delta} \quad (5.2)$$

$$\approx 10^{-14} \text{s}^{-1} \left(\frac{T_{core}}{10\,\text{K}}\right) \left(\frac{V_w}{400\,\text{km s}^{-1}}\right) \left(\frac{\mu_w}{1 \times 10^{-24}\,\text{g}}\right)^{-1} \left(\frac{\Delta}{0.003\,\text{pc}}\right)^{-1}$$

This is a rough upper bound to the rate of injection of H^+ per particle, ζ, at the coreward edge of the boundary layer.

Charnley et al. (1990) and Nejad and Hartquist (1994) followed the thermal and chemical evolution of gas maintained at constant pressure subjected to heating and the injection of H^+ at constant rates per particle; He^+ was assumed to be injected at a rate of 0.1ζ. The initial thermal and chemical conditions were taken to be amongst those thought to be appropriate for dense core gas. In the calculations of Nejad and Hartquist (1994) a magnetic contribution to the pressure assumed to scale as the inverse of the density squared and at the coreward side of the boundary layer taken to be either one times or ten times the thermal pressure.

In the Nejad and Hartquist (1994) models each boundary layer was assumed to be exposed to the radiation field resulting from a radiation field with an intensity of one tenth the standard unshielded interstellar radiation field passing through material causing two magnitudes of visual extinction. At the coreward

Table 1 Parameters for models of chemistry in dense cores–stellar wind boundary layers.

Model	Gas phase depletion	Γ_H 10^{-24}erg s^{-1}	ζ 10^{-15}s^{-1}	β_0
1	depleted	20.0	20.0	1.0
2	depleted	20.0	20.0	10.0
3	undepleted	20.0	20.0	1.0
4	undepleted	20.0	20.0	10.0
5	depleted	20.0	0.0	1.0
6	depleted	20.0	0.0	10.0
7	undepleted	20.0	0.0	1.0
8	undepleted	20.0	0.0	10.0
9	depleted	7.0	7.0	1.0
10	depleted	7.0	7.0	10.0
11	undepleted	7.0	7.0	1.0
12	undepleted	7.0	7.0	10.0
13	depleted	2.0	2.0	1.0
14	depleted	2.0	2.0	10.0
15	undepleted	2.0	2.0	1.0
16	undepleted	2.0	2.0	10.0

side of the boundary layer $n(H_2) = 1 \times 10^4$ cm^{-3} and $T = 10$ K. Two different sets of chemical conditions were assumed to obtain at the coreward side of the boundary layer; for one set, elements more massive than helium were taken to be depleted in the gas by roughly an order of magnitude while for the other set they were assumed to be undepleted. Table 1 gives the input parameters specifying the different models of Nejad and Hartquist (1994). β_o is the ratio of the magnetic pressure to thermal pressure at the coreward side of the boundary layer. (The total pressure was assumed to be constant throughout the boundary layer.) Table 2 gives the model temperature and density and chemical fractional abundances in gas that has moved about 0.05 pc along the boundary layer (due to momentum input by the wind) if the upstream wind speed is 300 km s^{-1}. Results for more species can be found in the paper by Nejad and Hartquist (1994). The values of $n(H_2)$, T and chemical fractional abundances at the coreward side of the boundary layer assumed for the models for which elements more massive than helium are depleted and undepleted are given at the top of Table 2.

The fractional abundances of OH and H$_2$O increase in these models relative to those for cold clouds. This is largely due to the fact that the injection of H$^+$ into the core gas leads to the formation of O$^+$ by

$$H^+ + O \rightarrow O^+ + H \tag{5.3}$$

which initiates the formation of OH and H$_2$O. The fractional abundance of C$^+$

Table 2 Fractional abundances for models of chemistry in dense core–stellar wind boundary layers.

Model	n(H$_2$) (cm^{-3})	T (K)	C$^+$	C	CO	CH	OH	H$_2$O	C$_2$	C$_2$H	H$_2$CO
depleted	10000	10	3.2(−08)	6.0(−08)	3.1(−05)	4.2(−09)	2.3(−07)	6.2(−07)	5.1(−09)	7.0(−09)	6.8(−09)
undepleted	10000	10	7.3(−09)	7.3(−10)	3.7(−04)	4.1(−11)	9.0(−08)	4.8(−06)	4.8(−11)	9.2(−10)	4.2(−09)
1	180	1100	2.6(−07)	3.0(−05)	2.8(−07)	2.3(−07)	1.7(−05)	2.7(−06)	9.1(−11)	5.2(−11)	4.8(−13)
2	1600	680	2.6(−06)	2.2(−05)	4.2(−06)	1.3(−06)	2.1(−05)	1.1(−05)	9.6(−09)	9.0(−09)	1.2(−11)
3	200	990	6.8(−06)	9.5(−05)	2.3(−04)	1.4(−05)	1.2(−04)	2.3(−04)	1.3(−06)	1.2(−06)	2.6(−08)
4	2500	420	9.9(−06)	8.1(−06)	3.4(−04)	3.0(−04)	3.2(−05)	1.8(−05)	1.0(−07)	4.9(−08)	3.0(−11)
5	180	1100	1.4(−08)	4.5(−07)	2.9(−05)	2.2(−07)	5.2(−07)	1.6(−05)	6.9(−09)	4.1(−09)	4.6(−10)
6	1600	650	2.1(−08)	4.7(−07)	2.9(−05)	1.7(−07)	2.3(−07)	1.6(−05)	7.4(−09)	5.2(−09)	1.3(−09)
7	210	970	1.8(−08)	5.1(−07)	3.6(−04)	1.4(−07)	6.3(−06)	1.9(−04)	8.6(−10)	1.1(−09)	4.4(−09)
8	2400	440	1.3(−08)	1.5(−08)	3.6(−04)	3.7(−10)	8.2(−07)	1.6(−05)	9.2(−12)	2.9(−12)	8.7(−10)
9	320	620	4.9(−06)	2.2(−05)	2.7(−06)	8.3(−07)	2.0(−05)	5.5(−05)	3.3(−09)	2.9(−09)	2.0(−12)
10	2700	380	8.2(−06)	3.3(−06)	1.9(−05)	2.6(−07)	1.3(−05)	3.2(−06)	6.9(−09)	6.5(−09)	6.1(−13)
11	470	420	1.5(−05)	9.5(−06)	3.4(−04)	4.5(−05)	4.7(−05)	1.9(−05)	1.4(−07)	6.9(−08)	2.0(−11)
12	3800	250	4.3(−06)	4.2(−06)	3.6(−04)	3.2(−08)	1.2(−05)	7.0(−06)	1.4(−08)	5.4(−09)	1.6(−11)
13	660	300	1.0(−05)	3.6(−06)	1.7(−05)	9.2(−08)	1.3(−05)	2.5(−06)	2.3(−09)	2.1(−09)	1.7(−13)
14	4100	220	3.6(−06)	1.7(−06)	2.5(−05)	1.5(−07)	7.3(−06)	1.4(−06)	5.1(−09)	5.0(−09)	1.2(−12)
15	1100	170	5.4(−06)	4.6(−06)	3.6(−04)	2.8(−08)	1.1(−05)	5.1(−06)	8.5(−09)	4.2(−09)	8.9(−12)
16	5900	130	1.8(−06)	3.1(−06)	3.7(−04)	8.2(−09)	3.3(−06)	2.9(−06)	3.1(−09)	1.3(−09)	2.6(−11)

Model	n(H$_2$) (cm^{-3})	T (K)	NH$_3$	HCN	CN	HCO$^+$	N$_2$H$^+$	CS	SO	HCS$^+$
depleted	10000	10	3.2(−08)	4.4(−09)	2.6(−08)	1.3(−09)	1.1(−10)	3.8(−08)	6.7(−09)	4.2(−11)
undepleted	10000	10	6.0(−08)	5.2(−09)	1.4(−09)	4.1(−09)	4.7(−10)	4.3(−07)	5.3(−07)	2.6(−10)
1	180	1100	1.4(−10)	1.3(−10)	3.5(−10)	8.0(−11)	3.7(−13)	1.3(−12)	2.0(−13)	3.1(−13)
2	1600	680	5.7(−10)	8.3(−09)	4.5(−09)	1.1(−08)	2.1(−12)	4.8(−10)	3.1(−11)	7.7(−11)
3	200	990	7.1(−07)	3.4(−06)	4.2(−07)	1.0(−06)	2.4(−08)	1.4(−07)	3.3(−08)	2.6(−08)
4	2500	420	4.5(−08)	3.0(−07)	2.8(−07)	4.6(−07)	2.8(−09)	2.1(−08)	1.7(−08)	1.9(−09)
5	180	1100	2.4(−07)	3.9(−08)	1.9(−10)	4.6(−07)	2.8(−08)	1.4(−08)	4.3(−10)	1.8(−09)
6	1600	650	4.2(−08)	4.9(−08)	1.3(−10)	1.2(−07)	6.7(−09)	1.3(−08)	6.4(−08)	7.2(−10)
7	210	970	3.4(−07)	6.3(−08)	4.1(−10)	1.6(−07)	1.1(−08)	1.4(−09)	1.3(−07)	1.1(−08)
8	2400	440	2.2(−08)	2.8(−08)	1.7(−10)	1.7(−07)	3.6(−09)	3.3(−10)	1.8(−07)	4.5(−11)
9	320	620	5.3(−11)	1.5(−09)	2.4(−09)	6.7(−09)	7.8(−13)	9.6(−11)	6.1(−12)	1.5(−11)
10	2700	380	3.8(−10)	2.9(−09)	8.1(−09)	2.5(−08)	2.4(−12)	4.4(−10)	5.2(−11)	4.3(−11)
11	470	420	4.3(−08)	3.6(−07)	5.6(−07)	4.7(−07)	4.3(−09)	1.9(−08)	1.3(−08)	1.8(−09)
12	3800	250	5.1(−08)	2.8(−08)	2.2(−07)	2.0(−07)	1.5(−09)	1.2(−08)	1.4(−08)	6.7(−10)
13	660	300	2.8(−10)	9.0(−10)	1.2(−08)	1.7(−08)	8.6(−12)	1.0(−10)	3.1(−11)	8.5(−12)
14	4100	220	1.3(−09)	1.0(−09)	3.7(−08)	1.3(−08)	1.1(−11)	1.3(−09)	1.2(−10)	8.2(−11)
15	1100	170	2.9(−08)	7.2(−09)	2.6(−07)	9.1(−08)	1.0(−09)	1.1(−08)	8.4(−09)	4.8(−10)
16	5900	130	2.9(−08)	3.8(−09)	6.0(−08)	3.3(−08)	4.3(−10)	2.0(−08)	7.4(−09)	5.7(−10)

increases as injected He$^+$ reacts with CO. The C$^+$ sometimes recombines creating C^0 or reacts with H$_2$ initiating the formation of CH and other hydrocarbons. In some models $T \geq 500$ K which is sufficient for the reactions

$$C^+ + H_2 \rightarrow CH^+ + H \tag{5.4}$$

$$O + H_2 \rightarrow OH + H \tag{5.5}$$

$$OH + H_2 \rightarrow H_2O + H \tag{5.6}$$

to become much more rapid (cf. Chapter 8) than they are at low temperatures. Following the formation of CH$^+$ the basic carbon chemistry can proceed. In some models $T \geq 10^3$ K allowing the reaction

$$C + H_2 \rightarrow CH + H \tag{5.7}$$

to be much more rapid than at low temperatures (cf. Chapter 8) which initiates interesting carbon chemistry. In model 3 the combination of a high gas phase abundance of CH_3 (a consequence of reactions that are much slower at low temperatures) and a high gas phase abundance of C^+ leads to a high fractional abundance of C_2H; an inspection of Fig. 11 causes one to suspect that the fractional abundance of HC_3N (which was not calculated) in such a model would also be high.

If the thickness of the boundary layer is a few per cent of the obstacle and the number density of the hot boundary layer gas is only a few per cent that of the core (as for model 3), the hot boundary layer gas has a total H_2 column density of only 10^{-3} of the core; thus, for the hot boundary layer gas to contain a substantial fraction of the total column density of species the fractional abundance of that species in the boundary layer must be at least 10^2 higher than it is in the core.

In addition, because the emissivity per unit volume of each of many species is proportional to the product of that species' density with $n(H_2)$, the low density of hot gas in the boundary layer may make the detection of boundary layer emissions of some species difficult even if the species have high fractional abundances there. Thus, in searches for boundary layer emission the focus should be on boundary layers around the highest density cores, around which the density in the hot parts of the boundary layers may not be greatly below the critical density above which the collisional de-excitation rate of a level exceeds the radiative de-excitation rate. Of course, significant effects on line profiles can be caused by absorption due to species having high column densities in parts of a boundary layer where the density is far below the critical value. Because in TMC-1 $n(H_2) \gg 10^4$ cm^{-3} and due to its proximity to a star, TMC-1 is especially suitable for the study of boundary layer emission. Turbulent driven transport of species formed in the hot part of a boundary layer from the hot part down to denser, colder parts may, in some cases, result in species formed only in hot regions being present in colder denser regions where excitation conditions are more favourable for the detection of emission from them.

Rawlings and Hartquist (1997) have constructed models of the chemistry of some simple species in boundary layers in which diffusion has been included. In their models OH provides a good example of a molecule with an abundance that diffusive transport from low density hotter parts of the boundary layer enhances in low temperature denser parts. Table 3 gives the steady state number density of OH as a function of (Z/Δ) for each of two model boundary layers; Δ is the boundary layer thickness, and Z is the distance from the coreward edge of the boundary layer. The chemical abundances at the coreward edge were assumed to be those appropriate for undepleted gas in a chemical equilibrium affected by a radiation field but unaffected (except for the production of H_2) by the presence of grains; there $n(H_2) = 3 \times 10^5$ cm^{-3} and $T = 10$ K. The chemical abundances at the outer edge of the boundary layer were estimated in the following way. It was assumed that the shocked wind flowed past the core at constant pressure and temperature (8000 K). Recombination for the flow-past time (10^4 years) was assumed to occur in this gas; the resulting abundances were those used.

Table 3 OH number density (in cm^{-3}) as a function of position in each of two models of steady state diffusive boundary layers.

Z/Δ	$T(K)$	Δ^2/κ	
		∞	3×10^{12} s^{-1}
0.0	10	5.5(−5)	5.5(−5)
0.505	20	5.3(−5)	5.5(−5)
0.747	40	5.3(−5)	1.8(−4)
0.838	62	5.5(−5)	1.1(−3)
0.879	82	5.6(−5)	3.7(−3)
0.909	110	5.7(−5)	1.8(−2)
0.929	140	5.7(−5)	6.9(−2)
0.939	160	5.7(−5)	1.3(−1)
0.949	200	6.1(−5)	2.4(−1)
0.970	320	3.4(−3)	6.1(−1)
0.980	450	9.7(−2)	7.9(−1)
0.990	760	1.1(−1)	6.7(−1)
1.0	8000	5.3(−5)	5.3(−5)

The radiation field throughout the boundary layer was assumed to be equal to that produced by the passage of the standard interstellar radiation field through material causing one magnitude of visual extinction. The diffusion coefficient, κ, in each model was taken to be constant and, consequently, the single free parameter that determines the effect of diffusion is Δ^2/κ. (If a boundary layer is translucent to optical or ultraviolet radiation, both κ and Δ must be specified, but here we give results for optically thin cases only.)

The rapidity of the reaction of O with H_2 at temperatures above 500 K and the drop off of the density with increasing temperatures cause OH to have its highest number densities in the temperature range of 500 K to 1000 K. One sees clearly from Table 3 that in the absence of diffusion ($\Delta^2/\kappa = \infty$) the abundance maximum is very localized, but with more rapid diffusion (decreased Δ^2/κ) the maximum broadens. It remains to be seen whether any realistic boundary conditions in models of diffusion modified chemistry in boundary layers result in the abundances of HC_3N, C_2H and a number of other species being at maxima near the windward edge of a boundary layer but diffusive transport carrying significant enough quantities of those species inwardly to sufficiently dense parts of the boundary layer to cause their emissions to be observationally detectable.

Though most of the contents of this section are relevant to boundary layers between cores and the winds of high-mass stars as well, low-mass star forming regions have been mentioned more directly. In fact, the existing molecular observations that are most clearly likely to result in additional insight into the natures of wind-core boundary layers are of objects in the region of high-mass star formation in Orion. Tedds et al. (1995) have observed H_2 infrared radiation in a number of lines emitted in the interaction of wind material with dense molecule gas in objects that are smaller than dense cores and were *perhaps* ejected from a young stellar object; the observations were made at sufficiently high angular resolution and velocity resolution to apparently resolve the emission spatially and to obtain detailed line profiles, but at this time the natures of the profiles remain unexplained. Wiseman and Ho (1996) have reported data from observations made in the NH_3 (1, 1) and NH_3 (2, 2) inversion transitions towards a roughly 0.5 pc long, narrow molecular streamer interacting with the wind from the star forming region; the data imply that the temperature in general increases towards the edges of the streamer and almost certainly contain information about the energy dissipation occurring in a boundary layer.

6 Attempts to Observe the Collapse of Dense Cores in Regions of Low-mass Star Formation

Because dense cores are thought to be the progenitors of protostars, efforts to detect signatures in molecular emission line profiles of the collapse of such cores have been made. Figure 12 shows the shape of an NH_3 emission line feature arising in L1498, a dense core (Myers and Benson 1983). Because the source was observed in two separate lines, a temperature could be derived; the NH_3 emission profile shows very little deviation from that expected from a static core with very subsonic turbulence and there are no broad wings like those that might be expected to arise, during some stages of the gravitationally induced collapse of a core in which magnetic effects are negligible, in the fastest (supersonically) infalling material.

In some cases, matter infalling transonically should give rise to asymmetric self-absorption nearer the centres of some emission features. If the excitation temperature of a level is higher in the central part of the core, the feature is optically thick and infall is symmetric with respect to the plane perpendicular to the line of sight, infalling material on the side nearer the observer will absorb emission from the centre of the core and depress the redward part of the emission profile while the infalling material on the far side of the core contributes somewhat to the blueward part of the emission profile.

Early observational detections of this sort of blue-red asymmetric self-absorption were made in CO emission line profiles formed in a number of clouds (Snell and Loren 1977) and in a CS emission line study of IRAS 16293-2422 (Walker et al. 1986). However, the uniqueness of the infall explanation of the origin of the asymmetries of these features was challenged by Leung and Brown (1977) in the case of the CO observations and by Menten et al. (1987) in that of IRAS 16293-2422. Menten et al. (1987) explained the observations of that source with a model in which the emission from a rapidly rotating core undergoes foreground absorption. Given the possible existence of ambiguities, Zhou (1992) performed radiative transfer calculations based on simplifying approximations

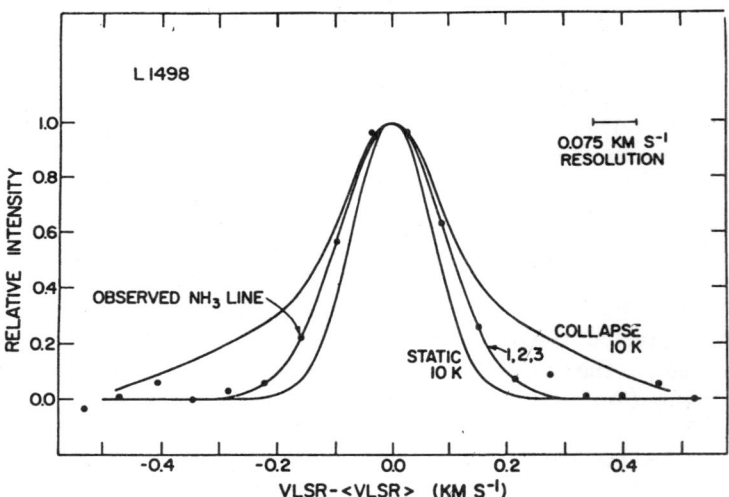

FIG. 12. The NH$_3$ line profile of L1498. The curve associated with the dots is the observed profile. That marked static is the one expected from 10 K gas that is not turbulent and experiences no systemic motion. The curve marked collapse is the profile expected if the fractional abundance of NH$_3$ is constant and the core is undergoing collapse governed by a particular solution of the class that Shu (1977) investigated. From Myers and Benson (1983).

in an attempt to establish more thoroughly the characteristics of spectral line profiles formed in collapse models including one studied analytically by Shu (1977). Shu (1977) showed that an initially static, singular isothermal sphere will undergo self-similar collapse with the collapse wave propagating out from the centre where the collapse velocity and the density are infinite. (Collapse in which the infall speed decreases with radius and the outer radius of the collapsing region, r_{out}, increases with time has come to be known as "inside-out" collapse.) Computations like those of Zhou (1992) and the more accurate Monte-Carlo calculations of Choi et al. (1995) give results that imply that for cores undergoing inside-out collapse and in which the temperature decreases with radius: 1) optically thick lines show the blue-red self-absorption asymmetry; 2) for fixed angular resolution and equal optical depths the width of a line of a transition for which the critical density is large is greater than the width of a line of a transition for which the critical density is small; 3) the self-absorption and linewidth of a line appear to increase with increasing angular resolution of the central parts of a core.

With the above three points in mind, Zhou et al. (1993) demonstrated that H$_2$CO and CS emission profiles originating in B335, a low-mass star forming core with an embedded 3 L$_\odot$ infrared source and a collimated outflow, show the characteristic shapes associated with infall. High resolution (3″ to 5″) aperture synthesis maps of ^{13}CO and C^{18}O emissions also support the hypothesis that B335 is still undergoing collapse (Chandler and Sargent 1993).

In order to confirm these encouraging results there is a need to: a) observe more molecules and transitions; b) consider the effects of the outflow on the profiles; c) detect outflows in more sources. Points (a) and (c) have recently been addressed by Zhou et al. (1994), Mardones et al. (1994), Myers et al. (1995), Velusamy, Kuiper and Langer (1995), Wang et al. (1995) and Zhou, Evans and Wang (1996); new infall candidates have been discovered (in particular, the low-mass cores L1527 and L483 which are associated with very young embedded sources) and characteristic infall signatures have been observed in a variety of molecular lines: C^{18}O(2-1), C$_3$H$_2$($2_{1,2}$-$1_{0,1}$) and C$_2$S(2_1-1_0), besides H$_2$CO and CS. Point (b) is a difficult one to address because it complicates already sophisticated and involved models. Myers et al. (1996) developed a simple analytic model of radiative transfer in which the contribution of outflowing gas to spectral line profiles from contracting clouds is also considered. This model provides a simple way to quantify characteristic infall speeds, and its use to interpret data strongly suggests that the inward motions derived from the line profiles are gravitational in origin.

Another way to avoid complications due to stellar outflows is to study starless cores on the assumption that some of them are already collapsing but appear starless because of the insufficient development of the embedded protostar. So far, only one starless core has shown evidence of infall asymmetry profiles which strongly suggest infall motions: L1544 (Myers et al. 1996; Tafalla et al. 1998). In this core, measured linewidths are extremely small ($\sim$0.3 km s^{-1}) meaning that thermal pressure is playing an important role in the dynamics; moreover, it is one of the most opaque cores in the Taurus Molecular Cloud, suggesting the presence of high column densities. Myers and collaborators have defined L1544 as the *most evolved starless core*. Another interesting starless core is L1498 which shows an intriguing double-peaked CS feature with the blue peak stronger than the red peak (Lemme et al. 1995). However, the physical and chemical structure of L1498 is quite complicated and hinders an easy interpretation of observational data (see Kuiper, Langer and Velusamy 1996).

High sensitivity and high spectral resolution interferometric observations are opening a new window to the innermost parts of star forming low-mass cores, and helping to disentangle infalling from outflowing motions. In particular, Ohashi et al. (1997) recently used the Nobeyama Millimeter Array to observe ^{13}CO (1–0) and C^{18}O(1–0) emissions towards the infrared source embedded in L1527. Their ^{13}CO maps show bipolar outflowing shells with a V-like shape as well as a circumstellar envelope surrounding the central source. On the other hand, the C^{18}O map shows a flattened envelope of $\sim$2000 AU radius overlaying the central source, and perpendicular to the outflow axis. The C^{18}O line profile is explained well with a rotating disc-like envelope with radial motions toward the central star.

Interferometers are also needed to identify which part of the cloud is traced by the chosen molecular species. For example, C$_2$S in B335 is tracing the outer parts of the collapsing envelope of the core (Velusamy et al. 1995), whereas H_2CO and CS emission is coming from deeper regions. A similar result has been found from high resolution observations toward L1498 (Kuiper et al. 1996). This starless core shows a chemically differentiated onion-shell structure, with the NH_3 in the inner and the C_2S in the outer part of the core; CS and C_3H_2 emission seem to lie in between NH_3 and C_2S. The chemical and physical properties of L1498 have been interpreted by Kuiper et al. in terms of a "slowly contracting" dense core in which the outer envelope is still growing.

Some questions may arise from the above summary on observational evidence of dark cloud core collapse: 1) How deeply in the core are infalling motions traced by H_2CO and CS observations? These are the species used most for this kind of study. 2) How strongly do stellar outflows affect the abundance and the excitation of the above molecular species? 3) Why does NH_3 not reveal infall signatures? 4) Why do CS and C_2S seem to trace "envelope" material, even though they are both high density tracers? They trace densities higher than NH_3 but their emission seems to be external to ammonia emission. 5) What is happening to molecular material in a region immediately surrounding the accreting young star, and can depletion on to grains occur in spite of the proximity of a central source?

More high sensitivity and high resolution observations are required to answer these questions and to better understand the physics of gravitational collapse in cloud cores and the chemical processes in star forming regions.

7 Theoretical Studies of the Chemistry of Dense Core Collapse in Regions of Low-mass Star Formation

Variations in dense core chemical composition have been measured and have effects on the profiles of lines observed to study core collapse. To obtain the maximum information from the profiles about the dynamics of core collapse the variations in chemical composition must be understood theoretically so that their effects on the profiles can be reliably deconvolved from those of the dynamics.

As mentioned in the previous section, one key problem in the observations of dense cores concerns the absence of infall signatures in NH_3 line profiles. Menten et al. (1984) suggested that depletion in the infalling gas may be so high that NH_3 is not observable in it. Rawlings et al. (1992) took this suggestion as the starting point for a theoretical examination of chemistry in dense core collapse and, following Hartquist and Williams (1989), attempted to identify gas phase species that have non-diminishing or at most slowly diminishing fractional abundances during some stages of depletion (cf. the discussion of this topic in Section 4). Rawlings et al. (1992) proposed that the lines of such species would be suitable ones to observe in efforts to discover unambiguous spectral signatures of ongoing collapse, which up to that time had been unsuccessful.

For gas taking part in a collapse described by the similarity solution for the inside-out collapse of a singular isothermal sphere due to Shu (1977), Rawlings et al. (1992) performed a point-by-point integration of the time-dependent chemical rate equations for each radius on a grid of radii. Apart from small differences arising from differences between the thermal speeds of species and corrections for positively charged ions having collision rates with grains affected by the negative charge carried by each grain, the same depletion rate was used for all species more massive than helium, other than sodium atoms and ions. It was assumed that Na and Na^+ returned immediately to the gas after striking a grain, in the form of Na. The depletion rate was adjusted so that the model NH_3 line profile is in harmony with that observed towards L1498 (Myers and Benson 1983). The gas was assumed to be in a very dark region so that the effects of the external radiation field were negligible.

The initial chemical conditions affect the results of any study like that of Rawlings et al. (1992). They worked within the framework of the cyclic dynamical and chemical model described in Section 3 (cf. Fig. 7). They started the integration of the time-dependent chemical equations for each parcel of gas just at the instant (point D in Fig. 7) after it emerged from the postshock region of the shock through which the ablated material–wind mixture had been decelerated. The chemical composition at that time was taken to be that given at the corresponding time in one of the cycles in one of the cyclic models studied numerically by Nejad, Charnley and Williams (1990). Shock processing was assumed to return depleted elemental O, C, N and S to the gas phase primarily as H_2O, CH_4, NH_3 and H_2S except for that material that was in very stable species such as CO and N_2 when it froze out of the gas phase. Therefore, the initial fractional abundance of H_2O was considerably higher than would obtain in purely gas phase chemical equilibrium; this high H_2O fractional abundance was found to have consequences in the subsequent collapse phase which are described below.

The chemical evolution of a parcel of gas was then followed as it underwent free-fall collapse until a fixed density, which differed from parcel to parcel, was reached; then the chemistry was followed as the parcel remained at fixed density. The results of the integrations for all the parcels were then used to construct the time-varying and position-varying chemical structure of a truncated, singular isothermal sphere as it formed from material at one initial density and as it remained static for a fixed length of time. This stage of the integration ended when selected average chemical fractional abundances (including those of NH_3 and CH) matched measured values appropriate for the region around core B of TMC-1.

During the next stage of the integration the densities and velocities of the parcels were taken to evolve in a way governed by the solution due to Shu (1977) appropriate for a 0.96 $M_\odot$ truncated singular isothermal sphere of temperature 10 K. From the calculated chemical evolution and the assumed dynamical evolution, Rawlings et al. (1992) obtained time-varying profiles for optically thin lines of a number of species on the assumption that for each line the critical de-excitation number density for the excited level is small compared to the number densities in the truncated isothermal sphere. The line profiles were calculated for angular resolutions obtainable with existing single-dish telescopes and for an object at the distance of L1498. Species found to have noticeably broader line profiles than NH_3 included HCO, HNO, N_2H^+, HCO^+, HS, CH and H_2S. For a number of these species the primary cause for their greater widths was that as depletion occurs the reduction of the high gas phase H_2O fractional abundance (an artifact of the initial high fractional abundance mentioned several paragraphs before in the consideration of initial conditions) decreases the rate of the primary removal mechanism of the species itself or a species that is a progenitor of it. Unfortunately, Rawlings et al. (1992) did not follow the behaviour of CS. (As mentioned in Section 2 the behaviour of the sulphur depletion is a key unanswered question in star formation, and it must be addressed before the evolution of the CS abundance can be understood.) However, it was concluded that the suggestion of Menten et al. (1984) is plausible, and that the proposal may be further tested by study of the line profiles of the additional species listed above.

Various models of the ways in which magnetic fields and ambipolar diffusion (cf. the final paragraph of Section 2) affect dense core collapse exist (e.g. Ciolek and Mouschovias 1995), and rather than always perform chemical calculations for detailed dynamical models it is sometimes useful to adopt a simple description of the dynamics or even assume a fixed density and explore the effects of depletion in models constructed for various chemical initial conditions, visual extinctions, and assumptions about the depletion processes themselves.

Nejad, Hartquist and Williams (1994) studied the chemical evolution in a single parcel of gas undergoing cycling in one cyclic model. Unlike in many other cyclic model calculations, the effects of an external radiation field were included. Its strength was assumed to be about one tenth that of the typical interstellar background field where the visual extinction was zero; the visual extinction to the parcel was taken to vary between 2.0 and 5.8 depending on at which point in its dynamical evolution it was. Due to the inclusion of a finite external radiation field, Nejad et al. (1994) found that the attainment of agreement between some models and TMC-1 measured fractional abundances required an input value of the cosmic ray ionization rate of 1×10^{-16} s^{-1}, about an order of magnitude in excess of typically assumed values. Depletion was treated in much the same way as in the Rawlings et al. (1992) work as described above. Species found to have fractional abundances increasing with time or at least remaining fairly level with time at model times when some important TMC-1 fractional abundances were reasonably well matched by the model fractional abundances included CH, OH, C_2H, H_2CO, HCN, HNC and CN. These species might be good ones to observe in studies of infall, as H_2CO has, in fact, proved to be (Zhou et al. 1993). CS and other sulphur-bearing species show rapid decreases in fractional abundances with time at key times due primarily to S^+ containing a sufficient fraction of the sulphur (a consequence of the high value of the cosmic-ray-induced ionization rate) for its collisions with grains to dominate the depletion of gas phase elemental sulphur. (Positive ions collide with negatively charged grains over an order of magnitude more frequently than neutrals of the same mass; grains are negatively charged in dark regions because electrons move so much more rapidly than ions.)

8 Some Dynamics and Chemistry of Regions of High-mass Star Formation

Ultracompact HII regions (UCHIIRs) are manifestations of newly formed massive stars that are still embedded in their natal molecular clouds (e.g. Churchwell 1991). UCHIIRs are associated with small (~0.1 pc), dense ($n(H_2) \sim 10^7$ cm^{-3}), and "hot" ($T \sim 100$–200 K) molecular cloud cores which are thought to be the type of condensations from which the OB stars powering the HII regions formed. Systematic motions in such cloud cores may reflect the dynamical conditions relating to the formation and very early evolution of the OB stars. However, direct evidence of collapse motions is hard to find because of their confinement to the innermost part of the core and because of the large distance of massive star formation sites (e.g. Walmsley 1995). Interferometry is needed. During the past decade, radio interferometric techniques have given several indications that the molecular gas surrounding each of a few well-observed UCHIIRs is collapsing toward a central mass concentration.

Zheng et al. (1985) used the Very Large Array (VLA) to observe the $(J, K) = (1, 1)$ NH$_3$ inversion line toward the UCHIIR ON1 which is an isolated source, not associated with other HII regions and probably represents a site of first-generation star formation. They resolved dense NH$_3$ condensations showing a systematic shift in velocity with position. This trend may be due to rotation, and the large velocity gradient (~11 km s^{-1} pc^{-1}) observed at the 0.1 pc scale was interpreted by Zheng et al. as the "spin-up" motion expected from partial conservation of the angular momentum during the collapse process. They suggested that the spin-up motion during condensation may provide a new means to identify collapse as it is much easier to detect than actual collapse.

Another rapidly rotating core was soon found toward the UCHIIR G10.6-0.4 by Ho and Haschick (1986). In this work, more evidence of spin-up and collapse motions was collected. The envelope in which the rapidly rotating core is embedded was seen to rotate with a gradient of ~1 km s^{-1} pc^{-1} in the same sense as the core; redshifted absorption of the NH$_3$(1, 1) line is observed in the direction of the continuum source. From the spatial distribution, Ho and Haschick suggested that infalling material consists of only a small portion of the central core. They estimated an accretion rate of ~2×10^{-4} $M_\odot$ yr^{-1} at a radius of 0.025 pc, equivalent to a pressure of 6×10^8 cm^{-3} K which may be sufficient to restrain the rapid expansion of UCHIIRs. Higher spatial resolution observations of the same region G10.6-0.4 in the $(J, K) = (1, 1)$ and $(3, 3)$ lines of NH$_3$ allowed Keto, Ho and Haschick (1987) to confirm the previous model of gravitational collapse and spin-up of the infalling gas and determine density and temperature profiles. They found that $n(H_2) \propto R^{-2}$, $T \propto R^{-1/2}$, and that the highest temperatures in the core occur at the most redshifted velocities, indicating that the gas with the highest collapse velocity is closest to the central heating source, as would be expected if the region were gravitationally collapsing. Later, Keto, Ho and Haschick (1988) resolved the accretion flow of the molecular gas onto the central HII region and showed that the flow field is characterized by differential rotation and accelerating infall. They inferred that infall is localized near the central source and not distributed over the entire molecular cloud.

Other well-known regions containing UCHIIRs and showing characteristics similar to G10.6-0.4 are W3(OH) and G34.3+0.2 (Keto, Ho and Reid 1987; Pahre et al. 1993). Welch et al. (1987) studied the W49A region, which contains a recently formed cluster of massive O and B stars. They used the Hat Creek Interferometer to map the HCO^+(1-0) line. Variations in the HCO^+ profiles across the region have been interpreted as due to free-fall of the inner envelope of the giant molecular cloud toward the 2 pc ring of the central HII regions. HCO^+ radial velocities fit a simple model of inside-out gravitational collapse of a once magnetically supported cloud. So, Welch et al. (1987) inferred the existence of a large-scale collapse of giant molecular clouds and speculated that this could be the mechanism to produce massive star clusters.

Analogous conclusions were reached by Rudolph et al. (1990) who studied W51, another well-known region of massive star formation. In this work, inverse P Cygni profiles of HCO^+ lines were detected toward the compact HII region W52 e2 and W51 IRS 2, which contains an infrared source, a collection of H_2O masers and a compact HII region (Genzel et al. 1982). Redshifted and blueshifted absorption features have been observed toward W51 IRS1, a long curving structure where free–free emission dominates and which does not contain much mass. Their data are fitted well with a model of large-scale dynamical collapse of a large section of the W51 cloud, supporting the suggestion of Welch et al. (1987). However, recent VLA observations of the same region in the NH_3(1, 1) and (2, 2) transitions give evidence of radial contraction extending only over a few tenths of a parsec near W51 e2, instead of over the entire W51 molecular cloud complex (Ho and Young 1996), and the W51 collapse is then analogous to that of G10.6-0.4 (Keto, Ho and Haschick 1988).

Recently, Cesaroni, Walmsley and Churchwell (1992) conducted a survey of "hot" ammonia emission from a sample of UCHIIRs with the Effelsberg 100 m telescope. One of the sources in this sample, G45.47+0.05, showed redshifted absorption and blueshifted emission in the NH_3(4, 4) and NH_3(5, 5) lines, characteristic of collapse. Wilner, Ho and Zhang (1996) used the OVRO millimeter array to observe HCO^+(1-0) and SiO(2-1) lines and study the kinematics of this region. They found no evidence for the spherical infall suggested by the NH_3 inverse P Cygni profiles. Instead, complex structure in both position and velocity, indicative of multiple velocity components, is resolved. They suggested that emission and absorption from separate sites of activity in the region may mimic the infall signature detected by the single-dish experiment, confirming again that interferometric observations are key in this

The measured chemical compositions of a number of the "hot cores" associated with UCHIIRs and having physical properties described in the first paragraph of this section differs markedly from the compositions of cool (10 K to 30 K) dark cores (e.g. Walmsley and Schilke 1993; Millar 1993).

Other so-called hot cores, not directly associated with UCHIIRs, also give some information about high-mass star formation. However, they may differ from those associated with UCHIIRs in that they may be shock heated, whereas the former are more likely to be radiatively heated. The method and rate of heating are important for the propagation of star formation.

The Orion Hot Core, which is associated with the cluster of luminous infrared sources of the Kleinmann–Low Nebula, is one of the best studied hot cores. The measured gas phase fractional abundances of NH_3, HCN and H_2S for this hot core exceed those for the cooler surrounding gas by more than one, roughly two, and about three orders of magnitude, respectively. When a molecule of one of many (but not all) unsaturated species sticks to a dust grain in hydrogen-rich environment, surface reactions add hydrogen to it to produce a molecule of a saturated species; thus, N, CN and S are converted by freeze-out on to a grain surface to NH_3, HCN and H_2S ices. Such freeze-out and ice formation probably occurred in the material now in the Orion hot core when it was cooler. Heating of the hot core material then later induced the evaporation of ices on the grains and injected the molecules into the gas phase.

Amongst other species having gas phase abundances greatly affected by the evaporation of ices on grains are the deuterated forms of many common molecules. Fractionation is described in Chapter 4; the ratios of the fractional abundances of a deuterated species and the protonated form can greatly exceed the cosmic abundance ratio of deuterons to protons in gas phase equilibrium if the temperature is well below 100 K and some other conditions are met. The high abundance ratios of NH_2D to NH_3, CH_3OD to CH_3OH and D_2CO to H_2CO found in the Orion Hot Core have been attributed to fractionation occurring in cold gas, followed by freeze-out on to grains on which these gas phase ratios are unaltered and eventual melting leading to much higher gas phase ratios than would be appropriate for 100 K gas in chemical equilibrium.

As mentioned near the end of Section 5, there is also much hotter ($\approx$2000 K) molecular gas in the region of high-mass star formation in Orion. Though the nature of the wind–obstacle interaction that heats the gas remains unclear at this time, the chemistry in this gas is similar to that in shock-heated gas, the subject of Chapter 8.

Bibliography

1. Arons, J. and Max, C. E. (1975). *ApJ*, **196**, L77.
2. Bertoldi, F. and McKee, C. F. (1992). *ApJ*, **395**, 140.
3. Cesaroni, R., Walmsley, C. M. and Churchwell, E. (1992). *A&A*, **256**, 618.
4. Chandler, C. J. and Sargent, A. I. (1993). *ApJ*, **414**, L29.

5. Charnley, S. B., Dyson, J. E., Hartquist, T. W. and Williams, D. A. (1988). *MNRAS*, **235**, 1257.
6. Charnley, S. B., Dyson, J. E., Hartquist, T. W. and Williams, D. A. (1990). *MNRAS*, **243**, 405.
7. Chièze, J. P., Pineau des Forêts, G. and Herbst, E. (1991). *ApJ*, **373**, 110.
8. Choi, M., Evans II, N. J., Gregersen, E. M. and Wang, Y. (1995). *ApJ*, **448**, 742.
9. Churchwell, E. (1991). In *The physics of star formation and early stellar evolution*, eds C. J. Lada and N. D. Kylafis. Kluwer Academic Publishers, Dordrecht, p.221.
10. Ciolek, G. E. and Mouschovias, T. Ch. (1995). *ApJ*, **454**, 194.
11. Ciolek, G. E. and Mouschovias, T. Ch. (1996). *ApJ*, **468**, 749.
12. Draine, B. T. and Sutin, B. (1987). *ApJ*, **320**, 803.
13. Elmegreen, B. G. and Lada, C. J. (1977). *ApJ*, **214**, 725.
14. Genzel, R., Becklin, E. E., Wynn-Williams, C. G., Moran, J. M., Reid, M. J., Jaffe, D. T. and Downes, D. (1982). *ApJ*, **255**, 527.
15. Goldsmith, P. F., Langer, W. D. and Wilson, R. (1986). *ApJ*, **303**, L11.
16. Hartquist, T. W. and Dyson, J. E. (1988). *Ap&SS*, **144**, 615.
17. Hartquist, T. W. and Dyson, J. E. (1993). *QJRAS*, **34**, 57.
18. Hartquist, T. W. and Dyson, J. E. (1996). *Ap&SS*, **245**, 263.
19. Hartquist, T. W. and Williams, D. A. (1989). *MNRAS*, **214**, 417.
20. Hartquist, T. W., Williams, D. A. and Caselli, P. (1996). *Ap&SS*, **238**, 303.
21. Hartquist, T. W., Rawlings, J. M. C., Williams, D. A. and Dalgarno, A. (1993). *QJRAS*, **34**, 213.
22. Hirahara, Y., Suzuki, H., Yamamoto, S., Kawaguchi, K., Kaifu, N., Ohishi, M., Takano, S., Ishikawa, S.-I. and Masuda, A. (1992). *ApJ*, **394**, 539.
23. Ho, P. T. P. and Haschick, A. D. (1986). *ApJ*, **304**, 501.
24. Ho, P. T. P. and Young, L. M. (1996). *ApJ*, **472**, 742.
25. Howe D. A., Taylor, S. D. and Williams, D. A. (1996). *MNRAS*, **279**, 143.
26. Keto, E. R., Ho, P. T. P. and Haschick, A. D. (1987). *ApJ*, **318**, 712.
27. Keto, E. R., Ho, P. T. P. and Haschick, A. D. (1988). *ApJ*, **324**, 920.
28. Keto, E. R., Ho, P. T. P. and Reid, M. J. (1987). *ApJ*, **323**, L117.
29. Kuiper, T. B. H., Langer, W. D. and Velusamy, T. (1996). *ApJ*, **468** 761.
30. Kulsrud, R. and Pearce, W. A. (1969). *ApJ*, **156**, 445.
31. Langer, W. D. (1987). Private communication.
32. Langer, W. D., Velusamy, T., Kuiper, T. B. H., Levin, S., Olsen, E. and Migenes, V. (1995). *ApJ*, **453**, 293.
33. Lemme, C. M., Walmsley, C. M., Wilson, T. L. and Muders, D. (1995). *A&A*, **302**, 509.

34. Leung, C. M. and Brown, R. L. (1977). *ApJ*, **214**, L73.
35. McKee, C. F. (1989). *ApJ*, **345**, 782.
36. Mardones, D., Myers, P. C., Caselli, P. and Fuller, G. (1994). In *Clouds, cores, and low mass stars*, eds. D. P. Clemens and R. Barvainis. Astronomical Society of the Pacific, San Francisco, p.192.
37. Menten, K. M., Walmsley, C. M., Krügel, E. and Ungerechts, H. (1984). *A&A*, **137**, 108.
38. Menten, K. M., Serabyn, E., Güsten, R. and Wilson, T. L. (1987). *A&A*, **177**, L57.
39. Millar, T. J. (1990). In *Molecular astrophysics—a volume honouring Alexander Dalgarno*, ed. T. W. Hartquist. Cambridge University Press, Cambridge, p.115.
40. Millar, T. J. (1993). In *Dust and chemistry in astronomy*, eds. T. J. Millar and D. A. Williams. Institute of Physics Publishing, Bristol, p.249.
41. Mouschovias, T.Ch. (1987). In *Physical processes in interstellar clouds*, eds. G. E. Morfill and M.Scholer. D. Reidel Publishing Company, Dordrecht, pp.453 and 491.
42. Mouschovias, T.Ch. and Psaltis, D. (1995). *ApJ*, **444**, L105.
43. Myers, P. C. (1986). In *IAU symposium no. 115—star forming regions*, eds. M. Peimbert and J. Jugaku. D. Reidel Publishing Company, Dordecht, p.307.
44. Myers, P. C. (1990). In *Molecular astrophysics—a volume honouring Alexander Dalgarno*, ed. T. W. Hartquist. Cambridge University Press, Cambridge, p.328.
45. Myers, P. C. (1995). *ApJ*, **442**, 186.
46. Myers, P. C., Bachiller, R., Caselli, P., Fuller, G. A., Mardones, D., Tafalla, M. and Wilner, D. J. (1995). *ApJ*, **449**, L65.
47. Myers, P. C. and Benson, P. J. (1983). *ApJ*, **266**, 309.
48. Myers, P. C. and Khersonsky, V. K. (1995). *ApJ*, **442**, 186.
49. Myers, P. C., Mardones, D., Tafalla, M., Williams, J. P. and Wilner, D. J. (1996). *ApJ*, **465**, L133.
50. Nejad, L. A. M., Charnley, S. B. and Williams, D. A. (1990). *MNRAS*, **246**, 183.
51. Nejad, L. A. M. and Hartquist, T. W. (1994). *Ap&SS*, **220**, 253.
52. Nejad, L. A. M., Hartquist, T. W. and Williams, D. A. (1994). *Ap&SS*, **220**, 261.
53. Norman, C. A. and Silk, J. (1980). *ApJ*, **238**, 158.
54. Ohashi, N., Hayashi, M., Ho, P. T. P. and Momose, M. (1997). *ApJ*, **475**, 211.
55. Osterbrock, D. E. (1961). *ApJ*, **134**, 270.

56. Pahre, M. A., Ho, P. T. P., Reid, M. J., Keto, E. R. and Proctor, D. (1993). In *Massive stars: Their lives in the interstellar medium*, eds. J. P. Cassinelli and E. B. Churchwell. Astronomical Society of the Pacific, San Francisco, p.126.
57. Rawlings, J. M. C. and Hartquist, T. W. (1997). *ApJ*, **487**, 672.
58. Rawlings, J. M. C., Hartquist, T. W., Menten, K. M. and Williams, D. A. (1992). *MNRAS*, **255**, 471.
59. Rudolph, A., Welch, W. J., Palmer, P. and Dubrulle, B. (1990). *ApJ*, **363**, 528.
60. Ruffle, D. P., Hartquist, T. W., Taylor, S. D. and Williams, D. A. (1997). *MNRAS*, **291**, 235.
61. Savage, B. D. and Mathis, J. S. (1979). *ARA&A*, **17**, 73.
62. Schilke, P., Keene, J., LeBourlet, J., Pineau des Forêts, G. and Roueff, E. (1995). *A&A*, **294**, L17.
63. Shalabiea, O. M. and Greenberg J. M. (1995). *A&A*, **296**, 779.
64. Shu, F. H. (1977). *ApJ*, **214**, 488.
65. Shu, F. H., Adams, F. C. and Lizano, S. (1987). *ARA&A*, **25**, 84.
66. Snell, R. L. and Loren, R. B. (1977). *ApJ*, **211**, 122.
67. Tafalla, M., Mardones, D., Myers, P. C., Caselli, P., Bachiller, R. and Benson, P. J. (1998). *ApJ*, submitted.
68. Taylor, S. D., Morata, O. and Williams, D. A. (1996). *A&A*, **313**, 269.
69. Tedds, J. A., Brand, P. W. J. L., Burton, M. G., Chrysostomou, A. and Fernandes, A. J. L. (1995). *Ap&SS*, **233**, 39.
70. Velusamy, T., Kuiper, T. B. H. and Langer, W. D. (1995). *ApJ*, **451**, L75.
71. Walker, C. K., Lada, C. J., Young, E. T., Maloney, P. R. and Wilking, B. A. (1986). *ApJ*, **309**, L47.
72. Walmsley, C. M. (1995). Rev Mex AA, **1**, 137.
73. Walmsley, C. M. and Schilke, P. (1993). In *Dust and chemistry in astronomy*, eds T. J. Millar and D. A. Williams. Institute of Physics Publishing, Bristol, p.37.
74. Wang, Y., Evans II, N. J., Zhou, S. and Clemens D. P. (1995). *ApJ*, **454**, 217.
75. Welch, W. J., Jackson, J. M., Dreher, J. W., Terebey, S. and Vogel, S. N. (1987). *Science*, **238**, 1550.
76. Williams, D. A. and Hartquist, T. W. (1991). *MNRAS*, **251**, 351.
77. Williams, D. A., Hartquist, T. W. and Caselli, P. (1996). *MNRAS*, **282**, 900.
78. Williams, D. A., Hartquist, T. W. and Whittet, D. C. B. (1992). *MNRAS*, **258**, 599.
79. Williams, J. P., Blitz, L. and Stark, A. A. (1995). *ApJ*, **451**, 252.

80. Wilner, D. J., Ho, P. T. P. and Zhang, Q. (1996). *ApJ*, **462**, 339.
81. Wiseman, J. J. and Ho. P. T. P. (1996). *Nature*, **382**, 139.
82. Zheng, X. W., Ho, P. T. P., Reid, M. J. and Schneps, M. H. (1985). *ApJ*, **293**, 522.
83. Zhou, S. (1992). *ApJ*, **394**, 204.
84. Zhou, S., Evans, II, N. J., Kömpe, C. and Walmsley, C. M. (1993). *ApJ*, **404**, 232.
85. Zhou, S., Evans II, N. J. and Wang, Y. (1996). *ApJ*, **466**, 296.
86. Zhou, S., Evans II, N. J., Wang, Y., Peng, R. and Lo, K. Y. (1994). *ApJ*, **433**, 131.

PART III

Young Stellar Objects and Herbig–Haro Objects

6
The Magnetohydrodynamics of Outflows from Low-mass Young Stellar Objects

T. W. Hartquist
Max-Planck-Institut für extraterrestrische Physik

and

D. P. Ruffle
Department of Physics and Astronomy, University College London

1 Introduction

Winds of high-mass main sequence and highly evolved stars are probably driven primarily by radiation pressure acting on the wind gas (e.g. Lucy and Abbot 1993), while the sustained winds of low-mass and intermediate-mass highly evolved stars may be affected substantially by radiation pressure acting on dust (cf. Chapter 12). However, radiation pressure is incapable of inducing the high mass-loss rates of 10^{-9} $M_\odot$ yr^{-1} to 10^{-4} $M_\odot$ yr^{-1} characteristic of low-mass ($\lesssim 4$ $M_\odot$) young stellar objects (YSOs) in some phases of their evolution. The highest mass-loss rates are often associated with clearly defined bipolar flows (e.g. Bachiller 1996); some low-mass YSOs are also seen to possess jets and Herbig–Haro objects (cf. Chapter 10).

Hartmann and MacGregor (1982) investigated whether YSOs rotating near the critical rotation frequency required for breakup could possess bipolar winds with the required mass-loss rates, driven by the rotating magnetic fields. However, the so-called T-Tauri stars, some of which have mass-loss rates of up to 10^{-6} $M_\odot$ yr^{-1}, rotate at frequencies that are about an order of magnitude below their critical frequencies (e.g. Bouvier *et al.* 1993), and the Hartmann and MacGregor model required modification before it could be used to explain such high mass-loss rates from stars rotating at such frequencies. The subjects of centrifugally driven magnetized stellar and disc winds have had many important contributions, but we will focus on the X-wind model developed by Shu *et al.* (1994a,b), Najita and Shu (1994), Ostriker and Shu (1995), Shu *et al.* (1995) and Li and Shu (1996).

In Section 2 we present some of the critical conceptual ideas underlying the X-wind model. Section 3 contains the basic axisymmetric equations that govern X-winds. In Section 4 the boundary conditions that are specified to study the winds in the regions where they are supersonic but sub-Alfvénic are described and a solution for one wind in such a region is presented. The remaining two sections contain considerations relevant to the derivation of the properties of an X-wind at very large distances from a star. Some of the boundary conditions on an X-wind solution depend on the nature of the magnetic field structure, where material flows from the disc to the star. Thus, we address that structure in Section 5 before showing in Section 6 that jets can form in X-winds at large distances from the star.

2 Conceptual Basis of the X-wind Model

2.1 Qualitative Picture of the Flow Structure

We consider a highly idealized and somewhat unrealistic picture of the evolution of an accretion disc around a YSO, in order to clarify the relationship of the magnetic field structure in an X-wind to a simple dipole structure that one might use to model, in a rough way, the magnetic field of a star without a disc. Imagine that initially the inner edge of the disc is at a radius much greater than R_x, the radius at which the Keplerian orbital frequency is equal to the rotation frequency of the star Ω_*; imagine also that the magnetic poles of the star lie on its rotation axis, and that axis is also the symmetry axis of the disc. The structure of the field lines nearest to the star would resemble that depicted in Fig. 1a. (We will not concern ourselves in detail with the behaviour of field lines at greater distances from the star than those shown; their behaviours depend to a large degree on the conductivity of the disc, and if the disc were a perfect conductor and had no intrinsic field, at large distances from the star the magnetic field would fall to zero like a quadrupole field.) Now assume that viscous coupling within the disc leads to the outward transport of angular momentum, allowing some of the disc material to move nearer to the star, so that the disc extends inwardly to distances from the star less than R_x. The disc has a finite magnetic diffusivity, which implies that some disc material will become threaded with stellar magnetic field lines. In the limit of a very tenuous disc, these field lines would rotate at nearly the stellar rotation frequency and cause disc material which they thread to also rotate at nearly the stellar rotation frequency. Material near the disc, at distances to the star less than R_x would then be rotating at frequencies less than the local Keplerian frequencies and would flow towards the star, whereas material near to the disc at distances from the star greater than R_x would then be rotating at frequencies greater than the local Keplerian frequencies and be driven into an outflow. If the ram pressure of the outward flow is great enough, it will stretch the field lines on which it is occurring out and away from the star and a field line structure like that depicted in Fig. 1b would obtain. In many cases, the field lines and flow stream lines are essentially parallel. In the theoretical work on X-winds, the assumption that the flow becomes steady has been made and solutions are for equations in which all derivatives with respect to time have been set to zero.

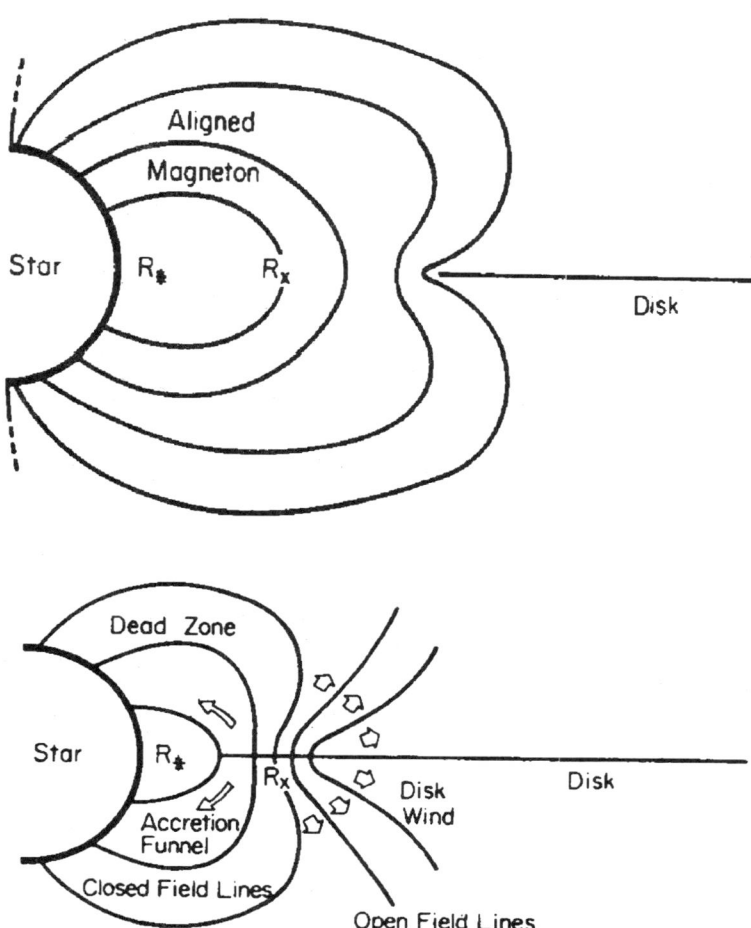

FIG. 1. Time evolution of the magnetic field structure in an idealized picture of the development of an X-wind. Only the structures of field lines nearest to the star in the equatorial plane are shown. a) The structures of field lines before the inner edge of the disc approaches R_x. b) The structures of field lines after the flow has become steady within the region shown. Arrows depict the flow directions in the accretion funnel and the wind. Adapted from Shu et al. (1994a).

The qualitative characteristics of the steady flows are indicated in somewhat more detail in Fig. 2 than in Fig. 1b. To understand Fig. 2, we must first consider the effective gravitational potential in a frame rotating about the symmetry axis with frequency Ω_*. In a cylindrical coordinate system in which $\tilde{\omega}$ is the distance

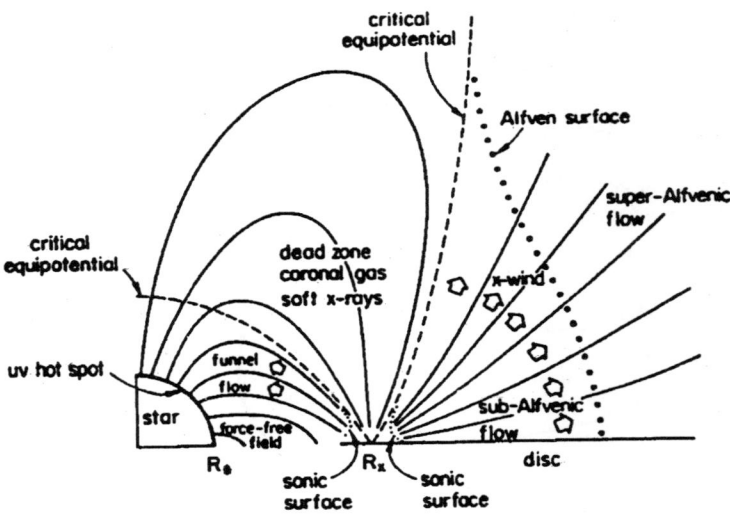

FIG. 2. A schematic diagram of a steady X-wind structure showing the locations of sonic and Alfvénic surfaces. From Shu et al. (1994a).

from the symmetry axis and z is the height above the mid-plane of the disc, that effective potential may be taken to be

$$V_{\text{eff}} = -\frac{GM_*}{(\tilde{\omega}^2 + z^2)^{1/2}} - \frac{1}{2}\Omega_*^2\tilde{\omega}^2 + \frac{3}{2}\Omega_*^2 R_x^2 \qquad (2.1)$$

so that $V_{\text{eff}} = 0$ at $z = 0$ and $\tilde{\omega} = R_x$; M_* is the mass of the star and G is the gravitational constant. In Fig. 2, the critical equipotential lines indicate where $V_{\text{eff}} = 0$. Figure 2 also shows two small dotted curves near R_x; they indicate the positions at which material at $\tilde{\omega} = R_x$ flowing from the disc begins to move at speeds measured in the frame co-rotating with the star to be greater than the ordinary isothermal sound speed. (Throughout this chapter we will assume gas to be isothermal.) At increasing distances from the X-point, these sonic surfaces asymptotically approach the $V_{\text{eff}} = 0$ surfaces. Material that flows between the two equipotential surfaces (the dead zone) always moves subsonically in the co-rotating frame and accretes onto the star. Material that flows through the inner sonic surface accretes onto the star in a supersonic funnel flow. Two types of waves propagate parallel to a uniform magnetic field (e.g. Krall and Trivelpiece 1973). A wave of one type is like an ordinary sound wave and involves the compression and relaxation of gas along the field lines. A wave of the other type is called an Alfvén wave and involves no compression of material; instead it has varying components of the velocity and of the magnetic field that are perpendicular to the background magnetic field and propagate at the Alfvén speed,

$$v_A = \frac{B_0}{(4\pi\rho)^{1/2}} \tag{2.2}$$

where B_0 is the strength of the background magnetic field and ρ is the gas mass density. In many YSO winds the Alfvén speed is much greater than the ordinary sound speed, and there is a region in which the wind is said to be supersonic but sub-Alfvénic. The surface at which the wind begins to flow at a speed measured (in the frame co-rotating with the star) to be greater than the Alfvén speed is indicated in Fig. 2.

2.2 The Relationship between the Disc Inner Truncation Radius and R_x

Consideration of where the torques exerted by the magnetic field of the star on the disc significantly affect the disc dynamics (Ghosh and Lamb 1979) lead to the result that the distance, R_t, of the inner edge of the disc from the star is given by

$$R_t = \Gamma_t R_* \tag{2.3}$$

$$\Gamma_t = \alpha_t \left(\frac{B_*^4 R_*^5}{GM_* \dot{M}_d^2} \right)^{1/7} \tag{2.4}$$

where B_* is the strength of the magnetic field at the stellar surface, $\dot{M}_d$ is the mass infall rate through the disc and α_t is a constant which depends on the magnetic diffusivity. These torques will enforce near co-rotation, at an angular frequency of about Ω_*, from $\tilde{\omega} = R_t$ to $\tilde{\omega} = R_t + \Delta_t$, called the T-region by Shu et al. (1994a), where Δ_t depends on the magnetic diffusivity. Possibly the T-region extends nearly to R_x; indeed, Shu et al. (1994a) have suggested that $(R_x - R_t)/R_x < 0.3$.

Shu et al. (1994a) have called the region in which the wind originates the X-region; it extends from $\tilde{\omega} \simeq R_x$ to $\tilde{\omega} = R_x + \Delta_x$. Δ_x can be estimated from considerations of the wind flow timescale. Take a_x to be the isothermal sound speed near R_x. (For discs around classic T-Tauri stars $\epsilon \equiv (a_x/\Omega_* R_x) = 0.02$.) Assume that mass input to the wind comes from the region where the wind speed is less than a_x. The acceleration of the wind is of order $\Omega_*^2 \Delta_x$; the wind is, thus, accelerated to a speed a_x on a length scale

$$\Delta_x = \frac{a_x}{\Omega_*}. \tag{2.5}$$

We now argue that R_t is proportional to R_x. To make this demonstration, we first note that the total mass loss rate of the wind, $\dot{M}_w$, is given roughly by

$$\dot{M}_w \simeq (2\pi R_x)(2\Delta_x)\rho_a a_x \tag{2.6}$$

where $4\pi R_x \Delta_x$ is a crude approximation to the area of the sonic surface and ρ_a is the mass density at the sonic surface. Hence, from the definition of ϵ, (2.5)

and (2.6), we have
$$\rho_a \simeq \frac{\epsilon^{-2} \dot{M}_w}{4\pi R_x^3 \Omega_*}. \tag{2.7}$$

After passing through the sonic surface, the wind diverges rapidly, so that the area of the surface through which it is flowing increases, to the order of $4\pi R_x^2$; also the speed quickly becomes of the order $\Omega_* R_x$. Thus, from the requirement that $\dot{M}_w$ is constant, it follows that the density ρ drops rapidly from ρ_a to about $\epsilon^2 \rho_a$. The divergence of the flow causes the strength of the magnetic field to drop from B_x, the value that it has near $\tilde{\omega} = R_x$, to a value of the order ϵB_x. If the X-region magnetic field is to fling gas into the escaping wind, the Alfvén speed in the diverged flow must be comparable to $\Omega_* R_x$; that is (cf. eqn (2.2))

$$\frac{\epsilon B_x}{(4\pi \epsilon^2 \rho_a)^{1/2}} \approx \Omega_* R_x. \tag{2.8}$$

In addition, ϵB_x must have a strength similar to that given by the stellar dipole field at R_x if there were a vacuum around the star; that is

$$\epsilon B_x \simeq B_* \left(\frac{R_*}{R_x}\right)^3. \tag{2.9}$$

Using (2.7), (2.8) and (2.9) we find that

$$\dot{M}_w \simeq \frac{B_*^2 R_x}{\Omega_*} \left(\frac{R_*}{R_x}\right)^6. \tag{2.10}$$

If we assume that $\dot{M}_w \propto \dot{M}_d$ and, furthermore, that $\dot{M}_d - \dot{M}_w$, $\dot{M}_d$ and $\dot{M}_w$ are all of the same order of magnitude and make use of the fact that

$$R_x = \left(\frac{GM_*}{\Omega_*^2}\right)^{1/3} \tag{2.11}$$

then (2.9) may be rearranged to express R_x as a function of $\dot{M}_d$ and other quantities,

$$R_x = \Gamma_x R_* \tag{2.12}$$

$$\Gamma_x = \alpha_x \left(\frac{B_*^4 R_*^5}{GM_* \dot{M}_d^2}\right)^{1/7} \tag{2.13}$$

where α_x is a constant.

Shu *et al.* (1994a) have suggested that the similarity of (2.12) and (2.13) to (2.3) and (2.4) shows that if a stellar magnetic field is strong enough to truncate an accretion disc beyond the stellar surface, it is strong enough to drive a substantial wind by picking up material around $\tilde{\omega} = R_x$.

2.3 The Relationship between $\dot{M}_\text{w}$ and $\dot{M}_\text{d}$

We now show that

$$\dot{M}_\text{w} = f\dot{M}_\text{d} \tag{2.14}$$

where f is a fraction less than but of order unity.

The angular momentum of a star is given by $bM_*R_*^2\Omega_*$, where b is a dimensionless measure of its moment of inertia and usually has a value between 0.136 and 0.205 if the star is fully convective. The time rate of change of that quantity is governed by

$$\frac{\text{d}}{\text{d}t}(bM_*R_*^2\Omega_*) = (1-\tau)\dot{M}_\text{d}R_x^2\Omega_* - \dot{M}_\text{w}\bar{J}R_x^2\Omega_* \tag{2.15}$$

where $\dot{M}_\text{w}\bar{J}R_x^2\Omega_*$ is the total angular momentum carried in the wind and

$$\tau = -\frac{\Im_x}{\dot{M}_\text{d}R_x^2\Omega_*} \tag{2.16}$$

with

$$\Im_x = \left(2\pi\tilde{\omega}^2\nu_\text{d}\sigma\frac{\text{d}\Omega}{\text{d}\tilde{\omega}}\right)_x. \tag{2.17}$$

Here ν_d is the coefficient of viscosity, σ is the surface density of the disc and Ω is the disc's rotation frequency at $\tilde{\omega}$. The expression is evaluated at $\tilde{\omega} = R_x$ and $\Im_x$ is the viscous torque exerted by material beyond $\tilde{\omega} = R_x$ on material interior to it.

Mass conservation requires that

$$\dot{M}_* = \dot{M}_\text{d} - \dot{M}_\text{w}. \tag{2.18}$$

We follow Shu et al. (1994a) in making the bold assumption that Γ_x in (2.12) is a constant and also assume that

$$\frac{\dot{R}_*}{R_*} = \frac{\dot{M}_*}{M_*} \tag{2.19}$$

(e.g. Palla and Stahler 1991). Here, as elsewhere in this chapter, the dot above a quantity implies that the derivative with respect to time should be taken. Use of (2.11), (2.12) and (2.19) gives

$$\frac{\dot{\Omega}_*}{\Omega_*} = -\frac{\dot{M}_*}{M_*}. \tag{2.20}$$

If b in (2.15) is taken to be constant, (2.15), (2.12), (2.19) and (2.20) yield

$$2b\dot{M}_* = (1-\tau)\dot{M}_\text{d}\Gamma_x^2 - \bar{J}\dot{M}_\text{w}\Gamma_x^2. \tag{2.21}$$

Then, (2.18), (2.21) with (2.14) lead to

$$f = \frac{1 - \tau - 2b/\Gamma_x^2}{\bar{J} - 2b/\Gamma_x^2}. \tag{2.22}$$

We may take $b/\Gamma_x^2 \ll 1$ for most situations, but the value of τ is not known, because the origin of the viscosity of an accretion disc is not known. Shu et al. (1994b) and Najita and Shu (1994) assumed that $\tau = 0$ to obtain estimates for f from their solutions to the two-dimensional, time independent, single fluid MHD equations, which we describe in the next section.

3 The Equations Governing the X-wind

3.1 Axisymmetric MHD flow in a Frame Rotating with the Star

The conservation of mass in a flow viewed in an inertial frame is ensured by the requirement that

$$\frac{\partial \rho}{\partial t} + \nabla \cdot (\rho \boldsymbol{v}) = 0 \tag{3.1}$$

where ρ, $\boldsymbol{v}$ and t are the mass density, velocity and time. It requires that the time rate at which the mass contained in a volume changes is the negative of the flux of the mass carried through the surface enclosing that volume. The ith component of the differential equation of motion in an inertial frame is

$$\frac{\partial (\rho v_i)}{\partial t} + \sum_j \nabla_j \cdot (\rho v_i v_j) = f_{di} \tag{3.2}$$

where $(\nabla_j \cdot)$ is used to indicate the jth component of divergence and f_{di} is the ith component of the force per unit volume acting on the fluid. Equation (3.2) states that the time rate of change of the ith component of the momentum carried by the fluid in a volume is the negative of the ith component of the momentum flux through the surface of that volume, plus the ith component of the force acting on the fluid in that volume.

$\boldsymbol{f}_d$ has various contributions. One arises from the variation of the thermal pressure P; that contribution is $-\nabla P$. Another arises if a nonuniform magnetic field is present; from the Lorentz force law it follows that that contribution to $\boldsymbol{f}_d$ is $(\boldsymbol{J} \times \boldsymbol{B})/c$ where $\boldsymbol{B}$ is the magnetic field, $\boldsymbol{J}$ is the electric current and c is the speed of light. If we consider sufficiently low frequency phenomena, we can neglect the displacement current and take $\nabla \times \boldsymbol{B} = 4\pi \boldsymbol{J}/c$; hence, the magnetic field gives rise to a contribution to $\boldsymbol{f}_d$ of $(\nabla \times \boldsymbol{B}) \times \boldsymbol{B}/4\pi$. We assume that the charge density of the system is zero, so that the electric field is zero. Gravity contributes a term $\rho \boldsymbol{g}$, where $\boldsymbol{g}$ is the gravitational field, to $\boldsymbol{f}_d$.

As we have assumed the gas to be isothermal, $P = \rho a_x^2$. For a perfectly conducting medium with no electric field

$$\boldsymbol{v} \times \boldsymbol{B} = 0. \tag{3.3}$$

This equation does not hold in the disc where mass is loaded onto the field lines. Gauss's law is
$$\nabla \cdot \boldsymbol{B} = 0. \tag{3.4}$$

Because we investigate steady flows, we will take the derivatives with respect to time to be zero. Also, it is convenient to work in a frame rotating about the symmetry axis, with an angular frequency of Ω_*. This choice of frame does not affect the equation of mass conservation, and from (3.1) we find that

$$\nabla \cdot (\rho \boldsymbol{v}) = 0. \tag{3.5}$$

"Fictional" contributions must be added to $\boldsymbol{f}_d$ due to the frame's rotation (e.g. Batchelor 1967); their contributions to $\boldsymbol{f}_d$ are

$$-2\rho\Omega_*(\hat{\boldsymbol{z}} \times \boldsymbol{v}) - \rho\Omega_*^2 \hat{\boldsymbol{z}} \times (\hat{\boldsymbol{z}} \times \boldsymbol{v}),$$

where the first term is the Coriolis force per unit volume and the second is the centrifugal force per unit volume.

Collecting the various real and "fictional" contributions to $\boldsymbol{f}_d$, making use of (3.5) and using

$$(\boldsymbol{v} \cdot \nabla)\boldsymbol{v} = \nabla(\boldsymbol{v} \cdot \boldsymbol{v})/2 + (\nabla \times \boldsymbol{v}) \times \boldsymbol{v},$$

we find that the time independent version of (3.2) may be written as

$$\nabla \frac{1}{2}|\boldsymbol{v}|^2 + (2\hat{\boldsymbol{z}} + \nabla \times \boldsymbol{v}) \times \boldsymbol{v} = -\frac{\epsilon^2}{\rho}\nabla\rho - \nabla V_{\text{eff}} + \frac{1}{\rho}(\nabla \times \boldsymbol{B}) \times \boldsymbol{B} \tag{3.6}$$

where we have adopted R_x, $\Omega_* R_x$, $\dot{M}_w/4\pi R_x^3 \Omega_*$ and $(\Omega_* \dot{M}_w/R_x)^{1/2}$ as the fiducial units of length, speed, density and magnetic field strength, and

$$V_{\text{eff}} = -\frac{1}{r} - \frac{1}{2}\tilde{\omega}^2 + \frac{3}{2}. \tag{3.7}$$

3.2 The Stream Function and Recasting the Equations

Equations (3.3)–(3.7) govern the X-wind flow. Equation (3.3) is satisfied if we write
$$\boldsymbol{B} = \beta\rho\boldsymbol{v}. \tag{3.8}$$

Equations (3.4) and (3.5) require that
$$\boldsymbol{v} \cdot \nabla \beta = 0 \tag{3.9}$$

i.e. that β is constant on a streamline. It is useful to introduce a stream function, $\Psi(\tilde{\omega}, z)$, such that

$$\rho v_{\tilde{\omega}} = \frac{1}{\tilde{\omega}}\frac{\partial \Psi(\tilde{\omega}, z)}{\partial z} \tag{3.10}$$

and

$$\rho v_z = -\frac{1}{\tilde{\omega}}\frac{\partial \Psi(\tilde{\omega}, z)}{\partial \tilde{\omega}}. \tag{3.11}$$

As we take the flow to be axisymmetric, (3.9) and (3.10) imply that (3.5) is automatically satisfied. They also imply that

$$\boldsymbol{v}.\nabla\Psi = 0 \tag{3.12}$$

which means that the lines of constant Ψ define streamlines in the meridional plane. From (3.9) and (3.12) we see that β depends only on Ψ.

We now show that (because we are expressing quantities in terms of the fiducial units listed below eqn (3.6)) the requirement that the mass loss rate is $\dot{M}_w$ automatically implies that for $z \geq 0$, Ψ takes values from 0 to 1. For the fiducial units adapted, the mass loss rate is 4π. Thus, if $v_{\tilde{\omega}}(\tilde{\omega}, z) = v_{\tilde{\omega}}(\tilde{\omega}, -z)$, then for each $\tilde{\omega}$

$$\begin{aligned} 4\pi &= 2\int_0^{z_e(\tilde{\omega})} (2\pi\tilde{\omega})\rho v_{\tilde{\omega}} dz \\ &= 4\pi \int_0^{z_e(\tilde{\omega})} \frac{\partial \Psi}{\partial z} dz = 4\pi \Psi(\tilde{\omega}, z_e) \end{aligned} \tag{3.13}$$

where $z_e(\tilde{\omega})$ is the (limiting) streamline having the highest value of z. In fact, $\Psi(\tilde{\omega}, z)$ gives the fraction of the mass loss at $\tilde{\omega}$ occurring between $-z$ and z.

If we take the dot product of $\boldsymbol{v}$ and eqn (3.6), we obtain Bernoulli's theorem

$$\boldsymbol{v}.\nabla\left(\frac{1}{2}|\boldsymbol{v}|^2 + \epsilon^2 \ln \rho + V_{\text{eff}}\right) = 0 \tag{3.14}$$

from which it follows that

$$\frac{1}{2}|\boldsymbol{v}|^2 + \epsilon^2 \ln \rho + V_{\text{eff}} = H(\Psi) \tag{3.15}$$

where H depends only on Ψ.

Equations (3.5) and (3.14) may be used to find

$$(|\boldsymbol{v}|^2 - \epsilon^2)\nabla.\boldsymbol{v} = |\boldsymbol{v}|^3 \nabla.\left(\frac{\boldsymbol{v}}{|\boldsymbol{v}|}\right) - \boldsymbol{v}.\nabla V_{\text{eff}}. \tag{3.16}$$

If the subsonic to supersonic transition (ϵ is the isothermal sound speed in the system of fiducial units) takes place smoothly, the right hand side of (3.16) must be zero at the sonic surface.

Equations (3.6), (3.8) and (3.15) yield

$$\frac{dH}{d\Psi}\nabla\Psi + [z\hat{z} + \nabla\times\boldsymbol{v} - \beta\nabla\times(\beta\rho\boldsymbol{v})]\times\boldsymbol{v} = 0. \tag{3.17}$$

The $\hat{\phi}$ component (ϕ is the angular variable in the cylindrical coordinate system adopted) of (3.17) implies that

$$\boldsymbol{v}.\nabla J = 0 \tag{3.18}$$

where J (which throughout the rest of this paper should not be associated with the magnitude of the electric current) is defined by

$$J \equiv \tilde{\omega}^2 + \tilde{\omega}(1 - \beta^2 \rho) v_\phi \tag{3.19}$$

and is the sum of the $\hat{z}$-components of the specific angular momentum carried by the gas in an inertial frame and the specific angular momentum carried by the magnetic field (in a torsional Alfvén wave). (3.18) implies that J is a function of Ψ only. The $\hat{\tilde{\omega}}$ and $\hat{z}$ components of (3.17) require that

$$\frac{\omega}{\tilde{\omega}\rho} + \frac{dJ}{d\Psi}\frac{v_\phi}{\tilde{\omega}} + \frac{d\beta}{d\Psi}\beta\rho|v|^2 - \frac{dH}{d\Psi} = 0 \tag{3.20}$$

where

$$\omega \equiv \frac{\partial}{\partial z}[(1-\beta^2\rho)v_{\tilde{\omega}}] - \frac{\partial}{\partial \tilde{\omega}}[(1-\beta^2\rho)v_z] \tag{3.21}$$

which is the ϕ component of the vorticity of the gas, plus a similar contribution from the magnetic field. Equations (3.10), (3.11), (3.20) and (3.21) yield

$$\frac{1}{\tilde{\omega}}\frac{\partial}{\partial \tilde{\omega}}\left(\tilde{\omega} A \frac{\partial \Psi}{\partial \tilde{\omega}}\right) + \frac{\partial}{\partial z}\left(A \frac{\partial \Psi}{\partial z}\right) = Q \tag{3.22}$$

where

$$A \equiv \frac{\beta^2 \rho - 1}{\tilde{\omega}^2 \rho} \tag{3.23}$$

and

$$Q \equiv \rho \left[\frac{dJ}{d\Psi}\frac{v_\phi}{\tilde{\omega}} + \frac{d\beta}{d\Psi}\beta\rho|v|^2 - \frac{dH}{d\Psi}\right]. \tag{3.24}$$

From (3.23) and (3.19) we can write

$$\rho = \frac{1}{\beta^2 - \tilde{\omega}^2 A} \tag{3.25}$$

and

$$v_\phi = -\frac{1}{\tilde{\omega}}\xi(\beta^2 - \tilde{\omega}^2 A) \tag{3.26}$$

with

$$\xi \equiv \frac{1}{A}\left(\frac{J}{\tilde{\omega}^2} - 1\right). \tag{3.27}$$

From (3.15) and (3.21) through (3.26) we obtain

$$\nabla \cdot (A \nabla \Psi) = Q \tag{3.28}$$

and

$$(\beta^2 - \tilde{\omega}^2 A^2)[|\nabla \Psi|^2 + \xi^2] + 2\tilde{\omega}^2 \left[V_{\text{eff}} + \epsilon^2 \ln\left(\frac{\epsilon^2 h}{\beta^2 - \tilde{\omega}^2 A}\right)\right] = 0 \tag{3.29}$$

with Q rewritten as

$$Q = -\frac{\xi}{\tilde{\omega}^2}\frac{dJ}{d\Psi} + (|\nabla\Psi|^2 + \xi^2)\frac{\beta}{\tilde{\omega}^2}\frac{d\beta}{d\Psi} + \left(\frac{\epsilon^2}{\beta^2 - \tilde{\omega}^2 A}\right)\frac{1}{h}\frac{dh}{d\Psi} \qquad (3.30)$$

and h is defined by

$$H \equiv -\epsilon^2 \ln(\epsilon^2 h). \qquad (3.31)$$

If the three arbitrary functions $\beta(\Psi)$, $h(\Psi)$, $J(\Psi)$ and boundary conditions are specified, then equations (3.28) and (3.29) can be solved for $\Psi(\tilde{\omega}, z)$ and $A(\tilde{\omega}, z)$. These three specified and two calculated functions can then be used with the analysis given in this subsection to obtain the functions ρ, $\boldsymbol{v}$ and $\boldsymbol{B}$ that solve equations (3.3) through (3.6). In reality, physical processes occurring outside the regions where the adopted governing equations are valid determine what $\beta(\Psi)$, $h(\Psi)$ and $J(\Psi)$ are. For instance, the nature of the magnetic diffusivity in the disc between $\tilde{\omega} = R_t$ and $\tilde{\omega} = R_x + \Delta_x$ and near $z = 0$ governs the behaviour of $\beta(\Psi)$, which basically describes the amount of mass loaded on to different magnetic field lines.

In another approach, $h(\Psi)$ would be specified with the same appropriate boundary conditions mentioned above and the surface $\tilde{\omega}_A(z)$ on which $A = 0$ would be specified. The significance of that surface is realized if one notes that $\beta^2 \rho$ is the inverse of the square of the Alfvén Mach number (the ratio of the fluid speed to the Alfvén speed), and that $\beta^2 \rho = 1$ implies that $A = 0$. The specification of the surface on which $A = 0$ places restrictions on $J(\Psi)$, since from (3.19) we see that $J = \tilde{\omega}_A^2(z)$ on it. In this approach, $\beta(\Psi)$ as well as Ψ and A would be calculated through the solution of (3.28) and (3.29). In fact, this approach does not yield a unique solution for $\beta(\Psi)$; rather a range of $\beta(\Psi)$s lead to solutions that satisfy the boundary conditions on the Alfvén surface (Shu et al. 1995). Najita and Shu (1994), whose work is described in the next section, chose a numerical approach for obtaining the values of Ψ and β as functions of position on the Alfvén surface that yields a $\beta(\Psi)$ which is near the maximum allowed, ensuring that the resultant $\beta(\Psi)$ and other functions given by the solution are consistent with the flow making a smooth transition through the fast-mode magnetosonic surface (Shu et al. 1995).

4 The Supersonic, Sub-Alfvénic Region of an X-wind

4.1 Boundary Conditions

Three boundary conditions that can be applied, in either of the solution approaches mentioned in the last two paragraphs of the previous section, are:

$$\Psi = 0 \text{ for } z = 0 \qquad (4.1)$$

which can be used only if ϵ is very small, so that the thermal pressure of the disc is so small that it is thin,

$$\Psi = 1 \text{ where } \frac{1}{2}(\beta\rho|\boldsymbol{v}|)^2 = P_{\text{ext}} + \frac{1}{2}|\boldsymbol{B}_{\text{ext}}|^2 \qquad (4.2)$$

with P_{ext} and $\boldsymbol{B}_{ext}$ being the pressure and magnetic field associated with an ordinary wind or the dead zone, and

$$\nabla A.\nabla \Psi = Q \quad \text{on the surface where} \quad A = 0 \tag{4.3}$$

which guarantees that the flow through the Alfvén surface is smooth.

Equation (4.2) limits the angular extent of the X-wind regions. In reality, P_{ext} and $\boldsymbol{B}_{ext}$ are not always known. Hence, in practice a curve, $z_e(\tilde{\omega})$, on which $\Psi = 1$, was specified in each of several calculations.

In practice, Najita and Shu (1994) specified $z_e(\tilde{\omega})$ and the surface on which $A = 0$, as described in the last paragraph of the previous subsection. Shu et al. (1994b) introduced a special set of curvilinear coordinates, q and t (which from this point on will designate the spatial coordinate rather than time), to simplify the specification procedure. To define q and t we first define s and θ, such that

$$s \cos \vartheta = \tilde{\omega} - 1 \tag{4.4}$$
$$s \sin \vartheta = z. \tag{4.5}$$

$\vartheta_x(s)$ is the curve in this coordinate system on which $\Psi = 1$. In the (q,t) coordinate system, the $t = 1$ curve is defined to be the one on which $\Psi = 1$, and

$$t(s,\vartheta) \equiv \frac{\vartheta}{\vartheta_x(s)}. \tag{4.6}$$

The curves of constant q were chosen to be orthogonal to those of constant t; hence,

$$\nabla q.\nabla t = 0. \tag{4.7}$$

For $\vartheta = 0$, $q = s$.

Najita and Shu (1994) always took the Alfvén surface (the one for which $A = 0$) to be the one on which q is a constant, q_A, which was varied from calculation to calculation. The inner boundary was taken to be $q = 0.05$, which typically corresponds to a surface somewhat outside the surface at which the subsonic to supersonic transition occurs.

The boundary condition for Ψ at $q = 0.05$ was specified by the method of matched asymptotic expansions. For $\epsilon \ll 1$, analytic approximations for the solution in the subsonic region of the wind flow were extrapolated to large values of s/ϵ and analytic approximations for the solution in the supersonic but sub-Alfvénic region of the flow were extrapolated to small values of s. The requirement that the large s/ϵ extrapolation of one solution matches the small s extrapolation of the other yields constraints on the solution in the supersonic but sub-Alfvénic region, which can be expressed as a boundary condition on Ψ at a surface on which $q \ll 1$.

4.2 Solutions

Najita and Shu (1994) have solved equations (3.28) and (3.29) in the supersonic but sub-Alfvénic region, in the appropriate (q,t) coordinate systems defined

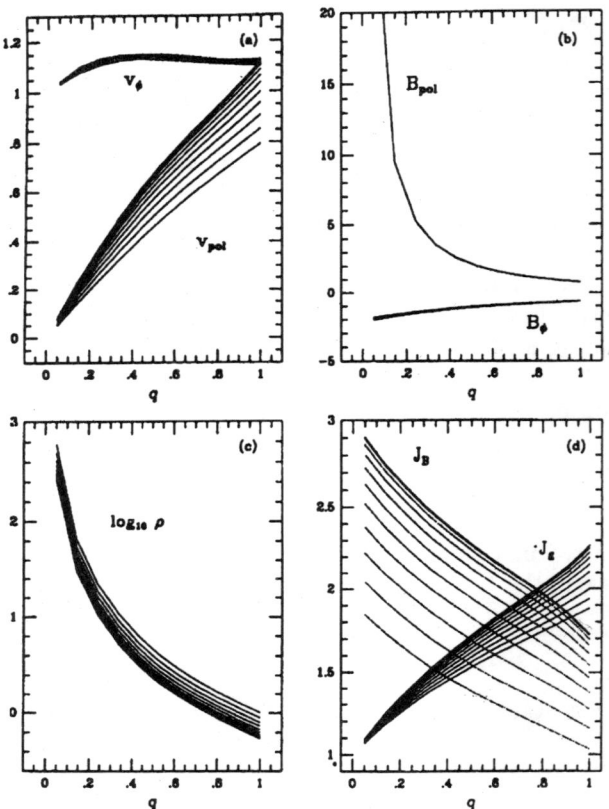

FIG. 3. The derived quantities for the flow in the supersonic but sub-Alfvénic region in a model of an X-wind. a) The poloidal and toroidal components of the velocity. b) The poloidal and toroidal components of the magnetic field. c) The density. d) The specific angular momenta in the gas and in the magnetic field. In a), b) and d) the curves are for $t = 0, 0.1, 0.2, \ldots, 1.0$. Quantities are expressed in the fiducial units defined after (3.6). From Najita and Shu (1994).

in subsection 4.1. They imposed the boundary conditions described in that subsection and restricted their studies to situations in which $\epsilon \ll 1$ and took $\epsilon^2 \ln[\epsilon^2 h/(\beta^2 - \tilde{\omega}^2 A)]$ and $[\epsilon^2/(\beta^2 - \tilde{\omega}^2 A)h][dh/d\Psi]$ to be negligibly small.

Figure 3 displays results for $v_{\text{pol}} \equiv (v_{\tilde{\omega}}^2 + v_z^2)^{1/2}$, $B_{\text{pol}} \equiv (B_{\tilde{\omega}}^2 + B_z^2)^{1/2}$, v_ϕ (as measured in an inertial frame), B_ϕ, ρ, J_g (the specific angular momentum carried by the gas) and J_B (the specific angular momentum carried by the magnetic field in a torsional Alfvén wave). The results are for $\vartheta_x(s) = 45°$, for all values of s (i.e. $\Psi = 1$ on the line on which $\tilde{\omega} - 1 = z$) and $q_A = 1$. The quantities are all expressed in fiducial units. The poloidal acceleration and the transfer of angular momentum from the magnetic field to the gas are clearly demonstrated. The value of $\bar{J}$ for this model is 3.607, which according to Shu et al. (1994a) and Najita and Shu (1994) can be used in eqn (2.22) to obtain the fraction of the mass that flows through the disc which is lost in an X-wind, for which the set of input model parameters are appropriate. Najita and Shu (1994) also presented results for other model parameters.

5 The Magnetic Field in the Funnel Flow and Dead Zone Regions

We are interested in the properties of the X-wind solution at large distances (relative to R_x) from the star. Before treating them, we must digress to discuss briefly the magnetic field in the dead zone and funnel flow region (cf. the last paragraph of subsection 2.1). As the funnel flow is likely to be very sub-Alfvénic everywhere and the density in the dead zone is very small, and everywhere thermal pressure is small compared to the magnetic pressure, eqn (3.6) yields that approximately

$$(\nabla \times \boldsymbol{B}_p) \times \boldsymbol{B}_p = 0 \qquad (5.1)$$

in the funnel flow region and the dead zone (Ostriker and Shu 1995). $\boldsymbol{B}_p$ is the poloidal component of the magnetic field, the sum of the $\hat{\tilde{\omega}}$ and $\hat{z}$ components of $\boldsymbol{B}$. Ostriker and Shu (1995) have argued further that more specifically

$$\nabla \times \boldsymbol{B}_p = 0 \qquad (5.2)$$

in the dead zone and funnel flow region, which is more restrictive than (5.1).

They introduced a flux function Φ related to the stream function by

$$d\Phi = -\beta d\Psi \qquad (5.3)$$

and from which $\boldsymbol{B}_p$ can be calculated,

$$\boldsymbol{B}_p = -\frac{\hat{e}_\phi \times \nabla \Phi}{\tilde{\omega}}. \qquad (5.4)$$

From (5.2) and (5.4) it follows that

$$\nabla \cdot \left(\frac{\nabla \Phi}{\tilde{\omega}^2}\right) = 0 \qquad (5.5)$$

in the dead zone and in the funnel flow.

They solved (5.5) for the boundary conditions

$$\Phi = 0 \text{ when } \tilde{\omega} = 0 \text{ or } z = \infty \qquad (5.6)$$

$$\frac{\partial \Phi}{\partial z} = 0 \text{ on } z = 0 \qquad (5.7)$$

$$\Phi = \Phi_e = \text{constant on } z = z_e(\tilde{\omega}) \tag{5.8}$$

$$\Phi = \frac{\Phi_{dx}\tilde{\omega}^2}{(\tilde{\omega}^2 + z^2)^{3/2}} \text{ as } \tilde{\omega}^2 + z^2 \to 0 \tag{5.9}$$

$$\Phi = \Phi_e(\vartheta/\vartheta_e) \text{ for } s \to 0 \text{ and } \vartheta_e \leq \vartheta \leq \pi. \tag{5.10}$$

Equation (5.6) sets the zero of the flux function along the z-axis; the field line leaving the pole of the star attaches itself to its open counterpart in the wind region only at $z = \infty$. Reflection symmetry across the $z = 0$ plane requires (5.7). Condition (5.8) is analogous to the requirement (4.2) on Ψ for the wind solution, while (5.9) requires that the poloidal components of the magnetic field match an unperturbed stellar dipole field at small distances from the star. (Φ_{dx} is the non-dimensional flux of the unperturbed dipole at the $\tilde{\omega} = 1$ and $z = 0$ point. Condition (5.10) requires that the field fans out between the upper wind streamline (where $\Psi = 1$ and $\vartheta = \vartheta_e(s)$) and the last flux tube on which $\Phi = \Phi_t \equiv (\pi/\vartheta_e)$.

Once the solution for Φ was obtained for these boundary conditions, $\beta(\Phi)$ could be somewhat arbitrarily specified, but it is subject to the restriction that

$$\Delta\Phi_{\text{fun}} = \int_0^{1-f} \beta d\Psi \equiv \beta_{\text{fun}}(1-f) \tag{5.11}$$

where $\Delta\Phi_{\text{fun}}$, β_{fun} and $(1-f)$ are, respectively, the flux, the streamline averaged value of β and the fraction of the disc mass accretion rate contained in the funnel (cf. eqn (2.14)).

From the calculation defined in this subsection one finds that for large distances from the star at the boundary between the wind and funnel flow/dead zone region, the magnetic pressure on the side of the boundary away from the wind is dominated by B_z and that

$$B_z \to \frac{2\bar{\beta}}{r^2 \sin^2 \theta_1} \tag{5.12}$$

where $\bar{\beta}$ is the average of β over streamlines; $\theta_1(r)$ gives the location of the $\Psi = 1$ streamline (the outer edge of the wind) in a spherical coordinate system (r, θ, ϕ), in which $\tilde{\omega} = r \sin \theta$ and $z = r \cos \theta$.

6 The X-wind at Large Distances from the Star

The reason for the digression that Section 5 represents is that we wish to determine the location of the boundary between the dead zone and the wind at large distances from the star, from the condition that the magnetic pressure on either side of the boundary equals that on the other. To complete this determination we must estimate the magnetic pressure in the wind at large distances from the star. Hence, we now return to the properties of the wind in the region where $r \gg 1$ and $r \sin \theta$ is at least moderately large as well. In this region, ρ varies as

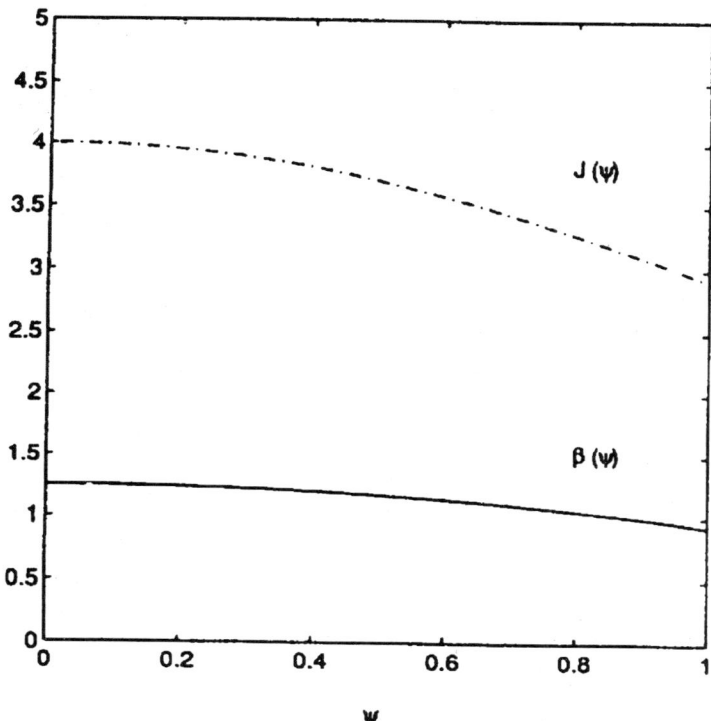

FIG. 4. Adopted forms for $J(\Psi)$ and $\beta(\Psi)$ used in the calculation of the flow properties in the super-Alfvénic region of the wind. From Shu et al. (1995).

r^{-2} times a function that varies only weakly with r and depends on θ. Also

$$A \approx -\frac{1}{r^2 \rho \sin^2 \theta} \qquad (6.1)$$

and varies only weakly with r and $\xi \approx -1/A$. In addition Ψ depends on θ, but only weakly with r. Hence, (3.28) is well approximated by

$$\frac{1}{r^2 \sin \theta} \frac{\partial}{\partial \theta}\left(A \sin \theta \frac{\partial \Psi}{\partial \theta}\right) = \frac{1}{r^2 \sin^2 \theta}\left(\frac{1}{A}\frac{\mathrm{d}J}{\mathrm{d}\Psi} + \frac{\beta}{A^2}\frac{\mathrm{d}\beta}{\mathrm{d}\Psi}\right). \qquad (6.2)$$

Multiplication of (6.1) by $2Ar^2 \sin^2 \theta \, \partial \Psi/\partial \theta$ yields

$$\frac{\partial}{\partial \theta}\left[\left(A \sin \theta \frac{\partial \Psi}{\partial \theta}\right) - 2J\right] = \frac{1}{A}\frac{\partial}{\partial \theta}(\beta^2). \qquad (6.3)$$

For large r, eqn (3.29) is well approximated by

$$A \sin \theta \frac{\partial \Psi}{\partial \theta} - 2J = -3 + \frac{2\beta^2}{A}. \qquad (6.4)$$

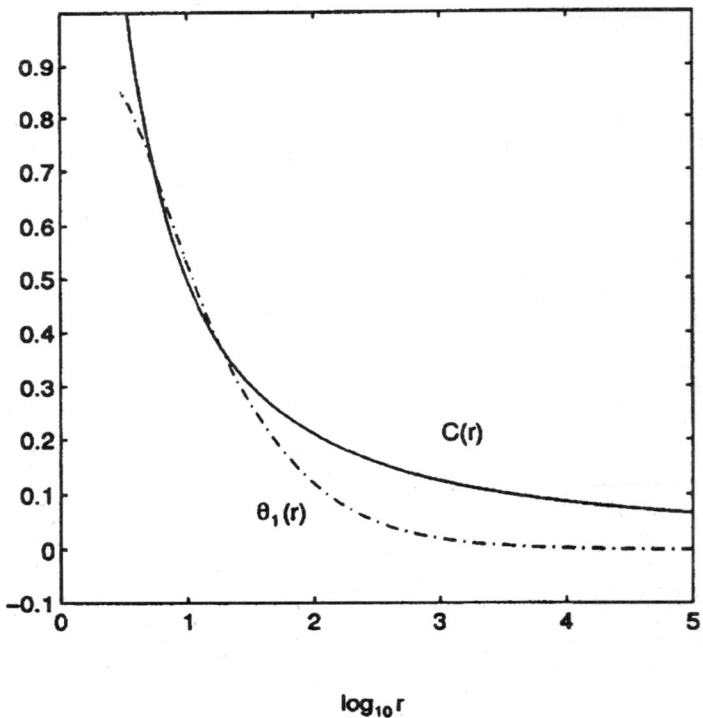

FIG. 5. Derived values of $C(r)$ and $\theta_1(r)$ for the adopted values of $J(\Psi)$ and $\beta(\Psi)$ shown in Fig. 4. From Shu et al. (1996).

Equations (6.3) and (6.4) give

$$\frac{\partial}{\partial \theta}\left(\frac{2\beta^2}{A}\right) = \frac{1}{A}\frac{\partial}{\partial \theta}(\beta^2) \tag{6.5}$$

which integrates to give

$$A = -\frac{\beta}{C} \tag{6.6}$$

where C is an arbitrary, but slowly varying positive function of r. Equations (6.1) and (6.6) give

$$\rho = \frac{C}{\beta r^2 \sin^2 \theta}. \tag{6.7}$$

The substitution of eqn (6.7) into eqn (6.4) and integration gives

$$F(C, \Psi) = -\ln[\tan(\theta/2)] \tag{6.8}$$

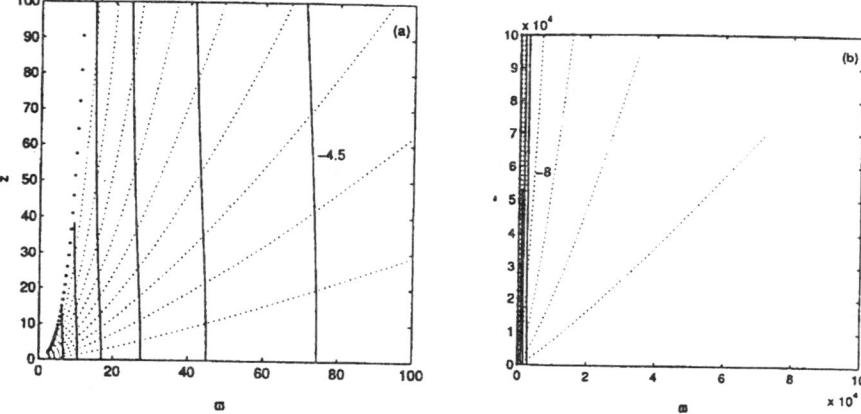

FIG. 6. Isodensity contours (solid curves) and streamlines (dotted curves) for the super-Alfvénic region of an X-wind. The first panel shows results for z and $\tilde{\omega}$ up to 100, while the second gives results for z and $\tilde{\omega}$ up to 10^4. The labels on the rightmost isodensity contour in each panel give $\log_{10}\rho$ where ρ is expressed in fiducial units. Successive isodensity contours correspond to increases of 0.5 in $\log_{10}\rho$. The $\Psi = 0$ streamline is not plotted and lies just above the $z = 0$ line. From right to left the $\Psi = 0.1, 0.2, 0.3, \ldots, 1.0$ streamlines are shown; 10 per cent of the mass loss in the region shown occurs between any two of the plotted streamlines. The $\Psi = 1$ streamline bounds the X-wind to the left.

where

$$F(C, \Psi) \equiv \frac{1}{C} \int_0^\Psi \frac{\beta(\Psi)}{[2J(\Psi) - 3 - 2C\beta(\Psi)]^{1/2}} d\Psi. \qquad (6.9)$$

From $\tan(\theta/2) = (1 - \cos\theta)/\sin\theta$, it follows that eqn (6.8) may be rewritten as

$$\sin\theta = \text{sech}[F(C, \Psi)]. \qquad (6.10)$$

In an inertial frame at $\tilde{\omega} \gg 1$, the super-Alfvénic wind hardly rotates, which implies that in the frame co-rotating with the star, $v_\phi \approx -\tilde{\omega}$, which along with (3.8) and (6.7) implies that in the wind

$$B_\phi \approx -\frac{C}{r\sin\theta}. \qquad (6.11)$$

(In the wind, B_ϕ drops off more rapidly than this.) Equating the squares of the right hand sides of (5.12) and (6.11), we find

$$\sin\theta_1 = \frac{2\bar{\beta}}{Cr}. \qquad (6.12)$$

Using (6.10) and (6.12), we find that $C(r)$ can be calculated from

$$\frac{1}{C}\cosh[F(C,1)] = \frac{r}{2\bar{\beta}}. \tag{6.13}$$

Figure 4 shows $\beta(\Psi)$ and $J(\Psi)$ for a model very similar to that for which Fig. 3 presents results. Figure 5 gives the functions $C(r)$ and $\theta_1(r)$, calculated for the $\beta(\Psi)$ and $J(\Psi)$ of Fig. 4.

From the derived $C(r)$, $\beta(\Psi)$ and $J(\Psi)$, the streamlines can be calculated with (6.8) and (6.9). From $\Psi(r,\theta)$, $\beta(\Psi)$ and (6.7), one can calculate $\rho(r,\theta)$. Figure 6 shows constant density contours and streamlines for the $\beta(\Psi)$ and $J(\Psi)$ of Fig. 4. It is clear from Fig. 6 that the density increases dramatically in the spatial region in which Ψ approaches 1.

Thus, the X-wind model suggests that a bipolar wind with a substantial opening angle at $\tilde{\omega} \approx 1$ and carrying mass at a non-negligible fraction of the accretion rate of the disc, narrows into a jet-like feature. The winds are important molecular sources themselves (cf. Chapter 7); they drive shocks into molecular regions (cf. Chapter 8) and collimate into jets like those associated with Herbig Haro objects (cf. Chapter 10).

Bibliography

1. Bachiller, R. (1996). *ARA&Ap*, **34**, 111.
2. Batchelor, G. K. (1967). In *An introduction to fluid dynamics*, Cambridge University Press, Cambridge.
3. Bouvier, J. Cabrit, S., Fernandez, M., Martin, E. L. and Matthews, J. M. (1993). *A&A*, **272**, 176.
4. Ghosh, P. and Lamb, F. K. (1979). *ApJ*, **232**, 259.
5. Hartmann, L. and MacGregor, K. B. (1982). *ApJ*, **259**, 180.
6. Krall, N. A. and Trivelpiece, A. W. (1973). In *Principles of plasma physics*, McGraw-Hill, New York.
7. Li, Z.-Y. and Shu, F. H. (1996). *ApJ*, **468**, 261.
8. Lucy, L. B. and Abbott, D. C. (1993). *ApJ*, **405**, 738.
9. Najita, J. and Shu, F. H. (1994). *ApJ*, **429**, 808.
10. Ostriker, E. and Shu, F. H. (1995). *ApJ*, **447**, 813.
11. Palla, F. and Stahler, S. W. (1991). *ApJ*, **375**, 288.
12. Shu, F. H., Najita, J., Ostriker, E., Wilkin, F., Ruden, S. and Lizano, S. (1994a). *ApJ*, **429**, 781.
13. Shu, F. H., Najita, J., Ruden, S. and Lizano, S. (1994b). *ApJ*, **429**, 797.
14. Shu, F. H., Najita, J., Ostriker, E. and Shang, H. (1995). *ApJ*, **455**, L155.
15. Shu, F. H., Najita, J., Ostriker, E. and Shang, H. (1996). *ApJ*, **459**, L43.

7
Chemistry in the Winds of Young Stellar Objects

A. E. Glassgold
Physics Department, New York University

1 Introduction

The realization that powerful winds accompany the formation of stars has played an important role in understanding how stars form (e.g. Shu, Adams and Lizano 1987; Lada and Shu 1990). Although considerable evidence for mass loss from young stellar objects (YSOs) existed before 1980, the discovery of bipolar molecular outflows (Snell, Loren and Plambeck 1980; the review by Lada 1985; Snell 1987; Fukui *et al.* 1993; Bachiller 1996) was a turning point in star formation research. Some of the striking phenomena related to the bipolar outflows are forbidden lines with asymmetric lineshapes, Herbig–Haro objects (cf. Chapter 10), and jets (reviewed by Edwards, Ray and Mundt 1993). When we also consider that low-mass stars (our primary concern here), form in dense cloud cores (cf. Chapter 5) and that much of the matter is transferred to one through an accretion disc, we can appreciate that the formation of a star involves a number of interacting systems with different size scales and physical properties.

Dynamics important in the formation of a low-mass star involve the inner accretion disc and associated flows and fields. This region is not much larger than $20\,R_\odot$, less than one milli-arcsecond for the nearest star-forming region (140 pc distant), and thus not susceptible to direct imaging with current techniques. In these circumstances, spectroscopic and chemical diagnostics of small scale dynamical phenomena play an important role in star formation studies. Understanding the physics and chemistry of the inner regions of a YSO is a prerequisite for interpreting such observations. Winds are especially interesting in this context since they carry information from the centre of activity out to distances where their effects can be observed with existing instrumentation of radio and infrared astronomy.

The central role of the winds of YSOs dictates that any viable theory of star formation be able to account for the occurrence of outflows. Shu and his collaborators have developed a promising theory of low-mass star formation that is based on a magnetocentrifugally driven wind called the *X-wind* (Shu *et al.* 1994a; also see Chapter 6). This model has the potential to explain all of the observed outflow phenomena, including jets and Herbig–Haro objects. It is therefore important to test this theory as close to the origin of the wind as possible.

The dynamical properties of the X-wind are largely independent of the physical conditions as long as a sufficient level of ionization exists. Thus the calculation of the physical, chemical, and diagnostic properties of the X-wind can be separated from the dynamics, which can be regarded as "given". Although I will often adopt the perspective of the X-wind model, most of the general conclusions about these properties apply to any theory of a wind that emanates from the central region of a YSO.

The physical conditions of YSO winds are perhaps more diverse than any other type of interstellar or circumstellar material, ranging from hot atomic to cool molecular regions. Bipolarity of course means that they are not spherically symmetric, although they are likely to have a high degree of cylindrical symmetry. One of their most important characteristics is that the particle density in the inner region is much higher than in interstellar clouds, and can even approach stellar densities. The wind velocity is also very high, of order several hundred kilometres per second, and shocks and boundary layers become important when the wind interacts with other parts of the star-forming complex or with ambient material. The winds are intrinsically magnetohydrodynamic, and are accompanied by processes characteristic of magnetically active stars, e.g. X-rays and high-energy particles (stellar "cosmic rays"). Only a small part of the fascinating physics and chemistry of YSO winds has actually been explored so far. I emphasize published work on relatively simple models, but I will also point out their limitations and try to indicate fruitful directions for further research. The main focus will be on the flows near the active centre of star formation, rather than on the effects that they generate at much larger distances.

In the next section I will discuss earlier chemical modelling of YSO winds, emphasizing the important roles of temperature and dynamics. In Section 3 I will sketch the next level of modelling, based on the X-wind dynamics of Shu and collaborators. The chapter ends with a brief set of conclusions.

2 Previous Studies

2.1 Early Calculations

Molecular synthesis in YSO winds proceeds in three or more stages: (1) the formation of molecular hydrogen, (2) the creation of simple heavy-atom radicals, notably OH, and (3) the formation of CO, other well-bound diatomics, and more complex species. The formation of H_2 in a dust-free environment can be accomplished by three pathways that involve the H^- negative ion, the H_2^+ ion, and three-body reactions (cf. Chapter 3):

$$e + H \rightarrow H + h\nu, \quad H^- + H \rightarrow H_2 + e; \tag{2.1}$$

$$H^+ + H \rightarrow H_2^+ + h\nu, \quad H_2^+ + H \rightarrow H_2 + H^+; \tag{2.2}$$

$$H + H + H \rightarrow H_2 + H. \tag{2.3}$$

The three-body process is only effective at very high densities, e.g. $\sim 10^{13}$ cm^{-3}; the inverse reaction can be an important destruction process. The subsequent stages of molecular synthesis will be discussed in subsection 2.3.

The first calculation of YSO wind chemistry was done by Rawlings, Williams and Cantó (1988, henceforth RWC) for the particular case of a T-Tauri wind. RWC assumed that the wind is spherically symmetric and starts off hot ($T \sim$ 20 000 K), following Hartmann, Edwards and Avrett (1982). RWC reasoned that the wind would cool at least as fast as adiabatic ($T_{ad} \propto r^{-3/4}$) and rapidly recombine, and they asked whether molecules would form under these conditions. Adopting a mass-loss rate $\dot{M} \sim 10^{-9} M_\odot$ yr^{-1} and an expansion (terminal) velocity $u_W = 230$ km s^{-1} for a dust-free wind, they obtained only weak molecularization, with the abundance of H_2 reaching a level of order 10^{-4} and heavy radicals (e.g. CH and OH) and CO reaching levels of order 10^{-9}. They also showed that including a 1% admixture of H_2 and dust (with the usual interstellar dust-to-gas ratio) promotes early molecule formation, and that lowering the initial wind temperature enhances the yield of CO. On the basis of their chemical model, RWC predicted that high velocity outflows from YSOs would not be detectable with the rotational line of CO at 2.6 mm in the absence of mass-loading.

At about this time, extremely high velocity (EHV) flows were discovered by Lizano et al. (1988) in both the CO 2.6 mm and the H 21 cm transitions. In the most common manifestation of bipolar outflows, the velocities sampled with the CO $J = 1$–0 line are relatively modest, ranging from a few to 30 km s^{-1} (e.g. Lada 1985). Lizano et al. detected atomic and molecular gas moving away from the embedded source SVS 13 in the Perseus cloud at velocities going up to 300 km s^{-1}, justifying the EHV appellation. The subsequent, extensive observations of EHV flows have been recently reviewed by Bachiller (1996).

Unaware of the RWC paper, our group was stimulated by the discovery of EHV flows to study molecule formation in YSO winds. We realized later (Glassgold, Mamon and Huggins 1989) that the reason RWC achieved only modest levels of synthesis was their restriction to small mass-loss rates $< 10^{-8} M_\odot$ yr^{-1}, appropriate for the most mature YSOs, the so-called "naked" T-Tauri stars. The mass-loss rates in less evolved (and also somewhat more massive YSOs) can be several orders of magnitude larger, going up to and even exceeding $\dot{M} = 10^{-5} M_\odot$ yr^{-1}. The basic concept of our first calculations was similar to RWC, except that we regarded the temperature phenomenologically, and we gave a more detailed calculation of the H^+ abundance that included excitation of the $n = 2$ level of atomic H and trapping of Lyman α. We modelled the temperature as: $T = T_o$ for $r < R_o$ and $T = T_o(R_o/r)^p$ for $r > R_o$, and defined a standard model by $T_o = T_* = 5000$ K, $R_o = 10^{12}$ cm, and $p = 1$ (a somewhat faster decline than adiabatic). The result was a cooler and less heavily ionized wind, with a CO abundance $\sim 10^{-5}$ for a mass-loss rate characteristic of SVS 13, $\dot{M} = 3 \times 10^{-6} M_\odot$ yr^{-1}. Perhaps the most important conclusion, in addition to the importance of the mass-loss rate, was that the results depend sensitively on how the temperature decreases in the wind. For example, if the size R_o of the inner, high-temperature region is large enough, then molecule formation is reduced because synthesis is postponed until the density is too low for it to be efficient.

2.2 Temperature

Ruden, Glassgold and Shu (1990, henceforth RGS) made a detailed study of the thermal properties of YSO winds, including excitation effects important for calculating the abundance of H^+. Although the theory was spherically symmetric, it included bipolarity in an approximate way by representing the *average* number density of hydrogen nuclei as

$$n(r)_H = \left(\frac{\dot{M}_H}{4\pi r^2 u_W}\right)\left[\frac{1}{\mathcal{A}\mathcal{U}}\right], \qquad (2.4)$$

where $\mathcal{A} \leq 1$ and $\mathcal{U} \equiv u(r)/u_W < 1$ are, respectively, *collimation* and *acceleration* factors that increase the density at small radii r. The idea behind $\mathcal{A}$ is that the area crossed by a bipolar wind is $\mathcal{A}(4\pi r^2)$, whereas $\mathcal{U}$ is the wind speed in units of the terminal or expansion velocity u_W. In accord with the early ideas underlying the X-wind model (Shu *et al.* 1988), RGS used the approximate form,

$$\mathcal{A} \approx \mathcal{U} = 1 - R_*/r, \qquad (2.5)$$

with essentially all of the acceleration and collimation occurring within a few R_*, as is appropriate for a *stellar* wind.

The RGS thermal theory included a full treatment of hydrogen ionization and chemistry. The H^+ and e abundances play a key role in the formation of H_2 as seen in subsection 2.1. The H^+ ion is produced primarily by collisional excitation of the $n = 2$ level of H followed by photoionization by Balmer continuum photons. Photodissociation of H^- and H_2^+ by the stellar radiation field are important destruction processes, and the proper treatment of the latter requires consideration of the vibrational excitation of the H_2^+ ions. The effects of heavy elements were represented by sodium ions, produced by photoionization and destroyed by radiative recombination.

In addition to including the effects on the wind temperature of the energy transfers in all atomic and molecular reactions, RGS considered two thermal processes intrinsic to the wind, adiabatic cooling and ambipolar-diffusion heating. The latter is caused by the frictional coupling between ions and neutrals when ions drift relative to the neutrals in response to magnetic forces. In all of the cases considered, adiabatic cooling dominates over all other thermal processes at small distances and initially produces a steep temperature decline. This promotes H_2 formation, which tends to heat the flow by the release of the molecular binding energy. As the temperature increases, however, collisional dissociation also increases, so that in models where H_2 is efficiently formed, the temperature is regulated by H_2 formation and destruction (cf. the discussion in Chapter 8 of the formation of H_2 behind shocks). At somewhat larger distances, where the ionization level is small, ambipolar-diffusion heating becomes important and the temperature declines only as $1/r$. The ability of H_2 and ambipolar-diffusion heating to counter the omnipresent adiabatic cooling, depends very much on the parameters of the YSO wind, i.e. mass-loss rate and wind speed. For parameters that correspond to the EHV flow in SVS 13, RGS found that, even though the temperature is orders of magnitude larger than given by the adiabatic law, the wind still becomes extremely cold (less than the cosmic microwave background) by the time it reaches distances where it might be observable with existing radio telescopes and arrays. Although significant amounts of H_2 are produced (and make it possible for heavy molecules to form), much of the atomic hydrogen remains. RGS concluded that spherically symmetric stellar winds can form H_2 (and by implication, heavier molecules), while retaining an atomic character, and that they become very cold in the absence of any interaction with their surroundings.

2.3 Chemistry

On the assumption that the YSO wind starts out from the source as an atomic (or ionic) wind, molecular synthesis proceeds in three stages that involve the successive formation of: (1) H_2, (2) heavy diatomic radicals or ions, and (3) diatomic and more complex molecules. This sequence is quite appropriate for "stellar" winds which originate close to the YSO, as assumed in the early work discussed in the previous subsections. The same sequence may be repeated if the wind is reheated enough at larger distances for wind-synthesized molecules to be destroyed. If the wind starts out with more of a "disc" character, some parts of it may initially have a significant molecular component, which again might be destroyed by reheating. Similar considerations apply to the dust content of YSO winds. Unless the initial temperature is $< 2500\,K$, the dust-free assumption is appropriate. As the wind cools, however, YSO winds will recycle their heavy atoms back into dust (RGS).

Because the physical properties of YSO winds vary over such a large range of temperature and density (mass-loss rate), the chemistry is more varied than interstellar cloud chemistry, which usually occurs under cold (i.e. 10–30 K) or warm (i.e. 100–200 K) conditions. Furthermore, molecular synthesis occurs quite close to the YSO so that stellar radiation is important. In terms of temperature and the effects of external radiation, YSO wind chemistry is more closely related to the chemistry of photo-transition regions (Sternberg and Dalgarno 1995; Hollenbach and Tielens 1997, Chapter 10), supernova ejecta (Lepp, Dalgarno and McCray 1990; Liu and Dalgarno 1996, Chapter 19), and the early Universe (Dalgarno and Lepp 1987, Chapter 3) than to shielded interstellar clouds. YSO chemistry is unique, however, in that densities in the inner wind are many orders of magnitude larger than in these other astrochemical applications. For example, eqn (2.4) assumes the numerical form,

$$n_{\rm H} = 3 \times 10^{35}\,{\rm cm}^{-1} \left(\frac{\dot{M}_{\rm H\,-6}}{u_{W\,7}}\right)\left[\frac{1}{\mathcal{AU}}\right]\frac{1}{r^2} \tag{2.6}$$

with the mass-loss rate in units of $10^{-6}\,M_\odot\,{\rm yr}^{-1}$ and the wind speed in units of $100\,{\rm km\,s}^{-1}$. Even at the large distance of $r = 100R_* \sim 1\,{\rm AU}$, the density is $n_{\rm H} \sim 10^9\,{\rm cm}^{-3}$.

We have already given the essence of the first stage of the chemistry, the formation of H_2, in the previous subsections. Some additional reactions were considered by Rawlings, Drew and Barlow (1993), especially the associative ionization reaction between ground and excited state hydrogen atoms,

$$H(n=2) + H(n=1) \rightarrow H_2^+ + e, \tag{2.7}$$

which is more important than the corresponding radiative association reaction proposed by Latter and Black (1991),

$$H(n=2) + H(n=1) \rightarrow H_2 + h\nu. \tag{2.8}$$

In contrast to reaction (2.8), reaction (2.7) is endothermic by about 1.1 eV, and so requires relatively high temperatures as well as a maximal population of the $n = 2$ level. Rawlings, Drew and Barlow (1993) showed that associative ionization of excited H atoms plays a role in the chemistry of the winds from massive YSOs, but it is unlikely to be important for low-mass winds, except in exceptional circumstances.

The next stage of YSO wind chemistry involves the production of radicals or ions that lead to more stable molecules. In order to properly treat this and the next stage of molecular synthesis, my collaborators and I developed an elaborate chemical kinetics program for spherical winds (Glassgold, Mamon and Huggins 1991, henceforth GMH) of the type introduced by RGS and discussed above. Among the novel features of the code written by Gary Mamon is the capability to treat simultaneously an arbitrary *stellar* as well as *interstellar* radiation field. Because the wind can easily be optically thick in radiation absorbed by one of the species, the chemical rate equations have to be solved by an iterative procedure.

An important source of far ultraviolet opacity is provided by the photoionization continua of heavy atoms, e.g. the threshold for photoionizing C atoms in the ground state is 1101 Å. The potentially abundant molecules CO and H_2 are also dissociated in the wavelength band from 912 to 1100 Å. If the opacity in this band is high, the C atom is mainly destroyed by photoionization of atoms in excited electronic levels, which have lower thresholds (starting at 1239 Å, 1444 Å, etc.). Thus the correct calculation of the photoionization of a heavy atom may require a multi-level population treatment, as GMH found in the case of the C atom.

Broadly speaking, there are two main routes to molecular synthesis in winds, once some H_2 has been formed, *ion–molecule reactions* and *neutral radical reactions*. As long as the wind is far from being completely ionized, neutral atoms will be sufficiently abundant so that the first heavy radicals and molecules are produced by reactions of the type,

$$A + H_2 \to AH + H \qquad (2.9)$$

and

$$AH + B \to AB + H, \qquad (2.10)$$

where A and B are heavy atoms, e.g. A,B = O, C, N, S, and Si. The reactions symbolized by eqn (2.9) are generally endothermic, and the production of OH is strongly favoured because of the small endothermicity in this case ($\sim$4000 K) and the high abundance of oxygen, cf. Chapter 8. The reactions represented by eqn (2.10), which lead to more stable species, are generally exothermic. Neutral reactions of the type (2.9) are also important *destruction* mechanisms for H_2.

The ion–molecule reactions of interest are generally similar to those familiar from interstellar chemistry (cf. Chapter 4), but YSO winds utilize a somewhat different subset of reactions due to the nature of the primary ionization. In the models developed so far, the main source of ionization is the stellar radiation field, either operating through the Balmer continuum in conjunction with collisional excitation of atomic hydrogen or by direct photoionization of heavy atoms by far ultraviolet radiation. For example, the initiator of ion–molecule chemistry in interstellar clouds is H_3^+, produced by

$$H_2^+ + H_2 \to H_3^+ + H. \qquad (2.11)$$

In the interstellar medium, H_2^+ is made by cosmic ray ionization of atomic and (especially) molecular hydrogen whereas, in the the wind models considered so far, H_2^+ is produced by the radiative association of H^+ and H or by reaction (2.7). The molecular ion CO^+ can be more important in these winds than H_3^+. It is produced by the reaction

$$C^+ + OH \to CO^+ + H, \qquad (2.12)$$

and leads to CO after charge exchange with atomic hydrogen

$$CO^+ + H \to CO + H^+. \qquad (2.13)$$

Since CO^+ is also destroyed (and HCO^+ produced) by the reaction

$$CO^+ + H_2 \to HCO^+ + H, \qquad (2.14)$$

the most abundant molecular ion after H_2^+ is generally OH^+, which is formed by the sequence

$$H^+ + O \to O^+ + H \qquad (2.15)$$

and
$$O^+ + H_2 \to OH^+ + H \tag{2.16}$$

and, somewhat more efficiently, by charge–exchange

$$H^+ + OH \to OH^+ + H. \tag{2.17}$$

Where there are enough ions for these processes to be efficient, the H_2 abundance is likely to be too small to make much H_2O^+ and H_3O^+ by further reactions of the same type as (2.16); OH^+ is also destroyed by dissociative recombination, which is a rapid process for the ionization levels in these winds. Thus reaction (2.15) is usually not very efficient in initiating heavy-molecule synthesis, especially since reaction (2.17) is so effective in destroying OH.

More generally, H^+ is able to charge exchange (or react) with many radicals and molecules, and is thus a molecular "poison". It does not react, however, with H_2, CO, or N_2. The cosmic-ray produced He^+ ion is an important destruction agent of molecules in interstellar clouds, including CO and N_2. It was ignored by GMH because YSOs are relatively cool, and produce few ionizing UV photons capable of ionizing He. In Section 3, I will argue for stellar production of cosmic rays, in which case He^+ reactions may be important.

In addition to charge exchange with molecules (which then leads to their destruction by dissociative recombination), charge exchange between atomic ions A^+ and atoms B,

$$A^+ + B \to B^+ + A, \tag{2.18}$$

is crucial for the correct calculation of the ionization of the wind. GMH emphasized the paucity of measured or reliably calculated theoretical charge-exchange rate coefficients. In the absence of any better information, they relied mainly on the educated theoretical guesses of Péquignot and Aldrovandi (1986). The situation has improved somewhat in recent years (see, e.g., Pradhan and Dalgarno 1994; Kimura et al. 1996). Charge exchange greatly reduces the abundance of H^+ and causes the ionization to be transferred to heavier atomic ions which can survive longer in the wind. This tendency is assisted by the reduction in the shielding of various ionizing far-ultraviolet continua that results from the ionization of the responsible atoms. Thus charge exchange ensures that the ionization level remains moderately high even for high-density winds.

The final results obtained by GMH are qualitatively similar to the early work of RWC and Glassgold, Mamon and Huggins (1989). For a high-velocity wind with density given by eqns (2.4) and (2.5) and temperature by RGS, molecule formation becomes quite efficient once $\dot{M} \geq 10^{-6} \, M_\odot \, \text{yr}^{-1}$. This is illustrated in Table 1, which shows the fraction of hydrogen, carbon, and oxygen in stable molecules for three values of the mass-loss rate. The abundance of species X is defined relative to the total number density of H nuclei, $x(X) = n(X)/n_H$; $f \equiv 2n(H_2)/n_H$; and x_C and x_O are the total fractional abundances of carbon and oxygen in all species. The wind speed in the table is $u_W = 150 \, \text{km s}^{-1}$.

Chemistry in the Winds of Young Stellar Objects 169

Table 1 Asymptotic Abundances

$\dot{M}$	3×10^{-7}	3×10^{-6}	3×10^{-5}
x_e	2.8×10^{-4}	8.5×10^{-5}	7.3×10^{-6}
$f(H_2)$	7.0×10^{-7}	1.6×10^{-5}	0.82
$x(CO)/x_C$	1.5×10^{-3}	1.0	1.0
$x(H_2O)/x_O$	0	0	0.45

The efficiency of molecule formation expressed in Table 1 is actually greater than obtained in the preliminary calculations discussed in subsection 2.1. Among the various improvements that are responsible, the new calculation of H^+ is most important, particularly its destruction by charge exchange and by neutral molecule reactions. The entries for x_e also indicate that the ionization does not get too small, even for the largest mass-loss rate. Again this is mainly due to the inclusion of a wide array of charge exchange reactions, which transfer the ionization held by H^+ charge to heavy atoms, where it is somewhat safer from destruction. The ionization levels in Table 1 range from a factor of 100 to 3 times larger than obtained by RGS, who used only Na^+ as a generic heavy ion. As a result, RGS obtained too much ambipolar-diffusion heating, which is inversely proportional to x_e. This led to reheating the wind at large radii and a reduction in molecule formation, an effect that is greatest for intermediate mass-loss rates ($\dot{M} \sim 10^{-6}\ M_\odot\ yr^{-1}$). Thus the molecule fractions in Table 1 are likely to be underestimates for this model; they also indicate that the thermal and chemical calculations have to be done consistently.

2.4 Disc Winds

The ionization and thermal properties of disc winds have been considered by Safier (1993a). His approach is analogous to that of RGS in that he starts from a specific dynamical model and treats the heating and cooling of the wind in the context of a pure hydrogen chemistry. His model is based on the magneto-centrifugally driven wind theory of Blandford and Payne (1982) and Königl (1989). The initial flow properties and the magnetic field depend on powers of the radial distance from the YSO, i.e. the entire disc participates in generating the outflow. The development and physical basis for this type of model are reviewed by Königl and Ruden (1993). Safier gives results for $\dot{M} = 10^{-7}\ M_\odot\ yr^{-1}$, a mass-loss rate intermediate between values characteristic of the most evolved (T Tauri stars) and the least evolved (class 0 and 1) YSOs. Questions have been raised about whether it is actually possible to set up the disc magnetic fields necessary to drive such a wind (e.g. see the discussion in Safier 1993a and in Königl and Ruden 1993).

Safier finds that, as the wind expands away from the disc, it makes a transition from a warm (or cool), low-ionization phase to a hot, high-ionization phase. Most of the wind is in the hot phase, with high temperature and electron fraction, $T \sim 10^4\ K$ and $x_e \sim 0.1$. According to Safier, the dominant thermal processes are adiabatic expansion and ambipolar-diffusion heating, similar to RGS for this level of mass-loss. The main difference is that adiabatic expansion is less important for a disc wind compared to a stellar wind, especially a disc wind generated at large radial distances from the YSO. In Safier's model, the high H^+ abundance is produced mainly by collisional ionization and destroyed by radiative recombination. Because ambipolar-diffusion heating decreases as $1/x_e$ and H^+ increases with T, the temperature is regulated to $\sim 10^4$ K, surprisingly large for ambipolar-diffusion heating at high x_e. As the wind expands to large vertical heights $\sim 10-100$ AU, the ionization fraction gets frozen out at a constant value.

Safier did not address the chemical properties of disc winds other than ionization. The initial state of his wind is probably molecular, except very close to the YSO. Of course he recognized that his high model temperatures would lead to the destruction of the molecules, and so he assumed that the wind was atomic. However, he might have gotten quite different results had he included charge exchange and molecule formation and destruction. Then the interesting question arises as to whether the much reduced H^+ abundance that probably ensues would produce even higher wind temperatures.

In a second paper, Safier (1993b) calculated forbidden line fluxes, and claimed qualitative success in explaining the observations. Although there is certainly evidence for regions of hot, highly ionized gas near YSOs, they probably do not pervade the entire YSO wind. Indeed there is a large body of observations that indicate that winds from YSOs are primarily neutral, with a significant average ionization level $x_e \ll 1$ (e.g. Natta et al. 1988a,b). The optical and radio observations of jets and EHV flows also demonstrate that the outflows are spatially inhomogeneous on the scale on which they are observed, and very likely on smaller scales as well. In general, there are two apparent outflow components that are generated by YSOs, a jet and an EHV outflow (e.g. Edwards, Ray and Mundt 1993; Königl and Ruden 1993; Bachiller 1996). Obviously, the physical conditions of real outflows cannot be adequately accounted for by either a spherical, stellar-wind model or an extended, disc-wind model.

3 Towards a Realistic YSO Wind Chemistry

In the previous section, we discussed two chemical models (RWC and GMH) for spherical winds with different mass-loss-rate ranges and two thermal-chemical models (RGS and Safier 1993a) with very different flows. The results in both cases show a sensitivity to the dynamics. Further progress in YSO wind chemistry requires more realistic dynamical models than those used in the past. Fortunately, self-consistent solutions of the X-wind (cf. Chapter 6) that can be used for this purpose (Shu et al. 1994a) have recently become available.

The first formulation of the X-wind model (Shu et al. 1988) described a limiting case where accretion spins the star up to breakup speed. The X-region, from which matter escapes by way of a magneto-centrifugally driven wind, is located at the stellar surface, which joins continuously to the accretion disc. This is a significant limitation, e.g. T Tauri stars are observed to rotate rapidly but considerably more slowly than breakup speed. In the more complete or generalized model (Shu et al. 1994a), the inward accretion flow is dramatically altered by the magnetic field of the protostar in the neighbourhood of the corotation radius R_X, defined by

$$\Omega_* = \left(\frac{GM_*}{R_X^3}\right)^{1/2}, \qquad (3.1)$$

where Ω_* and M_* are the angular velocity and mass of the YSO. At R_X, the accretion flow splits into a wind (solution in Shu et al. 1994b and Najita and Shu 1995) and into a flow onto the higher latitudes of the YSO (the "funnel flow" solution in Ostriker and Shu 1995). The wind asymptotically acquires the collimation of a jet (Shu et al. 1995; Li 1996), and sweeps up the surrounding molecular material into a bipolar outflow. Thus the new solution of the X-wind model contains two of the most dramatic aspects of YSO objects, bipolar outflows and jets.

In order to show that the X-wind can actually explain these phenomena, its physical and chemical properties must be established and then appropriate diagnostic line fluxes calculated. To accomplish this, something like the combined thermal-chemical program of RGS and MGH, described in Section 2 for spherical winds, must be executed for two-dimensional, axially symmetric winds. Detailed analysis of the X-wind solution (Shu et al. 1995; Shang and Shu 1996, private communication) shows that the stream functions of the X-wind can be simply approximated. For example, at large distances the velocity field asymptotically becomes radial, and the mass density assumes the form (Shu et al. 1995)

$$\rho = \frac{\mathcal{C}(r)}{\beta(r) r^2 \sin^2 \theta}, \qquad (3.2)$$

where r and θ are cylindrical coordinates. The function β is almost constant, and $\mathcal{C}$ varies smoothly with r. The density profiles are approximately surfaces of constant perpendicular distance, $r \sin\theta$, with a spacing (density stratification) that closely mimics a cylindrical jet.

Equation (3.2) fits the prescription for a radial wind given by eqn (2.4), but with the factor $\mathcal{A} = \sin^2\theta$. Gary Mamon and I have found that the chemical kinetics problem, i.e. solving the system of rate equations, can be achieved for this type of wind by minor reprogramming of the *one-dimensional* program used by GMH for spherically symmetric flows. The idea is to solve the rate equations with the radial streamline coordinate r replaced by the distance along each actual streamline, so that the rate equations are integrated along each streamline. We believe that this method will also work for more accurate representations of the X-wind solutions, which are needed close to the protostar. The solution of the thermal problem must be obtained simultaneously with the chemical solution because self-consistency is important, as demonstrated by the earlier experience of RGS (see subsection 2.3).

A large part of the wind chemistry program is devoted to the transfer of ionizing and dissociating radiation, as exemplified by the calculation of GMH. The problem is simplified if dust is the sole cause of the important far-UV opacity. In this case, if the dust properties are independent of distance, the opacity is completely determined because the total hydrogen (or mass) density is specified. This is true for both stellar and interstellar radiation sources. If, however, the opacity in some wavelength range is affected by atomic or molecular opacities, there is a significant difference between the two cases because the abundances are required for the opacity calculation. If the radiation emanates from an internal source, e.g. from the YSO, the problem is straightforward in that the opacity from "inside out" can be calculated step by step as the integration proceeds. However, the "outside to inside opacity", which is needed to calculate the transfer of external radiation, cannot be treated so simply and instead requires an iterative procedure (GMH). Fortunately, the radiation important for understanding the physical and chemical properties of the inner winds of YSOs is determined primarily by internal sources. For example, the method of solution sketched above for the X-wind could be carried out streamline by streamline starting with the innermost streamline, so that the opacity from the region inside a streamline is always known. Special attention has to be given to species for which line self-shielding is important, because the level population must also be calculated in this case, and this requires escape probabilities for all escaping photon directions, i.e. an iterative solution. Thus, despite the simplifications afforded by wind solutions like that given by eqn (3.2), solving for the thermal-chemical properties of any axial-symmetric wind involves a large computational burden, probably between one and two orders of magnitude greater than faced by GMH. Still, the advances in computers in the last half-decade should make the above program feasible on powerful workstations.

The development of a realistic wind chemistry requires broadening the physical as well as the dynamical basis of the model. The stellar dissociating and ionizing radiation will be much harder than that associated with the effective temperature of the YSO. For example, there is good evidence for a UV excess in T Tauri stars (Bertout 1989), which was formerly ascribed to an accretion shock. The X-wind model offers a specific scenario for the generation of UV radiation by the impact of the funnel flow onto the YSO that operates whenever accretion is significant. Approximate estimates based on the accretion luminosity yield a UV luminosity $\sim 1\,L_\odot$ and an equivalent temperature of the order of 8000 K. This radiation will play an important role close in, but it tends to be strongly attenuated, even if the wind is initially dust free.

On a larger scale, the X-rays emitted by a YSO can have an important effect on the physical and chemical properties of their immediate surroundings. YSOs have been known to be strong emitters of X-rays ever since the *EINSTEIN* observatory (e.g. Ku and Chanan 1979; Feigelson and Decampli 1981; Walter and Kuhi 1981; Montmerle *et al.* 1983). Little has been done, however, to investigate the effects of the X-rays on the circumstellar environment of YSOs, as has been emphasized by Feigelson, Montmerle, and their collaborators (e.g. Montmerle and Casanova 1995; Casanova et al. 1995). Meanwhile, new satellite observations are significantly extending the earlier observations. Those made with *ROSAT* provide increased sensitivity and spatial resolution, while those obtained with *ASCA* provide access to higher energies. The observations indicate that YSOs have a persistent hard X-ray component at all stages of evolution, with a typical X-ray luminosity fraction of $L_{XR} \sim 10^{-4}$–$10^{-3} L_{bol}$ and X-ray temperatures in the range $T_{XR} = 1$–10 keV.

Even before *Einstein*, theorists had studied how X-rays interact with the interstellar medium. In addition to investigating the microscopic processes involved in ionization and heating, idealized models were constructed for point sources inside isotropic clouds (e.g. Tarter et al. 1969; Tarter and Salpeter 1969). It was readily appreciated that, due to the dilution of the X-rays, the properties of the surrounding gas cloud change qualitatively with distance from the source, going from a highly ionized state close in to a molecular region far away (e.g. Halpern and Grindlay 1980; Lepp and McCray 1983). Chemical problems have been addressed by Langer (1978) and by Krolik and Kallman (1983) and, more recently and at much greater depth, by Neufeld, Maloney and Conger (1994), Maloney, Hollenback and Tillens (1996), Lepp and Dalgarno (1996), and Yan and Dalgarno (1997); also see Chapter 22. Broader issues raised by the interaction of YSO X-rays with their environment were also first addressed more than a decade ago. Krolik and Kallman (1983) examined the penetration of X-rays at large optical depths, and Silk and Norman (1983) suggested that the X-rays from YSOs influence the rate of star formation by affecting the ionization fraction of the nascent clouds. Because the rate at which the neutrals slip away from the charged particles varies inversely with the ionization fraction, a feed-back loop may be set up that "regulates" the rate of star formation.

My colleagues and I have begun a program to understand the role of YSO X-rays on the close-in components of the star-forming system, e.g. the wind and a disc. Preliminary results are available for the case of ionization of the accretion disc by hard, penetrating X-rays (Glassgold, Najita and Igea 1997), along the following lines. At a typical, intermediate location, the unattenuated ionization rate at the "top of the disc" is very large. Going towards the disc midplane, the rate decreases due to absorption but remains larger than the Galactic cosmic ray ionization rate ($\zeta_{CR} \approx 6 \times 10^{-18}$ s^{-1}, Spitzer and Tomasko 1968) to fairly large vertical column densities, $N_H \sim 10^{24}$–10^{25} cm^{-2}, depending on the spectrum and luminosity of the X-rays, testimony to the penetrating power of hard X-rays. The transfer of hard X-rays is determined by scattering as well as absorption, and scattering helps the X-rays diffuse into the disc.

Based on a simplified chemical model, we concluded that X-rays may be able to ionize the outer layers of protostellar accretion discs but not their interiors. The result is a layered disc ionization structure, similar to that discussed by Gammie (1996) on the basis of external cosmic ray ionization. In his picture, the Galactic cosmic rays penetrate ~ 100 g cm^{-2} into the disc (Umebayashi and

Nakano 1981) and produce an accreting surface layer that overlies a deeper, quiescent layer. Given the large range of observed protostellar X-ray properties, the disc ionization structure induced by X-ray irradiation implies a range in disc accretion rates, unlike the case of Galactic cosmic rays. It is unlikely that Galactic cosmic rays actually reach the inner accretion disc, because they will be excluded by YSO winds, in much the same way that the solar wind modulates the flux of Galactic cosmic rays. In the case of the Sun, the solar wind effectively excludes cosmic rays with energies <100 MeV from the inner solar system, precisely the energies that are responsible for most of the interstellar cosmic ray ionization rate (Spitzer and Tomasko 1968).

The ionization of the outer layers of the accretion disc illustrates how protostellar X-rays can affect the component flows involved in star formation. Because much of the X-ray spectrum is absorbed by the wind, we expect the soft X-rays to be important in ionizing the wind and in influencing its chemistry. The hard X-ray component can extend these effects to very large column densities, corresponding to several Thomson mean free paths, or more than a 1000 magnitudes of visual extinction (for a standard interstellar dust-to-gas ratio and dust size distribution). Some of the X-rays escape the protostellar system and ionize the nascent cloud, as envisaged in earlier studies (e.g. Silk and Norman 1983; Krolik and Kallman 1983). In addition to X-rays, YSOs (which are magnetically active stars) are also expected to accelerate particles to high energies, as our own Sun is observed to do. These stellar energetic particles dissipate most of their energy in exciting and ionizing hydrogen and helium, the main constituents of the circumstellar material. But they will also engage in nuclear reactions, as proposed long ago by Fowler, Greenstein and Hoyle (1962). Again the specific flow and field configurations of the X-wind model provide guidance in understanding how the energetic particles produced by a YSO affect the processes of star and planet formation. For example, Shu et al. (1997) have advanced new support for the stellar irradiation mechanism to synthesize short-lived radionuclides like ^{26}Al. They adopted a time-dependent version of the X-wind model which allows proto-chondritic solids to approach the region in the disc plane where magnetic reconnection *and* particle acceleration occur.

4 Conclusion

The main conclusion of this review of the inner wind chemistry of YSOs is that the most important work remains to be done! The early studies (discussed in Section 2) did identify important physical and chemical aspects of the problem, e.g. the dominance of adiabatic cooling, the critical role of ionization, and the need to treat thermal and chemical effects simultaneously. A serious defect of these models was that they were based on the assumption of spherically symmetric winds in a situation where the dynamical timescales are very short and where the chemistry is very sensitive to density. However, the future looks bright because the dynamical solutions required for definitive conclusions, which were lacking then, now exist. I suggest (recall Section 3) that the existence of the generalized

X-wind (Shu et al. 1994a, 1994b, 1995), not only challenges astrochemistry to deal with new regimes of density and temperature, but indicates the occurrence of important physical processes that were not considered previously. One area of special interest is the effects of protostellar X-rays and energetic particles on the ionization and chemistry of the winds.

Another aspect of previous studies of YSO wind chemistry is that the theory was limited to distances quite close to the YSO, i.e. the interactions of the wind with other parts of the system were ignored. The new dynamical solutions for the X-wind also improve this situation by making it possible to give a good account of the wind physical properties when it encounters the disc at low latitudes and infalling material at higher latitudes. Recent progress in the collapse dynamics (Li and Shu 1996, 1997; Shu and Li 1997) is also important in this regard, because it is now possible to envisage a more complete modelling of the star-forming region in which all of the crucial components can be included.

Acknowledgement

It is a pleasure to contribute to Alex Dalgarno's Festschrift. I have known Alex for more than 30 years and, even before we met, had the good fortune to encounter his research in basic atomic and molecular physics. When I began research in interstellar clouds, Alex had again anticipated my interest in this field. In the intervening years, I have benefited greatly from Alex's broad knowledge and friendly counsel. I wish him many more years of productive research.

The author would also like to acknowledge the indispensible help over the years of many collaborators in the area of protostellar wind physics and chemistry, especially Gary Mamon and Frank Shu.

Bibliography

1. Bachiller, R. (1996). *ARAA*, **34**, 111.
2. Bertout, C. (1989). *ARAA*, **27**, 351.
3. Blandford, R. D. and Payne, D. G. (1982). *MNRAS*, **199**, 883.
4. Casanova, S., Montmerle, T., Feigelson, E. D. and André, P. (1995). *ApJ*, **439**, 752.
5. Dalgarno, A. and Lepp, S. (1987). In *Astrochemistry*, eds. M. S. Vardya and S. P. Tarafdar. Reidel, Dordrecht, p.109.
6. Edwards, S., Ray, T. and Mundt, R. (1993). In *Protostars and Planets III*, eds. E. Levy and J. I. Lunine. University of Arizona, Tucson, p.567.
7. Feigelson, E. D. and de Campli, W. (1981). *ApJ*, **243**, L89.
8. Fowler, W. A., Greenstein, J. G. and Hoyle, F. (1962). *Geophys. J. Roy. Astron. Soc.*, **50**, 110.
9. Fukui, Y., Iwata, T., Mizuno, A., Ballt, J. and Lane, A. P. (1993). In *Protostars and Planets III*, eds. E. Levy and J.I Lunine. University of Arizona, Tucson, p.603.

10. Gammie, C. F. (1996). *ApJ*, **457**, 355.
11. Glassgold, A. E., Mamon, G. A. and Huggins, P. J. (1989). *ApJ*, **336**, L39.
12. Glassgold, A. E., Mamon, G. A. and Huggins, P. J. (1991). *ApJ*, **373**, 254.
13. Glassgold, A. E., Najita, J. and Igea, J. (1997). *ApJ*, **485**, 1010.
14. Halpern, J. P. and Grindlay, J. E. (1980). *ApJ*, **242**, 1041.
15. Hartmann, L., Edwards, S. and Avrett, E. (1982). *AJ*, **261**, 279.
16. Hollenbach, D. H. and Tielens, A. G. G. M. (1997). *ARAA*, **35**, 179.
17. Kimura, M., Sannigrahi, J. P., Gu. J. P., Hirsch, G., Buenker, R. J. and Shimamura, I. (1996). *ApJ*, **473**, 1114.
18. Königl, A. (1989). *ApJ*, **342**, 208.
19. Königl, A. and Ruden, S. P. (1993). In *Protostars and Planets III*, eds. E. Levy and J. I. Lunine. University of Arizona, Tucson, p.641.
20. Krolik, J. H. and Kallman, T.R (1983). *ApJ*, **267**, 610.
21. Ku, W.-K. and Chanan, G. A. (1979). *ApJ*, **234**, L59.
22. Lada, C. J. (1985). *ARAA*, **23**, 267.
23. Lada, C. J. and Shu, F. H. (1990). *Science*, **248**, 564.
24. Langer, W. D. (1978). *ApJ*, **225**, 860.
25. Latter, W. B. and Black, J. H. (1991). *ApJ*, **372**, 161.
26. Lepp, S. and Dalgarno, A. (1996). *A&A*, **306**, L21.
27. Lepp, S., Dalgarno, A. and McCray, (1990). *ApJ*, **358**, 262.
28. Lepp, S. and McCray, R. M. (1983). *ApJ*, **269**, 560.
29. Li, Z.-Y. (1996). *ApJ*, **465**, 855.
30. Li, Z.-Y. and Shu, F.H (1996). *ApJ*, **472**, 211.
31. Li, Z.-Y. and Shu, F.H (1997). *ApJ*, **475**, 237.
32. Liu, W. and Dalgarno, A. (1996). *ApJ*, **471**, 480.
33. Lizano, S., Heiles, C. F., Rodriguez, L. F., Koo, B.-C., Shu, F. H., Hayashi, S. and Mirabel, I. F. (1988). *ApJ*, **328**, 763.
34. Maloney, P. R., Hollenbach, D. J. and Tielens, A. G. G. M. (1996). *ApJ*, **466**, 561.
35. Montmerle, T. and Casanova, S. (1995). *Rev. Mex. Ser. Conf.*, **1**, 329.
36. Montmerle, T., Koch-Miramond, L., Falgarone, E. and Grindlay, J. (1983). *ApJ*, **269**, 182.
37. Najita, J. and Shu, F. H. (1995). *ApJ*, **429**, 808.
38. Natta, A., Giovanardi, C., Palla, F. and Evans, N. J., II. (1988a). *ApJ*, **327**, 817.
39. Natta, A., Giovanardi, C. and Palla, F., (1988b). *ApJ*, **332**, 921.
40. Neufeld, D. A., Maloney, P. R. and Conger, S. (1994). *ApJ*, **436**, L12.
41. Ostriker, E. C. and Shu, F. H. (1995). *ApJ*, **447**, 813.
42. Péquignot, D. and Aldrovandi, S. M. V. (1986). *A&A*, **161**, 169.

43. Pradhan, A. and Dalgarno, A. (1994). *Phys Rev A*, **49**, 960.
44. Rawlings, J. M. C., Drew, J. E. and Barlow, M. J. (1993). *MNRAS*, **265**, 968.
45. Rawlings, J. M. C., Williams, D. A. and Cantó, J. (1988). *MNRAS*, **230**, 695.
46. Ruden, S. P., Glassgold, A. E. and Shu, F. H. (1990). *ApJ*, **361**, 546.
47. Safier, P. N. (1993a). *ApJ*, **408**, 115.
48. Safier, P. N. (1993b). *ApJ*, **408**, 135.
49. Shu, F. H., Adams, F. and Lizano, (1987). *ARAA*, **25**, 23.
50. Shu, F.H and Li, Z.-Y. (1997). *ApJ*, **475**, 257.
51. Shu, F. H., Lizano, S., Ruden, S. P. and Najita, J. (1988). *ApJ*, **328**, L19.
52. Shu, F. H., Najita, J., Ostriker, E., Wilkin, F., Ruden, S. and Lizano, S. (1994a). *ApJ*, **429**, 781.
53. Shu, F. H., Lizano, S., Ruden, S. P. and Najita, J. (1994b). *ApJ*, **429**, 707.
54. Shu, F. H., Najita, J., Ostriker, E. and Shang, S. (1995). *ApJ*, **455**, L55.
55. Shu, F. H., Shang, S., Glassgold, A. E. and Lee, T. (1997). *Science*, **277**, 1475.
56. Silk, J. and Norman, C. (1983). *ApJ*, **272**, L49.
57. Snell, R. L., Loren, R. B. and Plambeck, R. L. (1980). *ApJ*, **239**, L17.
58. Spitzer, L. and Tomasko, M. G. (1968). *ApJ*, **152**, 971.
59. Sternberg, A. and Dalgarno, A. (1995). *ApJS*, **99**, 565.
60. Tarter, B. and Salpeter, E. E. (1969). *ApJ*, **156**, 953.
61. Tarter, B., Tucker, W. H. and Salpeter, E. E. (1969). *ApJ*, **156**, 943.
62. Umebayashi, T. and Nakano, T. (1981). *PASJ*, **33**, 617.
63. Walter, F. M. and Kuhi, L. V. (1981). *ApJ*, **284**, 194.
64. Yan, M. and Dalgarno, A. (1997). *ApJ*, **481**, 296.

8
Shock Chemistry

T. W. Hartquist
Max-Planck-Institut für extraterrestrische Physik

and

P. Caselli
Osservatorio Astrofisico di Arcetri

1 Introduction

The winds and jets of young stellar objects, the outflows of evolved stars, supernovae in normal galaxies and in starburst galaxies, and the winds and jets of the central objects of active galactic nuclei drive shocks into ambient gas. Thus, shock heated molecular gas exists in many types of sources.

One of the primary early applications of the sort of theory described in this chapter was to the molecular hydrogen emission from the Orion Kleinmann Low–Becklin Neugebauer region of star formation. However, detailed high velocity and high angular resolution observations of H_2 emission from the region show that it does not arise in the simple sorts of shock structures so far modelled (Tedds et al. 1995).

As seen in Chapter 4, the chemistries of oxygen and carbon are initiated in low temperature ($T \lesssim 10^2$ K) diffuse astrophysical media in which H_2 is by far the most abundant species by ion–neutral reactions even when most of each of those elements is in neutral species. The dominance of ion–neutral reactions as initiators is a consequence of some neutral–neutral reactions involving H_2 such as

$$O + H_2 \rightarrow OH + H \quad (1.1)$$
$$OH + H_2 \rightarrow H_2O + H \quad (1.2)$$

and

$$C + H_2 \rightarrow CH + H \quad (1.3)$$

being slow at such temperatures. Some ion–neutral reactions are also slow at low temperatures. For instance, the chemistry of sulphur in many $T < 10^2$ K media is initiated by neutral–neutral reactions involving species much less abundant than H_2 because

$$SH^+ + H_2 \rightarrow SH_2^+ + H \quad (1.4)$$

is slow in them.

In gas heated to $T > 10^3$ K all of the reactions (1.1)–(1.4) cease to be slow. Dissipation in shocks is one mechanism that in various regions heats gas sufficiently to induce rapidity in a variety of reactions that are slow at $T \lesssim 10^2$ K. Thus, the chemistry in shocked molecular gas often differs from that in low temperature material. Of course, under certain conditions too much heating can cause substantial molecular dissociation; thus, shocks progating too fast in some media cause the destruction of most molecules.

A number of mechanisms other than shocks can heat molecular gas to temperatures at which reactions that are slow at $T \lesssim 10^2$ K become important. These mechanisms include dissipation in boundary layers (cf. Chapter 10), absorption of ultraviolet radiation (cf. Chapter 9), and ionization by X-rays (cf. Chapter 22). Consequently, much of the chemistry important in shocked gas is of relevance to other types of regions as well.

In Section 2 we consider the physics and chemistry of hydrodynamic shocks in H_2-rich media; in that section we restrict attention to shocks that are too slow for a significant fraction of the H_2 in them to be dissociated by collisions. In Section 3 we address the dissociation of molecules by shocks and the chemistry in cooled downstream regions where H_2 is reforming. In fact, diffuse astrophysical sources are magnetized and Section 3 does contain a brief mention of some magnetic effects. Section 4 is an introduction to multifluid models of shocks in weakly ionized magnetized media. Section 5 concerns the important active dynamical role of dust grains in shocks in dark magnetized regions as well as the release of grain material into the gaseous neutral phase in shocked regions and the consequence of that release for the chemistry.

2 Nondissociating Hydrodynamic Shocks

We consider a semi-infinite nonmagnetized uniform medium into which a plane-parallel piston is moving. (The "piston" may be a boundary between higher pressure gas, such as a stellar wind, and the molecular medium.) We assume that the flow in the molecular medium is governed (except, as discussed below, in a very localized region) by one-dimensional equations of nonviscous hydrodynamics. For a medium with velocity $u\hat{z}$, density ρ, mean mass per particle μ, internal and thermal energy content per unit volume U, and pressure P these equations are:

$$\frac{\partial \rho}{\partial t} + \frac{\partial(\rho u)}{\partial z} = 0 \tag{2.1}$$

$$\frac{\partial(\rho u)}{\partial t} + \frac{\partial(\rho u^2)}{\partial z} + \frac{\partial P}{\partial z} = 0 \tag{2.2}$$

and

$$\frac{\partial}{\partial t}\left(\frac{1}{2}\rho u^2 + U\right) + \frac{\partial}{\partial z}\left(\frac{1}{2}\rho u^3 + uU + uP\right) = \rho Q \tag{2.3}$$

Q is the difference between energy input per unit mass (e.g. by photoabsorption) and energy loss per unit mass (e.g. by radiative emission). In addition,

$$P = \frac{\rho}{\mu} k_B T \tag{2.4}$$

where k_B and T are Boltzmann's constant and the temperature respectively.

$$U = \frac{3}{2}P + \frac{\rho}{\mu}\sum_j f_j \epsilon_j \tag{2.5}$$

where f_j is the fraction of particles having internal energy ϵ_j. In a general treatment, f_j is calculated through the solution of rate equations of the form

$$\frac{\partial f_j}{\partial t} + u\frac{\partial f_j}{\partial z} = F_j - D_j \tag{2.6}$$

where F_j is the rate per particle at which radiative and collisional processes cause a level with energy ϵ_j to be populated and D_j is the rate per particle at which processes depopulate that level. Often, the left hand side of (2.6) may be set to zero in which case a statistical equilibrium level population is said to obtain. In some situations, one can simply take

$$U = P/(\gamma - 1) \tag{2.7}$$

with γ being a constant. In gas that has no internal excitation, $\gamma = 5/3$. If in a pure H_2 gas the rotational levels of the lowest vibrational level have a thermalized population distribution but vibrationally excited levels have negligible populations, $\gamma = 7/5$.

If the piston moves at a speed such that sound waves carrying information upstream are overtaken, a shock develops in the flow. Even immediately ahead of the shock the upstream material is undisturbed. In the shock itself, over a length scale of a few collision mean free paths viscous dissipation occurs and material is accelerated up to a substantial fraction of the speed that the shock moves relative to the preshock ambient medium (e.g. Landau and Lifschitz 1959).

For simplicity, we will assume that upstream $(U - 3P/2) \ll U$ (i.e. the gas is rotationally cold, which is a reasonable assumption if upstream the gas is predominately H_2, and $T \ll 170$ K, which corresponds to the energy of the $J = 1$ level). Therefore, since momentum transfer cross sections are considerably larger than collision cross sections inducing internal excitation, we may take $\gamma = 5/3$ and use (2.7) in the thin region where the initial acceleration occurs. If we also assume that the timescales on which processes associated with Q heat and cool the gas are long compared to the timescale for the flow to travel a few collision mean free paths and work in a frame that moves with the shock front, we may take $Q = 0$ and set time derivatives to zero when deriving relationships between the preshock and postshock fluid parameters. Thus, from (2.1) through

(2.3) and (2.7)

$$\rho_2 u_2 = \rho_1 V_S \tag{2.8}$$
$$\rho_2 u_2^2 + P_2 = \rho_1 V_S^2 + P_1 \tag{2.9}$$
$$\frac{1}{2}\rho_2 u_2^3 + \frac{5}{2} P_2 u_2 = \frac{1}{2}\rho_1 V_S^3 + \frac{5}{2} P_1 V_S \tag{2.10}$$

where $V_S \hat{z}$ is the velocity of the upstream gas measured in the shock frame and the subscripts 1 and 2 indicate preshock and postshock values of quantities. As long as time derivatives in the shock frame can be neglected (2.8) and (2.9) are valid everywhere. The adiabatic acoustic speed is

$$c_a = \left(\frac{\gamma P}{\rho}\right)^{1/2} \tag{2.11}$$

We define the shock's Mach number as

$$M \equiv \frac{V_S}{c_{a1}} \tag{2.12}$$

Then for $\gamma = 5/3$ and $M \gg 1$, (2.4) and (2.8)–(2.12) yield

$$\frac{u_2}{V_S} = \frac{\rho_1}{\rho_2} \approx \frac{1}{4} \tag{2.13}$$

$$T_2 = \frac{3}{16}\frac{\mu}{k_B} V_S^2 \approx 4510\,\text{K} \left(\frac{\mu}{2\,\text{amu}}\right)\left(\frac{V_S}{10\,\text{km s}^{-1}}\right)^2 \tag{2.14}$$

In some cases (i.e. those in which instabilities do not develop) when the shock velocity relative to the uniform upstream gas remains constant for a time long compared to the various excitation and cooling timescales in the postshock region, the structure of the flow in a sizeable region behind the shock appears in the shock frame to be independent of time and can be studied in that frame on the assumption that the time derivatives are zero. The equations to be solved are (2.8), (2.9), and the time independent versions of (2.3) through (2.6).

In one simple case the density is so high that radiative decay rates of excited levels are small compared to collisional de-excitation rates; then, for an H_2 medium $\gamma = 7/5$ can be assumed as long as 170 K $\ll T \ll$ 6000 K, the temperature associated with the energy of the first vibrationally excited level of H_2. Usually (but not when radiative losses from species other than H_2 affect the thermal structure on a timescale short relative to the H_2 collisional de-excitation timescales) then for $M \gg 1$ shocks at a downstream distance of roughly $V_S t_D/4$ (where t_D is the typical de-excitation time)

$$\frac{u_2}{V_S} = \frac{\rho_1}{\rho_2} = \frac{1}{6} \tag{2.15}$$

$$T = \frac{5}{36}\frac{\mu}{k_B} V_S^2 \approx 3340\,\text{K} \left(\frac{\mu}{2\,\text{amu}}\right)\left(\frac{V_S}{10\,\text{km s}^{-1}}\right)^2 \tag{2.16}$$

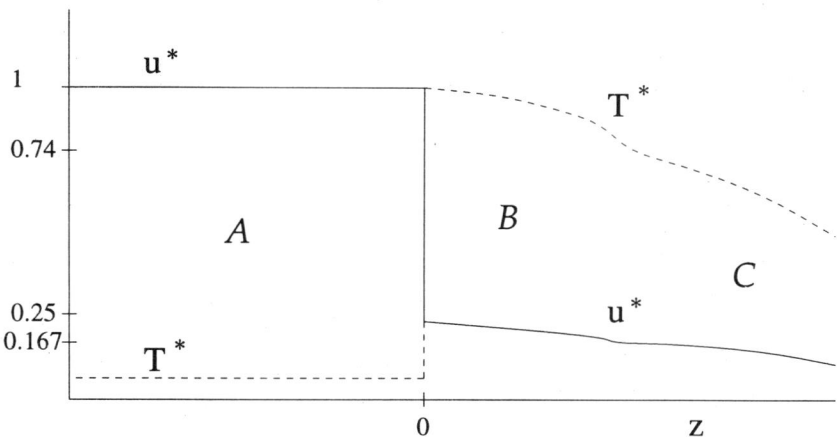

FIG. 1. Flow structure in a nondissociating hydrodynamic shock. It has been assumed that $M \gg 1$, the gas is predominantly H_2, the postshock gas is hot enough and dense enough that at least several excited rotational levels of H_2 become substantially populated with a thermalized population distribution, and excited vibrational levels are not substantially populated. The frame is comoving with the shock and the upstream gas moves with a velocity of $V_S \hat{z}$. $u^* \equiv u/V_S$ and $T^* \equiv 16\, k_B T/3\mu V_S^2$. Region A is upstream. Region B is where the excited rotational levels of H_2 become excited and region C is where radiative cooling occurs. At the transition between regions B and C the speed and temperature are given by (2.15) and (2.16); the ratio of the temperatures given by (2.16) and (2.14) is about 0.74.

Equations (2.15) and (2.16) follow from equations (2.8), (2.9) and (2.10) with 5/2 replaced by 7/2 as is appropriate for a $\gamma = 7/5$ gas. The gas will continue to behave as though $\gamma = 7/5$ until T drops too far, after which $\gamma = 5/3$ gives a better approximation. As long as the assumption that $\gamma = 7/5$ is reasonable, the time independent postshock evolution will be governed by (2.8), (2.9) and

$$\frac{\partial}{\partial z}\left(\frac{1}{2}\rho u^3 + \frac{7}{2}uP\right) = \rho Q \qquad (2.17)$$

which comes from (2.3) and (2.7). Figure 1 shows schematically the structure of a flow as described in this paragraph. One class of situations in which the adoption of $\gamma = 7/5$ is invalid is the topic of the next section.

For non-dissociative shocks cooling is usually dominated by CO line radiation at $T \lesssim 400$ K. The temperature range throughout which OH and H_2O line radiation dominates the cooling depends on the postshock density, but they tend to contribute substantially when 400 K $\lesssim T \lesssim$ 1000 K. H_2 line cooling tends to dominate when $T \gtrsim 1000$ K. H_2O cooling will dominate over H_2 cooling at T up to 2000 K or more if $\rho/\mu \gtrsim 10^6$ cm^{-3}. Tiné et al. (1997) have made a critical evaluation of available results for the collision cross sections for the de-excitation of H_2 rovibrational levels important in the calculation of level populations and radiative cooling by H_2. Sternberg and Dalgarno (1995) summarized the approach used to calculate radiative losses by CO, OH, and H_2O as well as losses due to the radiative decay of excited atomic fine structure levels. Recent important work in the area has been reported by Kaufman and Neufeld (1996). Before leaving the subject of cooling we point out that radiative cooling can for certain cooling rate coefficients induce thermal instability in a postshock region (e.g. Innes, Giddings and Falle 1987). Whatever its physical origin, instability introduces time dependence into the flow (even as viewed in the shock frame) and time derivatives cannot be neglected everywhere. It is often taken for granted that in the viscous dissipation region of a hydrodynamic shock time dependent effects can be neglected so that the jump conditions (2.8) through (2.10) are valid. However, in sections 4 and 5 we will consider hydromagnetic shocks for which the assumption of time independent flows in the main dissipation regions is sometimes valid. We return to the subject of radiative cooling in Section 4.

Clearly, the radiative cooling rate depends on the postshock chemical structure. Of particular relevance are the reactions (1.1) and (1.2) as are

$$OH + h\nu \rightarrow O + H \quad (2.18)$$

and

$$H_2O + h\nu \rightarrow OH + H \quad (2.19)$$

in regions exposed to strong enough UV radiation. The rate coefficients of (1.1) and (1.2) are 6×10^{-16} cm^3 s^{-1} $T^{1.53}$ exp(-4060 K/T) and 1.3×10^{-17} cm^3 s^{-1} $T^{1.95}$ exp(-1420 K/T) respectively (Wagner and Graff 1987). (The UMIST rate file updated by Millar, Farquhar and Willacy (1997) is a useful source of data for a wide range of studies of low temperature to high temperature and ultraviolet irradiated to dark regions. A good compilation of the rates of photoionization due to background photons as well as photons emitted in the interaction of H_2 with fast electrons produced by cosmic-ray induced ionization has been assembled by Sternberg and Dalgarno (1995)). The rates of (2.18) and (2.19) are roughly 9.4×10^{-11} s$^{-1}\chi$ and 1.7×10^{-10} s$^{-1}\chi$ respectively (Roberge et al. 1991) where χ is proportional to the intensity of the far UV radiation field and is unity for the typical interstellar background (cf. Chapter 9). In cases in which equilibrium obtains and (1.1), (1.2), (2.18) and (2.19) are the primary relevant reactions we find that for $T = 1\times 10^3$ K, $n(OH)/n(O) \approx 4\times 10^{-3}\, n(H_2)/\chi$ and $n(H_2O)/n(OH) \approx 1.3\times 10^{-2}\, n(H_2)/\chi$. $n(X)$ is the number density of species X. In dense dark $T \approx 10^3$ K regions in which equilibrium obtains, much of the oxygen not in CO is in H_2O; in diffuse clouds, substantial shock heating will result in the production of OH and H_2O but much of the oxygen will remain in O.

Even behind shocks propagating into very dark regions in which the postshock value of $n(H_2O)/n(OH)$ becomes very high, there is a hot region in which OH is abundant while (1.1) and (1.2) are converting O to OH and OH to H_2O. Such a region of high $x(OH)$, where $x(X)$ is the number density of species X relative to the number density of hydrogen nuclei, is chemically a very important one because OH reacts with a variety of neutral species. Important reactions involving OH include

$$O + OH \rightarrow O_2 + H \tag{2.20}$$

and

$$S + OH \rightarrow SO + H \tag{2.21}$$

CO is not removed in any special way in shocked gas, but high abundances of OH and O_2 behind shocks propagating into $n(H_2) \approx 200$ cm^{-3} cold clumps in giant molecular clouds (in which, as discussed in Chapter 5, a significant fraction of the carbon may be in C) result in the production of further CO by

$$C + OH \rightarrow CO + H \tag{2.22}$$

and

$$C + O_2 \rightarrow CO + O \tag{2.23}$$

In dark regions reactions (2.22) and (2.23) are most important in shocked gas in which 400 K $\lesssim T \lesssim$ 1000 K. In gas in which $T \gtrsim$ 1000 K reaction (1.3) becomes important and initiates the formation of carbon hydrides. In diffuse clouds in which most preshock carbon is in C^+, the reaction

$$C^+ + H_2 \rightarrow CH^+ + H \tag{2.24}$$

(which is endothermic by about 0.4 eV) is particularly important in initiating a sequence of hydrogen abstraction reactions. In diffuse clouds as well as $n(H_2) \approx 200$ cm^{-3} clumps in giant molecular clouds shock heating can result in the analogous reaction involving S^+.

$$S^+ + H_2 \rightarrow SH^+ + H \tag{2.25}$$

followed by (1.4) initiates a rapid sequence of hydrogen abstraction reactions. The rate coefficient of reaction (2.25) contains an $\exp(-9860/T)$ factor, so $T \gtrsim 10^3$ K for it to become important. The reaction

$$S + H_2 \rightarrow SH + H \tag{2.26}$$

has a similar exponential factor and is important in initiating sulphur chemistry in $T \gtrsim$ 1000 K dark dense regions.

3 Dissociating Shocks

The dissociation energy of H_2 is 4.48 eV and in dense enough ($n_H \gtrsim 10^4$ cm^{-3}) molecular gas with $T \gtrsim$ 6000 K dissociation of some of the H_2 proceeds on a timescale shorter than the timescale (10^6 s) for vibrationally excited levels of H_2

to decay radiatively. We can make an estimate of the minimum speed of a shock propagating into a dense initially cold nonmagnetized H_2 medium required to cause all H_2 to dissociate by requiring $\mu V_S^2/2 \geq 4.48$ eV which yields $V_S \geq 21$ km s^{-1}. As discussed in the next section the presence of a magnetic field raises this minimum value of V_S.

Dissociation by collisions lowers the temperature of a gas. The rate coefficient for dissociative cooling depends on density and is smaller when the density is far lower than the critical density required for collisions to nearly thermalize the vibrational level population distribution and is independent of density for densities much exceeding the critical density (Roberge and Dalgarno 1982; Lepp and Shull 1983) because most dissociation occurs due to a step-by-step excitation process through the vibrational levels rather than in a direct process from the ground vibrational level to the vibrational continuum. For some choices of collisional excitation rate coefficients the differences between the low density and high density collisional dissociation rate coefficients are not large (Hollenbach and McKee 1989).

In gas that is dense enough and hot enough for dissociation to proceed on a timescale short compared to any radiative timescales, one may calculate the steady flow structure in the shock frame by setting time derivatives and Q equal to zero and using (2.8), (2.9), (2.3), (2.5), (2.6) and an equation governing the variation of μ due to the dissociation of molecules. One obtains boundary conditions at $z = 0$ for an $M \gg 1$ shock by using (2.13) and (2.14) and by setting $U = 3k_B \rho_2 T_2/2\mu$, $f_j = 1$ for the lowest rovibrational level of H_2 and $f_j = 0$ for all others and by taking μ appropriate for a gas in which all of the hydrogen is in H_2. In practice, rate coefficients used in the calculation of F_j, D_j and the dissociation rate which determines the evolution of μ are uncertain. For situations in which the radiative rates are comparable to the collisional de-excitation rates Q cannot be neglected as radiative processes affect the values of F_j and D_j.

Neufeld and Dalgarno (1989a,b) and Hollenbach and McKee (1989) have considered the chemistry in shocks that dissociate essentially all H_2 that passes through them.

In shocks just fast enough for collisional dissociation of H_2 to be nearly complete, other molecules are removed in reactions with atomic hydrogen. For instance, the reverses of (1.1) and (1.2) remove H_2O and OH. Even CO which is so stable under many circumstances is removed effectively at $T \gtrsim 3000$ K by the endothermic reaction

$$H + CO \rightarrow OH + C \tag{3.1}$$

Qualitative differences exist between the structures of shocks that are just fast enough for collisional dissociation of H_2 to be important and much faster shocks. (Figure 2 shows the thermal structure of the gas around a fast shock.) For instance, in shocks behind which 10^4 K $\lesssim T \lesssim 10^5$ K after dissociation is complete, most of the remaining energy is emitted in Lyman α. The fraction of energy lost through radiation shortward of Lyman α is about 0.27 behind a 90 km s^{-1} shock and increases with increasing V_S (Neufeld and Dalgarno 1989a).

Radiation propagating upstream and downstream from a fast shock dissociates and ionizes species. For $V_S > 100$ km s^{-1} the radiation propagating upstream ionizes a substantial fraction of the gas before it actually reaches the shock. The emission of this far and extreme ultraviolet radiation in the postshock region cools the gas down to about 10^4 K at which time recombination of ionized hydrogen with electrons is important; neutral gas absorbs extreme ultraviolet radiation propagating downstream to become reionized and as a consequence the zone in which substantial recombination is occurring can be extended much further than it would be if atoms remained neutral after being formed by recombination.

Until now we have neglected the magnetic fields, which we will consider more thoroughly in the next two sections but which should now be introduced. For sufficiently high values of V_S the presence of a magnetic field makes no significant difference in the flow immediately behind the shock; (2.13) and (2.14) can be used as jump conditions and the hydrodynamic treatment that we have described until now provides a good description. However, after gas behind a fast shock has cooled sufficiently the magnetic pressure becomes important. For simplicity consider a shock that propagates in a direction perpendicular to the upstream magnetic field. In a fast "perpendicular shock" in which the ionization structure in the upstream gas is significantly altered by the absorption of radiation emitted in the shock (too low a fractional ionization can introduce complexities in shock structure as described in the next two sections), the magnetic field strength B increases as ρ and the magnetic pressure, $B^2/8\pi$, increases as ρ^2. The thermal pressure behind a hydrodynamic shock varies only fractionally. Thus, if cooling to 10^4 K occurs the magnetic pressure eventually dominates over thermal pressure behind most fast interstellar shocks; no matter how cold the gas gets its compression is limited by the fact that the postshock magnetic pressure will not exceed the shock ram pressure, ρV_S^2. The effects of magnetic pressure on postshock structure have been included in investigations of the recombination of ions and the formation of molecules far downstream of fast dissociating shocks.

The downstream chemistry begins with the formation of H_2. Where the fractional ionization, x_i, relative to n_H is above about 0.02 the reaction sequence

$$H + e \rightarrow H^- + h\nu \tag{3.2}$$

$$H^- + H \rightarrow H_2 + e \tag{3.3}$$

is an effective source of H_2 (cf. Chapter 3); an H_2 fractional abundance of order 10^{-3} is attained behind an 80 km s^{-1} shock propagating through a medium with a preshock number density of 10^5 cm^{-3} and magnetic field strength of 3×10^{-4} G before the temperature drops much below about 3000 K (Neufeld and Dalgarno 1989a). In Fig. 3 the fractional abundances of various species are shown as functions of N_H, the column density hydrogen nuclei between the shock front and the point considered. The temperature drops substantially below about 5000 K for the first time when $N_H \approx 10^{20}$ cm^{-2}. H_2 formed by (3.2) and (3.3) enters reactions (1.1) and (1.2) to make OH and H_2O which are removed by the

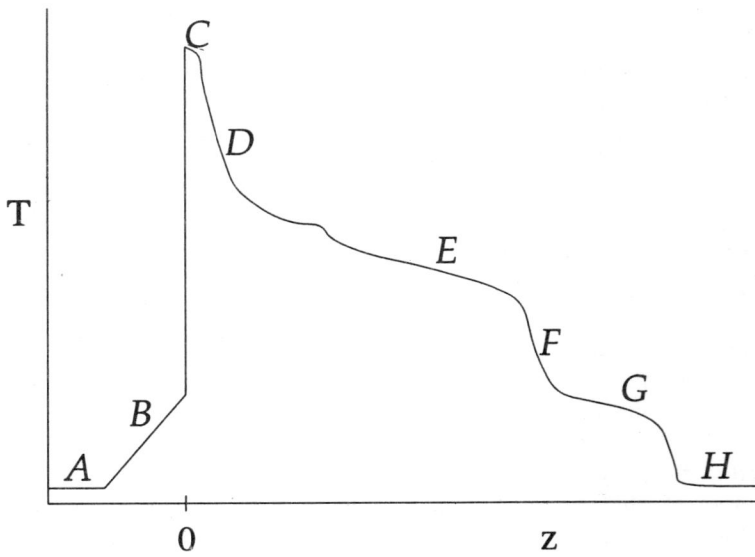

FIG. 2. Schematic representation of the thermal structure of gas near a fast dissociating and ionizing shock. Region A is the undisturbed preshock medium. In region B absorption of far ultraviolet and extreme ultraviolet photons emitted in region D dissociates molecules and produces ions. The shock front is at $z = 0$. In region C the remaining molecules are dissociated and the remaining neutrals ionized primarily by collisions. Region D is where most of the radiative cooling occurs. Usually somewhere in region D magnetic pressure begins to dominate over thermal pressure; beyond that point subsequent compression is by no more than a factor of $\sqrt{2}$. Region E is where ions recombine with electrons to form neutrals; absorption of extreme ultraviolet radiation from region D is important there. Subsequent cooling by atomic fine structure line emission lowers the temperature further in region F. The heat released by the formation of H_2, which occurs in region G, causes a temperature plateau to exist. After most of the hydrogen is in the form of H_2, cooling through molecular line emission further reduces the temperature until equilibrium is attained in region H.

reverse reactions and photodissociation. The reaction sequence

$$C^+ + OH \rightarrow CO^+ + H \tag{3.4}$$
$$CO^+ + H_2 \rightarrow HCO^+ + H \tag{3.5}$$
$$HCO^+ + e \rightarrow H + CO \tag{3.6}$$

is an important source of CO.

The temperature drop from 5000 K to about 1000 K occurs as N_H increases by considerably less than a factor of 2. At $T \gtrsim 10^3$ K, x_i is too low for (3.2) and (3.3) to be relevant. The formation of H_2, if it occurs, must take place on grain surfaces. The efficiency, ϵ_f, of H_2 molecules formed per every two hydrogen atoms striking grain surfaces is uncertain for $T \gtrsim 10^3$ K. ϵ_f has usually been assumed to be of order unity in studies of H_2 formation behind dissociating shocks. If ϵ_f is of order unity and H_2 is injected into the gas phase with translational and internal energies summing up to a substantial fraction of 4.48 eV, the gas is heated, as H_2 is forming, sufficiently to maintain its temperature at about 400–600 K. This temperature is high enough for reactions (1.1) and (1.2) to lead to substantial abundances of OH and H_2O in the H_2 formation region, and Elitzur, Hollenbach and McKee (1989) have argued that the excitation conditions are exactly correct in the H_2 formation region behind a dissociative shock in a dense cloud to produce H_2O maser line emission like that observed in the interstellar gas in regions of high-mass star formation.

After H_2 formation is nearly complete, the temperature drops to tens of degrees. When it does, reactions (1.1), (1.2), and (2.24) which are important as H_2 forms cease to be rapid. As these reactions become slow, many sequences that produce a rich chemistry in the H_2 formation zone become ineffective and photodissociation, caused by the far ultraviolet photons emitted upstream, leads to a large reduction in the fractional abundances of many species.

4 Multifluid Models of Perpendicular Shocks in Magnetized Weakly Ionized Media

In the previous section we noted that the presence of a magnetic field can limit the compression of cooled postshock gas, but in a weakly ionized medium a magnetic field can have other major effects on shock structure. In this section we restrict attention to the so-called perpendicular shocks, plane parallel shocks that propagate in directions perpendicular to the upstream magnetic fields.

In a fully ionized purely gaseous medium, information can be carried upstream perpendicular to the magnetic field at the fast-mode magnetosonic wave speed which for a wave vector perpendicular to **B** is given by

$$c_{f1} = \left(V_A^2 + c_a^2\right)^{1/2} \tag{4.1}$$

where c_a is given by (2.11) and the Alfvén speed is

$$V_A = \frac{B}{(4\pi\rho)^{1/2}} \tag{4.2}$$

A fast-mode MHD shock develops if disturbances are driven rapidly enough into a medium that fast-mode waves propagating in the upstream direction are overtaken. The Alfvénic Mach number of the shock can be defined as

$$M_A \equiv \frac{V_S}{V_{A1}} \tag{4.3}$$

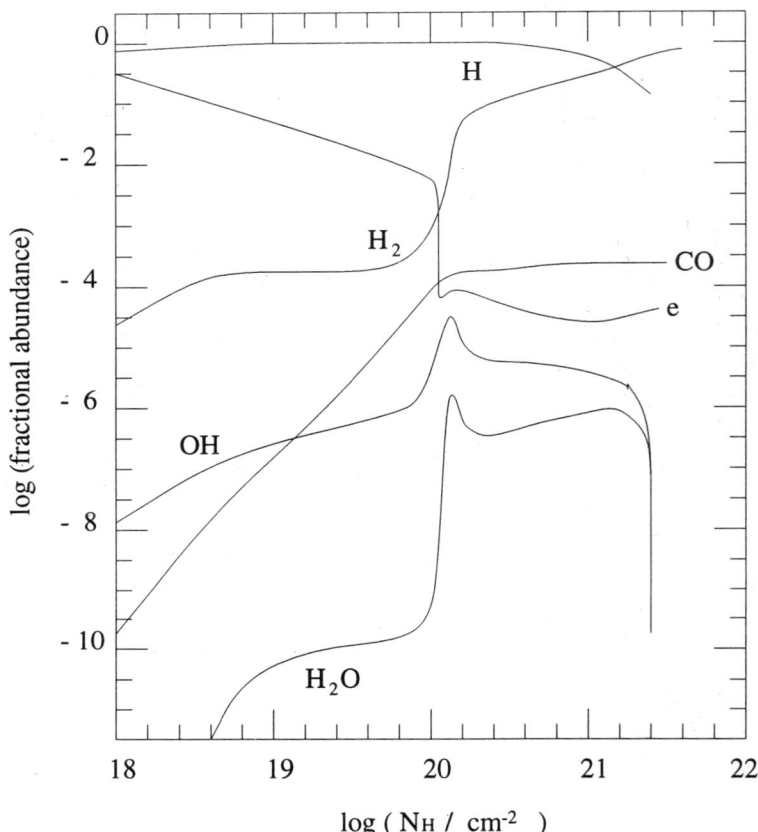

FIG. 3. Fractional abundances of major species as functions of N_H behind a 80 km s^{-1} perpendicular shock propagating into a medium with $n_H = 10^5$ cm^{-3} and $B = 3\times10^{-4}$ G. Results are from Neufeld and Dalgarno (1989a).

For $M_A \gg 8, M \gg 1$ and $\gamma = 5/3$, shock jump conditions are well approximated by (2.13), (2.14), and

$$B_2 = 4B_1 \qquad (4.4)$$

For lower M_As and Ms the jump conditions are more complicated (Field et al. 1968).

There are many weakly ionized, magnetized astrophysical media. Ions and electrons respond directly to electric and magnetic forces in such media, whereas neutral particles respond only indirectly to them through frictional coupling with the charged particles in the media. The timescales associated with the frictional coupling are often longer than other timescales of interest (e.g. that associated with radiative losses). Thus, multifluid models are sometimes used to describe flows in weakly ionized media. Henceforth, the subscripts i, e, and n will be used to indicate ion, electron, and neutral gas phase components. We define two Alfvén speeds

$$V_{AC} \equiv \frac{B}{(4\pi(\rho_i + \rho_e + \rho_n))^{1/2}} \quad (4.5)$$

and

$$V_{AU} \equiv \frac{B}{(4\pi(\rho_i + \rho_e))^{1/2}} \quad (4.6)$$

V_{AC} is the speed at which Alfvén waves with frequencies that are very small compared to the inverse of the timescale for neutrals subjected to no other forces to become coupled to ion motion, ν_{ni}, propagate; V_{AU} is the speed at which Alfvén waves with frequencies very large compared to the inverse of the timescale for ions subjected to no other forces to become coupled to neutrals, ν_{in}, propagate. We have

$$\frac{\rho_i}{\rho_n}\nu_{in} = \nu_{ni} = 2.41\left(\frac{m_i m_n}{m_i + m_n}\right)^{1/2} \frac{\rho_i}{m_i m_n} e\alpha_p^{1/2}$$

$$\approx 2 \times 10^{-9} s^{-1}\left(\frac{\rho_i}{m_i}/cm^{-3}\right) \quad (4.7)$$

(Osterbrock 1961) where m_i and m_n are the masses of each ion and neutral, e is the elementary charge, and α_p is the polarizability of the neutral species, which for H_2 is 8.04×10^{-25} cm^3. We now define two additional shock Alfvénic Mach numbers

$$M_{AC} \equiv \frac{V_S}{V_{AC}} \quad (4.8)$$

and

$$M_{AU} \equiv \frac{V_S}{V_{AU}} \quad (4.9)$$

We will assume that all fluids of the preshock media are sufficiently cold that the coupled and uncoupled fast-mode wave speeds are well approximated by V_{AC} and V_{AU} respectively; thus, M_{AC} and M_{AU} are the Mach numbers relevant to the consideration of shock propagation.

If $M_{AU} > 1$, no information is carried upstream in front of the disturbance; hence, a well-defined thin shock front, also called a jump-type shock or a J-type shock (Draine 1980; Draine, Roberge and Dalgarno 1983) will develop in all fluids. However, if $M_{AU} < 1$ information can be carried upstream for a certain distance. The information is carried upstream by the response of the charged species to the piston. In the absence of neutrals the structure of the upstream disturbance would get broader with time. However, the coupling between ions and neutrals over sufficient time and the fact that $M_{AC} > 1$ results in the upstream response being limited to a maximum distance Δ given approximately (Draine 1980) by

$$\Delta \approx \frac{V_S}{2\nu_{ni}M_{AC}^2} \quad (4.10)$$

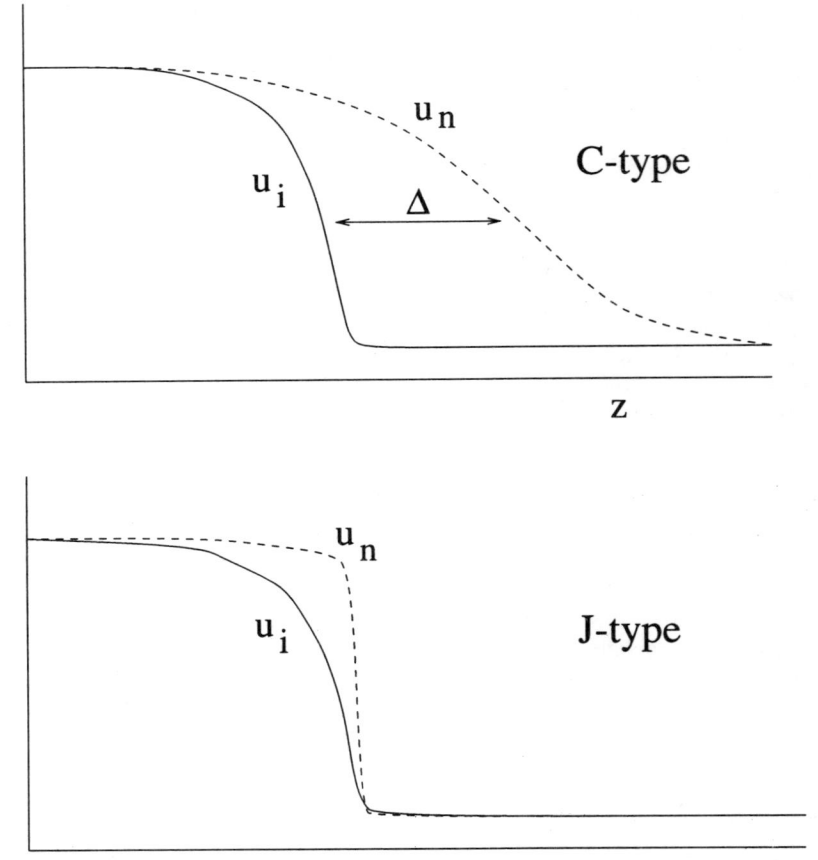

FIG. 4. Possible flow structures when $M_{AU} < 1$ and $M_{AC} > 1$. The flow is viewed in the frame comoving with the shock and the upstream flow is in the $\hat{z}$ direction. The charged fluids respond to the piston further upstream than the neutrals do, but ion–neutral coupling limits that distance. If heat generated by ion–neutral friction is efficiently radiated away, the flow remains smooth as depicted in the top panel. If radiative cooling is not efficient, a jump will develop in the flow of the neutrals (bottom panel).

Figure 4 shows schematically two possible steady shock structures that result when $M_{AU} < 1$, but $M_{AC} > 1$. The charged species move relative to the neutrals generating heat by friction. If the heat is radiated away sufficiently rapidly that $u_n > c_{an}$ (where c_{an} is the adiabatic hydrodynamic sound speed in the neutral fluid) everywhere, then no jump develops in the neutral fluid; the flow structure is described as a C-type shock (Draine 1980). If heating results in $u_n \leq c_{an}$ then a jump develops in the neutral flow and a J-type shock exists.

Chemistry is extremely important in governing the shock structure (e.g. the review by Hartquist, Flower and Pineau des Forêts 1990). The presence of some coolants can qualitatively alter the flow structure as mentioned in the previous paragraph, and the chemistry controlling the ionization structure determines the degree of coupling, the value of Δ, and the heating rate due to friction. Flower, Pineau des Forêts and Hartquist (1985) pointed out that in shocks with $V_S \geq 10$ km s^{-1} in typical diffuse interstellar molecular clouds (with $n_H \approx 10^2$ cm^{-3} and $B \approx 5 \times 10^{-6}$ G ahead of the shock) in which photoionization results in C$^+$ being the dominant ion (cf. Chapter 4) the removal of C$^+$ by (2.24) results in a significant drop in fractional ionization and subsequent broadening of the region in the shock where ions and neutrals are not effectively coupled; a major cause of the drop in fractional ionization is that dissociative recombination of molecular ions is much more rapid than radiative recombination.

Until now, we have neglected the fact that diffuse astrophysical media are often dusty. In the interstellar medium, the standard relationship between visual extinction (roughly the same as optical depth) and gas column density (Savage and Mathis 1979) implies that if we restrict attention to grains not much smaller than about 10^{-5} cm

$$\frac{\langle n_g \sigma_g \rangle}{n_H} \approx 5 \times 10^{-22} \text{ cm}^2 \qquad (4.11)$$

where n_g and σ_g are the number density and cross section of grains and $\langle \ \rangle$ implies that the appropriate size averaging has been taken. If grains were to move exactly as the ions do, use of (4.7) and (4.11) shows that grain–neutral friction would be of comparable importance for accelerating neutrals to ion–neutral friction if

$$x_i \approx 10^{-8} \left(\frac{V_S}{\text{km s}^{-1}} \right) \qquad (4.12)$$

We return to grains in the next section and note here only that grains are usually charged in diffuse astrophysical media and their effects on the coupling between charged species and neutrals have been included in various ways in many theoretical studies of shock structure.

Figure 5 shows the temperature profile of a 40 km s^{-1} shock obtained in one theoretical study by Kaufman and Neufeld (1996). The preshock medium was taken to have $n(H_2) = 10^5$ cm^{-3} and $B = 4.47 \times 10^{-6}$ G. Very small grains with radii of only 4 Å were assumed to have an upstream fractional abundance of 2×10^{-7} and to remain well coupled to the ions and electrons as well as to provide a source of drag on the neutrals. (The assumption of such a population of very small grains might be considered controversial but does not alter by large factors the effects of grain–neutral coupling from those of other models.) Note that the maximum value of T is greatly below that given by equation (2.14).

By lowering the maximum temperature in a shocked region the presence of a magnetic field significantly alters the value of the minimum shock speed required for dissociation to be important. This minimum shock speed depends

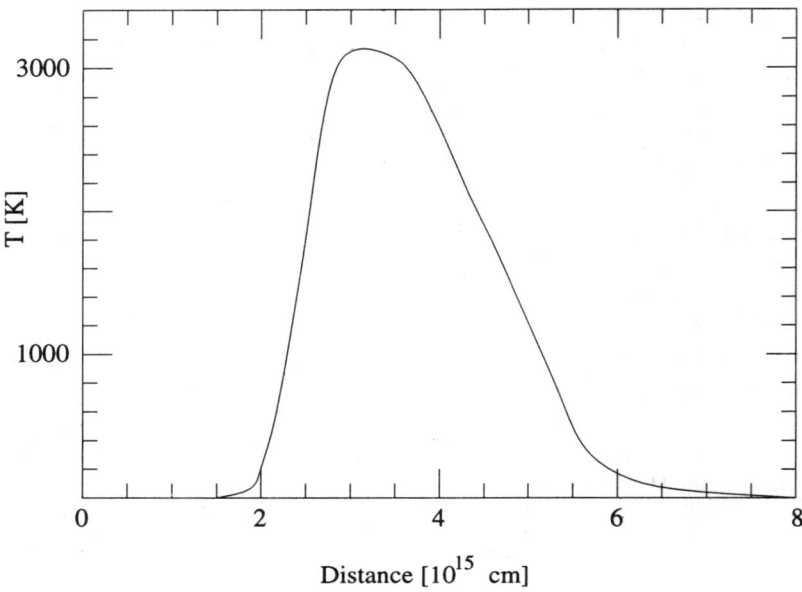

FIG. 5. The temperature profile of a 40 km s^{-1} perpendicular shock propagating into a postshock medium with $n(H_2) = 1\times10^4$ cm^{-3} and $B = 4.47\times10^{-6}$ G.

on the preshock magnetic field strength but is estimated to be around 40–50 km s^{-1} for typical interstellar shocks (e.g. Draine, Roberge and Dalgarno 1983).

As grains contain about one per cent of the interstellar mass their density should not be ignored in the calculation of V_{AU}. If grains were extremely well coupled to the magnetic field (which, as discussed in the next section, is not always the case), their density would lower V_{AU} to about 40 km s^{-1} for typical interstellar magnetic fields. As yet proper account of grain mass has not been taken in studies of shocks, but it will make little difference for shocks with $V_S \ll V_{AU}$ with V_{AU} calculated with the inclusion of grains.

Table 1 contains results from Kaufman and Neufeld (1996) for the fraction of the shock mechanical energy flux ($\rho V_S^3/2$) radiated by several different species. The preshock magnetic field strength was taken to be 1×10^{-6} G $(n(H_2)/cm^{-3})^{1/2}$. Their paper should be consulted for detailed results for individual lines observable in the far infrared with ISO and other instruments.

The numerical results given in this section are for models based on the assumption that the flows are steady. However, Wardle (1991) has shown that what is essentially a magnetic Rayleigh–Taylor or Parker instability develops when the $M \gg 1$ and $M_{AC} \geq 5$. Consider small amplitude periodic perturbations with associated magnetic field disturbances parallel to the shock propagation direction. The friction between charged particles and neutrals retards the

Table 1 Fraction of mechanical energy flux radiated by various coolants

Upstream $n(H_2)$ cm^{-3}	V_S km s^{-1}	H_2O	$H_2(J)$*	$H_2(V)$†	CO
1×10^4	10	0.7	24.7	0.0	27.4
	20	22.7	45.0	0.1	7.2
	30	14.2	63.1	3.5	3.9
	40	9.1	68.6	9.9	2.4
1×10^5	10	10.4	9.1	0.0	28.9
	20	44.5	20.3	0.4	10.2
	30	32.1	41.1	13.7	6.6
	40	21.5	42.3	31.3	4.2
1×10^6	10	27.1	2.2	0.0	13.9
	20	70.0	3.5	0.2	7.4
	30	55.9	11.3	14.6	6.4
	40	41.0	13.5	28.3	4.6

* Contribution from decay of excited $v = 0$, J levels
† Contribution from decay of vibrationally excited levels

charged particle motion along the perturbation magnetic field component so that the charged particle density increases in regions where the field bows in the upstream direction but decreases in regions where the field bows in the downstream direction. The charged particles in the regions of increased density experience a greater drag force per unit volume towards the upstream direction than the charged particles in the regions of decreased charged particle density; this tends to increase the amplitude of the bowing.

The lack of detailed knowledge of grain properties and the existence of instabilities means that detailed results (like those given in Table 1) must be considered as indicative rather than precise.

5 More on Grains, Their Possible Role in Lowering the Speed Limits of Steady Shocks, and Their Effects on Shock Chemistry

Because electrons have much higher thermal speeds than ions, a neutral grain in a plasma in which the gas phase ions and electrons have nearly equal number densities is more likely to encounter an electron than an ion; thus, grains in such circumstances carry an average negative charge of

$$Z_g e = -1.5e \left(\frac{T}{100\,\mathrm{K}}\right)\left(\frac{a}{10^{-5}\,\mathrm{cm}}\right) \quad (5.1)$$

if the ions are H^+, a is the radius of the grain, and $Z_g \ll -1$ (e.g. Draine and Sutin 1987). In shocks in dense enough ($n_H > 10^7 \text{cm}^{-3}$) dark Galactic molecular regions the electron temperature (resulting from the balance between electron–neutral frictional heating and energy loss from electron impact excitation of H_2) is typically several thousand degrees and Z_g becomes sufficiently negative that, if (4.11) is assumed, ion recombination onto grains is so rapid that

$$\frac{\mid n_i - \mid n_g Z_g \mid \mid}{n_i} \ll 1 \tag{5.2}$$

(Pilipp, Hartquist and Havnes 1990; Hartquist, Pilipp and Havnes 1997). Thus, as in the cold, dark molecular regions of the Galaxy (cf. Chapters 4 and 5) in shocks (including those with $n_H < 10^7 \text{cm}^{-3}$) grains are very important for establishing the ionization balance.

We have already given a rough criterion (eqn (4.12)) for neutral–grain friction to be important for the shock structure. That criterion is based on the assumption that grains are well coupled to the magnetic field, i.e.

$$\frac{\mid u_{gz} - u_{ez} \mid}{\mid u_{ez} - u_{nz} \mid} \ll 1 \tag{5.3}$$

where the subscript z denotes the $\hat{z}$ component of a vector, and the shock is assumed to propagate in the $\hat{z}$ direction. In fact, (5.3) does not always hold; in the regime in which

$$\left(\frac{\nu_{gn}}{\Omega_g}\right)^2 \gg 1 \tag{5.4}$$

(5.3) is invalid unless

$$\frac{\nu_{ng}}{\nu_{ni}} \left(\frac{\Omega_g}{\nu_{gn}}\right)^2 \gg 1 \tag{5.5}$$

(Hartquist, Pilipp and Havnes 1997) where Ω_g is the grain gyrofrequency and ν_{ng} and ν_{gn} are the inverses of the timescales for a neutral gas particle and for a grain to slow down if acted on by no forces other than grain–neutral drag.

In short, though grains are charged, a fluid of them does not behave like a perfectly conducting fluid due to the importance of grain–neutral friction. The imperfectly conducting nature of a grain fluid proves to be of particular importance for oblique shocks in weakly ionized dusty, dark clouds (Pilipp and Hartquist 1994; Hartquist, Pilipp and Havnes 1997). An oblique shock is one that propagates in a direction that is neither parallel nor perpendicular to the direction of the upstream magnetic field. We assume that in the shock frame the upstream fluid is moving with $u = V_S \hat{z}$ and has a constant magnetic field $\mathbf{B}_o = B_{ox}\hat{x} + B_{oz}\hat{z}$. We will consider steady shock models which implies that charge conservation requires that the $\hat{z}$ component of the current $\mathbf{J}$ is zero. In the shock frame the electric field is nonzero; $E_x = 0$ but $E_y = -V_S B_{xo}/c$ where c is the speed of light. For a massless, cold, perfectly conducting fluid $\mathbf{E} + (\mathbf{u} \times \mathbf{B})/c$

= 0 which implies that $\mathbf{u} \propto \mathbf{E} \times \mathbf{B}$. In the shock frame $(\mathbf{E} \times \mathbf{B})_z \propto E_y B_x$. If all charged fluids were perfectly conducting, they would all move with the same $(\mathbf{E} \times \mathbf{B})$ drift velocity. However, the grain–neutral friction is sufficient for the grains to drift differently than the other charged fluids in the $\mathbf{E} \times \mathbf{B}$ direction; since $\mathbf{E} \times \mathbf{B}$ has a $\hat{\mathbf{z}}$ component the imperfect conductivity of the grains causes a finite contribution to the $\mathbf{E} \times \mathbf{B}$ current's (or Hall current's) $\hat{\mathbf{z}}$ component. For $J_z = 0$, as we stated must be the case in a steady flow, the $\hat{\mathbf{z}}$ component of the $\mathbf{E} \times \mathbf{B}$ current must be counterbalanced by the $\hat{\mathbf{z}}$ component of a current that is parallel to $\mathbf{B}$; as long as $B_x \neq 0$ the current parallel to $\mathbf{B}$ also has an $\hat{\mathbf{x}}$ component. From Ampere's law $dB_y/dz = -4\pi J_x/c$ and the $\hat{\mathbf{x}}$ component of the current parallel to $\mathbf{B}$ causes the magnetic field to rotate around the shock propagation direction. This rotation of the field around the shock propagation direction, induced primarily by the grain–neutral drag (though ion–neutral drag contributes to a lesser extent), has significant consequences for shock structure. In fact, Pilipp and Hartquist (1994) failed to find (but did not rigorously rule out the existence of) steady oblique shock solutions for typical dark interstellar conditions when $V_S^2 4\pi(\rho_n + \rho_i + \rho_e + \rho_g)/B_{ox}^2$ exceeded unity by more than a small fraction. Possibly grain–neutral drag establishes lower speed limits for steady shocks than the instability investigated by Wardle (1991) does.

The issue of the extent to which grains are coupled to the magnetic field or to the neutral flow is important for the question of how much material is removed from grains and introduced into the gas phase by grain–neutral gas collisions (Draine, Roberge and Dalgarno 1983; Draine 1995; Caselli, Hartquist and Havnes 1997; Schilke et al. 1997) and by grain–grain collisions (Caselli, Hartquist and Havnes 1997) in shocks. Given the questions about the existence of non-dissociating steady shocks in magnetized molecular clouds raised above and the huge uncertainties in grain properties there are limitations to the rigour with which such studies can be conducted. Improved sputtering data (Jurac, Johnson and Dunn 1997) have become available after the most recent studies of sputtering in C-type shocks. The existence of gas phase SiO with fractional abundances of 10^{-6} in outflows around young low-mass stars (e.g. Martín-Pintado, Bachiller and Fuente 1992; Avery and Chiao 1996) and of 10^{-9} to 10^{-8} in regions where the formation of high-mass stars has recently occurred is almost certainly due to grain erosion in shocks (Caselli, Hartquist and Havnes 1997; Schilke et al. 1997). Caselli, Hartquist and Havnes (1997) found that grain–grain collisions in ambient gas through which an outflow is driving a shock can produce gas phase abundances of elemental silicon in harmony with the SiO data for regions of high mass star formation but that a higher speed shock in which sputtering dominates grain erosion is required to produce the higher gas phase elemental silicon abundances in gas around young stellar outflows.

We conclude this section and this chapter with the observation that studies of dynamics as well as investigations of chemistry in astrophysical molecular sources often require very careful considerations of dust. The validity of this observation certainly receives support from our knowledge of shock physics and chemistry.

Bibliography

1. Avery, L. W. and Chiao, M. (1996). *ApJ* **463**, 642.
2. Caselli, P., Hartquist, T. W. and Havnes, O. (1997). *A&A*, **322**, 296.
3. Draine, B. T. (1980). *ApJ*, **241**, 1021.
4. Draine, B. T. (1995). *Ap&SS*, **233**, 111.
5. Draine, B. T., Roberge, W. G. and Dalgarno, A. (1983). *ApJ*, **264**, 485.
6. Draine, B. T. and Sutin, B. (1987). *ApJ*, **320**, 803.
7. Elitzur, M., Hollenbach, D. J. and McKee, C. F. (1989). *ApJ*, **346**, 983.
8. Field, G. B., Rather, J. D. G., Aannestad, P. A. and Orszag, S. A. (1968). *ApJ*, **151**, 953.
9. Flower, D. R., Pineau des Forêts, G. and Hartquist, T. W. (1985). *MNRAS*, **216**, 775.
10. Hartquist, T. W., Flower, D. R. and Pineau des Forêts, G. (1990). In *Molecular astrophysics—a volume honouring Alexander Dalgarno*, ed. T. W. Hartquist. Cambridge University Press, Cambridge, p.99.
11. Hartquist, T. W., Pilipp, W. and Havnes, O. (1997). *Ap&SS*, **246**, 243.
12. Hollenbach, D. and McKee, C. F. (1989). *ApJ*, **342**, 306.
13. Innes, D. E. Giddings, J. R. and Falle, S. A. E. G. (1987). *MNRAS*, **227**, 1021.
14. Jurac, S., Johnson, R. E. and Donn, B. (1997). *ApJ*, submitted.
15. Kaufman, M. J. and Neufeld, D. A. (1996). *ApJ*, **456**, 611.
16. Landau, L. D. and Lifshitz, E. M. (1959). *Fluid mechanics*, Pergamon Press, Oxford.
17. Lepp, S. and Shull, J. M. (1983). *ApJ*, **270**, 578.
18. Martin-Pintado, J., Bachiller, R. and Fuente, A. (1992). *A&A*, **254**, 315.
19. Millar, T. J., Farquhar, P. R. A. and Willacy, K. (1997). *A&A Suppl*, **121**, 139.
20. Neufeld, D. A., and Dalgarno, A. (1989a). *ApJ*, **340**, 869.
21. Neufeld, D. A. and Dalgarno, A. (1989b). *ApJ*, **344**, 251.
22. Osterbrock, D. E. (1961). *ApJ*, **134**, 270.
23. Pilipp, W. and Hartquist, T. W. (1994). *MNRAS*, **267**, 801.
24. Pilipp, W., Hartquist, T. W. and Havnes, O. (1990). *MNRAS*, **243**, 685.
25. Roberge, W. G. and Dalgarno, A. (1982). *ApJ*, **255**, 176.
26. Roberge, W. G., Jones, D., Lepp, S. and Dalgarno, A. (1991). *ApJ Suppl*, **77**, 287.
27. Savage, B. D. and Mathis, J. S. (1979). *ARA&Ap*, **17**, 73.

28. Schilke, P., Walmsley, C. M., Pineau des Forêts, G. and Flower, D. R. (1997). *A&A*, **321**, 293.
29. Sternberg, A. and Dalgarno, A. (1995). *ApJ Suppl*, **99**, 565.
30. Tedds, J. A., Brand, P. W. J. L., Burton, M. G., Chrysostomou, A. and Fernandes, A. J. L. (1995). *Ap&SS*, **233**, 39.
31. Tiné, S., Lepp, S., Gredel, R. and Dalgarno, A. (1997). *ApJ*, **481**, 282.
32. Wagner, A. F. and Graff, M. M. (1987). *ApJ*, **317**, 423.
33. Wardle, M. (1991). *MNRAS*, **251**, 119.

9
Photon-dominated Regions

Amiel Sternberg
School of Physics and Astronomy, Tel Aviv University

1 Introduction

Photon-dominated regions (PDRs) are produced in dense neutral hydrogen clouds exposed to far-ultraviolet (FUV) radiation fields (6–13.6 eV). PDRs are present at the edges of HII regions, in reflection and planetary nebulae, and are pervasive inside clumpy star-forming molecular clouds illuminated by direct and scattered stellar FUV radiation. PDRs are present in starburst galaxy nuclei and active galactic nuclei.

The physical and chemical properties of PDRs are controlled by the incident FUV photons which heat the gas and drive the cloud chemistry. PDRs are optically thicker and generally denser than the more transparent translucent or diffuse clouds (see Chapter 4), and are observable as sources of a wide range of infrared and millimetre atomic and molecular line emission, and thermal infrared dust continuum emission. The observational and theoretical study of PDRs has been stimulated by rapid advances in millimetre and submillimetre receiver and infrared detector technologies for ground-based, airborne and space-based telescopes.

In this chapter I discuss some of the basic physical and chemical processes which operate in PDRs. In §2 I review theoretical and observational studies of molecular hydrogen emission produced in PDRs. In §3 I discuss heating and cooling mechanisms, and in §4 I discuss the chemical structure and molecular diagnostics in dense PDRs. Other reviews of PDR observations and theory have been presented by Genzel, Harris and Stutzki (1989), Sternberg, Yan and Dalgarno (1997), and Hollenbach and Tielens (1997).

2 Molecular Hydrogen

2.1 H/H_2 Transition

The outer part of a PDR consists of a layer of atomic hydrogen and an H/H_2 photodissociation front across which the hydrogen becomes molecular. Additional distinct chemical layers (discussed in §4) exist at greater cloud depths. The atomic and H/H_2 transition layers are important because observable H_2 emission lines are produced in them, and because molecular chemistry is initiated as the hydrogen becomes molecular.

H_2 is efficiently photodissociated in PDRs by the absorption of FUV photons in discrete Lyman and Werner (dipole) transitions from the ground X $^1\Sigma_g^+$ to excited B $^1\Sigma_u^+$ and C $^1\Pi_u$ electronic states followed by spontaneous radiative decays to the continuum of the ground state (Stephens and Dalgarno 1972; Abgrall et al. 1992). For a "unit" interstellar photon flux $\phi = 2 \times 10^7$ photons s^{-1} cm^{-2} within the 11.2 to 13.6 eV H_2 FUV absorption band (e.g. Draine 1978) the total unattenuated H_2 dissociation rate, D_0, summed over all of the relevant absorption lines equals 5×10^{-11} s^{-1}. The parameter χ will be used as a linear scaling factor of the incident FUV intensity such that $\chi = 1$ corresponds to ϕ and D_0 having the above values.

H_2 is formed by surface catalysis at a rate per unit volume of $Rnn(H)$ cm^{-3} s^{-1} where n is the total density of hydrogen nuclei, $n(H)$ is the density of atomic hydrogen, and R is the H_2 formation rate coefficient. For normal interstellar dust-to-gas mass ratios $R = 3 \times 10^{18} T^{1/2} y_F$ cm^3 s^{-1} where T is the gas temperature (K), and y_F is the formation efficiency which depends on the surface properties and gas and grain temperatures (Jura 1975; Duley 1996; Levinson, Chernoff and Salpeter 1997).

In a plane-parallel geometry the local H_2 density at a cloud depth z (cm) is governed by the equation

$$\frac{\partial n(H_2)}{\partial t} = Rnn(H) - \chi Dn(H_2) - \frac{\partial(n(H_2)v)}{\partial z} \tag{2.1}$$

where t is the time, D is the local dissociation rate for a unit incident FUV flux, and v is the local flow velocity of gas through the PDR. The behaviour is simplest if the cloud is in equilibrium and no flows are present, as is assumed in most model computations. The time derivative and the advection term then vanish, and the atomic to molecular hydrogen density ratio is simply

$$\frac{n(H)}{n(H_2)} = \frac{\chi D}{Rn}. \tag{2.2}$$

For a fixed cloud density and a constant H_2 formation rate coefficient the molecular fraction grows as the dissociation rate diminishes with cloud depth. The dissociation rate decreases as the FUV radiation is attenuated by continuum dust absorption and the H_2 line absorptions.

The Doppler cores of the strongest H_2 absorption lines, with oscillator strengths ~ 0.01 (Allison and Dalgarno 1970), become optically thick at H_2 columns of $\sim 10^{14}$ cm^{-2}. This is much smaller than the hydrogen column density at which the dust opacity becomes significant, because for the standard assumed effective FUV dust absorption cross section per hydrogen nucleus $\sigma \approx 1.9 \times 10^{-21}$ cm^2 at 100 nm. The H_2 molecules therefore "self-shield", and for the strongest lines the FUV absorption occurs out of the line-damping wings throughout most of the PDR. At sufficiently large depths the absorption lines overlap, further reducing the dissociation rate. A limit to the ratio of line and dust opacites at the wavelengths at which the lines overlap is

$$\frac{\tau^l}{\tau_{dust}} < 1.15 \left(\frac{10^{-21} \text{ cm}^2}{\sigma}\right)\left(\frac{f}{0.01}\right)\left(\frac{A^*}{10^9 \text{ s}^{-1}}\right)\left(\frac{\lambda}{100 \text{ nm}}\right)^4 \left(\frac{0.05 \text{ nm}}{\Delta\lambda}\right) \quad (2.3)$$

where $f \approx 0.01$, $A^* \approx 10^9$ s^{-1}, and $\lambda \approx 100$ nm, are the oscillator strengths, radiative transition rates, and wavelengths of the absorption line transitions, and $\Delta\lambda \approx 0.05$ nm is half the typical absorption line separation (Roberge 1981). For typical values of the dust absorption cross section σ the ratio $\tau^l/\tau \sim 1$ at the overlap wavelength, so that line overlap contributes significantly to the FUV attenuation at cloud depths corresponding to $\tau_{dust} > 1$ (Draine and Bertoldi 1996; see also Chapter 24, eqn (2.7)).

Integration of eqn (2.2) yields the expression

$$N(H) = \frac{1}{\sigma}\ln\left[\frac{\chi D_0}{Rn}G + 1\right] \quad (2.4)$$

for the total equilibrium column density of atomic hydrogen in plane-parallel PDRs (Sternberg 1988). In this expression $\alpha \equiv \chi D_0/Rn$ and $G \sim 3 \times 10^{-5}$ is a cloud averaged self-shielding factor. When $\chi/n \ll 0.01$ cm^3 the atomic column is small, and the hydrogen becomes molecular close to the cloud surface. In this regime the dust opacity associated with the atomic gas $\sigma N(H) \ll 1$. When $\chi/n \gg 0.01$ cm^3 a large column density of atomic hydrogen is maintained, and the dust opacity $\sigma N(H)$ associated with the atomic hydrogen layer is large.

A flow of molecular gas from the molecular zone into the atomic layer can affect the location and structure of the H/H$_2$ dissociation front when the flow time across the atomic layer $t_{flow} \sim N_H/nv$ is comparable to the effective H$_2$ dissociation time $t_{dis} \sim 1/\chi D_0 G$ in the HI layer. It follows from equation (2.4) that in the strong field limit ($\chi/n \gg 0.01$ cm^3) flows become important for flow velocities $v > 8\chi/n$ km s^{-1}. Flows with velocities of this order may be set up in PDRs adjacent to rapidly photoevaporating ionization fronts, and for sufficiently rapid flows the dissociation and ionization fronts may merge (Bertoldi and Draine 1996; Störzer and Hollenbach 1998).

2.2 H$_2$ Emission Line Spectra

The absorption of FUV photons by H$_2$ molecules in PDRs is followed by the emission of a rich array of ultraviolet, optical and infrared H$_2$ emission lines. An ultraviolet fluorescent emission spectrum is produced as the electronically excited H$_2$ molecules decay in rapid discrete transitions to excited rotational and vibrational (vJ) levels of the ground X state (Sternberg 1989; Jansen et al. 1995a; Neufeld and Spaans 1996). An infrared and optical fluorescent emission line spectrum is produced as the vibrationally excited molecules decay in a cascade of rotational and vibrational quadrupole transitions (Black and Dalgarno 1976; Black and van Dishoeck 1987; Sternberg 1988; Sternberg and Dalgarno 1989; Draine and Bertoldi 1996; Neufeld and Spaans 1996). The quadrupole

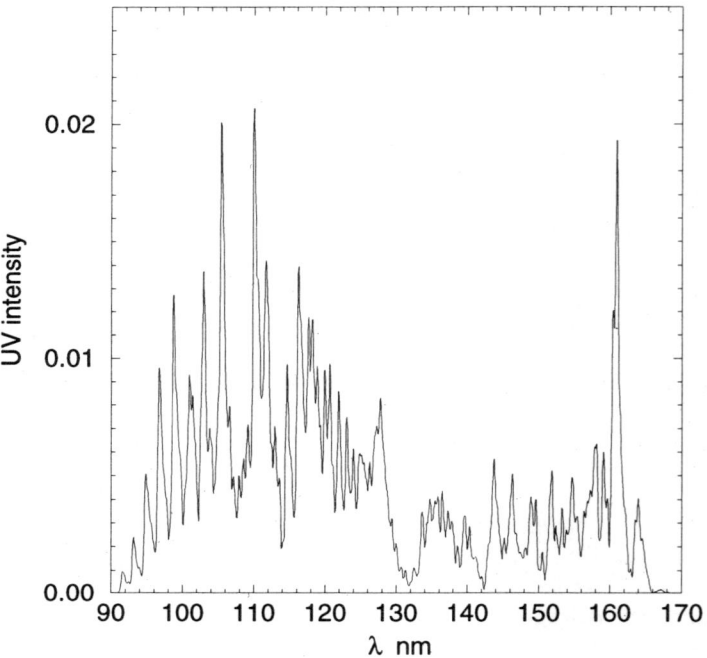

FIG. 1. FUV H_2 emission line spectrum.

transition probabilities are small, ($\sim 10^{-6}$ s^{-1} for $\Delta v = 1$ transitions) and the infrared/optical emission spectrum can be influenced by collisional processes in dense PDRs (Sternberg and Dalgarno 1989).

Figure 1 shows a computed FUV emission line spectrum of an isothermal PDR with a gas density $n = 10^3$ cm^{-3} and a temperature of 100 K, exposed to a $\chi = 10^3$ FUV field with an effective colour temperature of 10^4 K (Sternberg 1989). The assumed spectral resolution is $\lambda/\Delta\lambda = 500$. The UV emission line spectrum consists of a cluster of features between 90 and 130 nm, and a prominent blend of Lyman lines near 160 nm. The relative intensities of the FUV emission line spectrum are insensitive to n or χ, and depend slightly on the spectral shape of the illuminating radiation field. Using the IUE satellite Witt et al. (1989) observed FUV H_2 emission from the reflection nebula IC 63. They detected the intense 160 nm feature, and additional continuous emission between 140 and 160 nm which is produced during the molecular dissociation (Dalgarno, Herzberg and Stephens 1970; Jansen et al. 1995a). Ultraviolet H_2 emission has also been detected from the diffuse interstellar medium (Martin, Hurwitz and Bowyer 1990; Neufeld and Spaans 1996).

Figure 2 shows a computed infrared fluorescent emission spectrum ($\lambda/\Delta\lambda = 500$) produced in a PDR with $n = 10^3$ cm^{-3}, $\chi = 10^3$ and a uniform gas

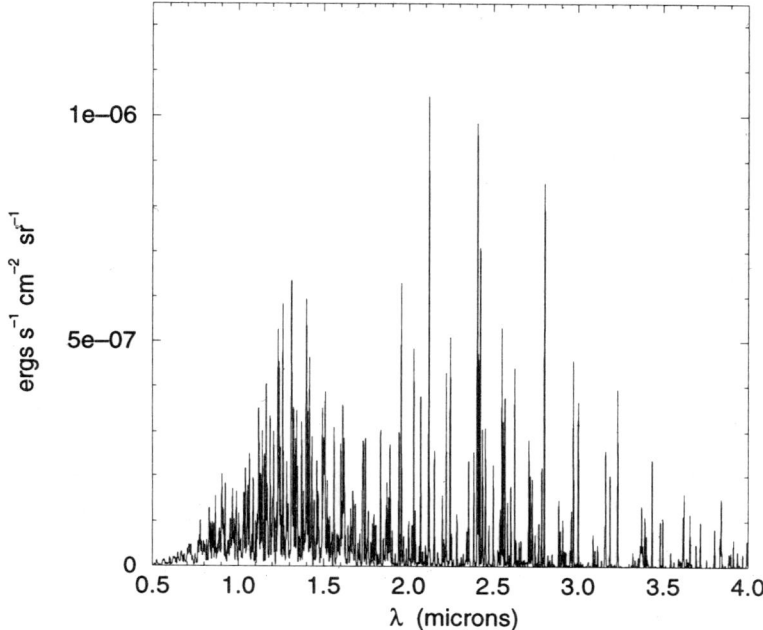

FIG. 2. Low density infrared/optical fluorescent H_2 emission line spectrum.

temperature of 500 K. In low density clouds ($n < 5 \times 10^4$ cm^{-3}) the relative intensities of the emission lines are determined primarily by the quadrupole transition probabilities (Turner et al. 1977; Wolniewicz, Simbotin and Dalgarno 1997) and are insensitive to n or to χ. The absolute intensities depend on the intensity of the incident radiation and the fraction of the FUV photons which are absorbed by H_2 rather than by dust grains. When χ/n is small a fixed fraction (~50%) of the FUV radiation is absorbed by the hydrogen molecules. The IR fluorescent emission intensity is then proportional to χ and independent of the cloud density. In this weak-field limit the specific intensity of the commonly observed $v = 1$–0 S(1) 2.12 µm line $I_{2.12\mu m} \approx 6 \times 10^{-8} \chi$ ergs s^{-1} cm^{-2} sr^{-1}. When χ/n is large, and the dust opacity associated with the atomic hydrogen layer becomes large, an increasing fraction of the FUV radiation is absorbed by the dust grains. The fluorescent emission intensity is then only weakly dependent on χ and is proportional to n. In this limit $I_{2.12\mu m} \approx 8 \times 10^{-9} n$ ergs s^{-1} cm^{-2} sr^{-1}.

Low density fluorescent H_2 emission has been detected in many ground-based near-IR observations in many objects including reflection nebulae (e.g. NGC 2023, NGC 7023, Parsamyan 18), planetary nebulae (Hubble 12, BD +303639) and in molecular cloud interfaces in Orion (Burton et al. 1992; Graham et al. 1993; Luhman and Rieke 1996; Luhman et al. 1997; Martini, Sellgren and Hora 1997).

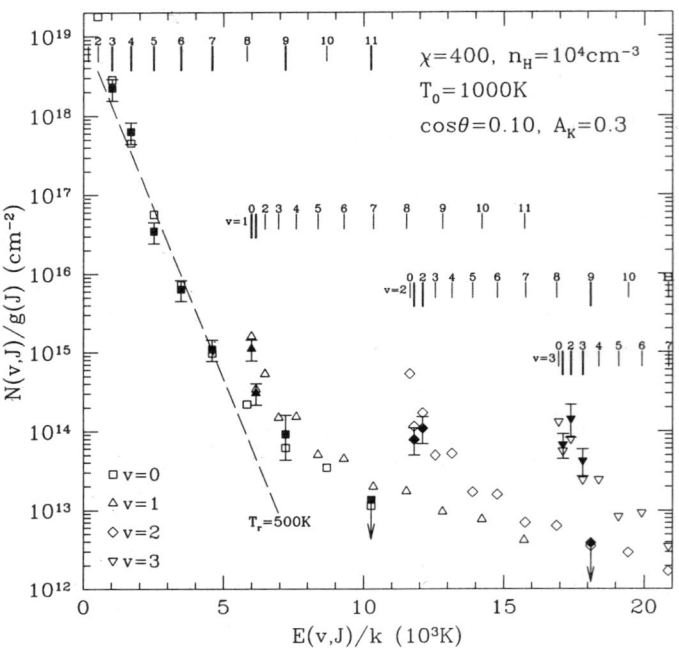

FIG. 3. ISO observations and model fit of rotational and vibrational H_2 emission in the S140 PDR (Timmermann et al. 1996).

H_2 emission from the S140 PDR was recently detected by Timmermann et al. (1996) with the Infrared Space Observatory (ISO). The S140 PDR is illuminated by a $\chi \approx 400$ FUV field produced by a B0.5V star located about 1.7 pc from the molecular filaments. The S140 data and a model fit are shown in Fig. 3. The filled symbols are the inferred column densities in the upper levels of the observed emission lines. The open symbols are the column densities for a (low-density) model cloud with $n = 10^4$ cm^{-3} exposed to a $\chi = 400$ radiation field. In the model the gas temperature is fixed at a maximum value of $T_0 = 1000$ K at the cloud edge, and decreases to less than 100 K through the H/H_2 transition zone. S140 is an edge-on system, and the model H_2 column densities were computed for a line-of-sight inclination angle θ with $\cos\theta = 0.1$. Several pure rotational mid-IR H_2 rotational lines were detected, as well as lines emitted from the first three excited vibrational levels. The populations in the $J \leq 5$ levels appear to be thermalized at a gas temperature of ~ 500 K, implying that the atomic hydrogen layer is warm with gas temperatures of at least 500 K. The model fit shows that the relative populations in the excited H_2 vibrational levels in S140 are nicely consistent with low-density fluorescent emission.

In dense ($n > 5 \times 10^4$ cm^{-3}) PDRs the quadrupole vibrational cascade is modified by collisions between the excited H_2 molecules and hydrogen atoms

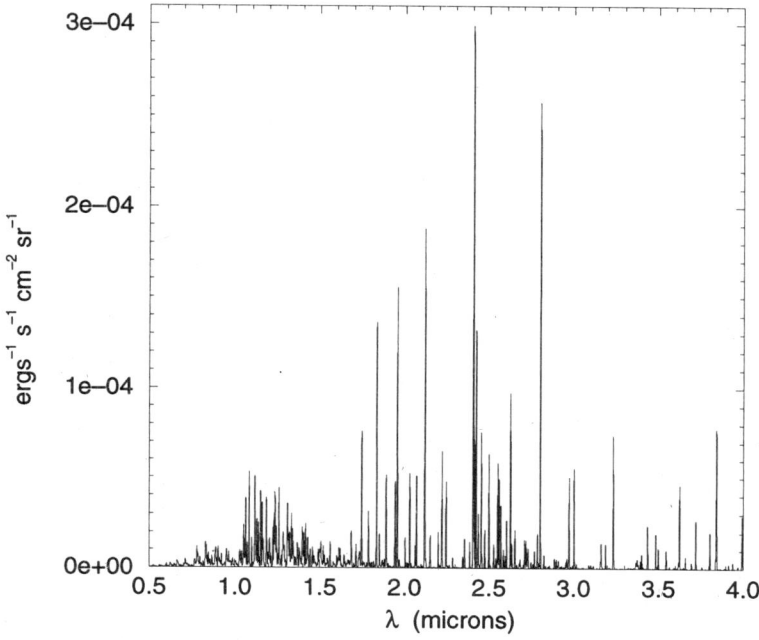

FIG. 4. High density fluorescent H_2 emission line spectrum.

(Sternberg and Dalgarno 1989; Burton, Hollenbach and Tielens 1990; Draine and Bertoldi 1996). Semiclassical trajectory computations of the rate coefficients for collisionally induced transitions, $H + H_2(vJ) \rightarrow H + H_2(v'J')$ have recently been carried out by Martin and Mandy (1995) and Lepp, Tiné and Dalgarno (1997). Figure 4 shows the collisionally modified H_2 emission spectrum for a PDR with $n = 10^6$ cm^{-3}, $\chi = 10^5$, and a gas temperature of 500 K. The Lepp, Tiné and Dalgarno rate coefficients for the H–H_2 collisions were used in this computation. A comparison with Fig. 2 shows that emission from the higher lying vibrational levels is more effectively quenched by the collisions than emission from the lower lying levels. For example, the near-red optical emission between 0.5 and 1 μm which is due to $\Delta v \geq 3$ transitions (Black and Dalgarno 1976; Neufeld and Spaans 1996) is suppressed relative to longer wavelength $\Delta v \leq 2$ transitions arising from low lying levels. Rotational excitation is enhanced by the collisional vibrational cascade since collisions are favoured in which the changes in the internal energies of the H_2 molecules are small.

In hot (≥ 800 K) gas, emission from the $v = 1$ levels is strengthened considerably by direct collisional excitation (Sternberg and Dalgarno 1989; Burton, Hollenbach and Tielens 1990; Draine and Bertoldi 1996).

Luhman et al. (1997) presented evidence for the transition from low density to high density fluorescent emission in quiescent PDRs along the Orion A molecular ridge and toward the Orion B molecular cloud near NGC 2024. Luhman et al. observed the 1–0 S(1) 2.12 µm, 2–1 S(1) 2.25 µm, and 6–4 Q(1) 1.60 µm emission lines. They found that the intensity ratios $I_{2.12}/I_{2.25}$ and $I_{2.12}/I_{1.60}$ increase from values of 2.2 and 3.0 to values of 6.1 and 42.0 as the emission is mapped across regions with varying hydrogen particle density as traced by the density-sensitive far-infrared [OI] 63 µm and [CII] 158 µm emission line ratio. For comparison, the ratios $I_{2.12}/I_{2.25}$ and $I_{2.12}/I_{1.60}$ increase from values of 2.0 and 3.8 in the model spectrum displayed in Fig. 2 to values of 3.9 and 27.7 in the spectrum displayed in Fig. 4.

Observations of extended H_2 2.12 µm vibrational emission from molecular cloud complexes in Orion and Galactic Center regions suggest that the total H_2 emission luminosity in the Milky Way is produced mainly in PDRs (Luhman et al. 1994; Pak, Jaffe and Keller 1996).

2.3 Time-dependent Emission

The timescale required for the development of an equilibrium H/H_2 dissociation front in a PDR is set by the H_2 formation time ($\sim 5 \times 10^8/n$ yr) which may be long compared to the turn-on times of the stellar FUV sources ($\sim 10^5$ yrs for OB stars, or $<10^3$ yrs for the evolving central stars in planetary nebulae). In clumpy molecular clouds new H/H_2 dissociation fronts may develop when fresh molecular gas is suddenly exposed to FUV radiation as embedded clumps move out from shadowed regions. When χ/n is large the fluorescent emission intensities are much more intense immediately after the onset of FUV radiation than the intensities produced after the dissociation fronts have reached their equilibrium configurations, because at early times the atomic hydrogen layers and their associated dust opacities are small (Goldshmidt and Sternberg 1995; Hollenbach and Natta 1995). The H_2 line intensities decrease with time as the H/H_2 fronts propagate into the clouds and the atomic layers become large. This behaviour is illustrated in Fig. 5 which shows how the total H_2 emission intensity varies with χ for several cloud densities and a range of FUV exposure times. For intermediate times which are longer than the effective dissociation time $1/\chi D_0 G \approx 10^7/\chi$ yrs but shorter than the H_2 formation time, the IR intensity decreases as $1/t$ for fixed values of χ and n.

H_2 emission in PDRs may also be enhanced by the rapid advection of molecular gas through the dissociation fronts in clouds containing flows (Bertoldi and Draine 1996; Störzer and Hollenbach 1998).

Time dependent behaviour probably explains the unusually intense fluorescent H_2 emission observed in several young planetary nebulae. For example, the intensity of the 2.12 µm line in the planetary nebula BD+303639 is $\sim 10^{-3}$ ergs s^{-1} cm^{-2} sr^{-1}. This is more than 100 times brighter than would be expected under conditions of equilibrium given the estimated FUV intensity ($\chi = 2 \times 10^4$) and density ($n < 3 \times 10^4$ cm^{-3}) at the edges of the molecular shells. However, this planetary nebula is probably only $\sim 10^3$ yrs old as inferred from its size and expansion velocity. At this age the predicted H_2 intensity is about 100 times

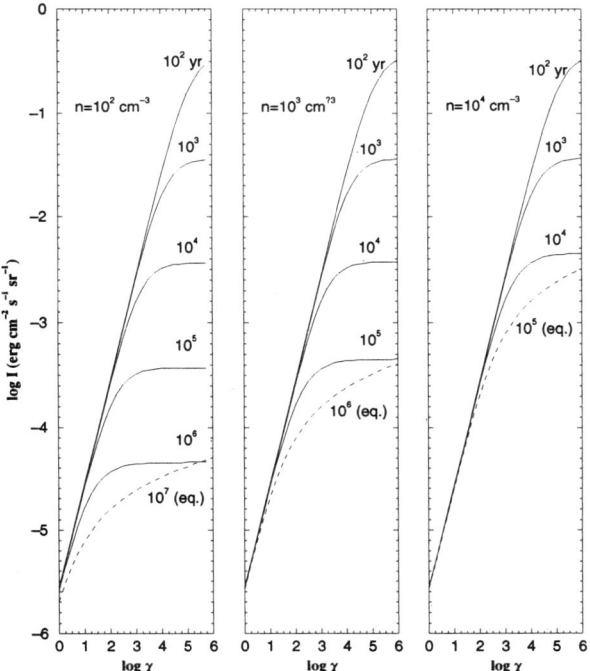

FIG. 5. Time dependent behaviour of the total H_2 fluorescent emission intensity $I \approx 50 I_{2.12\mu m}$.

larger than the equilibrium value, if it is assumed that the dynamical age is equal to the elapsed time since the central star became an FUV source (Goldshmidt and Sternberg 1995). A similar analysis of the H_2 emission in Hubble 12 has been presented by Luhman and Rieke (1996).

3 Heating and Cooling

PDRs are heated by photoelectric emission from grain surfaces (de Jong 1977; Draine 1978; Tielens and Hollenbach 1985), and by the photoionization of "large molecules" (LMs) or polycyclic aromatic hydrocarbons (PAHs) (Lepp and Dalgarno 1988; Leger and Puget 1989; Verstraete et al. 1996). The grain and LM photoelectric heating efficiencies vary with cloud depth as the radiation is attenuated and as the grain surface charge and LM ionization fractions vary. The grain photoelectric heating is dominated by small grains (with radii <15 Å) since they remain less positively charged than large grains for a given FUV intensity and electron density (Bakes and Tielens 1994).

Another major heating mechanism in PDRs is the collisional de-excitation of hydrogen molecules which have been vibrationally excited by the absorption of FUV photons. This mechanism operates most efficiently in the outer atomic hydrogen layers of dense PDRs where the H_2 FUV excitation is most efficient and where the excited molecules are de-excited by collisions with the hydrogen atoms (Sternberg and Dalgarno 1989).

PDRs are cooled by fine-structure emission lines of atoms and ions produced by photoionization and photodissociation. Major fine-structure cooling lines are the [CII] 158 μm, [OI] 63 and 146 μm, Si[II] 35 μm, and [CI] 370 and 609 μm emission lines. The gas is also cooled by rotational and vibrational emission from H_2 and rotational emission from less abundant species such as CO (Köster et al. 1994) OH and H_2O. The [CII] and [OI] emission lines are important diagnostics of PDRs (e.g. Herrmann et al. 1997), and typically emit ~1% of the incident FUV luminosity (most of which is reprocessed into far-infrared thermal dust emission).

Detailed computations show that gas temperatures of ~100 K can be maintained to cloud depths of a few visual extinctions. In dense PDRs the outer atomic hydrogen layers can become hot with temperatures exceeding 1000 K (Tielens and Hollenbach 1985; Sternberg and Dalgarno 1989; Burton, Hollenbach and Tielens 1990).

4 Chemistry in Dense PDRs

The chemical properties of PDRs are controlled by the incident FUV photons which heat the gas and drive the chemistry in sequences initiated by photoionization and photodissociation. The atomic and molecular abundances vary with cloud depth as the FUV radiation field is attenuated, and the effects of the photo-driven processes are diminished. Detailed chemical models of dense PDRs have been presented by Sternberg and Dalgarno (1995, henceforth SD95), and Jansen et al. (1995b). The models show that PDRs consist of several distinct chemical zones each of which are controlled by specific sets of gas-phase reactions. The sizes and locations of the different (and partially overlapping) zones depend on the cloud density, the FUV intensity, and the gas-phase elemental abundances.

In the SD95 model $n = 10^6$ cm^{-3} and $\chi = 10^5$ cm^{-3}, and the assumed gas phase abundances of oxygen, carbon, nitrogen, sulphur, and silicon are equal to 6×10^{-4}, 3×10^{-4}, 1×10^{-4}, 1×10^{-5}, and 1×10^{-6} relative to hydrogen.

4.1 HI Zone and H/H$_2$ Transition Layer

As discussed in §2 an HI and H/H$_2$ transition layer exists at the outer edges of PDRs. In this zone the hydrogen is mostly atomic, the carbon, sulphur, and silicon are singly ionized, and the oxygen and nitrogen are atomic. In dense clouds the gas can become hot (>1000 K) in the HI layer and the outer part of the H/H$_2$ dissociation front. In hot gas the oxygen, carbon, nitrogen, and sulphur chemistries are initiated by the endothermic reactions (cf. Chapter 8)

$$O + H_2 \rightarrow OH + H \qquad (4.1)$$

$$C^+ + H_2 \rightarrow CH^+ + H \qquad (4.2)$$

$$N + H_2 \rightarrow NH + H \qquad (4.3)$$

and
$$S^+ + H_2 \rightarrow SH^+ + H \tag{4.4}$$
where the atoms O, and N, and ions C^+ and S^+ are maintained by rapid photodissociation and photoionization. The internal energies of FUV-excited H_2 molecules may also be used to overcome the endothermicities of the above reactions. The neutral molecules OH, and NH are removed by the reverses of (4.1) and (4.3) and by photodissociation. The molecular ion CH^+ is removed by the reverse of (4.2) and by dissociative recombination
$$CH^+ + e \rightarrow C + H \tag{4.5}$$
and photodissociation. The SH^+ ions are removed by dissociative recombination and by the reverse of (4.4). Large abundances of OH, NH and CH^+ may be produced in the H/H_2 transition zone where the H_2 density becomes large but where the gas remains sufficiently warm for the endothermic hydrogen abstraction reactions to proceed efficiently.

In the SD95 model the OH abundance reaches a peak value of 6×10^{-6} at the cloud depth where $n(H_2)/n = 0.2$ and $T = 800$ K. The CH^+ abundance reaches a maximum value of 6×10^{-8} near the location of the OH density peak. At this cloud depth the OH and CH^+ column densities are equal to 1.5×10^{15} cm^{-2} and 3.5×10^{13} cm^{-2}.

The abundant OH radicals react with the abundant ions and atoms S^+, C^+, O, and N. Molecular ions SO^+, CO^+ and HCO^+ are formed via
$$S^+ + OH \rightarrow SO^+ + H \tag{4.6}$$
$$C^+ + OH \rightarrow CO^+ + H \tag{4.7}$$
$$CO^+ + H_2 \rightarrow HCO^+ + H \tag{4.8}$$
and are removed by dissociative recombination, photodissociation and reactions with H_2. Neutral molecules NO and O_2 and H_2O are produced by
$$N + OH \rightarrow NO + H \tag{4.9}$$
$$O + OH \rightarrow O_2 + H \tag{4.10}$$
$$H_2 + OH \rightarrow H_2O + H \tag{4.11}$$
and are removed by photodissociation. Water is also removed by the reverse of (4.11).

CO^+ and HCO^+ may be used as diagnostic probes of hot PDRs. Observational evidence for the production of CO^+ in hot PDRs has been provided by the detection of rotational emission lines of CO^+ at 236.06 and 235.79 GHz in the planetary nebula NGC 7027, and molecular cloud boundaries in the Orion Bar and M17 star-forming regions (Latter, Walker and Maloney 1993; Hogerheijde, Jansen and van Dishoeck 1995; Jansen et al. 1995b; Störzer, Stutzki and Sternberg 1995). The observations show that the CO^+ emissions peak at the edges of the molecular clouds close to the ionization fronts, and decrease rapidly with cloud depth. Further evidence for chemistry in hot H/H_2 transition layers is provided by interferometric observations of HCO^+ rotational emission at 89.19 GHz in the Orion Bar (Tauber et al. 1994) which show that the HCO^+ emission peaks close to the locations of the H_2 2.12 μm and CO^+ emission line maxima near the molecular cloud/HII region interface (van der Werf et al. 1996).

4.2 CII and SII Zones

Photoionization of atomic carbon

$$C + \nu \rightarrow C^+ + e \qquad (4.12)$$

by the incident FUV photons maintains a cloud layer in which most of the carbon is singly ionized. This layer is generally more extended than the HI zone due to the effective self-shielding of the H_2 molecules. Thus, a CII zone exists in which the carbon is ionized but where the hydrogen is fully molecular. Photoionization of atomic sulphur

$$S + \nu \rightarrow S^+ + e \qquad (4.13)$$

and charge transfer

$$C^+ + S \rightarrow C + S^+ \qquad (4.14)$$

maintain a cloud layer in which most of the sulphur is singly ionized. This layer is generally more extended than the layer of ionized carbon. Therefore an SII zone exists in which the sulphur is singly ionized but in which the carbon is neutral. Model computations of the gas temperature show that the CII and SII zones should usually be less than a few 100 K, so that the endothermic reactions (4.1)–(4.4) become much less efficient in these zones. In the SD95 model the CII zone is present at cloud depths between visual extinctions A_V of 0.7 and 1.7. The SII zone ranges from an A_V of 1.7 to 3.7.

Carbon bearing radicals such as CH and CH_2 are preferentially produced in the CII zone by a sequence initiated by

$$C^+ + H_2^* \rightarrow CH^+ + H \qquad (4.15)$$

where H_2^* are FUV-pumped H_2 molecules, and by

$$C^+ + H_2 \rightarrow CH_2^+ + \nu. \qquad (4.16)$$

followed by

$$CH^+ + H_2 \rightarrow CH_2^+ + H \qquad (4.17)$$

$$CH_2^+ + H_2 \rightarrow CH_3^+ + H \qquad (4.18)$$

$$CH_3^+ + e \rightarrow CH_2 + H \qquad (4.19)$$

$$CH_3^+ + e \rightarrow CH + H_2. \qquad (4.20)$$

The CH and CH$_2$ molecules are removed by photodissociation. The abundances of CH and CH$_2$ rise near the inner edge of the CII zone where the C$^+$ density is still large, and where the photo-destruction rates are diminished. The CH and CH$_2$ densities decrease rapidly through the SII zone and deeper cloud layers as the C$^+$ density becomes small. The carbon radicals are also removed by insertion reactions of the form

$$CH_n + C^+ \to C_2H_{n-1}^+ + H \quad (4.21)$$

near the inner edge of the CII zone. These reactions lead to the efficient production of complex hydrocarbons (e.g. C$_2$H) in this part of the cloud.

The density of neutral atomic carbon becomes large at the inner edge of the CII zone and through the SII zone. In these layers carbon is formed by radiative recombination of the C$^+$ ions, charge transfer (4.14), and photodissociation of CH and CO. Most of the carbon is driven into CO molecules in the transition from the CII to the SII zones via the neutral reactions

$$CH + O \to CO + H \quad (4.22)$$

$$CH_2 + O \to CO + H_2. \quad (4.23)$$

Large C/CO density ratios are maintained in these cloud layers. For example, in the SD95 model the atomic carbon abundance reaches a value of 3×10^{-5} near the inner edge of the CII zone. At this location the C/CO density ratio equals 0.25, and the ratio of column densities, N(C)/N(CO), equals 0.4. The column density ratio N(C)/N(CO) then decreases with increasing cloud depth as the carbon is incorporated fully into CO molecules.

Observations of the CI 492.2 and 809.3 GHz fine-structure emission lines in the Orion Bar, M17 and W51 star-forming regions suggest that the CI emission is produced in PDRs on the surfaces of dense clumps which pervade an interclump medium filled with FUV radiation. The observed N(C)/N(CO) ratios of $\sim$0.1 in the Galactic PDRs are generally consistent with the models. In lower density translucent clouds the C/CO ratio may be much larger. For example, in the high latitude translucent cloud toward the star HD 210121 the N(C)/N(CO) ratio is between 3 and 6 (Stark and van Dishoeck 1994). CI emission has been observed in the starburst galaxies IC 342, M82, and NGC 253 (Büttgenbach et al. 1992; Schilke et al. 1993; White et al. 1994; Harrison et al. 1995; Stutzki et al. 1997). The CI emission in these objects probably originates in dense PDRs. However, the emission appears to be unusually intense with the finding that the N(C)/N(CO) ratios are quite large (>0.5) compared to Galactic PDRs. A possible explanation is that the CI emission is enhanced by rapid recombination of C$^+$ ions in slowly cooling interclump gas suddenly shielded from the FUV sources by intervening dense clumps (Störzer, Stutzki, and Sternberg 1997).

4.3 SiII and SI Zones

Photoionization of atomic silicon

$$Si + \nu \to Si^+ + e \quad (4.24)$$

and charge transfer

$$S^+ + Si \to S + Si^+ \qquad (4.25)$$

maintain a cloud layer in which the silicon is singly ionized. This layer extends to larger depths than the C^+ or S^+ layers because of the generally lower Si abundance. Thus, an SiII zone exists in which the silicon is ionized but in which the carbon and sulphur are neutral. A broad SI zone may also exist in which most of the sulphur is present in neutral atomic form. The SI zone is maintained by photo-destruction of the sulphur bearing molecules, recombination of the S^+ ions and charge transfer (4.25). In the SD95 model the SiII zone is present from an A_V of 3.7 to 6, and the SI zone ranges from an A_V of 3.7 to 7. The FUV radiation is severely attenuated in the SiII and SI zones, and the chemistry is driven by a combination of photoionization and cosmic-ray ionization. The electron density becomes small, while the density of atomic oxygen remains large, so that reactions of the molecular ions with the oxygen atoms become competitive with dissociative recombination.

CS molecules may be preferentially produced in SI zones compared with more shielded regions. CS is produced by the sequence

$$S^+ + H_2 \to SH_2^+ + \nu \qquad (4.26)$$

$$SH_2^+ + e \to SH + H \qquad (4.27)$$

$$SH + O \to SO + H \qquad (4.28)$$

$$SO + C \to CS + O, \qquad (4.29)$$

where the carbon atoms are produced by the charge transfer reaction (4.14). The CS density becomes large near the inner edge of the SI zone where the photodissociation rates are diminished, but where the atomic carbon density remains large. At larger cloud depths the CS density decreases as the the density of S and C decrease and the sulphur is incorporated into SO and SO_2. In the SD95 model, the CS abundance reaches a peak value of 4×10^{-6} at the inner edge of the SI zone, a factor of 10^3 larger than in the dark core.

4.4 Chemical Diagnostics and Ionization Structure

Table 1 lists a set of molecular density ratios which may be used as diagnostics of the different chemical zones in PDRs. The listed ratios are large in the outer regions and become small with increasing cloud depth. The easily observable CN/HCN ratio is a particularly valuable diagnostic. In the SD95 model the CN/HCN density ratio is ~ 10 in the HI, CI, and CII zones and decreases to a value of 2×10^{-4} in the dark core. Figure 6 shows computed values of the CN and HCN abundances, CN/HCN density ratios, and CN column densities as functions of visual extinction for PDRs exposed to a $\chi = 2 \times 10^3$ FUV field and hydrogen gas densities in the range 10^3 to 10^6 cm^{-3}. For a given cloud density the CN/HCN is large near the cloud surface and becomes small at large visual extinctions. Observations of HCN 88.6 GHz and CN 113.5 and

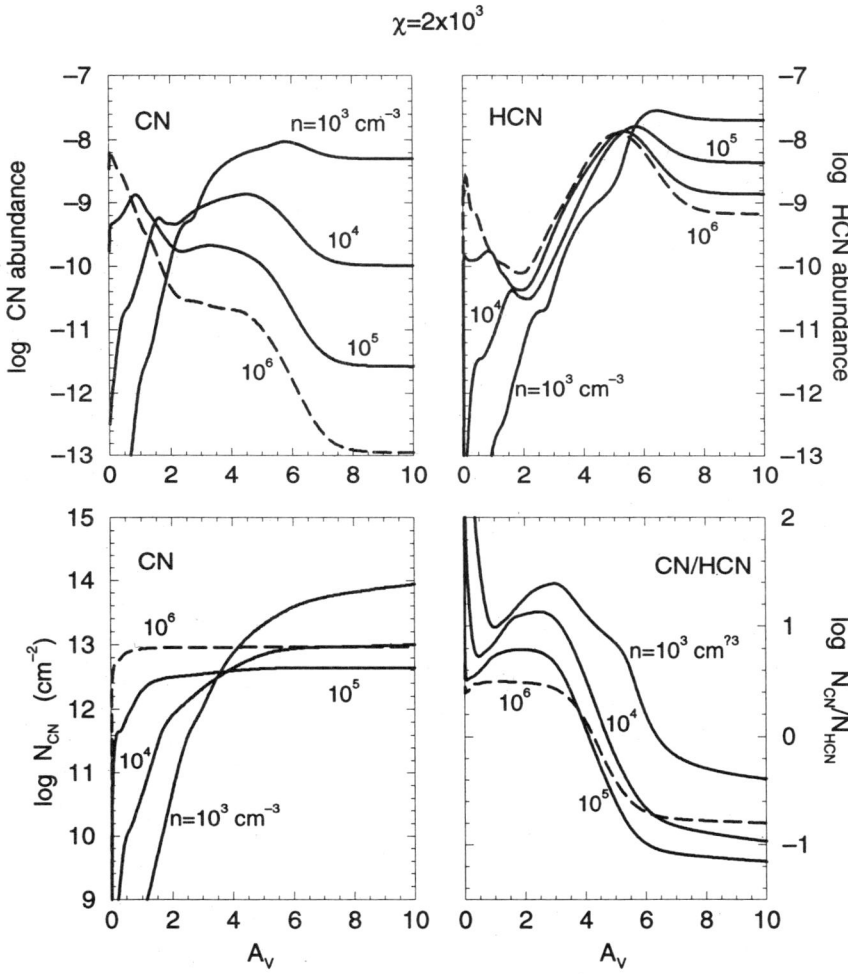

FIG. 6. CN and HCN abundances in dense PDRs.

226.9 GHz emission in the reflection nebulae NGC 2023 and NGC 7023 show that the CN/HCN density ratios are enhanced near the sources of FUV radiation (Fuente, Martin-Pintado and Gaume 1995, Fuente et al. 1993). The observed ratios decrease by a factor of $\sim$25 in NGC 2023, and $\sim$7 in NGC 7023 with increasing distance from the illuminating stars. In NGC 2023 the CN/HCN ratio is largest near the location of the 2.12 μm H_2 emission peak.

The fractional ionization in PDRs is controlled by the sequential photoionization of carbon, sulphur, and silicon as the FUV radiation is attenuated by dust absorption with increasing cloud depth. In the dark core the ionization

Table 1 Molecular Diagnostics

Density ratio	Hot HI H/H$_2$ $A_V = 0.6$	CII $A_V = 1.5$	SII $A_V = 3$	SiII $A_V = 5$	SI $A_V = 7$	Dark core $A_V > 10$
OH/H$_2$O	4.5(1)	2.3(1)	9.3	9.2(−2)	4.4(−3)	3.4(−4)
OH$^+$/H$_3$O$^+$	1.8	1.3	7.5(−2)	9.6(−3)	1.3(−3)	1.4(−5)
CO$^+$/HCO$^+$	5.1(−2)	3.5(−2)	1.5(−4)	1.7(−6)	2.9(−7)	2.0(−7)
SO$^+$/SO	1.7(4)	2.3(−1)	2.8(−3)	6.4(−4)	1.2(−4)	5.7(−4)
SiO$^+$/SiO	1.6(−1)	5.7(−3)	3.0(−5)	1.2(−7)	1.8(−9)	1.7(−10)
NH/NH$_3$	7.8(4)	2.2(5)	9.2(3)	9.3(−1)	1.7(−2)	4.9(−4)
CN/HCN	4.7	1.1(1)	8.4	3.3(−1)	1.3(−1)	1.8(−4)

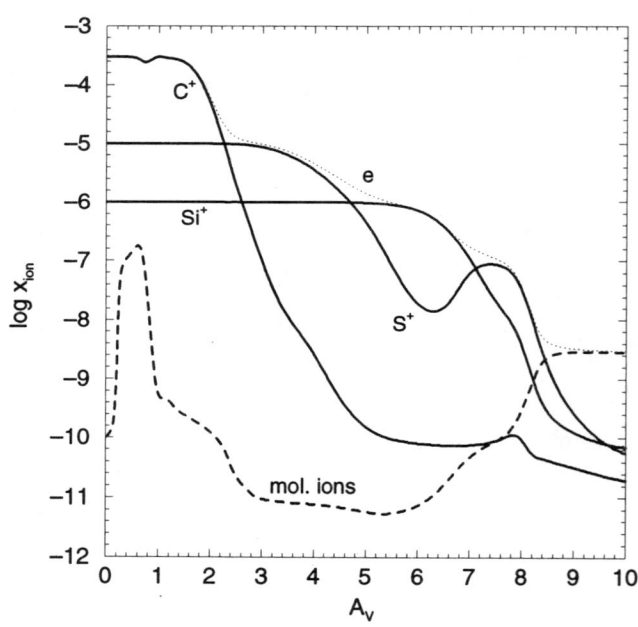

FIG. 7. Ionization structure in the $n = 10^6$ cm^{-3} and $\chi = 2 \times 10^5$ SD95 model.

is dominated by cosmic ray ionization of hydrogen and helium. The ionization structure in the SD95 model is displayed in Fig. 7. At the outer edge where C$^+$ is the dominant ion, the fractional ionization is $\sim 3 \times 10^{-4}$. The fractional ionization decreases to $\sim 3 \times 10^{-9}$ in the fully shielded core where the dominant ions are HCO$^+$ and H$_3$O$^+$. The ambipolar diffusion rates and the magnetically controlled star-forming cloud collapse times depend on the incident FUV flux and are sensitive to the cloud depth (McKee et al. 1989; Hartquist et al. 1993; Chapter 5).

Acknowledgements

I thank T. W. Hartquist, H. Störzer and E. F. van Dishoeck for comments on this chapter. I thank Alex for his scientific inspiration, leadership, and vision through the years.

Bibliography

1. Abgrall, H., Le Bourlot, J., Pineau des Forêts, G., Roueff, E., Flower, D. R. and Heck, L. (1992). A&A, **253**, 525.
2. Allison, A. C. and Dalgarno, A. (1970). Atomic Data, **1**, 289.
3. Bakes, E. L. O. and Tielens, A. G. G. M. (1994). ApJ, **427**, 822.
4. Bertoldi, F. and Draine, B. T. (1996). ApJ, **458**, 222.
5. Black, J. H. and Dalgarno, A. (1976). ApJ, **203**, 132.
6. Black, J. H. and van Dishoeck, E. F., (1987). ApJ, **322**, 412.
7. Burton, M. G., Hollenbach, D. J. and Tielens, A. G. G. M. (1990). ApJ, **365**, 620.
8. Burton, M. G., Bulmer, M., Moorhouse, A., Geballe, T. R. and Brand, P. W. J. L. (1992). MNRAS, **257**, 10.
9. Büttgenbach, T. H., Keene, J., Phillips, T. G. and Walker, C. K. (1992), ApJ, **397**, L15.
10. Dalgarno, A., Herzberg, G. and Stephens, T. L. (1970). ApJ, **162**, L49.
11. de Jong, T. (1977). A&A, **55**, 137.
12. Draine, B. T. (1978). ApJS, **36**, 595.
13. Draine, B. T. and Bertoldi, F. (1996). ApJ, **468**, 269.
14. Duley, W. W. (1996). MNRAS, **79**, 591.
15. Field, D., Gerin, M., Leach, S., Lemaire, J. L., Pineau des Forêts, G., Rostas, F,. Roueff, E. and Simons, D. (1994). A&A, **286**, 909.
16. Fuente, A., Martin-Pintado, J. and Gaume, R. (1995). ApJ, **442**, L33.
17. Fuente,A., Martin-Pintado, J., Gernicahro, J. and Bachille, R. 1993, A&A, **276**, 473.
18. Genzel, R., Harris, A. I., Jaffe, D. T. and Stutzki, J. (1988). ApJ, **332**, 1049.
19. Genzel, R., Harris, A. I. and Stutzki, J. (1989). In *Infrared Spectroscopy in Astronomy*, ed. M Kessler, ESA Publications, Dordrecht. ESA SP-290, p.115.
20. Goldshmidt, O. and Sternberg, A. (1995), ApJ, **439**, 256.
21. Graham, J. R., Herbst, T. M., Matthews, K., Neugebauer, G., Soifer, B. T., Serabyn, E. and Beckwith, S. (1993). ApJ, **408**, L105.
22. Harrison, A., Puxley, P., Russel, A. and Brand, P. W. J. L. (1995). MNRAS, **277**, 413.
23. Hartquist, T. W., Rawlings, J. M. C., Williams, D. A. and Dalgarno, A. (1993). QJRAS, **34**, 213.

24. Herrmann, F., Madden, S. C., Nikola, T., Geis, N., Townes, C. H. and Stacey, G. (1997). *ApJ*, **481**, 343.
25. Hogerheijde, M. R., Jansen, D. J. and van Dishoeck, E. F. (1995). *A&A*, **294**, 792.
26. Hollenbach, D. J. and Natta, A. (1995). *ApJ*, **455**, 133.
27. Hollenbach, D. J. and Tielens, A. G. G. M. (1997), *ARA&A*, in press.
28. Jansen, D. J., van Dishoeck, E. F., Black, J. H., Spaans, M. and Sosin, C. (1995a). *A&A*, **302**, 223.
29. Jansen, D. J., Spaans, M., Hogerheijde, M. R. and van Dishoeck, E. F. (1995b). *A&A*, **303**, 541.
30. Jura, M. (1975). *ApJ*, **197**, 575.
31. Köster, B,. Störzer, H., Stutzki, J. and Sternberg, A. (1994). *A&A*, **284**, 545.
32. Latter, W. B., Walker, C. K. and Maloney, P. R. (1993). *ApJ*, **419**, L97.
33. Lepp, S. and Dalgarno, A. (1988). *ApJ*, **335**, 769.
34. Lepp, S., Tiné, S. and Dalgarno, A. (1997). in prep.
35. Leger, A. and Puget, J. L. (1989). *ARA&A*, **27**, 161.
36. Levinson, A., Chernoff, D. F. and Salpeter, E. E. (1997). *Phys. Rev. B.*, submitted.
37. Luhman, K. L. and Rieke, G. H. (1996). *ApJ*, **461**, 298.
38. Luhman, M. L., Jaffe, D. T., Keller, L. D. and Pak, S. (1994). *ApJ*, **436**, L185.
39. Luhman, M. L., Jaffe, D. T., Sternberg, A., Herrmann, F. and Poglitsch, A. (1997). *ApJ* **482**, 298.
40. McKee, C. F. (1989), *ApJ*, **345**, 782.
41. Martin, C., Hurwitz, M. and Bowyer, S. (1990). *ApJ*, **354**, 220.
42. Martin, P. G. and Mandy, M. E. (1995). *ApJ*, **455**, L89.
43. Martini, P., Sellgren, K. and Hora, J. L. (1997). *ApJ*, **484**, 296.
44. Neufeld, D. and Spaans, M. (1996). *ApJ*, **473**, 894.
45. Pak, S., Jaffe, D. T. and Keller, L. D. (1996). *ApJ*, **457**, L43.
46. Roberge, R. (1981). PhD thesis, Harvard University.
47. Schilke, P., Carlstrom, J. E., Keene, J. and Phillips, T. G. (1993). *ApJ*, **417**, L67.
48. Stark, A. and van Dishoeck, E. F. (1994) *A&A*, **286**, L43.
49. Stephens, T. L. and Dalgarno, A. (1972) *J.Quant.Spect.Rad.Tran.*, **12**, 569.
50. Sternberg, A. (1988). *ApJ.*, **332**, 400.
51. Sterberg, A. (1989). *ApJ*, **347**, 863.
52. Sternberg, A. and Dalgarno, A. (1989). *ApJ*, **338**, 197.
53. Sternberg, A. and Dalgarno, A. (1995). *ApJS*, **99**, 567.

54. Sternberg, A., Yan, M. and Dalgarno, A. (1997). In *Molecules in Astrophysics: Probes and Processes*, IAU 178, ed. E. F. van Dishoeck.
55. Störzer, H. and Hollenbach, D. J. (1998). *ApJ*, in press.
56. Störzer, H., Stutzki, J. and Sternberg, A. (1995). *A&A* **296**, L9.
57. Störzer, H., Stutzki, J. and Sternberg, A. (1997), *A&A*, **323**, L13.
58. Stutzki, J., et al. (1997). *ApJ*, **477**, 33.
59. Tauber, J. A., Tielens, A. G. G. M., Meixner, M. and Goldsmith, P. F. (1994). *ApJ*, **422**, 136.
60. Tielens, A. G. G. M. and Hollenbach, D. (1985). *ApJ*, **291**, 722.
61. Timmermann, R., Bertoldi, F., Wright, C. M., Drapatz, S., Draine, B. T., Haser, L. and Sternberg, A. (1996). *A&A*, **315**, L281.
62. Tiné, S., Lepp, S,. Gredel, R. and Dalgarno, A. (1997). *ApJ*, **481**, 282.
63. Turner, J., Kirby-Docken, K. and Dalgarno, A. (1977). *ApJS*, **35**, 281.
64. van der Werf, P. P., Stutzki, J. Sternberg, A. and Krabbe, A. (1996). *A&A*, **313**, 633.
65. Verstraete, L., Puget, J. L., Falgarone, E., Drapatz, S., Wright, C. M. and Timmermann, R. (1996). *A&A*, **315**, 337.
66. White, G. J., Ellison, B., Claude, S., Dent, W. R. F. and Matheson, D. N. (1994). *A&A*, **284**, L23.
67. Witt, A. N., Stecher, T. P., Boroson, T. A. and Bohlin, R. C. (1989). *ApJ*, **336**, L21.
68. Wolniewicz, L. Simbotkin, I. and Dalgarno, A. (1998). *ApJ*, in press.

10
Molecular Hydrogen Emission from Herbig–Haro Objects

John C. Raymond
Harvard-Smithsonian Center for Astrophysics

1 Introduction

The faint emission line knots known as Herbig–Haro objects (Herbig 1951; Haro 1952) have been shown to be shock waves driven by (and in) outflows from young stars (Schwartz 1977; Dopita 1978; Raymond 1979). They are typically 0.1 pc long, with densities of order 10^4 cm^{-3}. Doppler shifts and proper motions of 100–400 km s^{-1} are common, and the exciting stars are most often classical T Tauri stars.

Though the flows are sometimes too complicated to sort out, there are four basic emission structures. The first is associated with knots along a highly collimated jet, which are typically very slow shock waves in cool, largely neutral atomic gas. Typical shock speeds are 20 km s^{-1}, and the shocks are generally attributed to variations in the speed or direction of the jet (Falle and Raga 1993; Hartigan and Raymond 1993; Hartigan, Morse and Raymond 1994; Biro and Raga 1994). Earlier models based on jet instabilities generally ran into difficulty with the spacing of the knots (Falle 1994). The second is characterized by bright emission at the end of the jet and arises when the jet material encounters the high pressure region of the working surface, creating a shock at the Mach disc. The third is seen in bright emission at the end of the jet and arises from the bow shock driven into the interstellar medium. Finally, some HH jets have cleared out cavities in the parent clouds, and the walls of these cavities may be heated by slow shocks or turbulent dissipation (e.g. Cantó and Rodriguez 1980) In principle, any of these four structures might produce H$_2$ emission.

In spite of the complicated flow structures created by Kelvin–Helmholtz, Rayleigh–Taylor and thermal instabilities (Blondin, Königl and Fryxell 1989; de Gouveia Dal Pino and Benz 1993; Stone and Norman 1993), simple bow shock models do a remarkably good job of predicting relative intensities and the line profiles of the the optical and UV emission lines (Hartmann and Raymond 1984; Raga and Böhm 1986; Hartigan, Raymond and Hartmann 1987; Morse et al. 1992, 1994). In particular, the relationship between line widths and excitation levels (e.g. [O III]/Hβ ratios) and the character of the line profile as a function of viewing angle work out quite well. Recent HST observations show that HH objects which are fairly simple at 1″ resolution are collections of far more complex structures when seen at 0.1″ resolution (Schwartz et al. 1993; Heathcote et al. 1996). The dynamical complexity makes spectral diagnostics for the nature of the H_2 excitation and the physical processes in the exciting shock waves especially important. The relationship between the high velocity material seen at optical wavelengths and the slower, denser outflows seen in CO has been controversial for many years (e.g. Chernin and Masson 1995), and H_2 emission offers a direct look at the interaction.

An excellent review of many of the topics covered here is given by Brand (1995), including shocked regions other than HH objects. Draine and McKee (1993) provide a comprehensive review of astrophysical shock waves. This paper will focus on the discrepancies between observations and current models of H_2 emission in HH objects.

2 Observational Aspects of H_2 Emission

The very presence of molecular hydrogen in flows faster than 100 km s^{-1} is somewhat surprising in view of the fact that shocks with speeds above 40 or 50 km s^{-1} should dissociate H_2 (e.g. Draine and McKee 1993; see Chapter 8). With the advent of array detectors for the near IR, excellent images of a large number of HH objects have become available in the 2.12 µm $v = 1$–0 S(1) line, the brightest vibrational transition. Striking examples include HH1/2 (Davis, Eislöffel and Ray 1994; Noriega-Crespo and Garnavich 1994), Cepheus A (Hartigan et al. 1996), and HH90, 110 and 111 (Davis, Mundt and Eislöffel 1994). The emission is highly structured. It is usually associated with the bow shock at the working surface, rather than the jet.

The intensities of the infrared vibrational lines generally resemble the emission from a thermal equilibrium population at a temperature of 2000–3000 K. The bright high excitation lines which could be produced by fluorescent pumping or by formation pumping are not generally seen. However, as emphasized by Brand (1995), the relative intensities do not always fit a single thermal distribution when a large number of lines is available. The highest and lowest excitation lines are systematically too bright. This is most easily seen in extremely bright objects such as Orion (Brand et al. 1988), but it is also apparent in some HH Objects (e.g. HH111/121 Gredel and Reipurth 1993). The ISO spectrum of Ceph A West (GGD37) shows high excitation attributed to pumping (Wright et al. 1996) like that occurring in PDRs (see Chapter 9).

The H_2 line profiles have been measured in a smaller sample of objects. Zinnecker et al. (1989), Carr (1993), Davis and Smith (1996) and Davis, Eislöffel and Smith (1996) presented line shapes for HH1/2, HH7-11, HH32, HH40, and L1448. In some cases the line widths are around 50 km s^{-1} and the shapes resemble those expected on the basis of straightforward bow shock models (e.g. Smith, Brand and Moorhouse 1991a). However, broader lines are seen as well. Zinnecker et al. (1989) show a 45 km s^{-1} FWHM peak which appears to stand upon a still broader component in HH7. The large line widths seen in Orion (Brand et al. 1989) and some HH objects present a severe problem for C-shock models of the molecular emission, as discussed below.

A third important characteristic of the infrared H_2 emission is its spatial distribution. Comparison with narrow band images of the optical lines shows a great deal of variety and complexity. Observations with NICMOS on the HST will provide far greater detail. For now, the salient feature is the close association of H_2 with optical features, even with very high excitation knots such as HH2A' and HH2H (Davis, Eislöffel and Ray 1994). In some cases the H_2 seems to lie along the wings of the bow shock, as might be expected if it is produced only at effective shock velocities below 50 km s^{-1} or if it arises after molecules reform after the shocked gas cools. In other cases, it coincides with the optical knots at the resolution level (Schwartz et al. 1988) and with the bow shock tip (HH7—Hartigan, Curiel and Raymond 1989), or even lies significantly ahead of the bow shock (one feature in Cepheus A—Hartigan et al. 1996). Figure 4 of Hartigan et al. (1996) is a useful set of schematics for the expected relative distributions of H_2 and the optical lines.

The H_2 ultraviolet emission from HH objects has received relatively less attention due to the difficulty of observing reddened objects in the UV. However, the less reddened HH objects have been detected with IUE and the GHRS. The first UV observations of HH1 and HH2 showed a short wavelength quasi-continuum which long remained unidentified (Böhm et al. 1987). The IUE spectrum of HH43 showed strong emission in eight transitions pumped by Lyα (which is emitted by hotter shocked gas in the HH object), along with a quasi-continuum extending up to about 1600 Å (Schwartz 1983). Lyα pumping of transitions B–X $v = 1$–2 P(5) and B–X $v = 1$–2 R(6) lying +99 and +15 km s^{-1} from the rest wavelength of Lyα has also been seen in sunspots (Jordan et al. 1977; Shull 1978) and in Burnham's Nebula (Brown et al. 1981). Schwartz (1983) and Böhm, Scott and Solf (1991) also observed HH47A. It showed the same Lyα-pumped transitions, but at a lower level relative to the apparent continuum. Curiel et al. (1995) investigated the cause of the difference with a GHRS observation of the 1263–1298 Å band in HH47A. It showed the three expected fluorescent transitions along with a fourth unexpected transition. Thus fluorescence is clearly important in this object as well, but either the fluorescent spectrum is more complex or some other process makes a contribution as well. Most recently, the Hopkins Ultraviolet Telescope (Davidsen et al. 1992) was used to observe HH2 during the Astro-2 mission in March 1995 (Raymond, Blair and Long 1997). The spectrum shows clear H_2 features at 1060 and 1105 Å and the quasi-continuum seen by IUE between 1250 and 1600 Å. The spectral resolution is twice that of IUE, and some of the fluorescent features longward of Lyα are apparent, but they are far weaker relative to the continuum than in HH43. We will discuss the implications of this observation below.

An important clue to the nature of the molecular emission is the relative luminosity of the atomic (mostly Lyα) emission to the H_2 UV and IR emission. While the luminosity determinations are somewhat uncertain due to the reddening corrections and the fact that the Lyα luminosity must be inferred from the Hα luminosity and model calculations, it should be possible to estimate the relative values to factor of two accuracy. In HH47A, Curiel *et al.* (1995) found luminosities for Lyα, H$_2$ UV emission and H$_2$ IR emission of 5×10^{32}, 2×10^{32}, and 0.8×10^{32} erg s^{-1}, respectively. In HH2H, the luminosity of the UV molecular hydrogen emission also appears to be about 1/3 the Lyα luminosity inferred from the Hα luminosity and bow shock models (Raymond *et al.* 1997). As Lyα accounts for the largest portion of the energy radiated by the optically bright shocks, it is necessary either to channel a comparable amount of energy into molecular shocks or to convert a substantial percentage of the Lyα into molecular emission through fluorescence.

3 Emission Mechanisms

3.1 Formation Pumping

When H$_2$ molecules form by radiative attachment or on dust grains, they can be in excited states. Hollenbach and McKee (1989) and Wolfire and Königl (1991) computed the emission spectra of H$_2$ forming in the cooling region behind a dissociative shock. The spectral signature of the process is strong emission in high excitation IR lines, such as the $v = 6-4$ Q(0-4) group near 1.6 µm. In most HH objects, the high excitation IR lines are weak, but GGD37, for instance, shows high excitation lines likely to be non-thermal in nature (Wright *et al.* 1996). A major limitation of formation pumping is its modest efficiency, since it produces only one excitation per molecule formed, and the brightness of 1-0 S(1) is much smaller than that produced by non-dissociative shocks. One would expect IR emission due to formation pumping to appear downstream from any optical emission.

3.2 Continuum Fluorescence

Continuum fluorescence has been treated in great detail for photodissociation regions (Black and Dalgarno 1976; Black and van Dishoeck 1987; Sternberg 1989; see Chapter 9). Photons between about 1049 and 1104 Å can be absorbed in various transitions by molecules in the ground vibrational state, and the resulting electronically excited molecules either dissociate (producing a continuum between 1300 and 1700 Å; Dalgarno, Herzberg and Stephens 1970) or decay to an excited vibrational level, producing a UV photon between 1200 and 1650 Å and an IR cascade. Again, strong high excitation IR transitions are a predicted observational signature, and in most HH objects these lines are weak. It would also be interesting to search for the red H$_2$ transitions predicted by Neufeld and Spaan (1996) to result from continuum fluorescence. A major limitation of continuum fluorescence for HH objects is the lack of a strong continuum flux. Wolfire and Königl (1991) computed the intensity of the helium 2-photon continua from shocks. However, observation of HH2 with the Hopkins Ultraviolet Telescope showed no continuum below 1200 Å (Raymond, Blair and Long 1997).

In general, the He two-photon emission accounts for less than 10% of the energy radiated by the shock, the fraction in the 912–1104 Å band is small (17%), and even within that band only a modest fraction can be absorbed by H_2 molecules, so that no more than a percent of the energy dissipated by a shock can appear as molecular hydrogen fluorescent emission.

3.3 Lyα Fluorescence

Lyα fluorescence is clearly responsible for strong UV emission in some HH objects. Schwartz (1983) showed that HH43 displays strong emission features resulting from absorption of Lyα photons in the B–X $v = 1$–2 P(5) and various branches of radiative decay in reasonable agreement with the proportions predicted by Shull (1978). The pumping transition is located nearly 100 km s^{-1} to the red of Lyα line centre. These fluorescent features dominate the IUE spectrum of HH43. Some additional continuum or quasi-continuum is also seen, some of it presumably the dissociation continuum implied by the Lyα pumping. A surprising feature of the HH43 spectrum is the absence of transitions pumped by the B–X $v = 1$–2 R(6) transition located closer to the Lyα line centre. This suggests that the Lyα line profile is narrow and red-shifted relative to the molecular gas, or else that H I atoms shield the molecular gas from photons near the line centre (e.g. Shull 1978).

Schwartz's IUE spectrum of HH47A also showed some evidence for Lyα pumping, but it was much less clear. Some of the Lyα-pumped transitions appeared to be present, but they did not dominate as they do in HH43. Curiel et al. (1995) obtained HST GHRS and FOC observations to sort this out. They identified the expected Lyα-pumped transitions at 1270.98, 1271.83 and 1293.75 Å, but there was a fourth, unexpected, transition at 1293.35 Å. At HST spatial resolution, the UV emission coincides with the optical shock emission rather than the more extended IR H_2 emission. The velocity centroid lies between the ambient cloud velocity and the velocity of the gas just ahead of the HH47A bow shock. One possible reason for the difference between the UV spectra of HH43 and HH47 is that more pumping transitions are excited in HH47 than in HH43, producing a more continuous spectrum and the unexpected 1293.35 Å line. This might result from a greater Lyα width or shift, or from larger populations in more highly excited levels. A second possibility is that collisional excitation contributes to the HH47A UV spectrum at some level, again providing emission distributed over a greater number of lines.

The Hopkins Ultraviolet Telescope spectrum of HH2 further complicates the situation (Raymond, Blair and Long 1997). It shows that the "blue continuum" investigated by Böhm et al. (1987) is indeed molecular hydrogen emission below about 1660 Å, with a hydrogen two-photon continuum at longer wavelengths, and it shows the Lyman band transitions at 1060 and 1105 Å. While some of the H_2 transitions that are strong in HH43 can be seen, they barely reach above the continuum in HH2. Thus HH2 seems to be an even more extreme case than HH47A. Either a still greater number of fluorescent transitions are pumped by Lyα, or collisional excitation dominates. The fluorescent explanation is somewhat attractive in that the Hα line widths in HH2H are 140–190 km s^{-1}, as compared with 100 km s^{-1} in HH47 (Hartigan, Raymond and Meaburn 1990), and probably an even smaller line width in HH43, judging by its low excitation. Fluorescence is also attractive from an energetic point of view. The total luminosity in the H$_2$ emission is $\simeq$ 30% of the Lyα luminosity inferred from the Hα luminosity, so that fairly efficient fluorescent conversion could account for the H$_2$ emission.

However, there is some difficulty in identifying plausible pumping transitions. Black and van Dishoeck (1987, Table 5) listed 46 potential pumping transitions close to Lyα, but only six arise from vibrational levels 2–4, and one of those has a very small oscillator strength. As the IR spectra of HH objects show temperatures near 2500 K (e.g. Brand 1995), the populations of vibrational levels 5 and higher will be orders of magnitude smaller than that of the second vibrational level. From the intensity of the (1–0) S(1) line (Elias 1980) and the radiative decay rate of its upper level, the column density in the $v = 1$, $J = 3$ level is about 10^{16} cm^{-2}. The pumping transitions have oscillator strengths of 0.03 or smaller, so the absorption cross sections are of order 10^{-14} cm^2, and transitions arising in vibrational levels above $v = 4$ should be optically thin. Thus it is far from clear that Lyα pumping can produce the quasi-continuous H$_2$ spectrum observed.

Lyα pumping can also produce IR emission, and again the observational signature should be emission from fairly highly excited vibrational states (e.g. Wolfire and Königl 1991). In fact, Lyα pumping cannot produce the strongest IR lines, because the Lyα photons can only be absorbed by vibrationally excited molecules in the first place. Thus Lyα pumping will enhance high excitation lines, but not the (1–0) S(1) most commonly observed.

3.4 Collisional Excitation

Hydrogen molecules can be excited by collisions with neutral atoms, molecules or electrons. Excitation by neutrals or molecules can account for the IR emission spectrum. They can quite efficiently convert thermal energy to infrared H$_2$ radiation, because in a molecular gas at HH object densities there are few other cooling mechanisms at temperatures near 2500 K. For the reasons mentioned in the discussion above, continuum fluorescence, formation pumping and Lyα fluorescence can enhance the high excitation IR transitions, but most of the IR emission must come from collisional excitation.

In the UV the situation is less clear. Lyα fluorescence clearly accounts for most of the UV emission in HH43, much of the UV in HH47, and perhaps that in HH2 as well. Excitation by hot electrons is also appealing, however. Liu and Dalgarno (1996) have computed the UV emission spectra from H$_2$ excited by electrons under various conditions. The spectrum produced by 20 eV electrons in a 1000 K H$_2$ gas matches the observed HH2 spectrum remarkably well, as seen in Figure 1. The strong blends at 1060 and 1105 Å clearly stand out, and longward

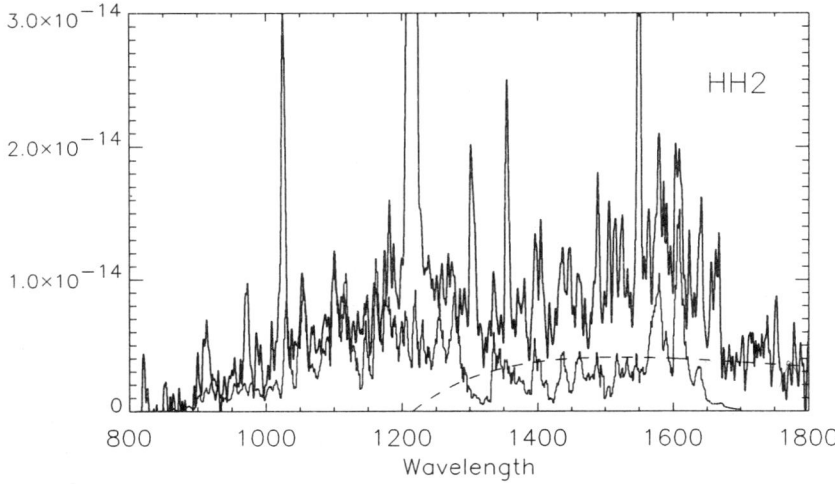

FIG. 1. UV spectrum of HH2 with the Liu and Dalgarno prediction for collisionally excited H_2 (thin line) and the hydrogen two-photon continuum (dashed lined).

of Lyα the spectrum is a quasi-continuum with a modest emission feature at 1610 Å. The reason for the quasi-continuous spectrum is that collisional excitation produces a large number of electronically excited levels, especially when the target molecules are distributed among many rotational and vibrational levels. The drawback of collisional excitation is that the molecules must not dissociate too quickly if they are to produce bright emission. Electron excitation in the Lyman and Werner bands meets this requirement, as the branching ratio for dissociation is only about 10%. However, competing dissociation processes must be smaller than the electron excitation rate, and this may be difficult to achieve. Excitation to triplet electronic states, in particular, leads to dissociation, so that the UV emission is unlikely to exceed two or three Lyman band or dissociation continuum photons per molecule entering the high temperature region.

4 Shocks, Turbulence, and Their Consequences

For most HH objects we must consider two types of shock wave models and the heating associated with turbulent mixing and dissipation. Stellar UV heating may be significant in some special cases, but that is a separate topic.

4.1 J-shocks

In classical fluid dynamics one treats a shock as an instantaneous jump in the fluid parameters—density, velocity and temperature. This approximation is widely used for astrophysical shocks in atomic gas, and it is an extremely good approximation for collisionless shocks. It is a risky approximation for a shock in molecular hydrogen, in that the collisions which thermalize the bulk speed of particles entering the shock will also excite the molecules. Therefore, the shock transition layer will both produce some of the emission lines which will be used as diagnostics and reduce the thermal energy content of the post-shock gas. The sharp transition approximation is acceptable if the dissociation rate is high enough that very little H_2 emission arises from the transition layer, but this may not be the case. Phan-Van-Diep, Erwin and Muntz (1989) have measured the particle distributions within the shock transition of a laboratory shock in H_2. The distributions were matched quite well with kinematic models, but these would be difficult to incorporate in a model for the H_2 emission spectrum. The Mott–Smith picture, in which the shock is described as a mixture of pre- and post-shock velocity distributions, and the fraction of each population varies between 0 and 1 over a few collisional mean free paths, might be a reasonable next approximation.

In any case, the H_2 is fairly quickly dissociated behind a J-shock faster than about 45 km s^{-1}. The shocked gas then cools by emitting Lyα, other hydrogen lines, and forbidden lines. If the shock velocity is above about 100 km s^{-1}, a significant amount of ionizing radiation is produced (Shull and McKee 1979; Raymond 1979). This might heat the upstream molecular hydrogen and produce IR emission with a fairly low temperature (depending on the relative density and ionizing flux; Wolfire and Königl 1991). Such an emission precursor seems to be present at one position in Cepheus A (Hartigan et al. 1996). As the gas cools, it can form molecules by gas phase reactions in the partially ionized zone above about 1000 K. This region would produce IR emission with a range of temperatures, accounting for the observations of excess emission in high and low excitation lines, and it could produce large enough column densities in the excited levels of H_2 to account for Lyα fluorescence (Curiel et al. 1995). As the gas cools further, formation on grains and the associated formation pumping produce additional IR emission.

From the point of view of HH object observations, J-shocks have several desirable features. They can account for optical emission lines, large line widths, strong Lyα and UV fluorescence, the close proximity of optical and molecular emission in some objects, and the range of temperatures inferred from the IR spectra of some objects. Their main drawback is their very low efficiency in producing H_2 IR emission. Most of the energy dissipated in the shock winds up in atomic hydrogen emission or optical forbidden lines. A second difficulty is that formation pumping and emission just behind the shock tend to produce too much of the emission in high excitation IR lines. If collisional excitation produces the UV emission in HH2, a J-shock would be the means to expose molecules to hot electrons. However, the luminosity estimates given, taken with plausible estimates for the flux of molecules into the HH2H shock, suggest that collisional excitation is not quite efficient enough to account for the H_2 UV emission.

4.2 C-shocks

A shock in molecular gas with a strong magnetic field and low ionization fraction can undergo a continuous transition from pre-shock to post-shock states (Mullan 1971; Draine 1980; see Chapter 8). The trick is that the magnetic field is tied directly only to the ionized component of the gas, and the Alfvén speed may be larger than the speed of the shock if the density of the ionized component is small enough. In that case, the field pushes ions and electrons though the molecular gas, and friction between the ions and neutrals heats both components. Provided that the cooling rate is large enough (which requires that hydrogen be molecular), the shock can produce a nearly constant temperature zone at 2000–3000 K where essentially all of the energy dissipated by the shock emerges as H_2 IR lines. C-shocks can exist only up to velocities of about 40 km s^{-1}, as faster shocks dissociate the molecules. As with the J-shocks, it may be prudent to re-examine the description of the shock as interpenetrating fluids with well-defined temperatures. Ions pushed by the magnetic field move through the background gas at speeds far above the thermal speeds, and they interact largely by charge transfer and (strongly forward peaked) elastic collisions. A fast neutral produced by charge transfer may excite several molecules before it slows down to join the thermal population. O'Brien (1995) shows that a detailed kinetic model can produce a different emission spectrum than the 2000–3000 K thermal spectrum given by the models. A second theoretical question is the stability of a C-shock. Wardle (1990) describes a dynamical instability due to accumulation of material in the trailing part of a ripple in the magnetic field, and it might disrupt C-shocks very quickly.

From the observational point of view, C-shocks are attractive in that they produce IR emission very efficiently, and the 2000–3000 K predicted spectrum is similar to those generally observed. Potential problems are the broad range of temperatures apparently seen in some objects, the impossibility of producing observed optical emission from the same shock, and line widths which exceed the theoretical maximum C-shock velocity (Brand et al. 1989; Zinnecker et al. 1989). A more general worry is the high magnetic field strength and low ionization state needed to ensure that the Alfvén speed exceeds the shock speed (e.g. Hartigan, Curiel and Raymond 1989). Field strengths of order 100 µG may well be common in clouds near young stars, but ionization fractions of order 10^{-5} or less may be difficult to maintain in the vicinities of fast shocks. An extreme case is HH2, where H_2 infrared emission appears virtually on top of the high excitation knots HH2A' and HH2H, where [O III] and C IV emissions imply high temperatures and strong ionizing photon fluxes (Davis, Eislöffel and Ray 1994; Noriega-Crespo and Garnavich 1994). In a bow shock geometry, the ionizing flux arises at the tip, and there may be enough attenuation to keep it from the wings. However, the models of Raymond, Hartigan and Hartmann (1988) show quite substantial irradiation of the gas which encounters the bow shock wings for suitable bow shock parameters.

4.3 Mixtures of Shocks

Some of the problems with the simple J- or C-shock models can be overcome by combining them in a bow shock. Models for optical and UV emission intensities and profiles use the normal component of the bow shock speed as the effective shock velocity at each position along the bow shock, the parallel velocity component being conserved. The range of shock parameters can then produce a range of excitation temperatures from C-shocks, or one can get J-shocks near the tip of the bow shock to produce optical emission and C-shocks in the wings to efficiently produce H_2 IR lines. Smith, Brand and Moorhouse (1991a) and Davis, Eislöffel and Smith (1996) have investigated this sort of model, and they could match morphologies and line profiles, though they had difficulty producing the very wide IR lines seen in some cases. Amplification of the magnetic field by passage of an earlier shock might alleviate the problem (Smith, Brand and Moorhouse 1991b). Fernandes and Brand (1995) presented a detailed model for HH7 which includes J-shocks near the tip of the bow shocks and C-shocks in the wings where the effective shock velocity falls below 40 km s^{-1} They find that the higher excitation IR lines require some contribution from UV pumping due to the UV emission from the bow shock tip. However, Gredel (1996) found that the IR lines can be matched by a thermal distribution.

A J-shock with an MHD precursor by itself possesses a mixture of characteristics usually associated with a C-shock as well as properties of a J-shock, and models of J-shocks with MHD precursors have been advanced to explain some Herbig–Haro object emission regions (Hartigan, Curiel and Raymond 1989; Curiel 1992). In such a shock there exists an extended region in which the ionized species move relative to the neutrals, but at some point the frictional heating due to the relative motion drives the neutral temperature high enough to dissociate H_2. The dissociation is followed by a catastrophic loss of cooling efficiency and a J-shock transition. An advantage of this picture is the possibility of IR and optical emission in very close proximity (as in HH7). The troublesome requirements of high field and low ionization state are exactly the same as for C-shocks. It is conceivable that MHD precursors might produce H_2 emission at high velocities. Draine and McKee pointed out that shocks faster than 50 km s^{-1} should be subject to ionization breakdown. However, the cross section for 100 km s^{-1} protons to ionize hydrogen is minute, and 100 km s^{-1} electrons are energetically incapable of ionizing. Thus the scale length for this breakdown is likely to exceed the precursor thickness unless the shock speed is very high. The possibility of close proximity of excited H_2 and the Lyα emission region just behind the J-shock is attractive for explaining the UV fluorescence, but the Doppler shift is in the wrong direction for pumping the B–X $v = 2, 1$ P(5) transition (Curiel et al. 1995).

4.4 Turbulent Mixing

The strong shear flow within the bow shock, between the edge of the jet and the backflowing cocoon, or at the outer boundary of the cocoon can generate

Kelvin–Helmhotz instabilities which lead to turbulence and turbulent mixing. They can heat molecular gas both by dissipation of bulk kinetic energy and by mixing hot and cool gas together (Cantó and Raga 1991; Slavin, Shull and Begelman 1993; Raga, Cabrit and Cantó 1995). The entrainment of dense gas by turbulent viscosity may be crucial in linking the fast atomic flow to the slower molecular outflows.

It is difficult to make detailed predictions for the spectral signature of turbulent mixing because of the difficulty in treating turbulence itself. Charnley et al. (1990) explored the chemical signatures of turbulent mixing in the context of a stellar wind interacting with a dense clump. Taylor and Raga (1995) presented a model of the thermal and chemical structure of a mixing layer in which much of the heat goes into H_2 IR emission. The spectral signature is relatively strong high excitation lines, similar to fluorescent and formation pumping predictions. Noriega-Crespo et al. (1996) find that the luminosity and the optical and H_2 morphologies of the HH110 jet agree with the Taylor and Raga predictions.

Interesting recent models for turbulent stripping of gas from a shocked cloud in a lower density medium were presented by Malone, Dyson and Hartquist (1994) and Raga et al. (1997). The cloud gas is initially heated by a slow shock, then fragments torn from the edge of the cloud are gradually accelerated by drag forces as they are exposed to the higher speed gas streaming by. This picture has the potential for explaining the acceleration of molecular gas to speeds above those obtainable in standard C-shocks. It should also produce close proximity of shocked molecular and atomic material. An intriguing question is whether the widths of different H_2 lines will depend on the excitation, since the gas should cool as it accelerates.

5 Summary

Observations of H_2 have developed rapidly over the past few years. While thermal collisional excitation accounts for most of the IR emission of molecular hydrogen, and Lyα fluorescence accounts for much of the UV, there is evidence that other excitation processes contribute. The nature of the shock waves or turbulent mixing zones which provide the energy for the molecular emission is not yet settled, with each observation seeming to point in a different direction.

Bibliography

1. Biro, S. and Raga, A. C. (1994). *ApJ*, **434**, 221.
2. Black, J. H. and Dalgarno, A. (1976). *ApJ*, **206**, 132.
3. Black, J. H. and van Dishoeck, E. (1987). *ApJ*, **322**, 412.
4. Blondin, J. M., Königl, A. and Fryxell, B. A. (1989). *ApJ*, **337**, L37.
5. Böhm, K.-H., Bürke, T., Raga, A. C., Brugel, E. W., Witt, A. N. and Mundt, R. (1987). *ApJ*, **316**, 349.
6. Böhm, K.-H., Scott, D. M. and Solf, J. (1991). *ApJ*, **371**, 248.

7. Brand, P. W. J. L. (1995). In *Shocks in Astrophysics*, eds. T. J. Millar and A. C. Raga. Kluwer, Dordrecht, p.27.
8. Brand, P. W. J. L., Moorhouse, A., Burton, M. G., Geballe, T. R., Bird, M. and Wade, R. (1988). *ApJ*, **334**, L103.
9. Brand, P. W. J. L., Toner, M. P., Geballe, T. R., Webster, A. S., Williams, P. M. and Burton, M. G. (1989). *MNRAS*, **237**, 929.
10. Brown, A., Jordan, C., Millar, T. J., Gondhalekar, P. and Wilson, R. (1981). *Nature*, **290**, 34.
11. Cantó, J. and Raga, A. C. (1991). *ApJ*, **372**, 646.
12. Cantó, J. and Rodriguez, L. F. (1980). *ApJ*, **239**, 982.
13. Carr, J. S. (1993). *ApJ*, **406**, 553.
14. Charnley, S. B., Dyson, J. E., Hartquist, T. W. and Williams, S. A. (1990). *MNRAS*, **243**, 405.
15. Chernin, L. M. and Masson, C. R. (1995). *ApJ*, **443**, 181.
16. Curiel, S. (1992). PhD thesis, Universidad Autonoma de Mexico.
17. Curiel, S., Raymond, J. C., Wolfire, M., Hartigan, P., Morse, J., Schwartz, R. D. and Nisenson, P. (1995). *ApJ*, **453**, 322.
18. Dalgarno, A., Herzberg, G. and Stephens, T. L. (1970). *ApJ*, **162**, L49.
19. Davidsen, A. F., *et al.* (1992). *ApJ*, **392**, 264.
20. Davis, C. J. and Smith, M. D. (1996). *A&A*, **309**, 929.
21. Davis, C. J., Eislöffel, J. and Ray, T. (1994). *ApJ*, **426**, L93.
22. Davis, C. J., Eislöffel, J. and Smith, M. D. (1996). *ApJ*, **463**, 246.
23. Davis, C. J., Mundt, R. and Eislöffel, J. (1994). *ApJ*, **437**, L55.
24. de Gouveia Dal Pino, E. M. and Benz, W. (1993). *ApJ*, **410**, 686.
25. Dopita, M. A. (1978). *ApJS*, **37**, 117.
26. Draine, B. T. (1980). *ApJ*, **241**, 1021.
27. Draine, B. T. and McKee, C. F. (1993). *Ann. Rev. Astr. Ap.*, **31**, 373.
28. Elias, J. H. (1980). *ApJ*, **241**, 728.
29. Falle. S. A. E. G. (1994). In *Kinematics and Dynamics of Diffuse Astrophysical Media*, eds. J. E. Dyson and E. B. Carling. Kluwer, Dordrecht, p.119.
30. Falle, S. A. E. G. and Raga, A. C. (1993). *MNRAS*, **261**, 573.
31. Fernandes, A. J. L. and Brand, P. W. J. L. (1995). *MNRAS*, **274**, 639.
32. Gredel, R. (1996). *A&A*, **305**, 582.
33. Gredel, R. and Reipurth, B. (1993). *ApJ*, **407**, L29.
34. Gredel, R., Reipurth, B. and Heathcote, S. (1992). *A&A*, **266**, 439.
35. Haro, G. (1952). *ApJ*, **115**, 572.
36. Hartigan, P. (1989). *ApJ*, **339**, 987.
37. Hartigan, P., Curiel, S. and Raymond, J. C. (1989). *ApJL*, **347**, L31.

38. Hartigan, P., Morse, J. A. and Raymond, J. C. (1994). *ApJ*, **436**, 125.
39. Hartigan, P. and Raymond, J. C. (1993). *ApJ*, **409**, 705.
40. Hartigan, P., Raymond, J. C. and Hartmann, L. (1987). *ApJ*, **316**, 323.
41. Hartigan, P., Raymond, J. C. and Meaburn, J. (1990). *ApJ*, **362**, 624.
42. Hartigan, P., Carpenter, J. M., Dougados, C. and Skrutskie, M. F. (1996). *AJ*, **111**, 1278.
43. Hartmann, L. and Raymond, J. C. (1984). *ApJ*, **276**, 560.
44. Heathcote, S., Morse, J. A., Hartigan, P., Reipurth, B., Schwartz, R. D., Bally, J. and Stone, J. M. (1996). *AJ*, **112**, 1141.
45. Herbig, G. H. (1951). *ApJ*, **113**, 697.
46. Hollenbach, D. and McKee, C. F. (1989). *ApJ*, **342**, 306.
47. Jordan, C., Brueckner, G. E., Bartoe, J.-D., Sandlin, G. D. and Van Hoosier, M. E. (1977). *Nature*, **270**, 326.
48. Liu, W. and Dalgarno, A. (1996). *ApJ*, **467**, 446.
49. Malone, M. T., Dyson, J. E. and Hartquist, T. W. (1994). In *Kinematics and Dynamics of Diffuse Astrophysical Media*, eds. J. E. Dyson and E. B. Carling. Kluwer, Dordrecht, p.143.
50. Morse, J. A., Hartigan, P., Cecil, G., Raymond, J. C. and Heathcote, S. (1992). *ApJ*, **399**, 231.
51. Morse, J. A. Hartigan, P., Heathcote, S., Raymond, J. C. and Cecil, G. (1994). *ApJ*, **425**, 738.
52. Mullan, D. J. (1971). *MNRAS*, **153**, 145.
53. Neufeld, D. A. and Spaan, M. (1996). *ApJ*, **473**, 894.
54. Noriega-Crespo, A. and Garnavich, P. (1994). *AJ*, **108**, 1432.
55. Noriega-Crespo, A., Garnavich, P., Raga, A. C., Cantó, J. and Böhm, K.-H. (1996). *ApJ*, **462**, 804.
56. O'Brien, I. (1995). *Ap&SS*, **233**, 185.
57. Phan-Van-Diep, G., Erwin, D. and Muntz, E. P. (1989). *Science*, **245**, 624.
58. Raga, A. C. and Böhm, K.-H. (1986). *ApJ*, **308**, 829.
59. Raga,A. C., Cabrit, S. and Cantó, J. (1995). *MNRAS*, **273**, 422.
60. Raga, A. C., Cantó, J., Curiel, S. and Taylor, S. (1997). preprint.
61. Raymond, J. C. (1979). *ApJS*, **39**, 1.
62. Raymond, J. C., Blair, W. P. and Long, K. S. (1997). *ApJ*, in press.
63. Raymond, J. C., Hartigan, P. and Hartmann, L. (1988). *ApJ*, **326**, 323.
64. Schwartz, R. D. (1977). *ApJ*, **212**, L25.4
65. Schwartz, R. D. (1983). *ApJ*, **268**, L37.
66. Schwartz, R. D., Cohen, M. and Williams, P. M. (1987). *ApJ*, **322**, 403.
67. Schwartz, R. D., Williams, P. M., Cohen, M. and Jennings, D. G. (1988) *ApJL*, **334**, L99.

68. Schwartz, R. D., Cohen, B. F., Jones, B. F., Böhmn K.-H., Raymond, J. C., Hartmann, L. W., Mundt, R., Dopita, M. A. and Schultz, A. S. B. (1993). *AJ*, **106**, 740.
69. Shull, L. M. (1978). *ApJ*, **224**, 841.
70. Shull, J. M. and McKee, C. F. (1979). *ApJ*, **227**, 131.
71. Slavin, J. D., Shull, J. M. and Begelman, M. C. (1993). *ApJ*, **407**, 83.
72. Smith, M. D., Brand, P. W. J. L. and Moorhouse, A. (1991a). *MNRAS*, **248**, 451.
73. Smith, M. D., Brand, P. W. J. L. and Moorhouse, A. (1991b). *MNRAS*, **248**, 730.
74. Sternberg, A. (1989). *ApJ*, **347**, 863.
75. Stone, J. M. and Norman, M. L. (1993). *ApJ*, **413**, 210.
76. Taylor, S. D. and Raga, A. C. (1995). *A&A*, **296**, 823.
77. Wardle, M. (1990). *MNRAS*, **246**, 98.
78. Wolfire, M. and Königl, A. (1991). *ApJ*, **383**, 205.
79. Wright, C. M., Drapatz, S., Timmerman, R., van der Werf, P. P., Katterloher, R. and de Graauw, Th. (1996). *A&A*, **315**, L301.
80. Zinnecker, H., Mundt, R, Geballe, T. R. and Zealey, W. J. (1989). *ApJ*, **342**, 337.

PART IV

Evolved Stars

11
Introduction to Stellar Evolution

Detlef Schönberner
Astrophysikalisches Institut Potsdam, Germany

and

Thomas Blöcker
Institut für Astronomie & Astrophysik der Universität Kiel, Germany

1 Setting the Stage
1.1 Principles of Stellar Evolution

Stars are an important constituent of galaxies and are exclusively responsible for the production, reprocessing and dissemination of the heavier elements. The existence of these elements provides the necessary prerequisite to establish chemical reactions under suitable conditions which then ultimately lead to the formation of organic molecules. Hence, it is of considerable importance to acquire a comprehensive and precise knowledge of all the complex physical processes that occur deep inside the stars, in their envelopes and also in the flow of matter from their surfaces into the interstellar medium.

The energy radiated by a star from its surface must be replenished either by potential (= gravitational) and/or nuclear energy. The energy balance is controlled by the virial theorem: one half of the energy released by contraction is radiated away, while the other half is converted into internal energy and heats up the interior. The contraction is virtually halted only when a nuclear fuel is available to cover the energy loss.

Once the contraction as a *protostar* is finished, a star evolves only very slowly as a *main-sequence star* burning hydrogen into helium. This continued element conversion leads to structural modifications, viz. increased central densities and temperatures, and also to a steady luminosity increase. Once hydrogen is exhausted in the central regions, the evolution accelerates as hydrogen is now burning in a shell surrounding the inert helium core. In turn, the core is increasing its density and temperature by continued contraction until the ignition of helium burning in its central regions with subsequent fusion of helium into carbon and oxygen. The cycle of core and shell burning with ever increasing densities and temperatures may continue until the most stable nuclei (those of the iron group) are formed. The further contraction (demanded by the photon and neutrino energy loss) is destabilized in the central high-temperature regions by nuclear dissociation caused by very energetic photons, which then leads immediately to core implosion followed by the blow-off of the entire envelope, observable as a *supernova*. The remnant is a very compact object: either a *neutron star* or a *black hole*.

It is important to realize that, once a fuel-exhausted core has formed, the matter above the new burning shell will greatly expand, with the consequence that only the core, bounded by the burning shell, is able to contract and to release potential energy for its further heating. In this sense it is the mass of the fuel-exhausted core (normally that part of a star within the hydrogen-burning shell) that controls how far the further evolution will continue. The influence of the total stellar mass is limited to the facts that i) the mass of the core scales with the total mass and ii) the matter outside the core (the envelope) provides the nuclear fuel for the shell source which in turn feeds the core with processed matter.

All the major burning phases (hydrogen through carbon) are well separated in time, or radial position if more than one burning region occurs within the star, since one of the decisive factors which controls the nuclear reaction rates, the *Coulomb barrier*, increases with the nuclear charges squared. The heavier the element, the larger the necessary fusion temperature, and a particular fusion reaction chain will start firstly in the central parts of a star where the temperatures are highest. Also, the available energy gain per nucleus decreases with nuclear charge, which leads to very short durations of the more advanced burning phases: central carbon burning may, for instance, only last for about 100 years!

Depending on the mass, it may happen during the course of evolution sketched above that the central densities increase to such high values that the electrons become degenerate and that their pressure dominates in stabilizing the stellar core, independent of temperature. The virial theorem is also valid for a degenerate electron gas, i.e. a loss of energy still requires contraction, releasing twice the energy lost by radiation (in the non-relativistic case.) The difference to the situation above, however, comes from the fact that the internal energy is now dominated by the (non-thermal) kinetic energy of the electrons It can be shown that virtually *all* the gravitational energy released by contraction is converted to the kinetic energy content of the electron gas (cf. the very detailed discussion in Kippenhahn and Weigert 1990).

The consequence for an isolated contracting object like a *white dwarf* in which the pressure support is mainly controlled by the degenerate electron gas is that the ions must lose about as much energy by cooling as is radiated away from the surface, since all the potential energy liberated by contraction is taken up by the electrons. Contraction is possible due to the small ion pressure contribution and continues until eventually a cold configuration (*black dwarf*) is reached where all of the internal energy is contained in the fully degenerate electron gas. In real objects the relativistic degeneracy in the stellar centre and the electrostatic interactions between the components of the stellar plasma change the energy distribution between the ionic and electronic component somewhat, but the results as outlined here remain essentially the same and form the basis of what is called in the literature the *cooling theory of white dwarfs*. It should be mentioned in this context that the interior of white dwarfs have a nearly uniform temperature due to the large heat conductivity of the degenerate electrons.

A stellar core with degenerate electrons evolves differently. It cannot simply cool by contraction since it is surrounded by a nuclear-active shell of very high temperature which heats up the core quite effectively by heat conduction. The shell also adds burned matter to the core, forcing it to contract (by the larger self-gravitation) and to increase the pressure support by the degenerate electron gas correspondingly. Since towards the outer parts of the core the degree of degeneracy is decreasing, the net effect of contraction and heating from above is a general temperature increase of the core, except in its central regions where cooling by neutrino emission may dominate over heating. With an equation of state insensitive to temperature any new fusion ignites (off-centre) under runaway conditions: heating by nuclear burning does not lead to expansion and cooling, but to ever increasing burning and heating until degeneracy is lifted and stable burning under non-degenerate conditions is established. A well-known example is the so-called *central helium flash* in the cores of *red giants*: for a few seconds the helium-burning luminosity reaches values up to $10^{11}\,L_\odot$, but all of the liberated energy is used to expand the core regions.

From the discussion above it is clear that the (self-)gravitation of a gaseous sphere is the force that drives the evolution through nuclear burning stages until an inert, very compact final stage is reached. However, not all stars will be able to reach during their lifetimes all possible nuclear-active phases. To get to the highest necessary burning temperatures, relatively large amounts of potential energy have to be converted into heat, i.e. only the more massive stars will have the chance to end by a supernova explosion. This fact is illustrated in Fig. 1 which gives for different masses the run of central density and temperature through the course of evolution. As a rule, lower initial masses attain lower central temperatures for a given density. One also sees from the path for the lowest mass shown in Fig. 1 ($M = 0.2\,M_\odot$) that in this case the helium-burning temperatures are not reached, and the centre finally cools off as described above for electron degenerated configurations. For larger masses (say, $M \gtrsim 2\,M_\odot$), helium burning can be achieved before electron degeneracy sets in, but carbon burning is delayed because of very efficient neutrino cooling. Shell helium and hydrogen burning continues to increase the core's mass (to about $1.4\,M_\odot$) and temperature (to about 10^9 K) until carbon burning sets in under highly degenerate conditions as a flash that will most likely disrupt the whole star ("carbon flash"). However, massive stars ($12\,M_\odot$ for instance, Fig. 1) are able to reach central temperatures sufficient for carbon burning under non-degenerate conditions since their hydrogen-exhausted cores grow in mass during the course of evolution to above $1.06\,M_\odot$, the minimum mass for carbon ignition under non-degenerate conditions.

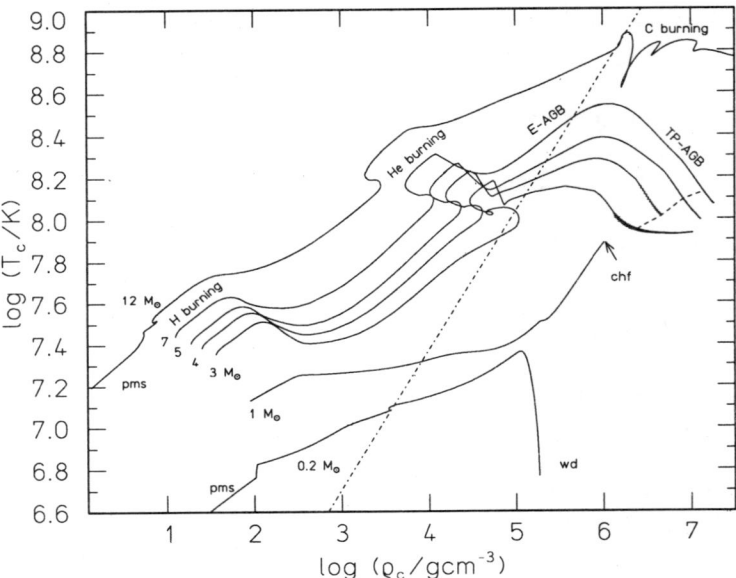

FIG. 1. Evolution of central densities and temperatures for different initial masses according to Blöcker (1995a). Burning stages and other important evolutionary phases are indicated. The pre-main sequence evolution (pms) is given for models of 0.2 and 12 $M_\odot$, the white dwarf (wd) cooling part for the 0.2 $M_\odot$ only. The track for 1 $M_\odot$ ends at onset of the central helium flash (chf). In the region on the right side of the diagonal line electron degeneracy is important. The dashed line indicates the end of the AGB evolution for sequences calculated with the mass-loss prescription of Blöcker (1995a). The wiggles along the TP-AGB correspond to the response of the core's centre to the occurrence of thermal pulses (cf. Section 2.1.1).

The structural changes accompanied with the nuclear evolution deep inside the stars lead to distinct paths in the *Hertzsprung–Russell Diagram* (HR diagram) as shown in Fig. 2. Once a hydrogen-exhausted core has developed, luminosities and radii increase while the surface temperatures, the *effective temperatures*, decrease. When the envelope expands and convective instabilities move inwards, matter that had experienced some CNO processing during the previous main-sequence phase will get stirred up and mixed with unprocessed surface material (*1st dredge-up*).

In particular, stars with electron-degenerate cores become very luminous and cool. Their envelopes are almost fully convective, and they evolve along the so-called *Hayashi limit*. These giant stars can either have a helium core surrounded by a hydrogen-burning shell (*red giants*) or a carbon–oxygen core surrounded by a helium-burning shell, a helium buffer layer, and hydrogen-burning shell. The

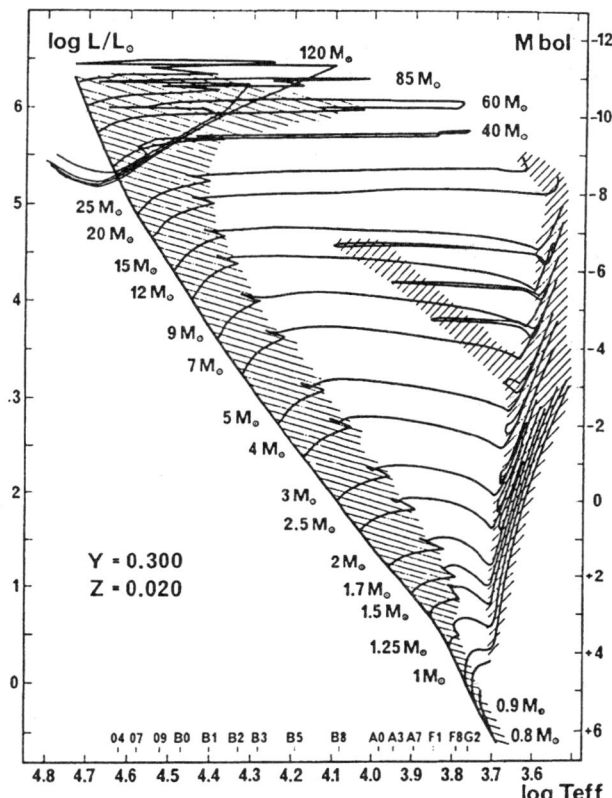

FIG. 2. Evolution of stellar models from 0.8 to 120 $M_\odot$ according to Schaller *et al.* (1992) with consideration of mass loss and overshooting. Hatched regions indicate phases of slow nuclear burning: the left is the main sequence, the right one is characterized by core helium burning (courtesy Springer Verlag, Heidelberg).

latter configuration is called an *asymptotic giant-branch star*, or AGB star for short, the evolution of which will be more thoroughly discussed in Section 2.

Giants close to the Hayashi limit have, at the upper part of the AGB, effective temperatures below 3000 K. The spectra of such cool atmospheres are characterized by the presence of CO and other molecular species, with their relative abundances depending on the C/O ratio. If the latter is smaller than unity, the star is said to be oxygen-rich (and classified as an *M star*), and the photosphere is dominated by oxygen-rich molecules like TiO, VO or H_2O. If s-process elements have already been produced and dredged up to the surface (cf. Section 2.1.1 and 2.1.4), oxides such as ZrO or YO can also be found (an *S star*). If the amount of dredged-up carbon leads to C/O larger than unity (a

C $star$), carbon-rich molecules like C_2, CN, HCN, C_2H_2, etc., dominate since almost all the available oxygen is locked up in CO.

1.2 Significance of Mass Loss

Stars, including our Sun, are not really static objects. In general, there is a perpetual flow of matter from the stellar surface into the interstellar medium, a so-called *stellar wind*, with a rate that depends on the stellar luminosity. For hot stars the wind material is highly ionized and causes broad emission and/or blue-shifted absorption lines against the photospheric light. The wind is driven by radiation pressure exerted on ions through the absorption of UV photons (Castor et al. 1975; Puls et al. 1993). The wind of very cool stars is neutral and consists mainly of molecular gas. At some distance from the stellar surface the temperature in the outflow falls below 1000 K so that even small solid particles (*dust*) can form out of the refractory elements. Although the mass fraction of condensable wind material is quite small ($\lesssim 1\%$), these newly formed grains are important for the outflow dynamics: they absorb stellar photons very effectively and distribute the gained momentum by collisions with gas particles. Calculations showed that such *dust-driven* winds very often need additional support by *shock waves* generated by pulsating stellar envelopes (Sedlmayr and Dominik 1995).

Other wind driving mechanisms exist (e.g. acoustic waves, cf. Lafon und Berruyer 1991), however it appears that only the two mechanisms mentioned above are able to generate mass-loss rates large enough to have significant consequences for the course of stellar evolution. Mass outflow rates up to about $10^{-4}\,M_\odot/\mathrm{yr}$ are observed and will change a star's evolution significantly: in the case of massive stars the still unprocessed envelope matter will be eroded until the products of hydrogen burning or even helium burning become visible (*Wolf-Rayet stars*, see Section 3). In giant configurations mass loss eventually stops the core's evolution and determines the final (= white dwarf) mass.

The efficiency of stellar mass loss and its significance for the stellar mass budget is witnessed by at least three observational facts:

(1) The mass distribution of white dwarfs is surprisingly narrow, with its maximum close to $0.6\,M_\odot$, well below the galactic turn-off mass;
(2) *planetary nebulae* (PNe) are created out of a substantial fraction of the former stellar envelopes;
(3) open clusters with turn-off masses as high as 5–6 $M_\odot$ still contain white dwarfs (see Koester and Reimers 1996).

The relation between initial and final masses, the so-called *initial–final mass relation*, based on white dwarfs in open clusters, is illustrated in Fig. 3 (Herwig 1997; cf. also Weidemann 1987). The determination of a reliable empirical initial–final mass relation is of the utmost importance to constrain mass-loss prescriptions used in evolutionary calculations. It is evident from Fig. 3 that there appears to be a general increase of the final with the initial mass. In

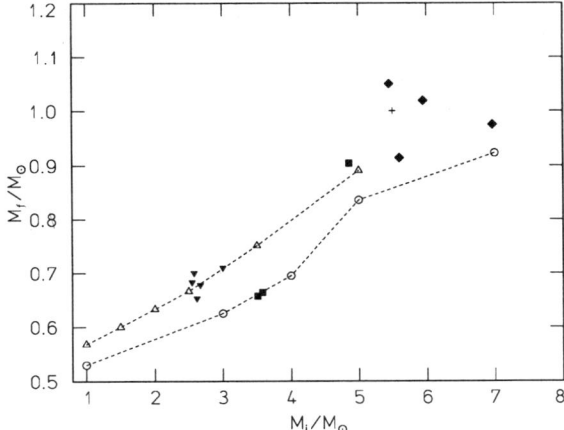

FIG. 3. Observationally determined combinations of initial and final masses for four open clusters according to Herwig (1997): the Hyades (filled triangles), NGC 3532 (filled squares), the Pleiades (cross), and NGC 2516 (filled rhombs). Two theoretical initial–final mass relations, based upon evolutionary calculations with specific mass-loss prescriptions, are also plotted: Vassiliadis and Wood (1993), $Z = 0.016$ (open triangles), Blöcker (1995a), $Z = 0.021$ (open circles).

the extreme case, up to more than 80% of a star's original mass is returned into interstellar space. Thus AGB stars are important suppliers which enrich the interstellar medium with nuclear-processed matter and solid particles (cf. Section 2 for more details).

1.3 Classification According to Stellar Mass

Despite the fact that the mass of the core created at the end of hydrogen burning determines the further evolution, it is customary to classify stars by their total masses

- *low-mass stars* go through hydrogen burning and ignite helium under degenerate conditions (helium flash);
- *intermediate-mass stars* ignite helium under non-degenerate conditions.

Both kinds of stars develop carbon–oxygen cores due to helium burning, but the onset of carbon burning (if possible) is influenced by two effects:

(1) neutrinos cool the central regions of the carbon–oxygen cores (cf. Fig. 1);
(2) if the hydrogen-exhausted core exceeds $\approx 0.8\,M_\odot$, the outer convection zone penetrates into the hydrogen-exhausted layers, reduces considerably the mass of the core, and again enriches the envelope with CNO-cycled matter (*2nd dredge-up*).

If the 2nd dredge-up reduces the mass of the hydrogen-exhausted core to below $1.06\,M_\odot$, carbon burning is for the moment avoided. During the following course of evolution along the AGB, effective neutrino cooling postpones carbon ignition until the core has reached $1.4\,M_\odot$. This event, however, is very unlikely to occur because of mass loss (see below).

- *Massive stars* are those that are able to ignite carbon under non-degenerate conditions and continue their evolution towards a supernova explosion, provided a core mass of about $1.37\,M_\odot$, the critical mass for neon burning, can be reached (e.g. Hashimoto et al. 1993). Otherwise a massive white dwarf ($\approx 1.2\,M_\odot$) consisting of O, Ne, Mg is formed (García-Berro and Iben 1994; Ritossa et al. 1996).

There still exists some ambiguity concerning the mass assignments to this classification scheme. In stars more massive than the Sun the central burning regions are convective, and the size of the hydrogen-exhausted core depends critically on the detailed treatment of the convective boundaries. If one makes the physically reasonable assumption that *overshoot* occurs, which means penetration of convective eddies into formally stable layers, the amount of matter that can be burnt is increased, and thereby also the mass of the hydrogen-exhausted core for a given stellar mass. Depending on the size of the assumed overshoot different stellar masses must be assigned to the classification introduced above. Also the opacity has some influence.

As an example, the recent calculations by D'Antona and Mazzitelli (1996) with the most recent opacities, but without considering overshoot, put the upper limit of intermediate-mass stars at $7\,M_\odot$, since this model develops a hydrogen-exhausted core of just $1.06\,M_\odot$ after the 2nd dredge-up, which then continues with quiet carbon burning. Consideration of overshooting would reduce this mass limit somewhat, and the computed relations of Fig. 1 have to be shifted to the left accordingly.

Rotation, which is normally not considered in stellar evolution calculations, acts in the same direction as overshooting. The effects are, however, only noticeable for rotation rates close to break-up values (cf. Langer 1997).

Together with Fig. 3 one may conclude that

i) there is consistency between observations and evolutionary calculations in the sense that stellar models produce cores of the correct size (e.g. $\approx 1\,M_\odot$ for $M \approx 6$ to $7\,M_\odot$);

ii) mass loss prevents any of these cores from growing to $1.4\,M_\odot$ necessary for the carbon flash to occur.

Stellar objects with $M \lesssim 0.8\,M_\odot$ cannot complete hydrogen burning within the Hubble time and are termed *very low-mass stars*. Finally, objects with masses below about $0.08\,M_\odot$ do not reach sufficiently high temperatures in their interiors to ignite hydrogen, and thus cannot cover their luminosity by nuclear processing at all. These so-called *brown dwarfs* constitute the stellar link to the giant planets.

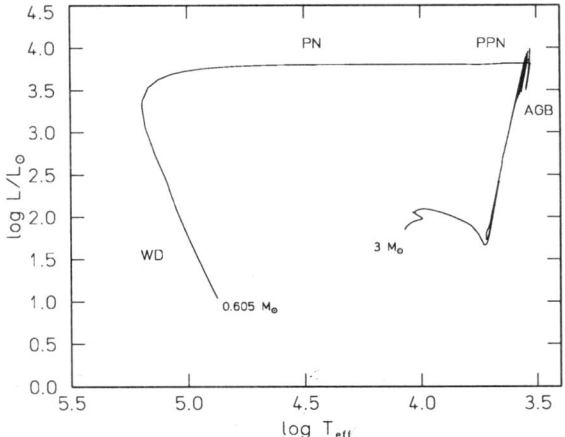

FIG. 4. Complete evolutionary path of an initially $3\,M_\odot$ star from the main sequence to a white-dwarf of $0.605\,M_\odot$ from Blöcker (1995a,b). The main evolutionary phases are indicated (AGB = asymptotic giant branch, PPN = proto-planetary nebula, PN = planetary nebula, WD = white dwarfs), and the last thermal pulses (from a total of 17) at the tip of the AGB are clearly visible. The model is burning hydrogen until the shell source is extinguished close to the turn-around point with $T_{\text{eff}} \approx 160000\,\text{K}$.

2 Low and Intermediate Mass Stars

In this section we will focus upon those stars which evolve after central helium burning through the AGB, become central stars of planetary nebulae and finally enter the white dwarf stage. The evolution along the AGB is characterized by three important features:

(1) thermal instabilities of the helium burning shell (*thermal pulses*);
(2) penetration of the convective envelope into the hydrogen burning shell for more massive objects (*hot bottom burning*).
(3) strong *mass losses*, most likely driven by shocks and dust.

An example of a complete evolutionary calculation of an initially $3\,M_\odot$ star from the main sequence through all following stages is shown in Fig. 4 (Blöcker 1995a, 1995b; without overshoot). Mass loss occurs mainly during the thermal-pulse phase, terminates abruptly the AGB evolution and determines the final mass, $0.605\,M_\odot$, of the central star (and white dwarf).

2.1 Evolution along the AGB

On the AGB the stellar structure can be described as follows: the very compact, hydrogen-exhausted core of mass M_H is surrounded by a very dilute, fully convective, hydrogen-rich envelope of mass M_e. The hydrogen-burning shell is the interface between core and envelope, and its radial position is only about

10^{-4} of that of the stellar surface! For a major fraction of the AGB evolution, hydrogen burning provides nearly all the luminosity that is radiated away from the stellar surface, and the luminosity itself increases with core mass, practically independent of the total stellar mass (Paczyński 1970; Kippenhahn 1981). The core increases by hydrogen burning at a rate given by

$$\dot{M}_{\rm H} = \frac{L_{\rm H}}{XE_{\rm H}},$$

where $E_{\rm H}$ ($= 6.3 \cdot 10^{18}$ erg/g) is the energy released per gram of hydrogen, X the hydrogen mass fraction in the envelope, and $L_{\rm H}$ the hydrogen-shell luminosity. For a typical AGB luminosity of $6000\,L_\odot$, corresponding to $M_{\rm H} \approx 0.6\,M_\odot$, one has $\dot{M}_{\rm H} \approx 10^{-7}\,M_\odot/{\rm yr}$.

For a hydrogen-exhausted core of about $0.6\,M_\odot$ the mass actually contained in the burning shell is very small, only about $5 \cdot 10^{-4}\,M_\odot$. Beneath this shell comes a layer of helium of about $2 \cdot 10^{-2}\,M_\odot$, which is at its bottom hot enough to burn helium to carbon and oxygen. The carbon–oxygen core itself is very dense (10^5–10^6 g/cm^3) and hot ($\approx 10^8$ K) and can be thought of as being a very hot white dwarf which grows in mass by accreting nuclear-processed matter from the envelope. The large densities within the core decrease over a small mass range of some $10^{-2}\,M_\odot$ occupied by the two burning shells towards the very low values of the envelope (typically 10^{-7} g/cm^3 or less). Helium burning contributes normally very little to the surface luminosity ($\approx 0.01\,L_{\rm H}$, but see Section 2.1.1); also the gravothermal contribution is very small. The energy loss by neutrino processes is also small, viz. $< 0.01\,L_{\rm H}$.

According to the general principles outlined in the previous chapter, the core evolves independently from the envelope as long as the latter contains sufficient mass to keep up the burning temperatures at its base by gravothermal energy release. The evolutionary path of an AGB star in the HR diagram, as shown in Fig. 4, is a consequence of the envelope's response to the growth in mass of the core: expansion of the envelope along the Hayashi border line in order to accommodate the increasing luminosity dictated by the core, but contraction at nearly constant luminosity when the envelope mass drops below a few percent of a solar mass, mainly by mass loss from the surface. When hydrogen burning cannot be sustained any longer, i.e. when $M_{\rm e}$ falls below $\approx 10^{-4}\,M_\odot$, the envelope shrinks rapidly to the (white dwarf) dimensions of the core, with a concomitant decrease of the luminosity (cf. Fig. 4).

2.1.1 Thermal Pulses

Early numerical calculations of AGB models showed that a structure with two burning shells is thermally unstable in the sense that the helium-burning shell enters repeatedly into runaway situations (Schwarzschild and Härm 1965; Weigert 1966). During these so-called *thermal pulses* or *helium-shell flashes* the luminosity by helium burning may exceed the surface luminosity by orders of magnitudes for a time span of about 100 years. The physics of thermal

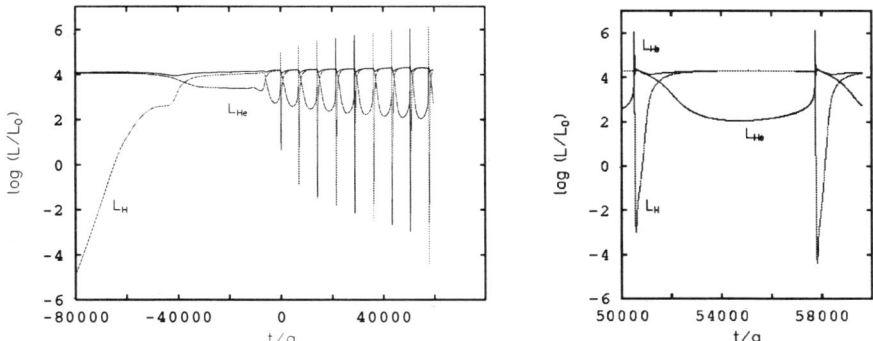

FIG. 5. Left: Surface luminosity and shell-source contributions on the AGB for a 5 $M_\odot$ model according to Blöcker (1995a). The core mass is $M_H = 0.823\,M_\odot$ at the first major pulse ($t = 0$ yrs) and increases to $0.834\,M_\odot$ until mass loss terminates the evolution after nine thermal pulses. Right: The last two pulses on an expanded time axis.

instabilities of nuclear shell sources is too complicated to be explained briefly here. The reader is, instead, referred to the book of Kippenhahn and Weigert (1990).

The situation is illustrated in Fig. 5 where the typical luminosity evolution of both shells is shown for a 5 $M_\odot$ model after completion of central helium burning. At first helium burning dominates (early AGB, or *E-AGB*), whereas later hydrogen burning takes over (thermally pulsing AGB, or *TP-AGB*). If one defines a thermal pulse cycle phase ϕ as the fraction of the time elapsed between two subsequent pulses, counted from the peak luminosities of the helium burning shell, it can be seen in Fig. 5 that immediately after a thermal pulse ($0 \leq \phi \lesssim 0.15$) hydrogen-burning is virtually extinguished, and that helium burning provides the surface luminosity. For the rest of the cycle hydrogen burning is restored and controls the evolution. The large energy excess liberated during a pulse is temporarily stored as potential energy and re-radiated later on a much longer time scale. Therefore, the changes at the star's surface, i.e. of its radius, effective temperature and luminosity, are much smaller than the many order-of-magnitude variations of the shell luminosities themselves, and are illustrated in Fig. 6. We note that all quantities calculated in terms of these parameters, viz. mass-loss rates or dynamical pulsational periods, will also vary during a pulse cycle. The strength of their variations depends, of course, on the parametrization used (see 2.1.3).

Thermal-pulse periods firstly increase and then (after the pulse strengths have reached "full" amplitudes) slowly decrease with core mass, a relation originally found by Paczyński (1975). For a typical core mass M_H of about $0.6\,M_\odot$ this period is of the order of 10^5 years, for $0.8\,M_\odot$ only of 10^4 years (see Fig. 5 with $M_H = 0.823\,M_\odot$ at the first pulse). More details how thermal-pulse cycle

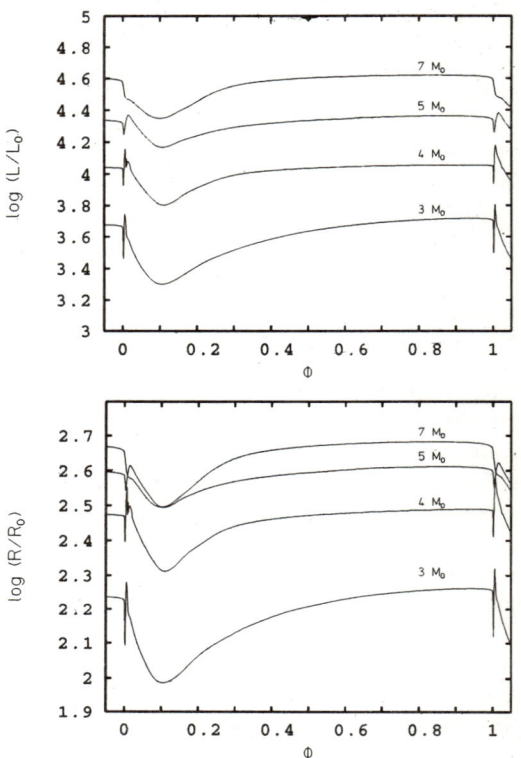

FIG. 6. Surface luminosity (upper panel) and stellar radius (lower panel) as a function of the thermal-pulse cycle ϕ for the 10th pulse of AGB models with different (initial) masses (Blöcker 1995a).

properties depend on stellar parameters are given, e.g., in Iben (1975), Sackmann (1977), Boothroyd and Sackmann (1988) or Wagenhuber and Weiss (1994).

An important consequence of the shell flashes is the possibility of mixing of processed or newly synthesized elements into the envelope and towards the observable surface. The large amount of energy liberated during the helium-shell instability drives a convective zone which mixes products of helium burning, i.e. carbon and oxygen, into the *intershell region* consisting of CNO-cycled matter. This leads to concomitant internal structural disturbances (cf. discussion of the luminosity variations above) during which, after hydrogen burning ceases, the envelope convection may penetrate into those carbon- and oxygen-enriched intershell layers (the so-called *3rd dredge-up*). For instance, in the example shown in Fig. 7 about $4 \cdot 10^{-3}\,M_\odot$ of carbon-rich matter ($X_c \approx 0.5$) are mixed into the envelope. With the declining disturbances the convection recedes, hydrogen re-ignites and burns quiescently until the next pulse and its mixing episode.

The 3rd dredge-up is a repeating phenomenon and may change the envelope's carbon-to-oxygen ratio from smaller than unity into larger than unity, i.e. an M star has converted itself into a C star.

2.1.2 Hot Bottom Burning

As the luminosity increases along the AGB, envelope convection extends further and further downwards during the quiescent hydrogen-burning phases. For models with $M \gtrsim 5\,M_\odot$ and $M_H \gtrsim 0.8\,M_\odot$, the lower boundary of envelope convection can even penetrate into the hydrogen burning regions, allowing burning at temperatures in excess of $50 \cdot 10^6$ K at the bottom *and* mixing of the newly processed isotopes throughout the stellar envelope (*hot bottom burning*, e.g. Iben 1975; Scalo *et al.* 1975). Such models become rapidly very luminous and do *not* obey the classical core-mass luminosity relationship any more (for more details see Blöcker and Schönberner 1991; Boothroyd *et al.* 1993; D'Antona and Mazzitelli 1996).

Hot-bottom burning has, in combination with the 3rd dredge-up, important consequences for the envelope's chemical evolution and the composition of stellar ejecta, viz. the abundances as observed in planetary nebulae. The recently found lithium-rich luminous AGB stars in the Magellanic Clouds (Smith and Lambert 1989, 1990) are explained as a direct consequence of hot bottom burning (e.g. Scalo *et al.* 1975; Blöcker and Schönberner 1991; Sackmann and Boothroyd 1992). Also the transformation of a C star into a nitrogen-rich S star (e.g. Iben 1975; Boothroyd *et al.* 1993) is due to hot bottom burning. Accordingly, the surface abundance of carbon depends on the competition between hot bottom burning and dredge up. More details can be found in Forestini and Charbonnel (1997), or Lattanzio and Boothroyd (1997). Planetary nebulae showing abundance patterns with the signature of former hot bottom burning are discussed in, e.g. Kaler and Jacoby (1989) and Clegg (1991).

2.1.3 Mass Loss

The importance for low and intermediate mass stars of mass loss by dust-driven winds has already been discussed in Section 1. Observations indicate rates of $10^{-7}\,M_\odot/\text{yr}$ for small-period *Mira stars* and up to $10^{-4}\,M_\odot/\text{yr}$ for luminous *long-period variables* (e.g. Wood 1997). At larger mass-loss rates the grains are responsible for complete obscuration of the underlying star. Such objects with optically thick, dusty envelopes are only visible at longer wavelengths (above 1 micron) because of the thermal emission of the grains, heated from the inside by the (invisible) star. The velocity of the outflowing matter is typically between 10 and 20 km/s, well below the escape velocity at the stellar surface. Further information can be gained from the thorough review of Habing (1996) which illuminates all the different aspects of mass loss along the AGB.

Since a rigorous theory of mass loss from AGB giants is still lacking, evolutionary calculations have to resort to empirical or semi-empirical descriptions of mass loss along the AGB (cf. Weidemann 1993). Currently two approaches have been used in model calculations along the AGB:

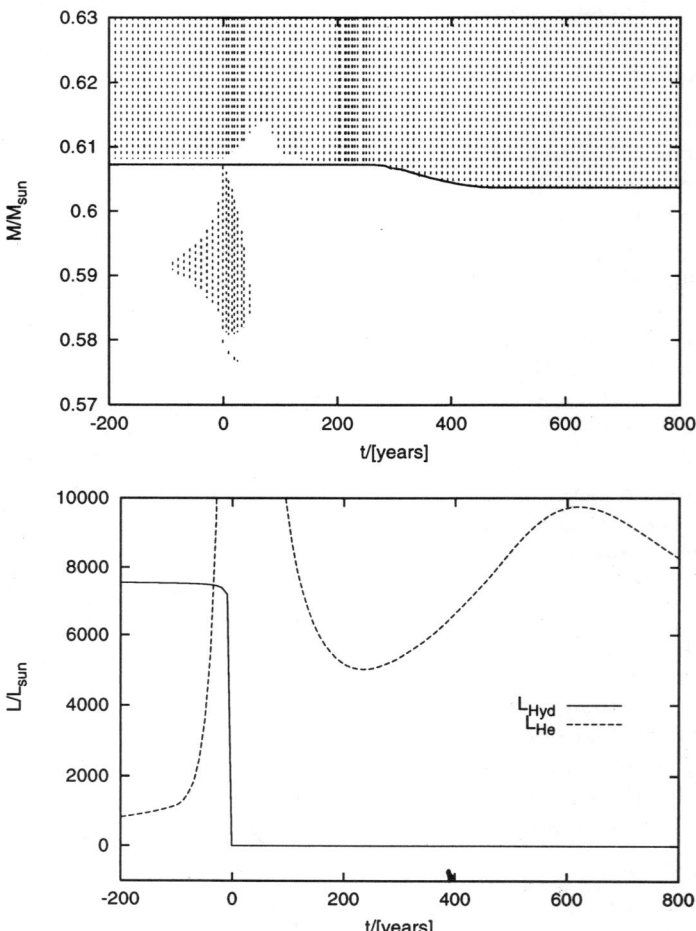

FIG. 7. The extension in mass of convective regions (top) and the luminosity contributions of the shell sources (bottom) vs. time for the 12th pulse of a $3M_\odot$ AGB star (Herwig et al. 1997). Time is set to zero at the peak luminosity of the He burning shell, which is about $10^6 L_\odot$. The solid line denotes the border of the hydrogen exhausted core and indicates also the bottom of the hydrogen-burning shell till time zero. The shaded regions indicate convective mixing. Helium-burning takes place between the mass coordinates $M_r = 0.585$ and $0.570\,M_\odot$. The pulse-driven convective zone exists for about 150 years. The dredge-up starts 250 years after this convective shell has disappeared and reduces the core's mass by about $0.004\,M_\odot$. Hydrogen re-ignites 5000 years later, and the next pulse commences after a further 60 000 years of evolution.

(1) Vassiliadis and Wood (1993) constructed a mass-loss formula by combining empirical relationships between mass-loss rates and outflow velocities on one hand and mass-loss rates and pulsational periods on the other.
(2) Blöcker (1995a) utilized the hydrodynamical calculations of pulsating Mira atmospheres performed by Bowen (1988) to construct a semi-empirical mass-loss formula.

It should be emphasized that both approaches predict so-called *accelerated* mass-loss rates that increase rapidly with luminosity and are considerably *larger* than those which follow from the popular Reimers formula (1975). Both sets of calculations predict initial–final mass relationships that are, within the observational errors, consistent with the observations (Fig. 3).

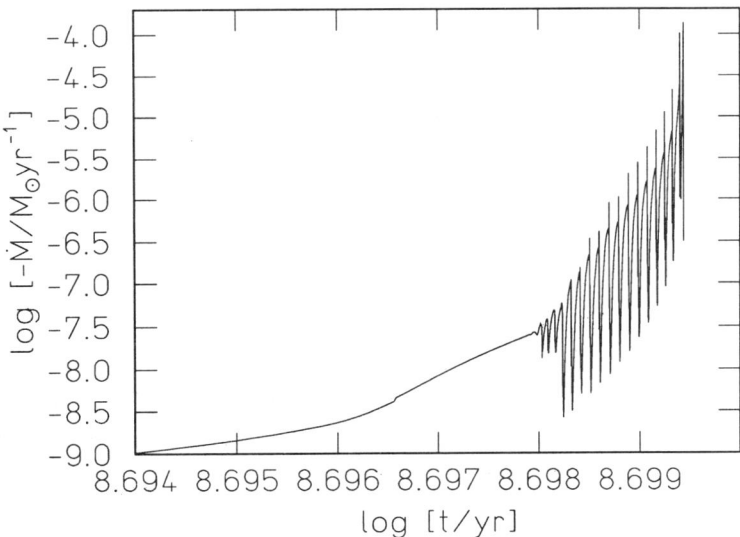

FIG. 8. Temporal evolution of the mass-loss rate along the AGB for a $3\,M_\odot$ model sequence (17 thermal pulses, Blöcker 1995a). The final mass is $0.605\,M_\odot$.

Both prescriptions predict huge mass-loss rate variations during the AGB evolution in general and thermal-pulse cycles in particular. This is demonstrated in Fig. 8 which shows as an example the whole mass-loss evolution along the AGB for the $3\,M_\odot$ model sequence already used in Fig. 4. One clearly sees how thermal pulses lead to order-of-magnitude modulations, or short interruptions, of the otherwise monotonically increasing mass-loss rate. Similar variations result from the Vassiliadis and Wood mass-loss prescription.

These (predicted) large mass-loss rate variations over a thermal-pulse cycle will have drastic consequences for the radial structure and the infrared emission of the wind envelopes, as is shown in Fig. 9. In these hydrodynamic calculations time-dependent values of stellar mass, luminosity and effective temperature (as shown in Fig. 6) as well as the resulting variable mass loss (as shown in Fig. 8) have been taken into account. As the consequence of a thermal pulse, objects on the upper AGB can develop *detached* dust shells and spend about 10 % of their lives away from the main IRAS two-colour relation for dusty AGB stars. The corresponding spectral energy distributions appear to be in good agreement with the observations, and mass-loss modulations of the computed size seem to account for the loops in the IRAS two-colour diagram as they have been observed for C stars as well as for M stars (cf. Olofsson *et al.* 1990; van der Veen and Habing 1988; Kwok *et al.* 1989; Zijlstra *et al.* 1992).

2.1.4 Comments on the s-Process Nucleosynthesis During the AGB Evolution

The intershell and helium-burning shell region of an AGB star is a site for the nucleosynthesis of *s-process* elements. Two possible reactions provide the necessary neutrons: $^{22}\text{Ne}(\alpha,\text{n})^{25}\text{Mg}$ and $^{13}\text{C}(\alpha,\text{n})^{16}\text{O}$. The temperatures reached in the intershell zone during thermal pulses appear to be too low to activate the $^{22}\text{Ne}(\alpha,\text{n})^{25}\text{Mg}$ reaction. Instead, the s-process is most likely driven by the $^{13}\text{C}(\alpha,\text{n})^{16}\text{O}$ neutron source. The problem with the latter reaction is how to mix just the right amount of protons into carbon-rich layers in order to produce sufficient amounts of ^{13}C via $^{12}\text{C}(\text{p},\gamma)^{13}\text{N}(\beta^+,\nu)^{13}\text{C}$. Iben and Renzini (1982) found that the protons can diffuse from the bottom of the convective envelope into the intershell zone due to semiconvection. However, this scenario is restricted to low metallicities (Iben 1983). To follow the s-process under typical conditions one has to ingest artificially a certain amount of ^{13}C at the proper time (e.g. Straniero *et al.* 1995).

The production of s-process elements, and of carbon as well, within a star is one problem, the other is to bring these elements to the stellar surface by dredge-up. The amount of dredge-up found in evolutionary calculations depends sensitively on metallicity, core and envelope masses (Wood 1981), and is in any case *too small* to account for the observations. Additionally, numerical details may play an important role (Frost and Lattanzio 1996). The cause of the large uncertainties concerning the efficiency of dredge-up is, of course, our inability to treat convection correctly.

Synthetic AGB calculations, therefore, utilize stellar parameters known from evolutionary AGB model sequences but take the minimum core mass for dredge-up, M_H^{min}, and the dredge-up parameter λ, as adjustable parameters (among others) in order to match the observations, i.e. the luminosity function of LMC carbon stars. For example, van den Hoek and Groenewegen (1997) find $M_H^{min} = 0.58\,M_\odot$ and $\lambda = 0.75$ as best fit values in contrast to the results of evolutionary calculations ($M_H^{min} \gtrsim 0/65\,M_\odot$ and $\lambda \approx 0.25$, see Wood 1997).

These intrinsic uncertainties of stellar evolution calculations and their apparent weakness in matching the observations led to the conclusion that mixing may take place outside the formally convective boundaries (e.g. Hollowell and Iben

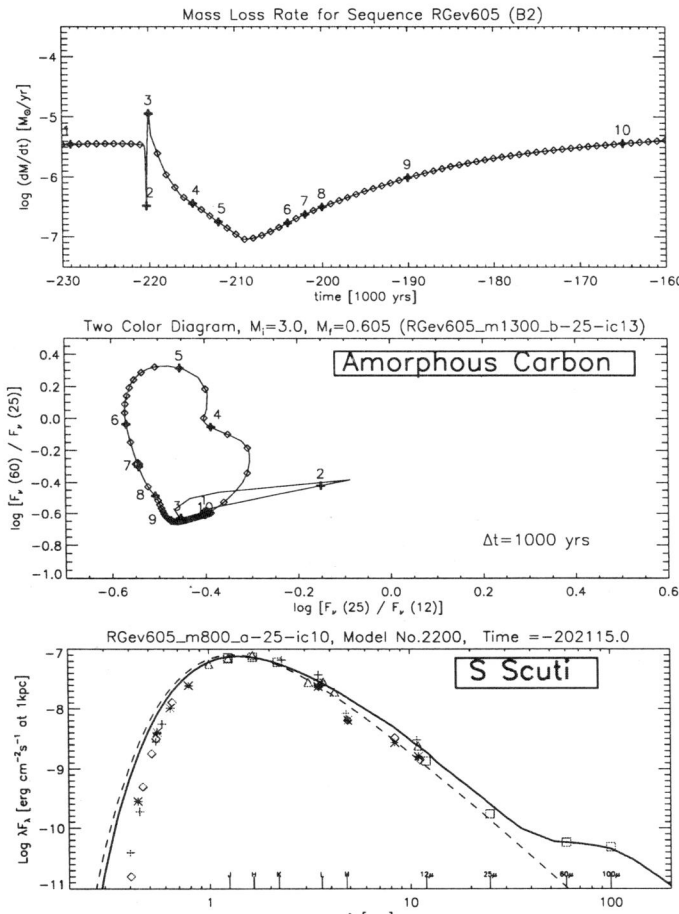

FIG. 9. Hydrodynamical simulation of an AGB star's dust shell evolution according to Schönberner et al. (1997). Top: Temporal change of the AGB mass-loss rate during a typical thermal-pulse cycle, according to Blöcker (1995a). Middle: Corresponding colour evolution of a model carbon star, with numbered + signs indicating the same times as in the top panel. Diamonds outline the time evolution in steps of $\Delta t = 1000$ years. The grains consist of amorphous carbon and have a radius of $0.05\,\mu$m. The adopted dust-to-gas mass ratio is $1.5 \cdot 10^{-3}$. Bottom: Observed spectral energy distribution of S Scuti (symbols; data from Groenewegen and de Jong 1994) compared with our theoretical spectrum (solid line) at the time labelled (7). The spectrum of the unobscured central star (dashed line) is that of a black body with $T_{\text{eff}} = 2760$ K at this moment. The infrared and IRAS pass bands are indicated on the wavelength scale

1988; D'Antona and Mazzitelli 1996; Wood 1997), or even that only a hydrodynamic approach of modelling the H/He interface will overcome the drawbacks of the local treatment of convection (Arlandini et al. 1996).

The latter conjecture seems to be confirmed by recent calculations of Herwig et al. (1997). They introduced overshoot based on the results of hydrodynamical simulations of convection by Freytag et al. (1996) to their treatment of convection, and applied it to AGB models. Indeed, these illustrative calculations showed that this method provides a sufficient amount of dredge-up *as well as* production of ^{13}C in AGB stars with rather *small* core masses of, say, about $0.6\,M_\odot$. It should be noted that a ^{13}C pocket formed during the dredge-up phase will not survive until the onset of the next pulse and, thus, will not be engulfed by the next flash-driven convection zone as often assumed (Straniero et al. 1995). Instead, due to the high temperatures reached in the intershell region in the course of evolution it will be burnt already during the interpulse phase, i.e. under radiative conditions. Accordingly, the s-process takes place in a radiative environment. For more details of s-processing, see Käppeler (1996) and Gallino et al. (1996).

All nucleosynthesis calculations should consider the influence of mass loss as well. Mass loss reduces the envelope mass and hence the dilution of newly made isotopes, but it reduces also the duration of the thermally pulsing AGB phase. Realistic mass-loss rates terminate this phase after at most some 10^6 years, and thus the number of thermal pulses is rather limited ($\lesssim 20$). The important interplay between mass loss and nucleosynthesis remains to be investigated in detail.

2.2 Post-AGB Evolution

Evolutionary calculations show that the models depart from the AGB when the envelope mass is reduced by mass loss and burning to the order of a percent of the total mass ($\approx 10^{-2}\,M_\odot$) (Schönberner 1979; Iben 1984). Since mass loss dominates so much over burning, the evolution off the AGB will be very rapid with only minute changes of the phase angle ϕ. Furthermore, since hydrogen burning dominates for about 80 % of the TP-AGB evolution, one also expects that most post-AGB stars will still burn hydrogen quiescently, and the existing mass-loss prescriptions used in model calculations confirm this view. The model shown in Fig. 4 started its post-AGB evolution with $\phi \approx 0.5$. For the most recent sets of post-AGB evolutionary calculations, see Vassiliadis and Wood (1994) and Blöcker (1995b).

The interpretation of observed central-star properties is not clear-cut: Schönberner (1981, 1986), Iben (1995) and Górny et al. (1994), for instance, found evidence that most central stars are burning hydrogen quiescently, but other interpretations are also possible (Dopita et al. 1997). A recent discussion on the evolutionary status of central stars has been given by Schönberner (1997).

2.2.1 Transition from the AGB Towards the Central-star Region

Observations indicate that the mass-loss rates of central stars of planetary nebulae are up to several orders of magnitude below those of the preceding AGB evolution (cf. Perinotto 1989). Therefore, the strength of the stellar wind has to decrease substantially during the transition from the AGB towards the central-star regime. However, up to now it is not known—either from observations or from theory—how and in which temperature range this mass-loss fading takes place. Accordingly, different prescriptions for the treatment of mass loss during this phase exist in evolutionary calculations (for comparisons see Blöcker 1995c; Steffen and Szczerba 1997).

The contraction from a cool AGB supergiant to a very much hotter object, and the consequences for the wind envelope, is presently of considerable interest (e.g. van Hoof et al. 1997, and references therein). This very brief evolutionary phase (see below) is determined by

- the thinning of the dust envelope due to expansion and decreasing mass-loss rate,
- photoionization and dissociation of molecules, and finally
- photoionization of atoms,

with a time development ruled by the evolutionary speed of the central stellar object. The physical processes involved are virtually identical to those in *photon-dominated regions* at the edges of HII regions around hot Pop. I stars (see Chapter 9). An object in this evolutionary stage is also called *proto-planetary nebula*, or PPN for short.

The evolutionary speed off the AGB depends highly on the reduction rate of the envelope mass, which, in turn, is given by the burning rate of the hydrogen shell and the mass-loss rate. In the vicinity of the AGB the ratio of mass loss to burning rate is very large (up to 10 000 for a remnant of $0.6\,M_\odot$), and mass loss completely controls the pace of evolution. Accordingly, the mass-loss rate itself determines also the temporal evolution of a cool dusty wind envelope and its transformation into an observable planetary nebula.

The thinning of the dust shell is observationally manifested by the reappearance of the optical radiation from the star itself and by far-infrared radiation from the detached dust shell. The IRAS satellite has detected many objects with spectral types F and G which are surrounded by cool (≈ 100 K) and optically thin dust shells (e.g. Parthasarathy and Pottasch 1986; Likkel et al. 1987; Pottasch and Parthasarathy 1988). Other typical members of this phase are the non variable OH/IR stars, according to their spectral appearance (cf. Habing et al. 1989).

Observations seem to indicate that the transition times should not be too long since the coolest post-AGB stars known have effective temperatures of about 5000 K, and kinematical ages of the youngest planetary nebulae are only of the order of 1000 years (van der Veen and Habing 1988; Hrivnak et al. 1989).

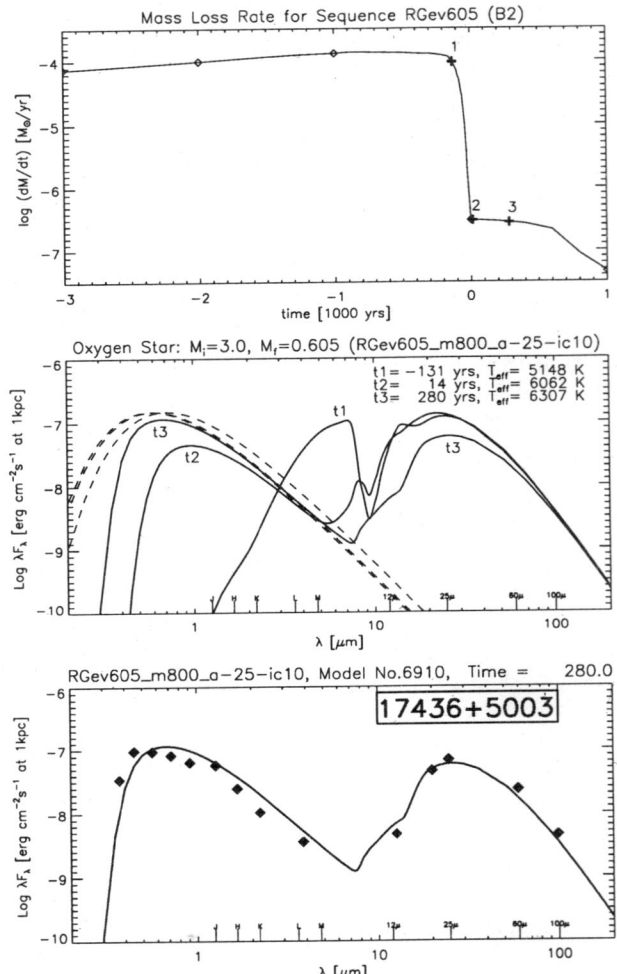

FIG. 10. Top: Adopted mass-loss rate during the transition from the AGB towards higher effective temperatures for an evolutionary track ending up with a remnant of 0.605 $M_\odot$ (Blöcker 1995a). $t = 0$ indicates the beginning of the post-AGB evolution with a significantly reduced mass-loss rate (compare also with Fig. 8 above). Middle: Spectral energy distributions for models with silicate grains ("Astronomical Silicates") at three selected times as indicated in the upper panel. The dashed lines are the corresponding intrinsic spectra of the central star. Bottom: Spectral energy distribution of the model at time $t3$ (solid line) compared to observations of $IRAS$ 17436 +5003 = HD 161 796 (diamonds; data from Hrvinak et al. 1989). Our central star ($T_{\rm eff} \approx 6300\,{\rm K}$) seems slightly cooler than the observed object ($T_{\rm eff} \approx 7000\,{\rm K}$).

Schönberner et al. (1997) continued their hydrodynamical simulations somewhat beyond the AGB in order to model also the envelope's detachment (Fig. 10). With the adopted mass-loss law, the star becomes visible in less than 100 years, and the comparison with the observation confirms that, at least in the case shown in Fig. 10, a large mass-loss rate must continue until the star's effective temperature has increased to about 5000 K, and must then decrease by orders of magnitude very rapidly. More simulations of this kind, together with detailed comparisons with observed spectral energy distributions of transition objects are necessary to constrain the mass-loss rate variations in the vicinity of the AGB tip.

2.2.2 From Central Stars to White Dwarfs

In the PN region proper mass loss due to radiation pressure on lines can compete with hydrogen burning only for luminous remnants, i.e. for those with $M \gtrsim 0.8\,M_\odot$ (Pauldrach et al. 1988). The total crossing time of the lighter central stars is uniquely given by the available fuel (i.e. the envelope mass) divided by the hydrogen luminosity. Because the envelope mass decreases and the luminosity increases with remnant mass, one arrives at the following figures for typical crossing times from, say, 10 000 K to the turn-around point in the Hertzsprung–Russell diagram: about 100 000 years for $0.55\,M_\odot$, 4 000 years for $0.6\,M_\odot$, and only 50 years for $0.94\,M_\odot$. Small differences for a given mass are due to the dependence of envelope mass and hydrogen luminosity from the phase angle ϕ. Also a metallicity dependence has been noted (Iben and MacDonald 1986).

When hydrogen burning cannot be sustained any longer because the envelope mass has become too small, the stellar luminosity drops very fast by at least one order of magnitude until it can be covered by gravothermal energy release. Helium burning is unimportant for most phase angles and dies away as well. The fading time of AGB remnants down to, for instance, 100 $L_\odot$ is thus controlled by the gravothermal energy release (and by neutrino energy losses as well). Both processes depend on the thermomechanical structure of the core, and thus on the *complete* evolutionary history. A more detailed discussion can be found in Blöcker (1995b) and Schönberner and Blöcker (1996).

A comprehensive set of evolutionary paths of hydrogen-burning post-AGB models is presented in Fig. 11. They are from the computations of Schönberner (1983) and Blöcker (1995b). The different evolutionary speeds across the Hertzsprung–Russell diagram and down to the white dwarf regime are illustrated by isochrones. Some observational data points indicating the positions of central stars of rather old planetary nebulae are also given. They have been taken from Jacoby and Kaler (1989) and Kaler and Jacoby (1989). From 82 listed galactic planetary nebulae we have selected those objects (29) which are optically thick and whose loci in the HR diagram can be determined by Zanstra temperatures ($T_{\rm HI} \approx T_{\rm HeII}$) and Shklovsky distances ($M_{\rm ion} = 0.2\,M_\odot$). As expected, nearly all of the objects are found in the regions where the fading speed of our models decreases considerably. For example, all very hot and presumably more massive

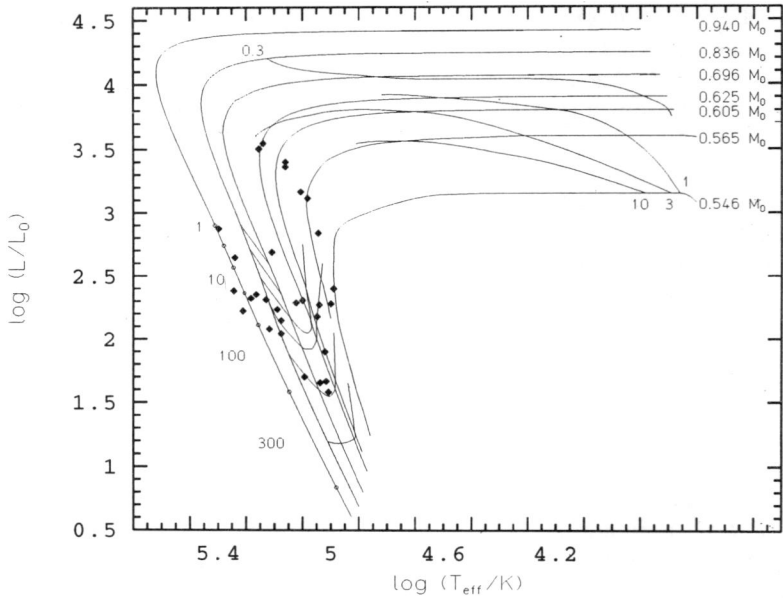

FIG. 11. Post-AGB evolutionary tracks of hydrogen-burning models taken from Schönberner (1983) and Blöcker (1995b) with isochrones and observational data of central stars (see text for details). Time marks are in units of 10^3 yrs.

central stars of this sample ($M \gtrsim 0.8\,M_\odot$) are located well above $\log(L/L_\odot) = 2$, some even above $\log(L/L_\odot) = 2.5$.

It should be noted that a significant fraction about 10 to 20 %) of post-AGB objects do not fit into this evolutionary scheme. These objects belong to the so-called *Wolf–Rayet central stars* and their descendants, the *O VI* and *PG 1159* objects, where about every second of the latter is without a detectable PN (for a spectroscopic definition of the Wolf–Rayet phenomenon see Section 3). A common property of all these stars is that their surfaces are virtually hydrogen-free, with carbon being the most abundant element, and that they follow similar evolutionary paths towards the white-dwarf regime as their "normal" counterparts (e.g. Hamann 1996; Werner et al. 1996). It is thought that thermal pulses may be responsible for these abundance anomalies, either at the very end of the AGB evolution or during the post-AGB phase. Thermally pulsing post-AGB models and their relevance for observed objects are discussed in, e.g., Iben and MacDonald (1995), Gorny and Stasinska (1995), Blöcker and Schönberner (1997), or Leuenhagen and Hamann (1997).

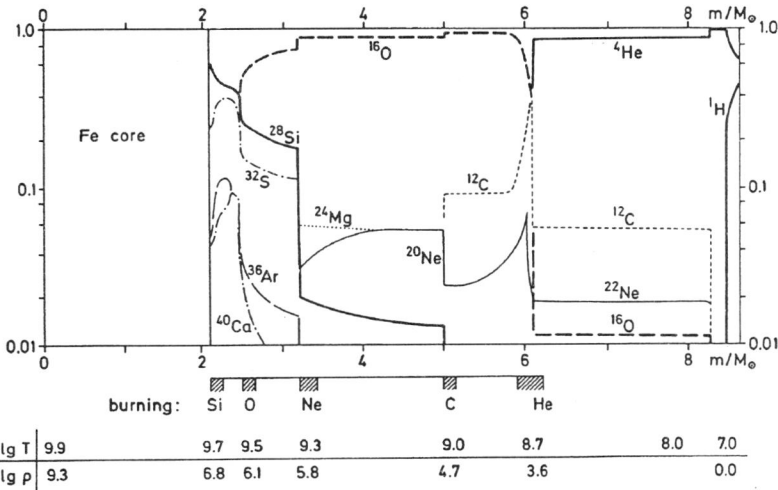

FIG. 12. The elemental composition of the interior of a highly evolved model of a 25 M$_\odot$ star. The mass concentrations of a few important elements are plotted against the mass m. Below the abscissa the locations of shell sources and typical values of temperature (in K) and density (in g/cm^3) are indicated (taken from Kippenhahn and Weigert (1990), courtesy Springer Verlag, Heidelberg).

3 Massive Stars

As outlined in Section 1, massive stars continue their evolution beyond helium burning, ignite carbon under non-degenerate conditions and pass then through all the following burning phases until finally an iron core is built up in the centre. The ultimate fate is then core collapse followed by a supernova explosion, during which a large variety of heavy elements, even those beyond iron, will be ejected into the interstellar medium. The mass range depends upon the physical assumptions made: the lower limit is determined by how convective overshooting is treated. With the currently accepted amount, this limit is definitively below 10 M$_\odot$. The upper mass limit is given by the onset of dynamical instabilities during hydrogen burning, either by pulsation or electron–positron pair creation, and is close to 100 M$_\odot$. For more details, see for instance Woosley and Weaver (1986, 1995). In addition, rotation plays an important role (e.g. Langer et al. 1997; Meynet and Maeder 1997).

Thus the massive stars constitute the most luminous, most short-lived, and hottest, albeit small fraction of stellar populations with large impacts on the chemical and dynamical evolution of their host galaxies. Their convective cores are very large and may contain 80 % of the total stellar mass. All the following burning phases take place within this former convective core and form what is called a onion-like layered structure (Fig. 12).

A typical property of massive stars is the occurrence of strong winds with high outflow velocities (>1000 km/s). Estimated mass-loss rates are as high as 10^{-5} to $10^{-4}\,M_\odot$/yr (de Jager et al. 1988; Castor 1993) and they must be considered in evolutionary calculations of hydrogen burning models. The HR diagram of a more recent calculation, including mass loss, is presented in Fig. 2 (Schaller et al. 1992). Models of stars that are not too massive evolve to the Hayashi limit as the intermediate mass stars, and contribute to a population of *massive supergiants* nearly indistinguishable from that formed by AGB stars, though their internal constitution is different. For instance, they do not experience the dredge-up episodes typical for the well-evolved AGB stars. The most massive stars even avoid becoming red supergiants but stay rather hot during core helium burning (cf. Fig. 2).

Although the lifetimes of these stars are extremely short, mass loss with the observed rates has profound impact on their mass budgets. The unprocessed envelope matter is effectively peeled off, and nuclear-processed matter originally contained in the former convective core exposed. From the spectroscopic point of view mass loss turns these stars from a normal appearance into what is called a *Wolf–Rayet (WR)* characteristic: hydrogen-poor surfaces with strong emission lines of nitrogen and helium (WN) or even carbon (WC) and oxygen (WO) when mass loss has uncovered also regions where once helium burning had taken place. Of course, only for the most massive and luminous stars will mass loss be sufficiently efficient to uncover layers deep inside. The Wolf–Rayet remnants may have masses as low as $10\,M_\odot$ (Langer 1991), and will evolve directly into supernovae (Woosley et al. 1993).

The observed properties of Wolf–Rayet stars (cf. van der Hucht 1992), such as spectral appearance, temperature, and luminosity, are the result of the competition between nucleosynthesis in the deeper regions and mass loss from the surface. The strength of mass loss during the main sequence (spectral type O) determines the surface chemistry and time of occurring of the Wolf–Rayet stage. Since winds driven by radiation pressure on spectral lines depend on the stellar metallicity, the spectral-type distribution of any Wolf–Rayet star population is determined by the metallicity of the host galaxy.

Investigations of the fundamental parameters and the surface compositions are not only important for our understanding of the evolution of Wolf–Rayet stars, but also for checking our knowledge of the nucleosynthesis. At present, however, the detailed comparison between predictions of stellar evolution theory and sophisticated spectral analyses is far from being gratifying: the observed objects are in general too cool and often also not luminous enough compared to the theoretical predictions from evolutionary calculations (Hamann 1996).

Bibliography

1. Arlandini, C., Gallino, R., Busso, M. and Straniero, O. (1996). Stellar evolution: What should be done. *32nd Liège International Astrophysical Colloquium*, eds. A. Noels et al., p.447

2. Blöcker, T. (1995a). *A&A*, **297**, 727.
3. Blöcker, T. (1995b). *A&A*, **299**, 755.
4. Blöcker, T. (1995c). White dwarfs. *Lecture Notes in Physics*, vol. 443, eds. D. Koester and K. Werner. Springer, Berlin, p.68.
5. Blöcker, T. and Schönberner, D. (1990). *A&A*, **240**, L11.
6. Blöcker, T. and Schönberner, D. (1991). *A&A*, **244**, L41.
7. Blöcker, T. and Schönberner, D. (1997). *A&A*, **324**, 991.
8. Boothroyd, A. D. and Sackmann, I. J. (1988). *ApJ*, **328**, 632.
9. Boothroyd, A. D., Sackmann, I. J. and Ahern, S. C. (1993). *ApJ*, **416**, 762.
10. Bowen, G. H. (1988). *ApJ*, **329**, 299.
11. Castor, J. I. (1993). Massive stars: their lives in the interstellar medium. eds. J. P. Cassinelli and E. B. Churchwell, *ASP Conf. Ser.*, **35**, 297.
12. Castor, J. I., Abbott, D. C. and Klein, R. I. (1975). *ApJ*, **195**, 157.
13. Clegg, R. E. S. (1991). Evolution of stars: the photospheric abundance connection, *Proceedings IAU Symp.145*, eds. G. Michaud and A. Tutukov, Kluwer, Dordrecht, 387.
14. D'Antona, F. and Mazzitelli, I. (1996). *ApJ*, **470**, 1093.
15. de Jager, C., Nieuwenhuijzen, H., van der Hucht, K. A. (1988). *A&AS*, **72**, 259.
16. Dopita, M. A., Vassiliadis, E., Wood, P. R., Meatheringham, S. J., Harrington, J. P., Bohlin, R. C., Ford, H. C., Stecher, T. P. and Maran, S. P. (1997). *ApJ*, **474**, 188.
17. Forestini, M. and Charbonnel, C. (1997). *A&AS*, **123**, 241.
18. Freytag, B., Ludwig, H.-G. and Steffen, M. (1996). *A&A*, **313**, 497.
19. Frogel, J. A., Mould, J. R. and Blanco, V. M. (1990). *ApJ*, **352**, 96.
20. Frost, C. A. and Lattanzio, J. C. (1996). *ApJ*, **473**, 383.
21. Gallino, R., Busso, M., Arlandini, C., Lugaro, M. and Straniero, O. (1996). *Mem. S. A.It.*, **67**, 761.
22. García-Berro, E. and Iben, I. Jr. (1994). *ApJ*, **434**, 306.
23. Górny, S. K. and Stasinska, G. (1995). *A&A*, **303**, 893.
24. Górny, S. K., Tylenda, R. and Szczerba, R. (1994). *A&A*, **284**, 949.
25. Groenewegen, M. and de Jong, T. (1994). *A&A*, **282**, 115.
26. Habing, H. J. (1996). *A&AR*, **7**, 97.
27. Habing, H. J., Tignon, J. and Tielens, A. G. G. M. (1994). *A&A*, **286**, 523.
28. Hamann, W.-R. (1996). Hydrogen-deficient stars, eds. C. S. Jefferey and U. Heber. *ASP Conf. Ser.*, **96**, 127.
29. Hashimoto, M., Iwamoto, K. and Nomoto, K. (1993). *ApJ*, **414**, L105.
30. Herwig, F. (1997). Priv. comm.

31. Herwig, F., Blöcker, T., Schönberner, D. and El Eid, M. (1997). A&A, **324**, L81.
32. Hollowell, D. and Iben, I. Jr. (1988). ApJ, **333**, L25.
33. Hrivnak, B. J., Kwok, S. and Volk, K. M. (1989). ApJ, **346**, 265.
34. Iben, I. Jr. (1975). ApJ, **196**, 525.
35. Iben, I. Jr. (1983). ApJ, **275**, 65.
36. Iben, I. Jr. (1984). ApJ, **277**, 333.
37. Iben, I. Jr. (1995). Physics Reports, **250**, 1.
38. Iben, I. Jr. and MacDonald, J. (1986). ApJ, **301**, 164.
39. Iben, I. Jr. and MacDonald, J. (1995). White dwarfs. Lecture Notes in Physics, vol. 443, eds. D. Koester and K. Werner. Springer, Berlin, p.48.
40. Iben, I. Jr. and Renzini, A. (1982). ApJ, **263**, L23.
41. Jacoby, G. H. and Kaler, J. B. (1989). AJ, **98**, 1662.
42. Kaler, J. B. and Jacoby, G. H. (1989). ApJ, **345**, 871.
43. Käppeler, F. (1996). Mem. S. A.It., **67**, 749.
44. Kippenhahn, R. (1981). A&A, **102**, 293.
45. Kippenhahn, R. and Weigert, A. (1990). Stellar Structure and Evolution. Springer, Berlin.
46. Koester, D. and Reimers, D. (1996). A&A, **313**, 810.
47. Kwok, S., Volk, K. M. and Chan, S. J. (1989). Evolution of peculiar red giant stars. Proceedings IAU Coll. 106, eds. H. R. Johnson and B. Zuckermann. Cambridge University Press, Cambridge, p.284.
48. Lafon, J.-P. J., Berruyer, N. (1991). A&A Rev, **2**, 249.
49. Langer, N. (1991). Wolf–Rayet stars and interrelations with other massive stars in galaxies. Proceedings IAU Symp. 143, eds. K. A. van der Hucht and B. Hidayat, Kluwer, Dordrecht, p.431.
50. Langer, N., Fliegner, J., Heger, A. and Woosley, S. E. (1997). Nuclei in the Cosmos IV, eds. M. Wiescher et al., Nuc. Phys. A, **621**, 457c.
51. Lattanzio, J. C. and Boothroyd, A. I. (1997). Astrophysical implications of the laboratory study of presolar materials, eds. T. Bernatowitz and E. Zinner, AIP Conf. Ser.
52. Leuenhagen, U. and Hamann, W.-R. (1998). A&A, **330**, 265.
53. Likkel, L., Omont, A., Morris, M. and Forveille, T. (1987). A&A, **173**, L11.
54. Meynet, G. and Maeder, A. (1997). A&A, **321**, 465.
55. Netzer, N. and Elitzur, M. (1993). ApJ, **410**, 701.
56. Olofsson, H., Carlström, U., Eriksson, K., Gustafsson, B. and Willson, L. A. (1990). A&A **230**, L13.
57. Paczyński, B. (1970). Acta Astr., **20**, 47.
58. Paczyński, B. (1975). ApJ, **202**, 558.

59. Parthasarathy, M. and Pottasch, S. R. (1986). *A&A*, **154**, L16.
60. Pauldrach, A., Puls, J., Kudritzki, R. P., Méndez, R. and Heap, S. R. (1988). *A&A*, **207**, 123.
61. Perinotto, M. (1989). Planetary nebulae. *Proceedings IAU Symp.131*, ed. S. Torres-Peimbert. Kluwer, Dordrecht, 293.
62. Pottasch, S. R. and Parthasarathy, M. (1988). *A&A*, **192**, 182.
63. Puls, J., Pauldrach, A. W. A., Kudritzki, R.-P., Owocki, S. P. and Najarro, F. (1993). *Rev. Mod. Astr.*, **6**, 273.
64. Reimers, D. (1975). *Mem. Soc. Sci. Liège*, **8**, 369.
65. Reimers, D. and Koester, D. (1988). *ESO Messenger*, **54**, 47.
66. Ritossa, C., García-Berro, E. and Iben, I. jr. (1996). *ApJ*, **460**, 489.
67. Sackmann, I. J. (1977). *ApJ*, **212**, 159.
68. Sackmann, I. J. and Boothroyd, A. I. (1992). *ApJ*, **392**, L71.
69. Scalo, J. M., Despain, K. H. and Ulrich, R. K. (1975). *ApJ*, **196** 805.
70. Schaller, G., Schaerer, D., Meynet, G. and Maeder, A. (1992) *A&AS* **96**, 269.
71. Schönberner, D. (1981). *A&A*, **103**, 119.
72. Schönberner, D. (1983). *ApJ*, **272**, 708.
73. Schönberner, D. (1986). *A&A*, **169**, 189.
74. Schönberner, D. (1997). Planetary nebulae. *Proceedings IAU Symp.180*, eds. H. J. Habing and H. J. G. L. M. Lamers, Kluwer, Dordrecht, in press.
75. Schönberner, D. and Blöcker, T. (1996). *Astr. & Sp. Sc.*, **245**, 201.
76. Schönberner, D., Steffen, M., Stahlberg, J., Kifonidis, K. and Blöcker, T. (1997). Advances in Stellar Evolution, eds. R. Rood and Renzini, A., Cambridge University Press, Cambridge, p.146.
77. Schwarzschild, M. and Härm, R. (1965). *ApJ*, **142**, 855.
78. Sedlmayr, E. and Dominik, C. (1995). *Sp. Sci. Rev.*, **73**, 211.
79. Smith, V. V. and Lambert, D. L. (1989). *ApJ*, **345**, L75.
80. Smith, V. V. and Lambert, D. L. (1990). *ApJ*, **361**, L69.
81. Smith, V. V., Lambert, D. L. and McWilliam, A. (1987). *ApJ*, **320**, 826.
82. Steffen, M. and Szczerba, R. (1997). *Ap&SS*, **251**, 131.
83. Straniero, O., Gallino, R., Busso, M., Chieffi, A., Raitieri, C. M., Salaris, M. and Limongi, M. (1995). *ApJ*, **440**, L85.
84. van den Hoek, L. B. and Groenewegen, M. A. T. (1997). *A&AS*, **123**, 305.
85. van der Hucht, K. A. (1992). *A&AR*, **4**, 123.
86. van der Veen, W. E. C. J. and Habing, H. J. (1988). *A&A*, **194**, 125.
87. van Hoof, P. A. M., Oudmaijer, R. D. and Waters, L. B. F. M. (1997). *MNRAS*, **289**, 371.
88. Vassiliadis, E. and Wood, P. R. (1993). *ApJ*, **413**, 641.

89. Vassiliadis, E. and Wood, P. R. (1994). *ApJS*, **92**, 125.
90. Wagenhuber, J. and Weiss, A. (1994). *A&A*, **286**, 121.
91. Weidemann, V. (1987). *A&A*, **188**, 74.
92. Weidemann, V. (1993). Mass loss on the AGB and beyond, ed. H. E. Schwarz. *ESO Conference and Workshop Proc. No. 46*, 59.
93. Weigert, A. (1966). *Z. Astrophys.*, **64**, 395.
94. Werner, K., Dreizler, S., Heber, U. and Rauch, T. (1996). Hydrogen-deficient stars, eds. C. S. Jefferey and U. Heber. *ASP Con. Ser.*, **96**, 267.
95. Wood, P. R. (1981). *Physical Processes in Red Giants*, eds. I. Iben Jr. und A. Renzini, Reidel, Dordrecht, p. 135
96. Wood, P. R. (1997). Planetary Nebulae, *Proceedings IAU Symp.180*, eds. H. J. Habing and H. J. G. L. M. Lamers, Kluwer, Dordrecht, in press.
97. Woosley, S. E. and Weaver, T. A. (1986). *ARAA*, **24**, 205.
98. Woosley, S. E. and Weaver, T. A. (1995). *ApJS*, **101**, 181.
99. Woosley, S. E., Langer, N. and Weaver, T. A. (1993). *ApJ*, **411**, 823.
100. Zijlstra, A. A., Loup, C., Waters, L. B. F. M. and de Jong, T. (1992). *A&A*, **265**, L5.

12
Dust Formation in Carbon-rich AGB Stars

Isabelle Cherchneff
Department of Physics, UMIST

1 Introduction

Stars ascending the Asymptotic Giant Branch (AGB) are characterized by high mass losses and extended circumstellar envelopes, especially when the stars reach the tip of the AGB. Typical values for mass loss rates range from 10^{-8} to 10^{-4} $M_\odot$ yr^{-1} for the most evolved objects and the extent of circumstellar envelopes can reach 1000 stellar radii. The strong wind developed by an AGB star is of crucial importance for the evolution of the object as the star expels about 40 % of the initial mass it has on the Zero-Age Main Sequence. Furthermore, the winds represent the means by which the interstellar medium (ISM) is replenished in chemical species and dust grains, therefore governing the elemental composition of the ISM in galaxies (for more details on late stages of stellar evolution, see Chapter 11).

All AGB stars show a strong excess emission in the infrared (IR) that has been ascribed to dust grains absorbing the visible and near-IR parts of the stellar radiation field and re-emitting as black bodies at IR wavelengths. The formation of dust particles is not too surprising as AGB stars have cool and dense photospheres. Typical values for the effective temperature and the gas number density at the outer edge of the photosphere are $T_{eff} \sim 3000$ K and $n_{phot} \sim 10^{16}$ cm^{-3}, respectively. Dust is of paramount importance as it participates both in the dynamics and chemistry of the stellar wind. Indeed, although other sources of mass loss are required to reproduce mass loss rates (see Section 2.2), the dust provides the ultimate acceleration to the stellar wind and permits high mass loss rates to be generated. Once the dust grains have formed they quickly interact with the gas not only dynamically but also chemically. When the dust leaves the hot inner layers close to the photosphere, the gas temperature decreases and the grains reach regions where the temperature becomes low enough to allow for gas–grain chemical interaction and the formation of certain molecules on the surface of dust particles. For example, ammonia, NH$_3$, methane, CH$_4$ and silane, SiH$_4$ are believed to form from "surface chemistry" in the intermediate region of carbon-rich circumstellar envelopes (Keady and Ridgway 1993). At even larger radii ($r > 100\ R_\star$, where $R_\star$ is the stellar radius), in the outer part of the envelope, dust grains shield molecules from photo-processes induced by the penetration of the interstellar radiation field and cosmic rays, and permit chemical species to survive and complex molecules to form (see Chapter 15 and references therein). Finally, the dust escapes the star-envelope bound system to reach the ISM where it is recycled and processed in the dense interstellar clouds.

This chapter reviews the various aspects of dust formation and mass loss processes in carbon-rich stars and is organized as follows: we discuss the dust composition in Section 2, the various processes proposed to drive mass loss are presented in Section 3, Section 4 covers the theories of dust nucleation and condensation, the chemical kinetics approach to dust nucleation is discussed in Section 5, and finally Section 6 outlines future directions for the study of circumstellar dust.

2 Dust Composition

The composition of circumstellar dust grains depends on the photospheric elemental composition of the star and whether the star has gone through thermal pulsing or not (see Chapter 11). During thermal pulses, the main source of nuclear energy in the stellar atmosphere swaps from the hydrogen shell surrounding the core to the helium shell, resulting in a helium shell flash and carbon dredge-up to the stellar surface (Boothroyd and Sackmann 1988). For all stars, either carbon-rich or oxygen-rich, the most stable molecule formed under thermal equilibrium in the photosphere is carbon monoxide, CO. If the star is oxygen-rich ($\equiv C/O < 1$), CO will lock up almost all the available carbon and silicate dust will form. If the star is carbon-rich ($\equiv C/O > 1$), the opposite occurs and carbon-bearing molecules are available in the flow to trigger the formation of carbonaceous grains.

Models of mass loss processes and dust formation have been developed for carbon stars (C-stars) rather than oxygen stars (O-stars) for two reasons essentially:

(1) The proximity of IRC+10216 (or CW Leo), an evolved C-star undergoing a strong mass loss ($\dot{M} \sim 4 \times 10^{-5}$ $M_\odot$ yr^{-1}). The star is at $\sim$200 pc and is one of the brightest objects in the IR. It has been extensively studied at IR and millimetre wavelengths and has no O-rich equivalent available for observation.

(2) Carbon chemistry has been studied in terrestrial laboratories, as carbon derivatives (soot, coal, graphite, diamond, etc.) are ubiquitous on Earth. The chemistry of refractory material is less developed due to the lack of experimental data on reaction rates (see Chapter 13).

As mentioned previously, the composition of circumstellar dust is deduced from fitting the IR excess emission appearing in the spectral energy distribution (SED) of the star. This procedure requires a complete treatment of the radiative transfer in the wind and assumptions of the composition and the size distribution of dust grains (Bagnulo et al. 1995; Groenewegen 1997). Obviously, a dust model needs also to reproduce extinction properties in the ultraviolet and visible spec-

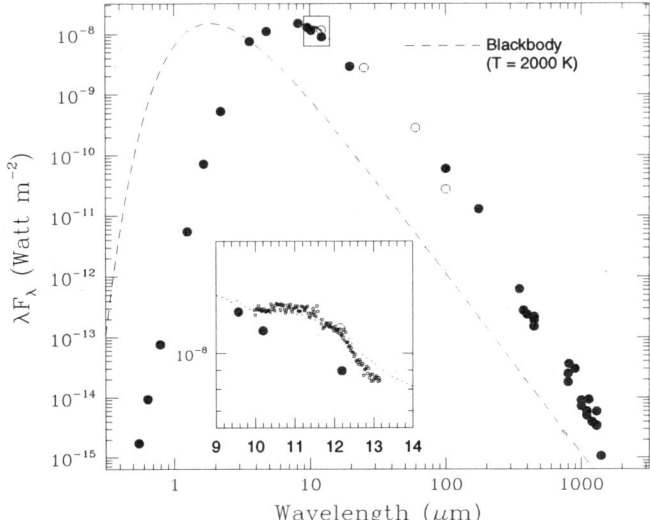

FIG. 1. Spectral energy distribution of IRC+10216 at IR wavelengths and a 2000 K black body curve (dashed line). The inset shows the 11.3 μm feature due to α-SiC grains. The observational data are: empty squares—UKIRT high resolution CGS3 spectrum (Bagnulo and Doyle, private communication), dotted line—IRAS low resolution spectrum, filled circles—ground-based photometry (Bagnulo et al. 1995).

tral range. Figure 1 represents the spectral energy distribution of IRC+10216 at IR wavelengths. Usually, SEDs for C-stars can be fitted well with a mixture of amorphous carbon (AC) and silicon carbide (SiC) grains. SiC exists in two forms, hexagonal α-SiC and cubic β-SiC, and the best fits are obtained using the hexagonal form. Optical properties of amorphous carbonaceous material are quite featureless in the visible and the IR while small α-SiC grains show surface mode-induced Fröhlich bands between 10.3 and 12.6 μm (Bohren and Huffman 1983). C-stars often show an emission band at 11.3 μm ascribed to small α-SiC grains or possibly small SiC grain cores coated with an AC mantle (Kozasa et al. 1996). Figure 1 represents the SED for IRC+10216 along with a 2000 K black body flux that mimics the stellar radiation field.

The 11.3 μm emission band is clearly evident and the IR excess is reproduced using a dust composition of ~93% AC and ~6% α-SiC (Bagnulo 1996). Therefore the dust in C-rich stars is represented by at least two grain populations (SiC and AC dust). It is likely that AC particles are not made of pure carbon but rather a composite material including SiC or other carbides and some amorphous carbon phase (Bernatowicz et al. 1996, and Section 7). Amorphous carbon is favoured over graphite or diamond as the outflow time scale is too small to allow for graphitization, especially in an environment devoid of strong UV radiation

(Sorrell 1990). Hence, we will consider in this review amorphous carbon and silicon carbide as the main constituents of dust grains in C-stars.

3 Mass Loss Processes

Several mechanisms have been proposed to explain the high mass loss rates and low outflow velocities of late-type stars. As dust has long been observed in AGB winds, it was suggested that radiation pressure acting on dust grains accelerates the grains which then drag the gas along as long as gas and grains remain coupled. This mechanism drives mass loss (Weymann 1960; Gehrz and Woolf 1971; Kwok 1975). However, the dust formation processes were not considered in early dust–driven wind models and it was assumed that dust condensation occurred under thermal equilibrium (TE) conditions, suggesting that dust condensation occurred at low temperatures (see Section 4.1) which were to be found at several stellar radii. The process was thought inefficient in driving mass loss as the gas density at the condensation radius would be very low, if it followed the gas hydrostatic scale height.

Alternative mechanisms were suggested to reproduce the mass loss rates or to provide a way to enhance the gas density close to the star in order to maximize the efficiency of dust–driven outflows. Magnetic (Alfvén) waves were studied in cool stars by Hartmann and MacGregor (1980). They assumed an input wave energy flux comparable to solar values and showed that the propagation of Alfvén waves led to terminal velocities that were too high, unless the waves underwent some dissipation processes close to the star. However, the wind terminal velocity was found very sensitive to the damping length of the dissipation processes assumed (Holzer et al. 1983), and AGB outflow velocities could be reproduced only for assumed dissipation lengths of the order of one stellar radius. Such a confinement of the damping length has no physical explanation and is difficult to reconcile with the fact that all AGB stars lose mass (Holzer et al. 1983). Finally, dissipation processes give rise to warm, extended chromospheres, a prediction which may be confirmed for stars like α Orionis but which cannot account for some cool AGB stars without chromospheres, such as IRC+10216.

Turbulent convection zones in the stellar atmosphere produce pressure fluctuations which generate acoustic waves. Stellar winds driven by acoustic waves were studied by Pijpers and Hearn (1989) and the model was applied to AGB stars by Pijpers and Habing (1989). They showed that both adiabatic and isothermal acoustic waves could drive mass loss and lead to terminal velocities in agreement with observational values. The wind velocities they predicted did not depend strongly on the choice of dissipation lengths as for Alfvén wave models, and acoustic waves are therefore good candidates as a mechanism for driving mass loss and stationary winds in evolved stars.

Late-type stars are large-amplitude pulsators with typical pulsation periods of a few hundred days. The pulsations result from the modulated absorption of radiation in the stellar atmosphere which leads to vibrational instabilities, the so-called κ-mechanism which is maximum in the helium second-ionization zones of the stellar atmosphere (Cox 1980). Several authors have investigated the propagation of periodic compression waves and their effects on the atmosphere of pulsating stars (Fedorova 1973; Willson and Hill 1979). The amplitude of compression waves propagating in an atmosphere with a decreasing gas density will grow because of conservation of energy and steepen into shock waves. There is observational evidence for shock waves in the outer atmosphere of late-type stars. For example, Hinkle (1978) studied the spectroscopic variability of CO and hydroxyl, OH, lines in the Mira variable R Leonis, and Gillet et al. (1983) studied the shock-induced variability of the hydrogen emission lines in Mira. The shocks deposit energy in the gas which then becomes hotter than its radiative equilibrium temperature. Depending on the local densities, the gas cooling processes are either radiative or mechanical, the latter process being equivalent to adiabatic expansion. The net result of shock propagation on the stellar atmosphere is to extend the gas layers near the photosphere and to change the gas hydrostatic scale height into an "extended" scale height (Willson and Hill 1979; Willson and Bowen 1984; Cherchneff et al. 1992).

All the above models provide energy dissipation in the regions close to the photosphere and result in an enhancement of the gas density over a few stellar radii. Compression waves driven by stellar pulsation appear to be the most "natural" mechanism to initiate the driving of winds as any AGB star pulsates and undergoes mass loss. The most satisfactory models of mass loss have combined the effect of compression waves and radiation pressure on dust grains and are able to reproduce observed mass losses and terminal wind velocities for AGB stars (see Kwok 1975 and Wood 1979 for stationary winds, Bowen 1988 for time-dependent winds). Bowen also shows that shock propagation results in the formation of "stationary layers" close to the photosphere where the gas is accelerated by the shocks but falls back onto the star due to gravity.

The remaining key problem in those models is the treatment of dust nucleation and condensation. Indeed, dust is usually assumed to form at a few stellar radii from the stellar surface, and the chemistry of dust formation is ignored. More comprehensive models of stationary and time-dependent winds, including dust formation, are discussed in Section 4.3.

4 Models of Dust Formation

4.1 Thermal Equilibrium Calculations

Early studies of dust formation in circumstellar environments concerned the transition between gas phase to solid phase under thermal equilibrium (TE) (Salpeter 1974; McCabe et al. 1979; McCabe 1982). This approach was motivated by the fact that high gas temperatures and densities were to be expected in stellar photospheres. However, a first drawback of the TE theory was that both AC and SiC have condensation temperatures at equilibrium which are low (~ 1700 K and ~ 1300 K, respectively) compared to the gas kinetic temperature in or just out of the photosphere. TE models were therefore constructed for assumed effective temperatures that were too low for the stars, or on the assumption that TE held too far away from the photospheres (Lafont et al. 1982). The dynamics of the inner stellar winds was also overlooked in the models although the pulsating properties of AGB stars were already known and their resulting effects on the gas dynamics already predicted (Fedorova 1973; Willson and Hill 1979).

Although it is accepted now that dust cannot form under TE conditions, TE remains a good description of the stellar photosphere. Furthermore, TE models are useful in predicting which chemical species are available in the photosphere for dust nucleation. For the gas parameters characterizing the photosphere of AGB stars (i.e. pressure $P_{phot} \sim 10^4$ dyn cm^{-2}, temperature $T \sim 2500$ K), acetylene, C_2H_2, is the dominant C-bearing molecule after CO in a C-rich environment and any nucleation process must use and convert acetylene to form dust (Tsuji 1973; Tarafdar 1987; Cherchneff and Barker 1992). We shall consider these processes in Section 5.

4.2 Homogeneous and Heterogeneous Nucleation of Dust

Nucleation theory has been extensively studied to model the formation of water droplets in the Earth's atmosphere (Abraham 1974; Friedlander 1977) and was first applied to the formation of stellar dust by Draine and Salpeter (1977); Gail et al. (1984); Gail and Sedlmayr (1985, 1987, 1988); see also Chapter 13. The theory describes the formation and growth of small liquid or solid particles from a gaseous phase on the assumptions that

(1) the gas is in thermal equilibrium;
(2) the gas becomes supersaturated;
(3) the growth occurs via the clustering of a "monomer" species whose size is small compared to the final sizes of the grains;
(4) the thermodynamic properties of monomers and small clusters can be extrapolated from the bulk material properties.

The theory predicts the existence of a "critical cluster size" N_c defining a region of unstable clusters with size $N_{cl} < N_c$ and a region of spontaneous stable growth of clusters with size $N_{cl} > N_c$. Homogeneous nucleation concerns the growth of clusters via a unique monomer. For example, for carbon clusters, the monomer can be atomic carbon and the growth is described by

$$C_N + C \longrightarrow C_{N+1}, \qquad (4.1)$$

where C_N represents a cluster containing N monomers. Heterogeneous nucleation describes the growth of clusters from collisions with other gaseous species. For example, (Gail and Sedlmayr 1988)

$$C_N + C_2H_2 \longrightarrow C_{N+2} + H_2. \qquad (4.2)$$

The final size distribution of grains is derived from a complex formalism based on the Zeldovich equation which links the time rate of change to the size rate of change of the size distribution function $f(N,t)$ where t is the time. Moments of the size distribution function are defined and correspond to various grain properties, such as the average particle size, the average number of monomers per grain etc., and moment equations are derived which form a set of first order differential equations that is solved for appropriate initial conditions (see Gail and Sedlmayr 1985, for details).

There are several drawbacks to nucleation theory applied to dust formation in AGB winds, and the theory was first seriously questioned by Donn and Nuth (1985) who favoured a stochastic, kinetic treatment when comparing experimental results of refractory material nucleation with classical nucleation theory. From the lists of assumptions above, condition 1 will be satisfied in the stellar photosphere or its vicinity, but not after the passage of strong shocks. Condition 3 is violated as the growth of very small clusters is likely to take place via a complex set of chemical reactions which involves several different gaseous species. Finally, condition 4 is not met, because extrapolation is not a reliable method to derive the thermodynamical properties of small clusters as small particles often show properties very different from the bulk material (Bohren and Huffmann 1983; Cherchneff *et al.* 1991). These points emphasize again the need to treat dust nucleation in a different way than classical nucleation theory (see also Chapter 13), and new approaches will be described in Section 5.

4.3 Hydrodynamical and Chemical Models of AGB Winds

The most complete models for dust-driven winds in AGB stars have been made by Höfner and Dorfi (1992), Höfner *et al.* (1995) and Fleischer *et al.* (1992, 1994). Höfner *et al.* considered stationary winds driven by radiation pressure on dust *only*, while Fleischer *et al.* have treated the combined effect of pulsation and radiation pressure, resulting in time-dependent wind models. The last study includes a treatment of the wind hydrodynamics, the effect of pulsation for a piston model for the outer photospheric boundary, and a time-dependent dust formation model based on classical nucleation theory, but does not treat explicitly the coupling between matter and radiation. The gas is assumed to be at its radiative equilibrium temperature (isothermal limit) or to follow an *ad hoc* cooling law. Höfner *et al.* proposed a model of stationary winds *excluding* the stellar pulsation and resulting shock wave propagation in the flow, included the same approach to dust condensation as Fleischer *et al.*, but did consider a proper treatment of radiative hydrodynamics.

Both groups reproduced with good agreement terminal velocities and mass losses for late-type stars, although stationary models are rather artificial as the outflow has to be time-dependent due to pulsation-induced shocks. Despite the different approaches of the two models, both studies provide evidence of the formation of dust-induced secondary shocks in the inner stellar wind: after the first condensation of dust grains at a few stellar radii, dust traps the stellar radiation field and this results in a back-warming of the gas located just before the condensation zone. Therefore dust can no longer condense in this hot gas,

Table 1 The dominant chemical species' TE abundances relative to the total gas density for parameters characterizing the region of pre-shock formation (from Willacy and Cherchneff 1998)

H_2	6.63(−1)	C_2H_2	1.56(−4)	Si	3.67(−5)
H	1.87(−1)	N_2	6.28(−5)	CS	1.27(−5)
He	1.51(−1)	C_2H	5.46(−5)	SiS	1.00(−5)
CO	9.46(−4)	HCN	5.11(−5)	C_3H	7.47(−6)

and this leads to a dust-free zone and a detached "first condensation" shell. When the gas cools again, a new condensation event occurs, and the net result is the formation of discrete shells of dust in the wind. The spacing between these shells is typically a few stellar radii and the shell periodicity is superimposed on the regular pulsation period of the star, implying that dust-induced shocks may be difficult to detect observationally.

The prediction of the formation of discrete dust shells in the wind of C-stars is interesting but it should be said that this process is highly sensitive to the dust formation mechanism. Both studies include dust condensation using nucleation theory, and—as already outlined in Section 4.2—this is not the appropriate approach to dust formation. As an example, the back-warming effect predicted by Fleischer et al. (1992, 1994) and Höfner et al. (1995) may not inhibit dust nucleation as much as is predicted in the post-shell gas. Indeed, the temperature enhancement in this region is only ∼1000 K over the local gas temperature (that is, ∼2000 K). Such a temperature is definitely high compared to the TE dust condensation temperature for amorphous carbon, but certain intermediate species involved in the dust formation process, such as polycyclic aromatic hydrocarbons (PAHs), can survive quite high temperatures due to an "inverse greenhouse" effect acting on them (Cherchneff et al. 1991; see Section 6.2). This effect will therefore counteract the formation of a dust-free zone in the post-shell gas and will work against the formation of discrete dust shells.

5 The Approach to Dust Formation Through Chemical Kinetics

As already mentioned in Section 4.1, the chemical composition of the photosphere of a C-star is given by TE and is rich in CO, C_2H_2, C_2H and HCN regardless of the carbon-to-oxygen ratio. Molecular hydrogen, H_2, is the dominant molecular species and is more abundant than atomic hydrogen. The TE chemical composition of the inner envelope of IRC+10216 is listed in Table 1 and corresponds to a gas temperature of 2062 K and a total number density of 3.68×10^{14} cm^{-3} for a solar elemental composition apart from a C/O ratio of 1.5. The dominant molecular species are CO and C_2H_2 but the strong binding energy of carbon monoxide prevents the molecule from being effectively destroyed by chemical

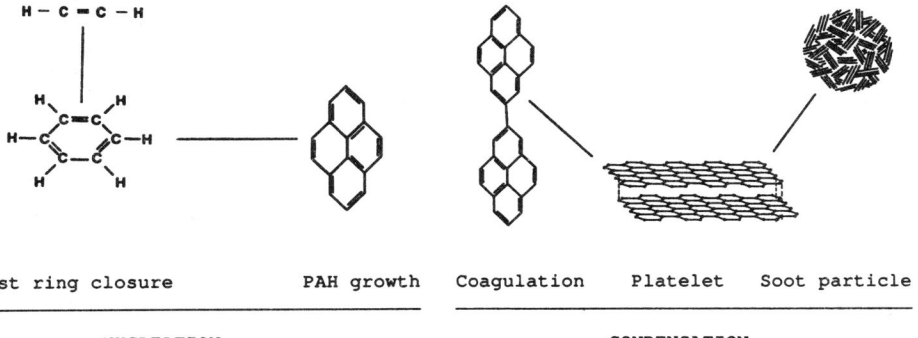

1st ring closure   PAH growth   Coagulation   Platelet   Soot particle

NUCLEATION   CONDENSATION

FIG. 2. Nucleation and condensation steps in soot formation in acetylenic flames (adapted from Tielens 1990).

reactions. Therefore, C_2H_2 is effectively the most abundant carbon-bearing species available in the flow, with C_2H second in abundances.

Terrestrial combustion systems such as shock-tube hydrocarbon pyrolysis and sooting premixed hydrocarbon flames are known to produce a large amount of soot particles and PAH molecules (Homann and Wagner 1967). Chemical modelling of shock-tube pyrolysis of acetylene and sooting acetylene flames have been studied by Frenklach *et al.* (1985) and Frenklach and Warnatz (1987). Apart from the presence of oxygen in flames, there exists a similarity between the flame and pyrolysis conditions and those of AGB photospheres. First, acetylene is the dominant molecular species in all these environments and, second, flame and pyrolysis temperatures are similar to those characterizing the inner envelope of AGB stars. Finally, both AGB star and pyrolysis environments are oxygen-free. Whether PAHs are intermediate species in the condensation of soot or by-products of the combustion is not entirely clear, but models using PAHs as intermediates can reproduce well the soot size distribution and formation yields, and demonstrate that PAHs act as building blocks in soot formation (Frenklach *et al.* 1985; Frenklach and Warnatz 1987; Frenklach 1989).

Models of flame chemistry are based on the formation sequence depicted in Fig. 2. The first step referred to as *nucleation* consists of forming large aromatic molecules from the acetylene-rich gas phase. The second step is the *condensation* stage where a three-dimensional, small solid particle formed from large aromatic precursors. We discuss in more details these various stages in the next sections.

5.1 Nucleation Processes

Formation of the first aromatic ring involves the isomerization of acetylene and the formation of larger hydrocarbons via thermolecular and bimolecular reactions which may have relatively high activation energies. However, these are overcome due to the high gas temperature in the flame, shock-tube or stellar environment. The nucleation step includes the closure of the first aromatic ring ($\equiv$ benzene

C_6H_6) and the subsequent growth of PAHs. Figure 3 represents the dominant routes to first ring closure (which represent the bottleneck in PAH formation) and PAH growth. First ring closure involves reaction of the propargyl (C_3H_3) radical with itself to form phenyl, and the reaction of C_4H_3 (1-buten-3-ynyl) with acetylene to form benzyne and phenyl.

Once benzene or phenyl are formed, the growth to PAHs can be simply represented by a set of chemical reactions which consist of (Frenklach 1989; Frenklach and Feigelson 1989; Cherchneff et al. 1992)

(1) acetylene addition;
(2) atomic hydrogen extraction to form an aromatic radical;
(3) second acetylene addition and closure of the ring.

Because of the high gas densities, the reverse reactions for all processes have to be considered. This set of reactions is of crucial importance as the thermodynamics of the processes determines a "temperature window" ($\equiv$ 900–1100 K) for which PAH growth occurs. Indeed, for gas temperatures larger than 1100 K, the H abstraction and the second acetylene addition are both reversible processes and the growth of PAHs cannot proceed. For gas temperatures between 900 and 1100 K, the second C_2H_2 addition become irreversible and growth is effective. At lower temperature than 900 K, the H abstraction is inhibited and the formation of aromatic radicals that trigger PAH growth stops. Then PAH growth will occur *only* in regions of the stellar outflow where the temperature window of 900–1100 K can be met.

5.2 Condensation Processes

The condensation stage comprises a series of chemical processes which will permit the growth of small solid condensates. Whereas nucleation treats the conversion of linear gas-phase species into planar, large aromatic molecules, condensation involves the growth of planar species into three-dimensional solid grains. Condensation is initiated with the formation of PAH dimers where the aromatics are linked by van der Waals forces. Condensation products form simultaneously and these differ from the dimers by the formation of six-membered or five-membered rings. For example, both dimers and condensation products of pyrene are observed in high-temperature pyrolysis of pyrene and the growth to soot particles was suggested to take place through aromatic polymerization and cyclodehydrogenation rather than through acetylene addition (Mukherjee et al. 1994). The growth would then continue through coagulation of condensation products of PAH molecules, deposition of carbon via acetylene surface reactions and surface condensation of free PAH molecules on the grains.

6 Model of Dust Nucleation in AGB Envelopes

For the last ten years, two groups have been involved in modelling the formation of PAHs and soot particles in AGB winds. A pioneering study by Frenklach and Feigelson (1989) applies chemical models developed for combustion systems to

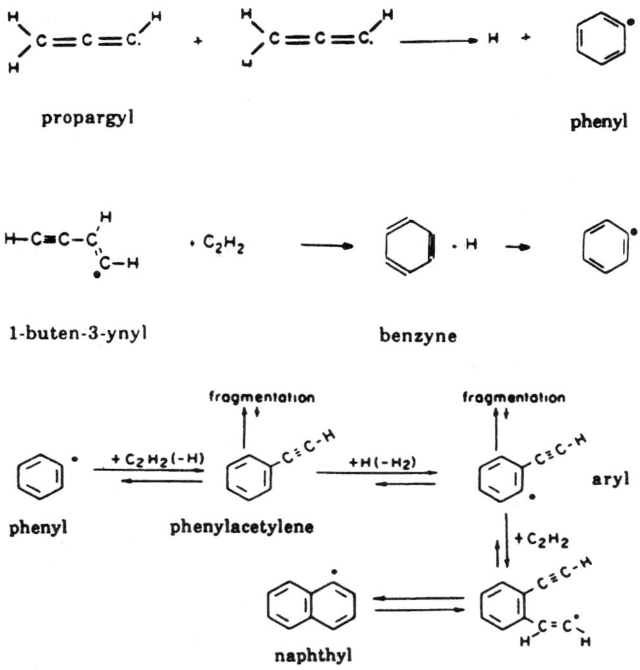

FIG. 3. Dominant pathways to PAH growth (adapted from Cherchneff et al. 1992).

the winds of C-stars. Rather than considering realistic wind models, the authors derived typical wind parameters for which the growth of PAHs and soot particles is possible. Their results show that only very cool, dense and slow winds were able to nucleate carbon dust. Based on temperature arguments, Frenklach et al. (1989) proposed that a dust grain could grow through amorphous carbon condensation on pre-existing SiC seeds and final PAH deposition on the grains. Finally, Cadwell et al. (1994) studied the chemistry of the induced-condensation of soot assuming an initial population of condensation nuclei in the form of SiC small particles. The model was able to generate optically thick dust shells for stationary winds.

Following Frenklach and Feigelson's 1989 study, Cherchneff et al. (1991, 1992) considered time-averaged stationary winds taking into account the effect of pulsation-driven shocks. They showed that PAH formation in AGB winds was difficult to generate. The basic problem is that the temperature window governing the growth of PAHs is found in stationary winds at ~ three stellar radii where the gas density is already low and small PAH formation yields are obtained.

A possible weakness in the approach to dust formation in AGB stars is the consideration of stationary winds instead of time-dependent outflows. As we mentioned in Section 3, the inner region close to the photosphere is subjected to the passage of periodic shocks. The net effect of the shock propagation is to form "stationary layers", that is, parcels of gas that are accelerated upwards but fall back on to the star because of gravity. These layers are shocked many times, span wide ranges of gas temperatures and densities, and may represent regions where molecules and dust precursors are formed and processed. We now investigate dust formation processes in this stationary region.

6.1 A Model for the Inner Envelope

The periodic shocks travelling through the inner layers of the envelope alter dramatically both the temperature and the density of the gas. Because the preshock gas is molecular and relatively cool (∼2000 K), the energy deposited by the shocks into the gas is lost mainly via the dissociation of molecular hydrogen, H_2, and adiabatic expansion, as shown by Fox and Wood (1985). Layers of gas are then accelerated upwards (the hydrodynamical postshock zone) but decelerate and eventually almost fall back to their initial position under the influence of stellar gravity. The cycle is then repeated with the next pulsation, and the gas parcels gradually move outwards and eventually escape from the shocked regions.

The dynamics of the shocked regions may be modelled in two stages (Willacy and Cherchneff 1998). We assume that a shock forms at the shock formation radius, r_s, outside the stellar photosphere and that its strength, Δv, is damped as the shock travels outward (Willson and Bowen 1986; Cherchneff et al. 1992). The effect of the shock on the postshock gas and its various cooling processes is studied adopting a Lagrangian formalism to describe the hydrodynamical cooling zone.

First, we derive the variation of the gas parameters (temperature, density and velocity) in the "chemical" cooling layer of the postshock region. Starting with the preshock gas parameters, we apply the Rankine–Hugoniot jump conditions to a diatomic gas to derive the parameters of the cooling region just after the shock front. We then assume thermal cooling by H_2 dissociation via collision with H atoms until the parameters have reached the values where hydrodynamical cooling starts. The length of the "chemical" cooling layer is defined by the collisional dissociation of H_2 and the dominant dissociative reaction for molecular hydrogen is

$$H + H_2 \rightarrow 3H \quad (6.1)$$

The rate for reaction (6.1) is

$$k_1 = 3.78\,10^{-8} \left(\frac{T}{300}\right)^{-0.5} \exp(-53280/T) \quad cm^3\ s^{-1} \quad (6.2)$$

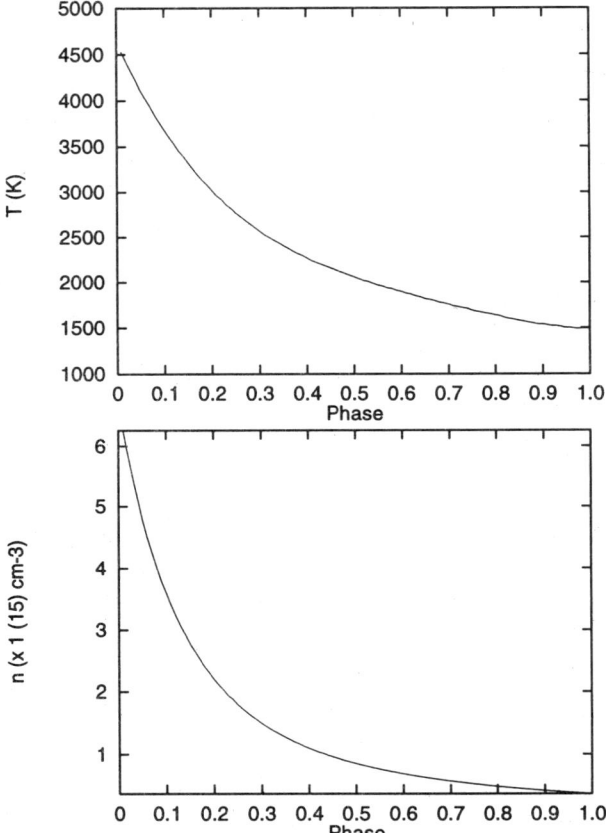

FIG. 4. The variation of the gas temperature T and number density n as a function of phase for the excursion induced by a 20 km s^{-1} shock in the inner wind of IRC+10216 (adapted from Willacy and Cherchneff 1998).

where T is the gas kinetic temperature. The H$_2$ dissociation length is defined as

$$l_D = \frac{1}{k_1 \, n(\mathrm{H})} \times \frac{v_{shock}}{n_{jump}} \qquad (6.3)$$

where $n(\mathrm{H})$ is the atomic hydrogen number density, v_{shock} is the shock velocity in the stellar rest frame and n_{jump} is the gas number density behind the shock front given by the Rankine–Hugoniot jump conditions (see Willacy and Cherchneff (1998) for details).

Second, the hydrodynamical postshock region and its quasi–ballistic trajectory is modelled following a formalism derived by Bertschinger and Chevalier (1985) who solved the continuity, momentum and energy equations for a periodically shocked, adiabatically expanding gas subjected to the stellar gravitational

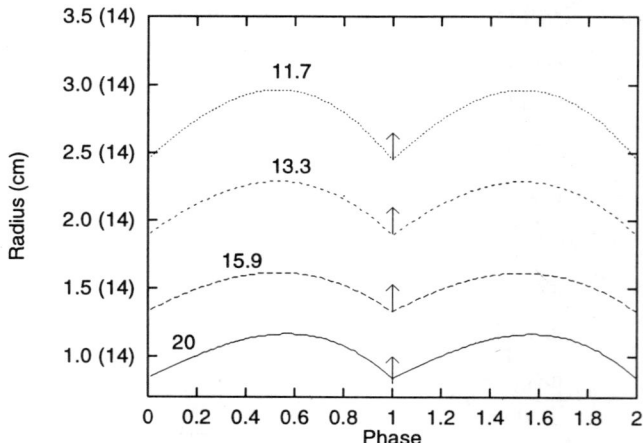

FIG. 5. Gas excursions induced by different shock strengths (in km s^{-1}) in the inner envelope of IRC+10216. The stellar radius is 7×10^{13} cm. The arrows show where the gas is shocked (adapted from Willacy and Cherchneff 1998)

field. The gas parameters (temperature, T, number density, n, and velocity, v) are calculated as functions of phase ($p = t/P$ where t is the time and P the stellar pulsation period) for Lagrangian parcels of gas. Strictly periodic motion of the gas (i.e. no mass loss) is assumed and this condition holds to a first approximation in the stationary region close to the photosphere.

The variation of T and n with p for the gas excursion induced by a 20 km s^{-1} shock in the wind of IRC+10216 is shown in Fig. 4. The adiabatic expansion is very efficient at cooling the gas (even below the gas radiative equilibrium temperature at large p values, as first mentioned by Bowen (1988)). Figure 5 represents the excursions of various shocked gas layers located at different positions in the envelope. The damped shock strengths inducing the excursions are also shown.

The model (consisting of the immediate postshock region and excursion for one shock strength at one position in the envelope) is run over two pulsation cycles to check that periodicity is obeyed. The output abundances for one model are used as the input to that for the next slowest shock strength and are rescaled according to the new local gas number density.

6.2 PAH Formation Yields

The chemical scheme for PAH formation of Cherchneff et al. (1992) has been extended to account for the formation of larger PAHs up to coronene ($C_{24}H_{12}$) (Cherchneff 1997). We consider the coagulation of PAHs and describe the formation of PAH dimers up to bi-coronene. The reaction rates for dimer formation from their PAH parents are not available and we base our estimates on the

Table 2 PAH and PAH dimer formation yields

Radius ($R_\star$)	Shock strength (km s^{-1})	PAH yield	PAH dimer yield
1.3	20.0	0	0
1.9	16.5	3×10^{-5}	10^{-8}
2.5	14.4	7×10^{-5}	2×10^{-8}
3.1	13.0	7×10^{-5}	2×10^{-8}

reaction of benzene with phenyl

$$C_6H_6 + C_6H_5 \longrightarrow C_6H_5C_6H_5 + H \tag{6.4}$$

for which Fahr et al. (1988) measured a rate of

$$k_2 = 5\times 10^{-13} \exp(-15.7/RT) \quad \text{cm}^3\ \text{s}^{-1} \tag{6.5}$$

where R is the perfect gas constant and T the gas temperature. The rates for the formation of PAH dimers larger than biphenyl are expected to be greater than k_2 because the van der Waals force between molecules increases with their size. This force can then act as a temporary glue which can hold the adduct together while the chemical bond forms (McKinnon 1989). Miller et al. (1985) have estimated the van der Waals forces for several PAHs and they found that the ratio between the force and the PAH number of carbon atoms is roughly constant. We use this result for PAH species present in this study and not considered by Miller et al. and estimate the dimer formation rate rescaling k_2 according to the van der Waals force values.

Small PAH molecules do not possess absorption bands in the visible and near-IR. This leads to radiative equilibrium temperatures for PAHs which are smaller than the gas kinetic temperature when the aromatics are collisionally decoupled from the gas (Cherchneff et al. 1991). This "inverse greenhouse" effect has been applied to PAHs but not to PAH dimers and the formation yield of PAHs and PAH dimers is defined as the total number of carbon atoms in PAHs and PAH dimers divided by the total number of carbon atoms initially in the form of hydrocarbons (Cherchneff et al. 1992). The yield is calculated for the cycles of shock + excursion described in Section 6.1.

Preliminary results are summarized in Table 2. In the 20 km s^{-1} shock and in the excursion following, no PAH/PAH dimers can form because the 900–1100 K temperature window necessary to the growth of PAHs cannot be reached in the postshock gas. PAH formation starts further out at radii of $\sim 2R_\star$, and the yields range between 10^{-5} and 10^{-4} at $r \geq 2R_\star$. The yields for the dimers follow the same trend as for PAHs but are three orders of magnitude smaller. No destruction processes have been considered for dimers and the PAH dimer formation yields are then upper limits. However, these compounds will be undoubtedly destroyed in the shocks.

An important conclusion is that the abundances calculated for each radius in the inner wind are governed by the chemistry occurring in the excursion (i.e. the hydrodynamically cooling part of the postshock region) and *not* in the immediate postshock region. This is due to the existence of a "very fast chemistry" region at the beginning of each excursion in which the inputs from the immediate postshock region are quickly changed (Willacy and Cherchneff 1997). Therefore, we expect the dimers to be destroyed in the shocks but to reform in the excursions and the yields listed in Table 2 depend only on the parameters characterizing the gas excursion associated with a particular shock strength.

7 Discussion and Conclusions

Dust contents of C-rich AGB envelopes have been derived by Knapp and Morris (1985) who obtained an average dust-to-gas number density ratio of 5.2×10^{-13} for C-stars. If we assume a population of dust grains with average radius $a = 500$ Å, the ratio of the number of C atoms in dust grains to the number of C atoms in acetylene is $\sim 5 \times 10^{-9}$. This value is in very good agreement with the PAH dimer formation yields found in Section 6.2 and supports the idea that PAH dimers and their condensation products (see Section 5.2) could act as condensation nuclei. The coagulation of PAH dimer condensation products can lead to small condensation nuclei of ~ 100 carbon atoms which then grow via acetylene and free PAH deposition.

However, it is unlikely that dust will form from pure carbon condensation nuclei. Frenklach et al. (1989) and Kozasa et al. (1996) have proposed that silicon carbide, SiC, particles condense first, act as condensation nuclei and grow through acetylene and PAH deposition. The ground for such a hypothesis is that SiC forms very efficiently at high temperatures ($T \sim 2000$ K) in IR pyrolysis of acetylene and silane (Lihrmann and Cauchetier 1994). The formation mechanism of solid SiC is not yet clear but Lihrmann and Cauchetier have proposed that reaction of atomic Si and SiC_2 triggers the nucleation of SiC grains via the process

$$Si + SiC_2 \longrightarrow 2\ SiC \tag{7.1}$$

and the subsequent coalescence of SiC molecules in solid particles. However, this reaction has an endothermicity of ~ 395 kJ mole^{-1} at 1000 K and is unlikely to occur except perhaps in very high temperature regions. Furthermore, Willacy and Cherchneff (1998) have shown that the amount of SiC_2 in the inner envelope of C-stars is not sufficient to account for the SiC content of dust derived from fitting spectral energy distributions. Other mechanisms must operate in the inner wind that have not been identified yet. However, it is well possible that SiC condensation takes place at the high temperatures met in the shock-induced gas excursions (see Fig. 4), preceding the formation of PAHs and PAH dimer species. A full treatment of the hydrocarbon/PAH and the silicon chemistry is necessary to shed light on this hypothesis.

New information on stellar dust can be derived from the isotopic and chemical analysis of carbon spherules extracted from meteorites, such as the Murchison meteorite (Bernatowicz et al. 1996). The small core nuclei extracted have been identified as carbides of Ti, Mo, and Zr, but SiC was not found in the spherules. This result implies that TiC, for example, would be favoured over SiC in the formation of condensation nuclei for AC grain growth, leaving free solid SiC particles as a second dust population in the flow.

Willacy and Cherchneff (1997) show that there exists in the flow a temporary formation of oxygen-bearing molecules in the chemical and hydrodynamical postshock regions. Although no oxygen molecules are formed in large amounts (apart from CO and SiO) and expelled in the outflow, it is possible that oxygen influences the dust nucleation processes. First, the formation of SiO, O, and O_2 may lead to the formation of refractory oxides as SiO_x in small amounts. Second, oxygen is known to inhibit the growth of PAHs because of oxygenation processes (McKinnon 1989) but also creates radical sites which can enhance PAH growth. Therefore, the net effect of oxygen on PAH nucleation and dust condensation needs to be investigated in the context of C-star winds.

Dust condensation in C-stars is far from been understood and requires the combined effort of many scientific communities, from scientists studying meteorites or combustion processes in the laboratory to astronomers and astrophysicists working on hydrodynamical and chemical models.

Acknowledgements

The author wishes to thank S. Bagnulo and K. Willacy for helpful discussions.

Bibliography

1. Abraham, F. F. (1974). *Homogeneous Nucleation Theory*. Academic Press, New York.
2. Bagnulo, S. (1996). *PhD Thesis "Modelling of Circumstellar Environments around Carbon and Oxygen Rich stars"*. Queen's University of Belfast, Northern Ireland.
3. Bagnulo, S., Doyle, J. G. and Griffin, I. P. (1995). *A&A* **301**, 501.
4. Bernatowicz, T. J., Cowsik, R., Gibbons, P. C., Lodders, K., Fegley, B., Amari, S. and Lewis, R. S. (1996). *ApJ* **472**, 760.
5. Bertschinger, E. and Chevalier, R. A. (1985). *ApJ* **299**, 167.
6. Boothroyd, A. I. and Sackmann, I. J. (1988). *ApJ* **328**, 671.
7. Bohren, C. F. and Huffman, D. R. (1983). *Absorption and Scattering of Light by Small Particles*. Wiley & Sons, New York.
8. Bowen, G. H. (1988). *ApJ* **329**, 299.
9. Cadwell, B. J., Wang, H., Feigelson, E. D. and Frenklach, M. (1994). *ApJ* **429**, 285.
10. Cherchneff, I. (1997). In: E. F. van Dishoeck (ed.) *Proc. IAU Symp. 178, Molecules in Astrophysics: Probes and Processes*. Kluwer, Dordrecht, p. 469.
11. Cherchneff, I. and Barker, J. R. (1992). *ApJ*, **394**, 703.
12. Cherchneff, I., Barker, J. R. and Tielens, A. G. G. M. (1991). *ApJ* **377**, 541.

13. Cherchneff, I., Barker, J. R. and Tielens, A. G. G. M. (1992). *ApJ* **401**, 269.
14. Cox, J. P. (1980). *Theory of Stellar Pulsation*. Princeton University Press, Princeton.
15. Donn, B. and Nuth, J. A. (1985). *ApJ* **288**, 187.
16. Draine, B. T. and Salpeter, E. E. (1977). *J. Chem. Phys.* **67**, 2230.
17. Fahr, A., Mallard, W. G. and Stein, S. E. (1988). *21st Symp (International) on Combustion*, p.825
18. Fedorova, O. V. (1973). *In Clusters*. Nauka, Alma-Ata, p. 55.
19. Fox, M. W. and Wood, P. R. (1985). *ApJ* **297**, 455.
20. Fleischer, A. J., Gauger, A. and Sedlmayr, E. (1992). *A&A* **266**, 321.
21. Fleischer, A. J., Gauger, A. and Sedlmayr, E. (1994). *A&A* **297**, 543.
22. Frenklach, M. (1989). *21st Symp. (International) on Combustion*, p.843.
23. Frenklach, M. and Feigelson, E. D. (1989). *ApJ* **341**, 372.
24. Frenklach, M. and Warnatz, J. (1987). *Combustion Sci. Tech.* **51**, 256.
25. Frenklach, M., Clary, D. W., Gardiner, W. C. and Stein, S. E. (1985). *20th Symp. (International) on Combustion*, p.887.
26. Frenklach, M., Carmer C. S. and Feigelson, E. D. (1989). *Nature* **339**, 196.
27. Friedlander, S. K. (1977). *Smoke, Dust and Haze: Fundamentals of Aerosol Behavior*. Wiley-Interscience, New York.
28. Gail, H. P. and Sedlmayr, E. (1985). *A&A* **148**, 183.
29. Gail, H. P. and Sedlmayr, E. (1987). *A&A* **171**, 197.
30. Gail, H. P. and Sedlmayr, E. (1988). *A&A* **206**, 153.
31. Gail, H. P., Keller, R. and Sedlmayr, E. (1984). *A&A* **133**, 320.
32. Gehrz, R. D. and Woolf, N. J. (1971). *ApJ* **165**, 285.
33. Gillet, D., Maurice, E. and Baade, D. (1983). *A&A* **128**, 384.
34. Groenewegen, M. A. T. (1997). *A&A* **317**, 503.
35. Hartmann, L. and MacGregor, K. B. (1980). *ApJ* **242**, 260.
36. Hinkle, K. H. (1978). *ApJ* **220**, 210.
37. Höfner, S. and Dorfi, E. A. (1992). *A&A* **265**, 207.
38. Höfner, S., Feuchtinger, M. U. and Dorfi, E. A. (1995). *A&A* **297**, 815.
39. Holzer, T. E., Flå, T. and Leer, E. (1983). *ApJ* **275**, 808.
40. Homann, K. H. and Wagner, H. G. (1967). *11th Symp. (International) on Combustion*, p.371.
41. Keady, J. J. and Ridgway, S. T. (1993). *ApJ* **406**, 199.
42. Knapp, G. R. and Morris, M. (1985). *ApJ* **292**, 640.
43. Kozasa, T., Dorschner, J., Henning, T. and Stognienko, R. (1996). *A&A* **307**, 551.
44. Kwok, S. (1975). *ApJ* **198**, 583.

45. Lafont, S., Lucas, R. and Omont, A. (1982). *A&A* **106**, 201.
46. Lihrmann, J. M. and Cauchetier, M. (1994). *J. European Ceramic Soc.*, **13**, 41.
47. McCabe, E. M. (1982). *MNRAS* **200**, 71.
48. McCabe, E. M., Smith, R. C. and Clegg, R. E. (1979). *Nature* **281**, 263.
49. McKinnon, J. T. (1989). PhD thesis MIT, Boston, USA.
50. Miller, J. H., Smyth, K. C. and Mallard, W. G. (1985). *20th Symp. (International) on Combustion*, p. 1139.
51. Mukherjee, J., Sarofim, A. and Longwell, J. P. (1994). *Combustion and Flame* **96**, 191.
52. Pijpers, F. P. and Habing, H. J. (1989). *A&A* **215**, 334.
53. Pijpers, F. P. and Hearn, A. G. (1989). *A&A* **209**, 198.
54. Salpeter, E. E. (1974). *ApJ* **193**, 585.
55. Sorrell, W. H. (1990). *MNRAS* **243**, 570.
56. Tarafdar, S. P. (1987). In: M. S. Vardya and S. P. Tarafdar (eds.) *Astrochemistry*. Kluwer, Dordrecht, p. 559.
57. Tielens, A. G. G. M. (1990). In: J. Tarter (ed.) *Carbon in the Galaxy*. NASA CP-3061, p.59.
58. Tsuji, T. (1973). *A&A* **23**, 411.
59. Weymann, R. (1960). *ApJ* **132**, 380.
60. Willacy, K. and Cherchneff, I. (1998). *A&A*, **330**, 676.
61. Willson, L. A. and Bowen, G. H. (1984). In: R. Stalio and J. B. Zirker (eds.) *Relation between Chromospheric Coronal Heating and Mass Loss in stars*. Triestre Workshop Series, Triestre, p. 127.
62. Willson, L. A. and Bowen, G. H. (1986). In: M. Zeilik and D. M. Gibson (eds.) *Cool Stars, Stellar Systems, and the Sun*. Springer, Berlin, p. 385.
63. Willson, L. A. and Hill, S. J. (1979). *ApJ* **228**, 854.
64. Wood, P. R. (1979). *A&A* **86**, 286.

13
Dust Formation in M Stars

Hans-Peter Gail
Institut für Theoretische Astrophysik, Universität Heidelberg

and

Erwin Sedlmayr
Institut für Astronomie & Astrophysik, Technische Universität Berlin

1 Introduction

Stars with initial masses between $\approx 2 M_\odot$ and $\approx 8 M_\odot$ on the main sequence develop first through the red giant phase until they ignite central helium burning. Later they start to climb up the Asymptotic Giant Branch (AGB) where they alternately burn helium and hydrogen in two thin shells over a degenerated carbon–oxygen core. In this stage, the thermal pulsing AGB stage (TP-AGB), they are cool ($T_{\text{eff}} \approx 2000\text{--}2500\,\text{K}$), luminous ($L \approx 10^4\, L_\odot$) and very extended ($R_* \approx 10^3\, R_\odot$) giant stars. For further discussion see Chapter 11. Observationally they are found to be emitters of a strong infrared emission and a rich zoo of molecular microwave emission lines which originate in a dense and very extended ($\gtrsim 10^4\, R_*$) circumstellar shell of cool gas and dust. The central stars are usually obscured by the circumstellar dust and in many cases they are even optically invisible and can only be identified by the IR-emission from warm dust and by microwave molecular line emission.

The most intense molecular lines are usually the thermally excited emission lines of CO and in many cases very strong maser emission lines of OH, H_2O and SiO are observed (e.g. the excellent review of Habing 1996; Chapter 14). A common characteristic of all the lines is the large velocity range of more than $20\,\text{km}\,\text{s}^{-1}$ which is simply interpreted as being the result of an expanding circumstellar envelope. Quantitative analysis of the line emission yields mass-loss rates ranging from less than $10^{-7}\, M_\odot\,\text{yr}^{-1}$ for optically visible objects with only a slight far-infrared excess emission by the circumstellar dust up to more than $10^{-4}\, M_\odot\,\text{yr}^{-1}$ for the most heavily obscured OH-IR-objects. The expansion velocity is usually found to be in the range between 10 and $25\,\text{km}\,\text{s}^{-1}$ (cf. Loup et al. 1993).

According to observations there exist two different classes of objects with circumstellar dust shells with distinctively different chemical composition: (i) stars of spectral type M (if the stellar atmosphere is visible) for which abundant oxygen bearing molecules and only trace amounts of carbon bearing molecules (besides the abundant CO molecule) are observed to exist in their atmospheres and their circumstellar shells, and (ii) stars of spectral type C which show a rich zoo of abundant carbon bearing molecules and only trace amounts of oxygen bearing molecules.[1] M stars mainly form silicate dust in their circumstellar shells, as can be seen by strong infrared features at $10\,\mu$m and $18\,\mu$m, usually ascribed to stretching and bending vibrational modes in an SiO_4 group, while C stars form carbon dust (see Chapter 12), as is inferred from the strong featureless continuous absorption of the main circumstellar dust component in such objects (see Dorschner and Henning 1995, and references therein). Minor dust components are likely to be formed besides the main components in both cases (e.g. Dorschner and Henning 1995; Waters et al. 1996), and exceptions to the above general relation between spectral type of the central star and nature of the main condensate are known (e.g. Sylvester et al. 1994).

These differences in the chemical composition of the mixture of molecules and dust are related to the relative element abundances of oxygen and carbon. In stars of spectral type M the oxygen abundance exceeds that of carbon ($\epsilon_O > \epsilon_C$) while in stars of spectral class C carbon is the more abundant ($\epsilon_C > \epsilon_O$) of the two elements. These differences in abundance result from a convective mixing between the central regions of the star subject to shell burning and the stellar surface in some fraction of the TP-AGB stars (e.g. Iben 1991). In this "third dredge up", operating at the top of the AGB, ashes of the He to C burning process are brought to the surface. The mixing then gradually increases the C abundance relative to the O abundance. As is well known the exceptionally high bond energy between C and O in the CO-molecule results in a chemical blocking of the less abundant of the two elements C and O in the unreactive CO molecule. This blocking of C in M stars and of O in C stars is responsible for the observed different chemical compositions of the gas phase and for the formation of silicate dust and probably some other oxide dust components in circumstellar shells of M stars and of soot, SiC and probably some additional non-oxide dust components in the case of C stars (e.g. Gail and Sedlmayr 1987).

The nature of the dust material formed and the possible chemical pathways for the nucleation and growth of circumstellar dust obviously depends crucially on the element mixture in the stellar outflow. This element mixture is fixed by the initial composition of the star at the time of its formation and by the nuclear burning processes operating deep within the star which results, by means of convective mixing processes, in certain compositional changes at the stellar surface in late stages of the star's evolution.

Intermediate mass stars in the range $2M_\odot \lesssim M \lesssim 8M_\odot$ have stellar lifetimes small compared to the age of our Galaxy. They are formed therefore from

---

[1] We ignore here the rare transition objects of spectral class S between spectral classes M and C.

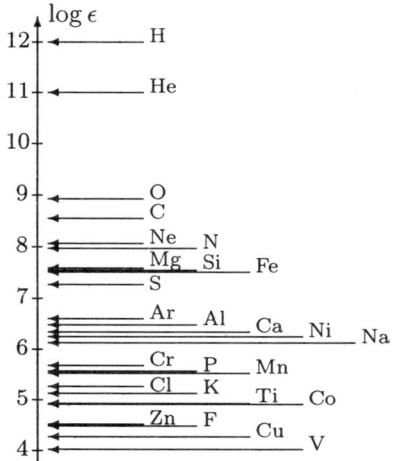

FIG. 1. Solar relative abundances of the most abundant elements according to Anders and Grevesse (1989) and Grevesse and Noels (1993).

material with a mixture of elements not much different from that observed for early type main sequence stars or that derived for the Solar System. This standard cosmic element mixture (Anders and Grevesse 1989; Grevesse and Noels 1993) is shown in Fig. 1 for the most abundant elements. The only elements shown in this figure are those likely to be involved in the dust formation process, and only the most abundant ones can be involved in the formation of the dust component(s) responsible for the strong extinction of the circumstellar dust shell.

The remaining elements have cosmic abundances that are much too low to form visible dust or to form significant numbers of seed nuclei for the growth of dust from the more abundant elements. Exceptions are possible only if very strong abundance excesses of such rare elements occur. The element abundances of M-giants have been studied, e.g. by Smith and Lambert (1990).

Despite its obvious astrophysical relevance no adequate descriptions of inorganic cluster and grain formation are available. The reason for this shortcoming lies in the enormous complexity of the various microphysical processes involved, in the lack of most of the required data necessary for a reliable modelling of the cluster structures, and in the heterogeneous nature of the condensates. Therefore, so far theorists have been forced to adopt strongly simplified descriptions like liquid droplet models, chemical homogeneity, and well defined mineralogical structures (e.g. Kozasa and Hasegawa 1987), yielding valuable insight into basic features of the dust formation process and into the complexity of the whole problem but of course not providing reliable quantitative predictions.

Through a broad discussion of the astrophysical situation, the applied theoretical description and some key condensates expected in this context, in this chapter we try to shed light on important problems and aim to establish a more secure foundation on which more realistic future developments can be based.

2 Dust Nucleation

At the starting point of the classical theory of formation of seed particles for dust growth is the formation of a cluster by the stepwise collection of a single molecular species A_1 (the monomer) from the gas phase:

$$A_1 \xrightarrow{A_1} A_2 \xrightarrow{A_1} A_3 \xrightarrow{A_1} A_4 \xrightarrow{A_1} A_5 \xrightarrow{A_1} \cdots.$$

More complicated processes involving cluster growth by addition of several different molecular species from the gas phase and by chemical surface reactions are not considered in homogeneous nucleation theory.

2.1 Equilibrium Condensation

In the following we characterize each cluster by the number N of the basic molecules required for its formation and denote its particle density by $f(N,t)$ and its partial pressure in the gas phase by $p(N)$. The distribution of the partial pressures of clusters with size N in a thermodynamic equilibrium state is given by the law of mass action

$$\overset{o}{p}(N) = p^N(1)\exp[-\Delta G(N)/RT] \tag{2.1}$$

where ΔG is the free enthalpy of formation of the cluster of size N from N monomers. This generally is rewritten in terms of the vapour pressure p_{vap} of the monomers over the condensate to obtain

$$\overset{o}{p}(N) = \exp[N \ln S' - \Delta(G(N) - NG(\infty))/RT] \tag{2.2}$$

where $\Delta G(\infty)$ is the free enthalpy of formation of the monomer from the condensate and S' is the supersaturation ratio

$$S' = p/p_{\text{vap}}. \tag{2.3}$$

p is the actual partial pressure of the monomers $N = 1$ in the gas phase; this is identical with $\overset{o}{p}(1)$. The monomer pressure is determined by the requirement

$$\sum_{N=1}^{\infty} N\overset{o}{p}(N) = f_\epsilon \epsilon P_{\text{H}}. \tag{2.4}$$

P_{H} is the fictitious pressure of hydrogen nuclei if they all were present as free particles, ϵ is the element abundance of the least abundant element required to form the monomer and $1 - f_\epsilon$ is the fraction of this element which is contained in other molecular species which do not participate in the condensation process.[2]

---

[2] In the case of carbon dust formation, for instance, $1 - f_\epsilon$ equals the fraction of the carbon bound in CO, because the C bound in CO cannot be condensed into dust. For condensation of dust in M stars such a blocking of some fraction of a condensible element in some strongly bound and unreactive molecular species seems not to occur, so that we have always $f_\epsilon = 1$ in this case.

Another way of writing (2.1) is

$$\overset{\circ}{p}(N) = p^N(1) \frac{U_N(T)}{U_1^N(T)} e^{-\Delta E_N/kT} (kT)^{-(N-1)} \qquad (2.5)$$

where $U_N(T)$ and $U_1(T)$ are the partition functions of a cluster of size N and the monomer, respectively, and ΔE_N is the energy of formation of the N-cluster from N monomers. The partition functions U are determined by the translational and internal energy states of the particle. A calculation of $\overset{\circ}{p}(N)$ from (2.5) requires besides the bond energy a complete knowledge of the vibrational and rotational energy spectrum of the clusters and in some cases also of the spectrum of their excited electronic states.

The difference of the free enthalpies

$$\Delta(G(N) - NG(\infty)) = \Delta H_f(N) - N \cdot \Delta H_f(\text{cond.}) + T\Delta S \qquad (2.6)$$

is essentially determined by the difference of the bond energy of the monomers in the finite cluster of size N and of that in the infinitely extended condensate. This energy difference is positive since the bond energy of monomers in the solid is higher than in a finite cluster. To some extent the difference of the Gibbs enthalpy depends also on the difference between the partition functions for the translational and internal degrees of freedom, which determines ΔS, but this contribution usually is neglected in *classical* nucleation theory.

To determine the energy defect $\Delta H_f(N) - N\Delta H_f(\text{cond.})$ in *classical* nucleation theory it is assumed that this can be calculated from the macroscopic surface energy of the bulk condensate which is given in terms of the surface tension σ and the surface area A_N of the particle by

$$\Delta H_f(N) - N \cdot \Delta H_f(\text{cond.}) = \sigma A_N. \qquad (2.7)$$

This basic assumption of *classical* nucleation theory is the source of the observed discrepancies (see Donn and Nuth 1985) between its predictions and experimental results. It takes for granted that the bond properties at the molecular level are essentially the same as for the infinitely extended condensate, but this is not true in real cases. Nevertheless, it is worthwhile to consider this simple approach further since the fundamental concepts of any nucleation theory can be demonstrated most easily for this model.

A second assumption usually introduced in classical nucleation theory is that the small clusters are spherical. This allows a calculation of the surface area of the cluster in a simple way and one obtains the following expression for the dependence of ΔG on cluster size N:

$$\Delta G = -RTN \ln S' + 4\pi\sigma \left(\frac{3V_0}{4\pi}\right)^{\frac{2}{3}} N^{\frac{2}{3}} - T\Delta S \qquad (2.8)$$

where V_0 is the volume occupied by one monomer in the condensed phase. The essential conclusions which can be drawn from this are:

(1) If the supersaturation ratio S' is less than unity then ΔG increases monotonically with increasing cluster size N. The size distribution of clusters according to (2.2) monotonically decreases with increasing N. Observable quantities of clusters exist only on the molecular level. No macroscopic sized dust grains exist.

(2) If the supersaturation ratio S' exceeds unity then ΔG depends in a more complicated way on N. The first term in (2.8) *decreases* linearly with increasing N while the second one *increases* somewhat more slowly with N, proportional to $N^{\frac{2}{3}}$. For sufficiently big N the first term dominates and ΔG decreases with increasing cluster size. The size distribution then *increases* with increasing N as

$$\overset{o}{p}(N) \propto S'^N \qquad (2.9)$$

In this case according to the constraint (2.4) most of the mass would be contained in a few very big clusters. The case $S' > 1$, thus, corresponds to complete condensation in thermodynamic equilibrium.

(3) For small N usually the second term dominates in (2.8). The size distribution then *decreases* with increasing N for small cluster sizes. Since for large N it has to satisfy (2.9) there necessarily exists some cluster size N_* where $\overset{o}{p}(N)$ has an absolute minimum. For very large supersaturation ratios, S', the first term in (2.8) dominates for all cluster sizes N. Then (2.9) holds for all $N \geq 1$ and the critical cluster size N_* where $\overset{o}{p}$ takes a minimum in this case is simply $N_* = 1$. The existence of this critical cluster size N_* where $\overset{o}{p}$ takes its absolute minimum is crucial for the whole nucleation problem, as we shall see below.

Realistic calculations of cluster abundances have to be based on detailed theoretical and experimental determinations of thermochemical functions for clusters up to sufficiently large size N (>10 at least) for the species which may be involved in the nucleation process. Examples for such calculations are presented in Section 3.

2.2 Classical Homogeneous Nucleation Theory

Nearly all attempts to calculate the formation of condensation nuclei in the outflows of cool stars up to now have been based on the approach of classical homogeneous nucleation theory, starting with Donn et al. (1968). The essential principles of applying this approach to dust condensation in circumstellar shells are outlined in (among many others): Salpeter (1974), Draine and Salpeter (1977), Yamamoto and Hasegawa (1977), Draine (1979), Deguchi (1980), Gail et al. (1984), Gail and Sedlmayr (1986) and Sedlmayr and Dominik (1995).

Basically the formation of clusters can be conceived as the product of a chain of chemical reactions resulting in a production rate of critical clusters (e.g. Gail and Sedlmayr 1988; Köhler et al. 1997)

$$J_* = \left[\sum_{i=1}^{N_{\max}} \frac{\tau_{\text{gr},i}}{\overset{o}{f}(i)}\right]^{-1}, \tag{2.10}$$

which in the case of a strong supersaturation, where $\overset{o}{f}(N)$ has a sharp minimum at $N = N_*$, can be approximated by the term

$$J_* = \frac{\overset{o}{f}(N_*)}{\tau_{\text{gr},N_*}}. \tag{2.11}$$

Thus we have the very plausible result that *the rate of dust particle formation is determined by the slowest growth step on the reaction chain along clusters of increasing size which occurs at the cluster size N_* where $\overset{o}{f}(N)$ is smallest.*

A calculation of the nucleation rate in circumstellar dust shells requires, then, the knowledge of the species which may be responsible for nucleation and the properties of all clusters formed from the relevant species. For a given thermodynamic state of the gas phase it is possible then to calculate the equilibrium abundances of all clusters and from this to determine the critical cluster size N_* and the nucleation rate J_*.

In the approximation of homogeneous classical nucleation theory, the equilibrium pressure $\overset{o}{p}$ of the clusters is given by (2.2) with ΔG given by (2.8). To evaluate the sum over N it is approximated by an integral which is then calculated by means of the saddle-point integration method. The result is (e.g. Feder et al. 1966)

$$J_* = \sqrt{\frac{2\sigma}{\pi m}}\, V_0\, n_1^2\, \exp\left[-\frac{16\pi\sigma^3 V_0^2}{3k^3 T^3 (\ln S')^2}\right]. \tag{2.12}$$

m is the mass and n_1 the particle density of the monomer, V_0 the volume occupied by the nominal monomer in the bulk condensate and σ is the surface tension. This equation allows a calculation of the nucleation rate from the properties of the condensate, its surface tension, and its vapour pressure. Up to now in astrophysics, considerations of condensation have been nearly exclusively based on this expression for J_*. The main objections against this approach, and its main shortcomings, are listed by Donn and Nuth (1985). Nevertheless (2.12) is generally used since no realistic theory of condensation in an oxygen rich environment is known at present.

An interesting application of (2.12) has been made by Nuth and Donn (1982b) to the condensation of SiO. The SiO molecule is one of the few molecules with an exceptionally high bond energy (cf. Fig. 3) and is stable even in the atmospheres of cool stars. It forms a pure SiO solid with a condensation temperature of approximately 1000 K in the laboratory. For this reason it has often been thought to be responsible for the initiation of dust condensation in circumstellar shells (e.g. Nuth and Donn 1982b; Gail and Sedlmayr 1986). Nuth and Donn (1982b) measured the nucleation rate J_* for SiO in the laboratory by evaporating and re-condensing SiO and fitted their results for J_* with the expression (2.12). The parameters $m = 7.3 \times 10^{-23}$ g, $V_0 = 2.4 \times 10^{-23}$ cm^3 are fixed by the properties of the condensate. The vapour pressure of SiO molecules over solid SiO is

$$p_{\mathrm{vap}} = 760 \exp\left[12.81\left(1 - \frac{2300.9}{T}\right)\right] \tag{2.13}$$

(in torrs). By fitting the experimental data with (2.12) Nuth and Donn (1982b) determined a fictitious surface tension $\sigma = 500$ dyn cm^{-2}. This σ in principle has nothing to do with a real surface tension but merely is that value of the parameter σ which yields the best approximation of the experimentally determined values of J_* by a relation of the type (2.12). Thus, a calculation of J_* for SiO nucleation from (2.12) with this value of σ yields results which are not subject to the usual errors of classical nucleation theory.

Figure 2 shows lines of constant nucleation rate in the P–T plane calculated by this approach for an element mixture of standard composition. An inspection of the figure shows that dust formation by nucleation of SiO does not occur until the temperature of the outflowing gas has fallen below 500 K. This is incompatible with a dust temperature of $\approx$1000 K in the dust condensation zone derived from models for the circumstellar IR-emission of optically thick dust shells. Therefore, SiO nucleation cannot be responsible for the condensation in these objects. There remains the possibility, however, that SiO triggers dust formation in *optically thin* dust shells where condensation temperatures as low as this are observed in many cases (Rowan-Robinson and Harris 1983).

2.3 Possible Nucleation Species

We are now able to state some rules which a particular species from the gas phase has to satisfy in order to be a favourable candidate for nucleation:

(1) The nucleation rate depends on the product of the number density of the monomer (via the collision rate) and the number density of the critical cluster. Thus, the least abundant element in the molecule should have an element abundance as high as possible in order that this product may become as large as possible.

(2) For nucleation to occur at a temperature as high as possible the monomer should already be formed at a temperature as high as possible. This requires an exceptionally high bond energy.

(3) Unlimited growth beyond some critical cluster size requires that the thermodynamic equilibrium size distribution $\overset{\circ}{p}(N)$ beyond the critical cluster size N_* increases indefinitely with increasing cluster size N. This is possible only if the gas phase is supersaturated with respect to the formation of the solid condensate from the monomer and this in turn requires that the solid

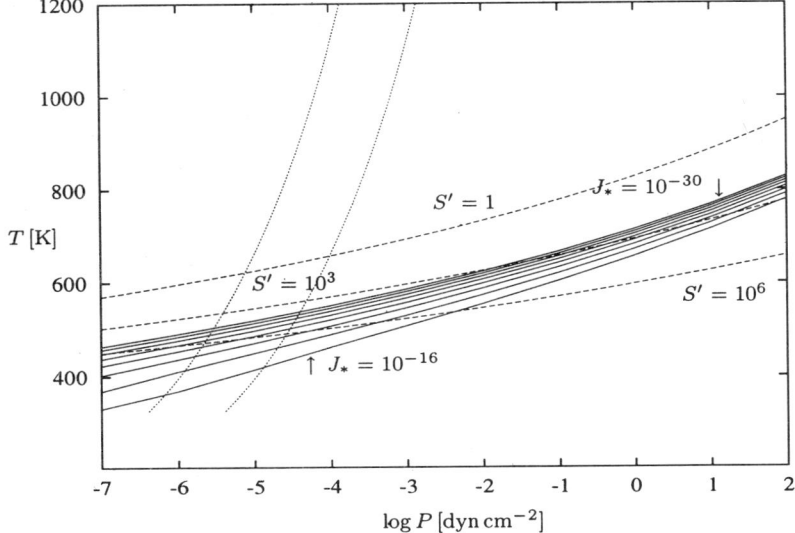

FIG. 2. Supersaturation ratio S' (dashed lines) and nucleation rate J_* (full lines) for SiO. J_* (seed nuclei per second and H-nucleus) is shown from $J_* = 10^{-30}$ (top) to $J_* = 10^{-16}$ (bottom) in steps of $\Delta \log_{10} J_* = 2$. The dotted lines show the P–T combinations of a stationary stellar wind at the sonic point for $\dot{M} = 10^{-6} \, M_\odot \, \mathrm{yr}^{-1}$ (left) and $10^{-5} \, M_\odot \, \mathrm{yr}^{-1}$ (right).

condensate formed from the monomer vaporizes at a temperature as high as possible.

(4) The nucleation rate depends on the probability that two particles stick together on collision. This probability should be as high as possible. This requires reactive molecules, radicals if possible, and an efficient way of removing the reaction energy, or alternatively the possibility of transferring (part of) this energy to internal degrees of freedom which later are de-excited, for instance by radiative transitions.

Figure 3 shows the bond energy of diatomic molecules with bond energies exceeding 4 eV for the more abundant elements. An inspection of this figure shows that there are only a few abundant molecular species with exceptionally high bond energies which are possible candidates for forming the first clusters in the outflow of M stars. These are listed in Table 1. Additionally, the two abundant elements Mg and Fe from which refractory condensates may be formed are present in the gas phase as free atoms. Other abundant species like H_2, CO, H_2O, and N_2 need not be considered since they condense at a very low temperature (some may condense as ices in the outermost parts of dense circumstellar shells).

At first glance, the SiO molecule seems to be the most likely candidate for nucleation since it satisfies both of the conditions: a bond energy as high as

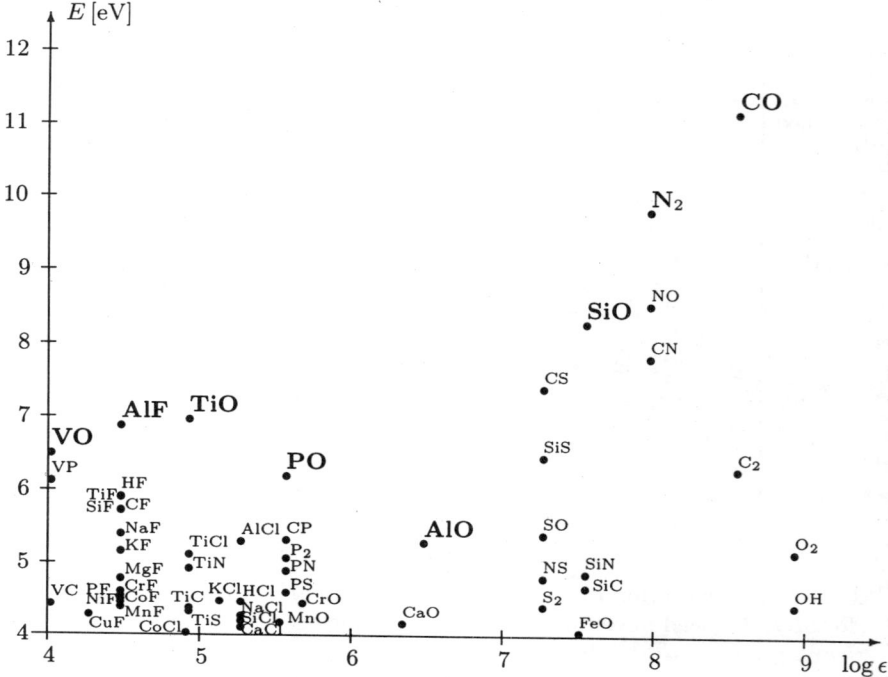

FIG. 3. Bond energies of particularly strongly bound diatomic molecules of abundant elements, plotted against the abundance of the less abundant of its constituents. (Bond energies from Lide 1995).

Table 1 Molecules with exceptionally high bond energy D and their electronegativity difference Δ.

Molecule	SiO	AlO	TiO	AlF	ZrO
D [eV]	8.28	5.29	6.97	6.88	8.04
Δ	1.7	2.0	2.0	2.5	2.1

possible, and a high abundance. It violates, however, the third requirement and, indeed, in Section 2.2 we showed that SiO is not a likely candidate for nucleation.

The next most abundant species in the gas phase which might be responsible for initiation of dust condensation are Fe and Mg. They are shown in Section 3.2 not to be responsible for circumstellar nucleation.

For other species from Table 1 no definite conclusion can be drawn at present with respect to the likelihood of their responsibility for nucleation. From the point of view of element abundance, aluminium- or titanium oxide clusters seem to be favoured. Due to the low abundance of F and Zr AlF and ZrO can be important only if nucleation of the former two species does not work.

3 Examples for the Calculation of Nucleation Rates

Realistic calculations of cluster abundances have to be based on detailed theoretical and experimental determinations of thermochemical functions for clusters up to sufficiently large size N (> 10 at least) for all of the species which may be involved in the nucleation process. Such calculations are just beginning to become available. We present here two examples relevant to the formation of MgO and Fe clusters.

3.1 MgO Clusters

The properties and thermodynamic functions for MgO clusters up to a cluster size of $N = 16$ have been studied by Köhler et al. (1997). This species was thought to be important for condensation in circumstellar shells at the time that work was begun, but since then the dissociation energy of MgO has been considerably revised downwards and now MgO is unlikely to be important for the circumstellar dust problem. The calculation of a nucleation rate is, however, a pretty example which demonstrates along which lines one has to proceed in the future to solve the problem of circumstellar dust nucleation.

The magnesium oxide solid is believed to be a pure ionic compound formed from Mg^{++} cations and O^{--} anions, respectively. The crystal structure of the solid is of the NaCl lattice type. The properties of such ionic crystals can well be described by the classical Born–Mayer potential model, as is well known from solid state physics. The total potential energy is assumed to be given by

$$\Phi = -\tfrac{1}{2} \sum_{\substack{i,j \\ i \neq j}} \frac{Z_i Z_j e^2}{r_{ij}} + \tfrac{1}{2} \sum_{\substack{i,j \\ i \neq j}} A e^{-r_{ij}/\rho}. \tag{3.1}$$

r_{ij} is the mutual distance between the ions i and j, Z_i and Z_j are their charges (± 2 for MgO), ρ_{ij} measures the steepness and A_{ij} the strength of the repulsive potential between the ions due to overlap of their outer electron shells. The summation is over all ions in the crystal. Usually, A_{ij} and ρ_{ij} are assumed to be independent of i and j. This model works quite well for the solid (e.g. Evans 1966) and allows a calculation of all of its properties, if the two parameters A and ρ are determined by a fit to two measured quantities. For the alkali halides this approach also works quite well for the free molecule with a single set of parameters A and ρ describing both the properties of the molecules and the solid. The potential model (3.1) has been used with great success, for example, in calculating the cluster properties for NaCl clusters (Martin 1983).

The same potential model may be applied to other compounds with chemical bonds of strong ionic character including the alkaline earths. Though their electronegativity difference is not as big as that of alkali halides, the solids form ionic crystals which are well described by Born–Mayer theory (e.g. Evans 1966). Application of the potential model (3.1) to the free molecule, however, yields unsatisfactory results. The main reasons for this shortcoming are:

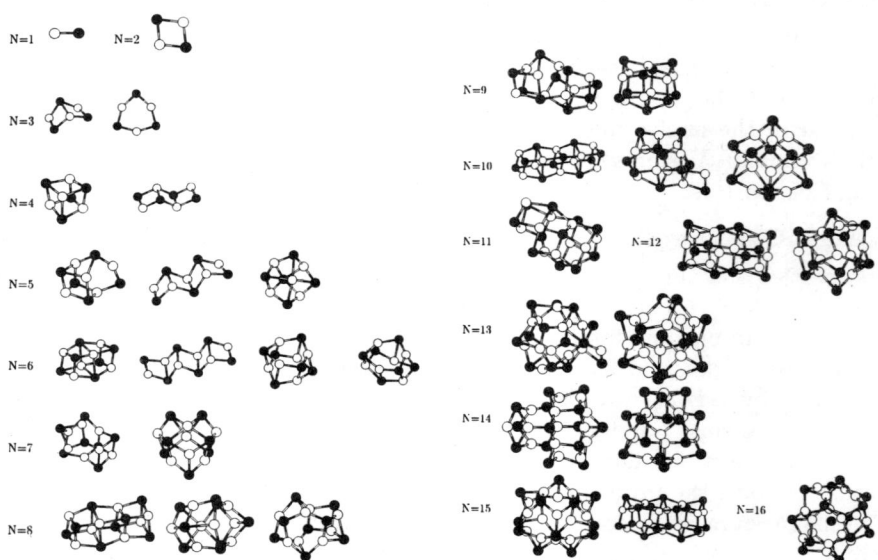

FIG. 4. Structures of MgO clusters for clusters of size $N = 1, \ldots, N = 16$ and their isomers. The ground state configuration is the leftmost one, the bond energy of isomers decreases from left to the right (for more details see Köhler et al. 1997).

- The ions are polarized in the local electric field due to the other ions. In the NaCl-type lattice of the solid there is no net polarization for symmetry reasons, but for small clusters without such a high symmetry the induced-dipole–monopole interaction contributes to the total energy.
- The bond in alkaline earth compounds has a considerable covalent contribution. The attraction cannot be described by Coulomb forces alone.
- Due to the lower electronegativity difference the ions carry only fractional charges.
- In the free alkaline earth molecule the ions are only singly charged, not double charged as in the solid.

In Köhler et al. (1997) the potential model (3.1) was modified by including the polarization of the ions and an additional Morse-potential like contribution to the particle–particle interaction potential and allowing for a size dependent fractional charging of the ions:

$$\Phi(N) = \sum_{\substack{i,j \\ i>j}} \left[A\,\mathrm{e}^{-2r_{ij}/\rho} - B(N)\,\mathrm{e}^{-r_{ij}/\rho} - \frac{Z_{\mathrm{eff},i} Z_{\mathrm{eff},j} e^2}{r_{ij}} \right] - \frac{1}{2} \sum_{\substack{i,j \\ i>j}} \frac{Z_{\mathrm{eff},i} e\,\vec{\mu}_j \cdot \vec{r}_{ij}}{r_{ij}^3}. \tag{3.2}$$

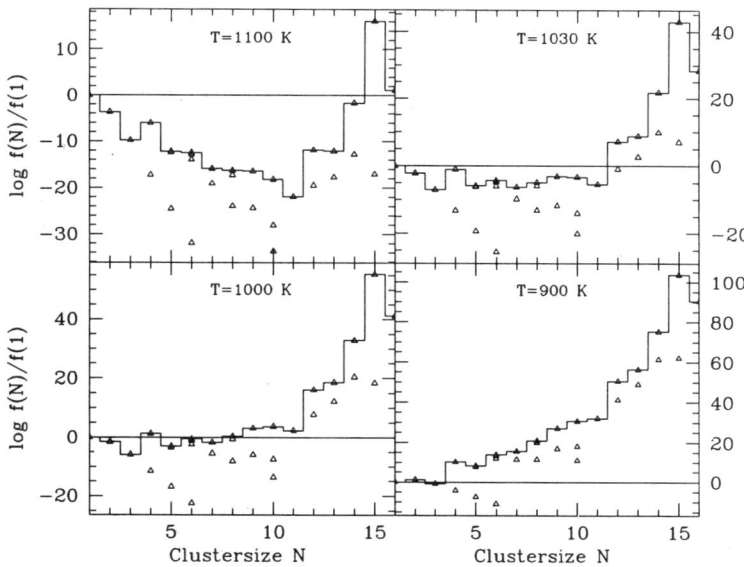

FIG. 5. Density of MgO clusters relative to the monomer density for four different temperatures and a typical gas density of $n_{H_2} = 10^{10}$ cm^{-3}. The full line connects the most abundant clusters of each size, triangles below this correspond to less tightly bound isomers. If the full curve decreases for small cluster sizes and increases for large cluster sizes, nucleation occurs. In this case, the least abundant cluster on the curve defines the critical cluster for nucleation (marked by a thick arrow).

$B(N)$ is an additional covalent attractive contribution to the potential, depending on cluster size N, $Z_{\text{eff},j}$ is the (size dependent) fractional charge, and

$$\vec{\mu}_j = \alpha_j \sum_{\substack{k \\ k \neq j}} \frac{Z_{\text{eff},k} e \vec{r}_{kj}}{r_{kj}^3} \tag{3.3}$$

is the induced dipole moment on particle j where α_j denotes its polarizability.

This potential model has been used to calculate the structure and properties of MgO clusters. The parameters of the potential were determined by a fit to (i) the bond energy and (ii) the vibrational frequency of the molecule and (iii) to the crystal lattice energy. The ground state of a cluster of size N with N cations Mg^{++} and N anions O^{--} each was calculated by determining the minimum of the potential (3.2) in the $6N$-dimensional configuration space $(\vec{x}_1, \ldots, \vec{x}_{2N})^T$ of all ion positions. Possible isomers were determined by searching for all deep local minima of Φ with an energy no more than a few electronvolts above the ground state. The structure of the first 16 MgO clusters and their isomers are shown in Fig. 4.

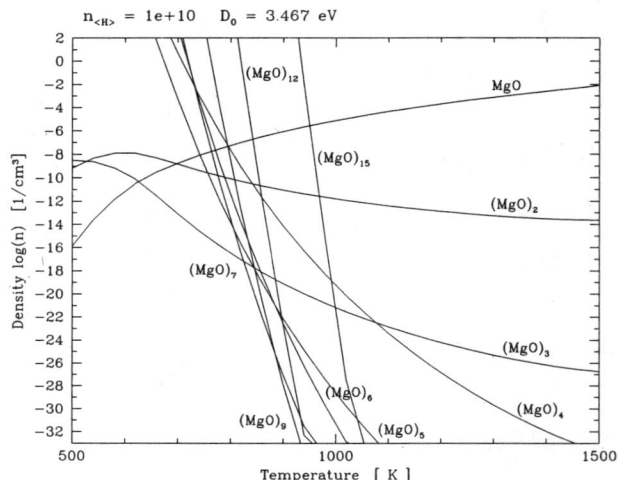

FIG. 6. Cluster densities for clusters of size $N = 1, \ldots, N = 15$ in chemical equilibrium with the monomer at a hydrogen density of $n_{H_2} = 10^{10}$ cm^{-3} which is typical for the condensation zone.

Once the cluster structures are known it is a simple task to calculate the moments of inertia and the vibrational frequencies from the eigenvalues of the force matrix. With these data and the calculated bond energies one readily calculates the free enthalpy of formation $\Delta G(N, T)$ of an (MgO)$_N$-cluster from N MgO molecules. The partial pressures $\overset{o}{p}(N)$ of clusters in thermodynamic equilibrium are given by (2.5). Results for the cluster densities relative to the monomer density for a characteristic hydrogen density of 10^{10} cm^{-3} in the dust condensation zone of a circumstellar shell are shown in Fig. 5. The full line corresponds to the ground states of the clusters. This line shows the expected behaviour for condensation to occur: a general decrease for small cluster sizes and a general increase for large cluster sizes. Considerable fluctuations around this general trend for individual clusters, however, are obvious. The cluster with least abundance along this line defines the critical cluster for nucleation which has to be used in eqn (2.10) for calculating the nucleation rate J_*. Figure 6 shows the results for the true cluster density. In this representation, the lower envelope of the family of curves shown in Fig. 6 defines the density of the critical cluster.

For the case of MgO the resulting nucleation rate obviously is much too low for MgO to become important for dust nucleation. The calculation of Köhler et al. (1997) demonstrates, however, the way in which a nucleation rate has to be determined for real situations.

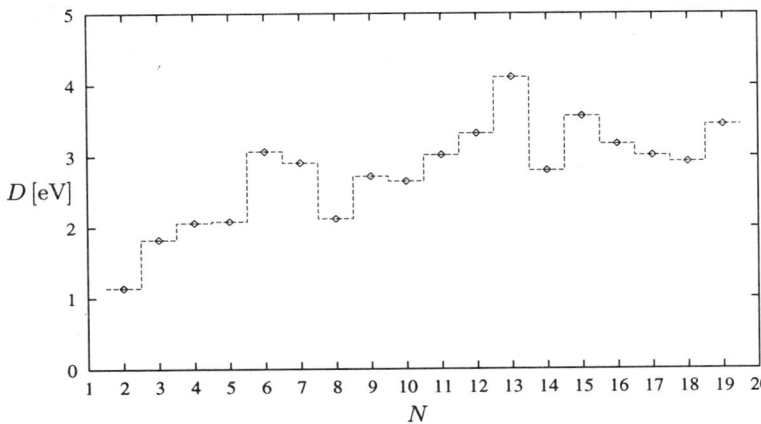

FIG. 7. Dissociation energy D (eV) of a single iron atom from an iron cluster of size N.

3.2 Iron Clusters

Iron atoms are one of the most abundant species in the gas phase which may be involved in circumstellar dust formation process, and the iron metal is one of the most stable condensates which may be formed (cf. Fig. 9). For this reason the clustering of iron atoms is a possible candidate for the process initiating dust formation. Following the above general concepts outlined for MgO, John (1995) has calculated the nucleation rate by clustering of Fe atoms. Obviously, the particle–particle interaction cannot be described by a potential model of the Born–Mayer type in this case. The bond energy of the iron clusters, instead, is available from the laboratory experiments of Lian et al. (1992) for cluster sizes up to $N = 19$. These are shown in Fig. 7. It is important to note that there is a very strong dependence of the bond energy of an iron atom within a cluster on the cluster size. The bond energy for small cluster sizes, particular for $N = 2, 3$, is very small. The equilibrium abundance of clusters with such a low dissociation energy clearly is extremely low at the temperature of ≈ 1000 K in the dust formation zone. They are stable only at much lower temperature and since during nucleation of iron each particle has to pass trough the stage $N = 2, 3$, we expect iron nucleation to be efficient only for temperatures $T \ll 1000$ K.

In order to derive the nucleation rate of Fe, John (1995) determined all the quantities required to calculate $\Delta G(N, T)$. The vibrational frequencies for the dimer and trimer are known from laboratory measurements, and for bigger clusters they are estimated from semi-empirical rules based on Debye theory. John (1995) derived the possible equilibrium configurations of Fe atoms in the clusters up to a cluster size $N = 19$ and calculated from this the moments of inertia for the clusters. The contribution of low lying electronic states to the partition function is particularly important in case of iron clusters. For the dimer these are known from quantum mechanical calculations (Shim and Gingerich 1982), and for bigger clusters the contribution of such states is estimated from

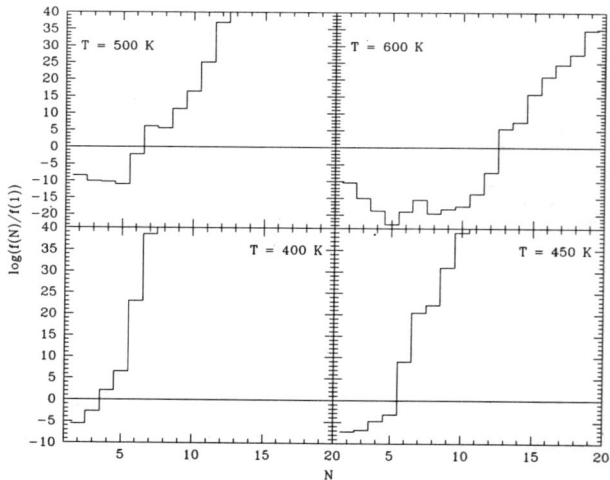

FIG. 8. Density of Fe clusters relative to the monomer density for four different temperatures and a typical iron atom density of $n_{Fe} = 4 \times 10^5\,\text{cm}^{-3}$ in the dust condensation zone. If the density of clusters decreases for small cluster sizes and increases for large cluster sizes, nucleation occurs. In this case, the least abundant cluster on the curve defines the critical cluster for nucleation.

a semiempirical interpolation procedure between the dimer case and the free electron model of the bulk condensate. From these data the free enthalpy of formation of clusters of size N from N free atoms is calculated. Details of this calculation will be published elsewhere (John and Sedlmayr 1998). Results for the cluster densities relative to the monomer density for a characteristic hydrogen density of $10^{10}\,\text{cm}^{-3}$ in the dust condensation zone of a circumstellar shell are shown in Fig. 8. The dependence of the normalized partial pressures on N shows the expected behaviour for condensation to occur: a general decrease for small cluster sizes and a general increase for large cluster sizes. Considerable fluctuations around this general trend for individual clusters are again observed.

An inspection of Fig. 8 shows that efficient nucleation of iron cannot be expected to occur until the temperature has dropped below 500 K since the density of the critical cluster is extremely low otherwise. The dust formation in a circumstellar shell, thus, cannot be triggered by the onset of iron nucleation because this requires a temperature much lower than the typical temperatures inferred from models for the IR-emission from circumstellar dust shells.

The abundant Mg atoms in circumstellar outflows also cannot be responsible for nucleation. The dimer has a bond energy of only 0.09 eV and the principal trend for the N-dependence of cluster bond energies should be similar to that for iron. Thus, the onset of nucleation of Mg also occurs at a very low temperature.

3.3 Future Prospects

The two examples above demonstrate the way in which one has to proceed in the future in order to arrive at a definite conclusion with respect to the primary nucleating species in circumstellar shells and in order to calculate the nucleation rate for circumstellar dust: one has to determine the structure and bond energies, the moments of inertia and the vibrational frequencies of all possible clusters of a given substance, and this has to be done for all substances which may be important for the nucleation problem. This is probably not such a difficult task as it might seem at first glance: only a few species can be important, see Section 2.3. For two of them we now have definite results:

- The most likely candidate for nucleation from the point of view of particle abundance and stability, SiO, is ruled out in most cases because the temperature for the onset of nucleation is much too low, as was shown in Section 2.2.
- The next most abundant possible species, the Fe atoms, are not responsible for dust nucleation (see Section 3.2).

According to our discussion in Section 2.3 the next likely candidates for nucleation are Al_mO_n clusters or TiO clusters, which remain to be investigated. These compounds have a strong ionic character of their bonding according to their big electronegativity difference (cf. Table 1) and, therefore, their structures and energies can be calculated by modelling their bond properties by a semiempirical potential model similar to that applied by Köhler et al. (1997) for MgO. Such potential models have been successfully applied in geophysics to calculate the properties of minerals (e.g. Catti 1986). Calculations based on such semi-empirical potentials for aluminium and titanium oxide clusters are currently under way at our institutes.

4 Iron–Magnesium-Silicate Compounds

4.1 Nature of the Condensate

For reasons of element abundance it is generally thought that the dominating dust component in circumstellar shells around M stars is some kind of magnesium–iron silicate (cf. Fig. 1 and Dorschner and Henning 1995). Such silicates occur with two different types of lattice structures and chemical compositions:

- The orthosilicates with the composition $Mg_{2x}Fe_{2(1-x)}SiO_4$. They form a continuous series of compounds with $0 \leq x \leq 1$. The two members at the endpoints of this series are fayalite ($x = 0$) and forsterite ($x = 1$). The intermediate case is known as olivine.
- The metasilicates with the composition $Mg_xFe_{1-x}SiO_3$. They form a continuous series of compounds with $0 \leq x \leq 1$. The two members at the endpoints of this series are gruenerite ($x = 0$) and enstatite ($x = 1$). The intermediate case is known as orthopyroxene.

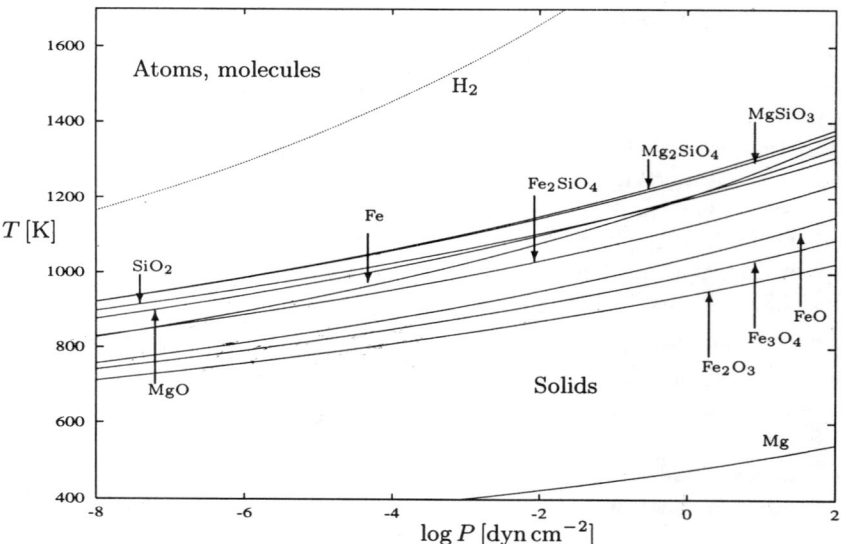

FIG. 9. Limit curves for destruction or growth of potential circumstellar condensates of the abundant refractory elements. Element abundances according to Anders and Grevesse (1989), thermodynamic data from Sharp and Huebner (1990). The dotted line is the limit curve for H_2 dissociation.

We now consider some aspects of the formation of the dominating dust component, which we believe sheds new light on the circumstellar dust formation process.

Let us consider the formation of forsterite (Mg_2SiO_4) from the gas phase in a thermodynamic equilibrium state. In view of the chemical composition of the gas phase in the dust formation zone of a circumstellar shell the corresponding chemical reaction is

$$2Mg + SiO + 3H_2O \longrightarrow Mg_2SiO_{4,\text{solid}} + 3H_2.$$

In chemical equilibrium the partial pressures of the gaseous species involved in this reaction have to satisfy the law of mass action

$$p_{H_2}^3 = p_{Mg}^2 \, p_{SiO} \, p_{H_2O}^3 \, K_p(T) \tag{4.1}$$

where

$$K_p(T) = \exp\left[-(\Delta G(Mg_2SiO_4) + 3\Delta G(H_2) - 2\Delta G(Mg) - \Delta G(SiO)\right. \\ \left. - 3\Delta G(H_2O))/RT\right]. \tag{4.2}$$

Let f be the fraction of the silicon bound in Mg_2SiO_4. If there exists no other dust condensate, the partial pressure of H_2 in the gas phase is $\frac{1}{2}P_H$, that of

SiO is $(1-f)\epsilon_{Si}P_H$, that of Mg is $(\epsilon_{Mg}-2f\epsilon_{Si})P_H$, and that of water vapour is $(\epsilon_O-\epsilon_C-(1+3f)\epsilon_{Si})P_H$. Inserting this into (4.1) yields

$$P^3 = \frac{(1+2\epsilon_{He})^3 K_p^{-1}(T)}{2^6(1-f)\,\epsilon_{Si}\,(\epsilon_{Mg}-2f\epsilon_{Si})^2\,(\epsilon_O-\epsilon_C-(1+3f)\epsilon_{Si})^3}. \qquad (4.3)$$

For given f this defines a curve in the P–T plane on which the degree of condensation f of silicon into forsterite has just this value. The limit curve $f=0$ for complete destruction of Mg_2SiO_4 is shown in Figs. 9 and 10. If Mg is replaced by Fe in this equation, one obtains the corresponding equation for fayalite Fe_2SiO_4. The limit curve $f=0$ for complete destruction of Fe_2SiO_4 is also shown in Fig. 9. We observe that forsterite is much more stable than fayalite. If the dust is formed in chemical equilibrium in a slowly cooling system, the magnesium silicate would be formed first. An inspection of Fig. 10 shows that circumstellar dust in equilibrium forms at a temperature of approximately 1000–1100 K, roughly consistent with the dust temperature derived for the dust formation zone of optically thick dust shells. Note that formation of dust grains at this temperature requires growth on some pre-existing seed particles.

4.2 Annealing

At a sufficiently high temperature the dust is subject to the process of annealing. During this process the lattice vibrations in the dust become sufficiently excited that activation energy barriers can be overcome causing atoms or groups of atoms to change their positions or their orientations within the lattice. The atoms in a poorly ordered or even amorphous dust material then start to rearrange and to migrate into energetically more favourable positions or orientations within the lattice, where they are more tightly bound and, then, become less mobile. By this process the dust material gradually develops some order in the local arrangement of the atoms and slowly changes its lattice structure to the locally ordered structure of a microcrystalline material. If amorphous dust grains contain a significant fraction of impurity elements within their lattice (for instance Al, Ca, and Na in a dust material of predominantly silicate composition) some kind of chemical fractionation may occur at the same time within the grain between the main component of the dust material and the impurities that were included during grain growth. These impurities may be driven out of the lattice and assemble at the surface or in separate inclusions. If annealing lasts sufficiently long complete crystallization of the grains may even occur.

The essential microscopic processes responsible for annealing are diffusion of vacancies and interstitials, and self diffusion. For the purpose of an order of magnitude estimation these processes can be described by a rather simple model: if the particles perform a 3D random walk with constant step size a one has for the diffusion coefficient

$$D = \tfrac{1}{3}a^2\nu e^{-E_a/kT} \qquad (4.4)$$

(e.g. Dekker 1963), where ν is the number of attempts per unit time for hopping to a neighbouring lattice site and E_a is the activation energy. The charac-

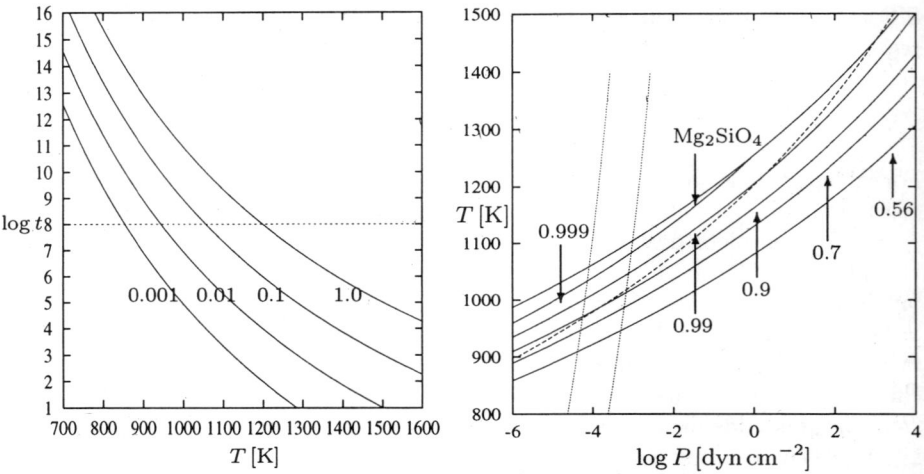

FIG. 10. Left part: Diffusion time scale (in s) for solid state diffusion in silicate dust particles of the indicated size (in μm). The horizontal dashed line indicates the characteristic timescale for temperature changes in the dust formation zone. Right part: Equilibrium composition of olivine. The numbers denote the value of the stoichiometric coefficient x for olivine with composition $Mg_{2x}Fe_{2(1-x)}SiO_4$. The uppermost full line is the stability limit of pure forsterite ($x = 1$). The dashed line shows the stability limit for condensed iron. The dotted lines shows the P–T stratification at the sonic point $v = c$ of a stationary stellar wind for $\dot{M} = 10^{-6}$ (left) and 10^{-5} $M_\odot$ yr^{-1} (right) for comparison.

teristic activation energy for silicate materials has been estimated by Lenzuni et al. (1995) and Duschl et al. (1996) to be $E_a = k \times 41\,000$ K based on the annealing experiments of Nuth and Donn (1982a) and Nuth and Donn (1986) for condensates from magnesium silicate smokes and on the assumption that the characteristic frequency ν equals the average vibrational frequency of the SiO_4 tetrahedron. Essentially the same E_a results if one uses the Debye temperature $\Theta_D = 763$ K for Mg_2SiO_4 (Kuskov and Galimzyanov 1986) for the determination of the characteristic frequency ν. We estimate the characteristic length a from the volume V of the basic molecule forming the lattice $V = Am/\rho = a^3$, where $A = 140.7$ is the molecular weight of Mg_2SiO_4, m the atomic mass unit and $\rho = 3.21$ g cm^{-3}, the mass density of Mg_2SiO_4. The coefficient of solid state diffusion within the silicate lattice is then

$$D = 1.2 \times 10^{-2} \, e^{-41\,000 \, K/T} \, [\text{cm}^2 \, \text{s}^{-1}]. \quad (4.5)$$

In a 3D random walk the average r.m.s. displacement d^2 within time t is given by

$$d^2 = 3 \cdot 2Dt. \quad (4.6)$$

The left part of Fig. 10 shows the typical time t required for a single atom to walk over some prescribed distance d. This may be identified with the time required for annealing the amorphous structure of interstellar grains and to form a local crystal structure extending at least over regions of size d. Since the characteristic time scale for changes of the temperature in the outflow is of the order of 10^8 s and since the time scale for dust particle growth may be somewhat shorter, 10^7 s for instance, we infer from Fig. 10 that for a dust temperature of about 900 K initially amorphous circumstellar grains would start to develop some degree of local order in their lattice structure. At about 1000 K the annealing of an amorphous silicate would take approximately one day (cf. Nuth and Donn 1982a) and at temperatures of about 1100 K and above a long range order would develop within no more than $\approx 10^6$ s, consistent with the finding in a laboratory experiment with striated orthopyroxene which showed annealing of the structure by heating for one week to 1100 K (Ashworth et al. 1984).

Recent observational findings (Goebel et al. 1994, Waters et al. 1996) indicate that the carriers of the circumstellar absorption bands show to some extent the structure which is compatible with at least a partial crystalline structure of the grain material and is less similar to that expected for an extremely amorphous structure, for instance like that proposed by Nuth and Hecht (1990). The grains are most likely hot enough in the dust formation zone ($T \gtrsim 900$ K) that solid state diffusion anneals part of the amorphous structure which probably first develops during surface growth from the gas phase (Gail and Sedlmayr 1984). The findings of the modelling of a large sample of circumstellar shells around M stars by Rowan-Robinson and Harris (1983) corroborates this, since they found, for most of their objects with dense circumstellar shells, condensation temperatures of ≈ 1000 K. It is unlikely, however, that big (> 0.1 μm) circumstellar grains develop a clear crystal structure since this requires heating to a temperature of at least 1 200 K, much higher than the stability limit of the silicates.

4.3 Equilibrium Chemistry of Mg–Fe Ortho-silicates

We now consider the formation of magnesium–iron ortho silicates (olivines) with a mixed composition $Mg_{2x}Fe_{2(1-x)}SiO_4$ with $x \in [0, 1]$ in a chemical equilibrium state. This is considered as a two step process where first pure forsterite Mg_2SiO_4 is formed and then converted into olivine by an exchange reaction with Fe atoms from the gas phase

$$Mg_2SiO_4 + 2(1-x)Fe \longrightarrow Mg_{2x}Fe_{2(1-x)}SiO_4 + 2(1-x)Mg.$$

We assume that the members of the continuous series of magnesium–iron silicates usually denoted as olivine form a solid solution of forsterite and fayalite. The free enthalpy of formation of one mole of the mixture from x moles of forsterite and $1-x$ moles of fayalite is given by the weighted mean of the free enthalpy of formation of both components and the mixing entropy term

$$\Delta G(Mg_{2x}Fe_{2(1-x)}SiO_4) = x\Delta G(Mg_2SiO_4) + (1-x)\Delta G(Fe_2SiO_4) + \Delta G_{mix} \quad (4.7)$$

where
$$\Delta G_{\text{mix}} = 2RT\left(x\ln x + (1-x)\ln(1-x)\right) \tag{4.8}$$
(Atkins 1994). In a state of chemical equilibrium between the gas phase and olivine, the partial pressures of Mg and Fe atoms in the gas phase must satisfy the law of mass action
$$p_{\text{Mg}}^{2(1-x)} = p_{\text{Fe}}^{2(1-x)} \cdot e^{-\Delta G_x/RT} \tag{4.9}$$
where
$$\Delta G_x = \Delta G(\text{Mg}_{2x}\text{Fe}_{2(1-x)}\text{SiO}_4) - \Delta G(\text{Mg}_2\text{SiO}_4) + 2(1-x)\left(\Delta G(\text{Mg}) - \Delta G(\text{Fe})\right) \tag{4.10}$$
is the change of free enthalpy in the conversion of forsterite into olivine. It follows that
$$\frac{p_{\text{Mg}}}{p_{\text{Fe}}} = K_x(T) = \exp\left[-\frac{\Delta G(\text{fay.}) - \Delta G(\text{for.})}{2RT} - \frac{\Delta G_{\text{mix}}}{2(1-x)RT}\right]. \tag{4.11}$$

If f is the fraction of silicon bound in the olivine, the partial pressures of Mg and Fe in the gas phase are $(\epsilon_{\text{Mg}} - 2xf\epsilon_{\text{Si}})P_{\text{H}}$ and $(\epsilon_{\text{Fe}} - 2(1-x)f\epsilon_{\text{Si}})P_{\text{H}}$, respectively. Then
$$\frac{\epsilon_{\text{Mg}} - 2xf\epsilon_{\text{Si}}}{\epsilon_{\text{Fe}} - 2(1-x)f\epsilon_{\text{Si}}} = K_x(T,x). \tag{4.12}$$
The degree of condensation f is determined by eqn (4.3) where the term $\epsilon_{\text{Mg}} - 2f\epsilon_{\text{Si}}$ has to be replaced in the present case by $\epsilon_{\text{Mg}} - 2xf\epsilon_{\text{Si}}$. Equations (4.12) and (4.3) determine the amount and the composition of the magnesium–iron silicate present in a chemical equilibrium state. We can solve (4.12) for f with the result
$$f = \frac{\epsilon_{\text{Mg}} - \epsilon_{\text{Fe}}K_x}{2\epsilon_{\text{Si}}(x - K_x(1-x))}. \tag{4.13}$$
For given x and T eqn (4.13) determines f and then from eqn (4.3) we can determine the total gas pressure corresponding to this state. For fixed x this defines a curve of constant composition x in the P–T plane. Such curves for selected values of x are shown in the right part of Fig. 10. Obviously, in a rather broad temperature strip below the stability limit of the olivine the chemical equilibrium composition of olivine would be that of nearly pure forsterite. Only at temperatures at least 100 K below the stability limit would substantial amounts of Fe be incorporated into the condensate. Inspection of Fig. 9 or Fig. 10 shows that the iron starts to condense as a separate dust material at similar but slightly higher temperatures as the olivine starts to incorporate substantial amounts of the iron. In constructing Fig. 10 we have assumed that iron is not condensed. The formation of solid iron would reduce the pressure of Fe atoms in the gas phase and this would favour, for given T, an even higher value of x.

Hence, the chemical equilibrium composition of the condensate does not correspond to an iron rich olivine which consumes most of the available magnesium and iron from the gas phase, as is commonly assumed, but instead it corresponds to a mixture of nearly pure forsterite and metallic iron particles. In Section 3.2 we saw that iron nucleation from the gas phase occurs at quite a low temperature. Thus, if iron condenses as a separate solid at a temperature of about 1000 K it condenses only on the surfaces of already existing grains, which means that small iron nuggets form on the surfaces of silicate grains. If there exists no process which detaches the iron particles from the surfaces they will be incorporated as small inclusions into the silicate grains.

4.4 Mg–Fe Meta-silicates

At temperatures only slightly less than the stability limit of forsterite the meta-silicate $MgSiO_3$ (enstatite) becomes stable (cf. Fig. 9). It can be shown that the conversion of forsterite into enstatite, for instance by the reaction

$$Mg_2SiO_4 + H_2 \longrightarrow MgSiO_3 + Mg + H_2O,$$

becomes thermodynamically favourable at a temperature roughly 30 K below the stability limit of the enstatite. Hence it is likely that part of the forsterite formed by condensation from the gas phase is later converted into enstatite. This process will be discussed elsewhere.

4.5 Iron Sulphide (Troilite)

Metallic iron grains may form with the abundant H_2S from the gas phase the quite stable mineral FeS, troilite. This is observed to be a component of the most primitive meteoritic material and it is also thought to be a component of the interstellar dust (Pollack et al. 1994). The formation of troilite from iron grains occurs according to the reaction

$$Fe_{solid} + H_2S \longrightarrow FeS_{solid} + H_2.$$

In chemical equilibrium the partial pressures of H_2 and H_2S must satisfy the law of mass action

$$p_{H_2} = p_{H_2S} \cdot K_p(T), \qquad (4.14)$$

where

$$K_p(T) = \exp\left[-(\Delta G(FeS) + \Delta G(H_2) - \Delta G(Fe) - \Delta G(H_2S))/RT\right]. \qquad (4.15)$$

If we assume that the fraction f of the total available iron has reacted to form FeS and if no other abundant sulphur bearing species exist then we have for the partial pressure of H_2S in the gas phase $p_{H_2S} = (\epsilon_S - f\epsilon_{Fe})P_H$. From (4.14) it follows that

$$f = \frac{1}{\epsilon_{Fe}}\left(\epsilon_S - \frac{1}{2K_p(T)}\right). \qquad (4.16)$$

The maximum fraction of the iron which can be converted into FeS equals the abundance ratio $\epsilon_S/\epsilon_{Fe} = 0.56$. If the quantity f calculated from equation (4.16)

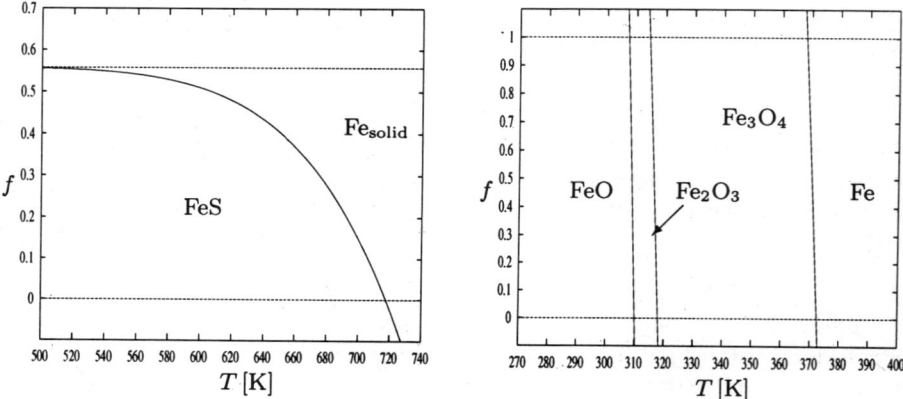

FIG. 11. Left part: Fraction f of the metallic iron which is converted by surface reactions with H_2S into FeS (troilite). Right part: Fraction of the iron converted by surface reactions with H_2O into the indicated iron oxides.

for a certain temperature T takes a value $0 \leq f \leq 0.56$, then for this temperature the iron metal is partially converted into solid FeS. If there results a value of $f < 0$, then all the iron is present as the free metal and we have to put $f = 0$ in this case. The left part of Fig. 11 shows the result for f for the standard element abundances. The conversion of FeS to Fe extends over a broad temperature region. It starts rather suddenly at a temperature $T \approx 720$ K, the conversion of solid Fe into FeS then gradually progresses with decreasing T and it is nearly completed at a temperature of about 500 K, where f approaches the maximum possible value of 0.56. Above the limit temperature defined by $K_p(T) = 1/2\epsilon_S$ the iron always is present as the free metal and no FeS can exist due to reduction of FeS by the abundant hydrogen. Below this temperature the iron would be converted to FeS by reactions with H_2S, but this can convert at most the fraction $f = \epsilon_S/\epsilon_{Fe}$ of the iron into FeS. If a fraction of the total iron abundance less than this is present as solid iron, the conversion of the metal to FeS is complete; if the abundance of iron metal exceeds this limit the excess remains as free metal and can be oxidized by water vapour (see Section 4.6).

The kinetics of the formation of FeS has been discussed for instance by Fegley and Prinn (1989) and more recently studied in detail in the laboratory by Lauretta et al. (1996). They found a rate coefficient for the formation of FeS from solid Fe of $k_f = 5.6 \exp(-3360/T)$ in units gram FeS per cm², hour, and atm. This holds for the linear kinetics regime where the rate is not controlled by solid diffusion. According to their findings diffusion only becomes important for particle sizes above 10 µm. This rate may be converted into a rate of radius change of spherical particles of radius a. One obtains

$$\frac{da}{dt} = k_f \cdot \rho_{Tr}^{-1} = 7.2\,10^{-9}\,e^{-3360/T}\,p_{H_2S} \qquad (4.17)$$

where $\rho_{\text{Tr}} = 4.7 \, \text{g cm}^{-3}$ is the density of troilite, the partial pressure of H_2S is now in units dyn cm^{-2}, and a and t are in units of cm and s, respectively. For a particle size of $0.1 \, \mu\text{m}$ and an estimated total pressure $P = 4 \times 10^{-5} \, \text{dyn cm}^{-2}$ in the circumstellar shell at the temperature of $T = 720 \, \text{K}$ for the onset of FeS formation, we obtain a typical time scale for the conversion of an iron grain of such size into FeS of $t \approx 2 \times 10^{14}$ s. Compared with a characteristic timescale for the expansion of the wind of 10^8 s in this region this time is much too long for conversion of iron grains into FeS to be possible.

Therefore, though troilite formation would be strongly favoured for chemical reasons it is kinetically inhibited in circumstellar outflows.

4.6 Iron Oxide

Iron oxides are thought to be a possible component of circumstellar dust (e.g. Henning et al. 1995). They may be formed by oxidizing iron grains by water vapour according to the reaction

$$\text{Fe}_{\text{solid}} + H_2O \longrightarrow \text{FeO}_{\text{solid}} + H_2$$

(e.g. Fegley and Prinn 1989). In chemical equilibrium the partial pressures of the molecules have to satisfy the law of mass action

$$p_{H_2} = p_{H_2O} \cdot K_p(T), \qquad (4.18)$$

where

$$\Delta G = \Delta G(\text{FeO}_{\text{solid}}) + \Delta G(H_2) - \Delta G(\text{Fe}_{\text{solid}}) - \Delta G(H_2O). \qquad (4.19)$$

If we assume that the fraction f of the totally available iron is oxidized from Fe into FeO and if we admit that some of the oxygen is bound in silicates (g O atoms per Si atom) then we have for the partial pressure of H_2O in the gas phase $p_{H_2O} = 2(\epsilon_O - \epsilon_C - g\epsilon_{\text{Si}} - f\epsilon_{\text{Fe}}) P_H$. From (4.18) it follows that

$$f = \frac{1}{\epsilon_{\text{Fe}}} \left(\epsilon_O - \epsilon_C - g\epsilon_{\text{Si}} - \frac{1}{2K_p(T)} \right). \qquad (4.20)$$

If the quantity f calculated from this equation for a certain temperature T takes a value in the interval $0 \leq f \leq 1$, then for this temperature the iron metal is partially oxidized to FeO. If there results a value of $f > 1$, then the oxidation is complete and we have to put $f = 1$. If there results a value of $f < 0$, then all Fe is present as free iron and we have to put $f = 0$ in this case. The right hand part of Fig. 11 shows the result for f for standard element abundances. It is assumed that silicon atoms have formed solid silicates in which case they have bound four oxygen atoms each ($g = 4$). An inspection of the figure shows that the solid iron metal is oxidized by the water vapour from the gas phase to solid FeO at $T \approx 310 \, \text{K}$ and the degree f of oxidation changes from zero to unity within a very narrow temperature interval. Above this temperature, the iron would be present as the free metal and no FeO can exist at such temperatures due to reduction of FeO by the abundant hydrogen. Below this temperature the iron would form FeO due to oxidation by water vapour.

Analogously one determines the limit temperatures for the oxidation or reduction of the two other possible iron oxides. For Fe_2O_3 (haematite) we obtain

$$f = \frac{2}{3\epsilon_{Fe}} \left(\epsilon_O - \epsilon_C - g\epsilon_{Si} - \frac{1}{2K_p^{\frac{1}{3}}(T)} \right) \qquad (4.21)$$

and for Fe_3O_4 (magnetite)

$$f = \frac{3}{4\epsilon_{Fe}} \left(\epsilon_O - \epsilon_C - g\epsilon_{Si} - \frac{1}{2K_p^{\frac{1}{4}}(T)} \right). \qquad (4.22)$$

The results for the fraction of the oxidized iron are also shown in the right hand part of Fig. 11. The temperature of formation of Fe_2O_3 is ≈ 330 K and Fe_3O_4 is formed at ≈ 370 K. Hence above 370 K all iron not bound in silicates, in a chemical equilibrium state would be present as the free metal. Below this temperature it would be oxidized by the water vapour to Fe_3O_4; at temperatures around 315 K this would be converted into Fe_2O_3 and below ≈ 310 K into FeO. Free metallic iron cannot exist below ≈ 370 K in chemical equilibrium with the gas phase while the oxides cannot exist above this temperature.

The time required for "rusting" of the iron grains is determined by the reaction kinetics of this process. The activation energy for wustite (FeO) formation in a H_2/H_2O atmosphere, for instance, has been found by Turkdogan et al. (1965) to be $E_a/k = 9\,600$ K. In view of this high energy barrier, the low particle density in the wind where rusting becomes chemically favourable, and the findings of the previous Section we can safely conclude that formation of iron oxides, though favourable for chemical reasons, again is kinetically inhibited in the stellar outflow. If iron grains are formed in the condensation zone of the wind, they will not be subject to subsequent chemical modification by surface reactions with species from the gas phase.

5 Concluding Remarks

The methods and the illustrative applications presented here yield important insights into grain formation in oxygen rich stellar environments, and provide basic information about the condensates to be expected under conditions encountered in circumstellar shells. Of course, these approaches can be considered only as first steps towards a realistic description of inorganic astrophysical grain condensation.

Bibliography

1. Atkins, P. W. (1994). *Physical Chemistry*, 5th edn, Oxford University Press, Oxford.

2. Anders, E. and Grevesse, N. (1989). *Geochimica et Cosmochimica Acta*, **53**, 197.
3. Ashworth, J. R., Mallinson, L. G., Hutchinson, R. and Biggar, G. M. (1984). *Nature*, **308**, 259.
4. Catti, M. (1986). In *Chemistry and Physics of Terrestrial Planets*, ed. S. K. Saxena, Springer, New York, p.224.
5. Dekker A. J. (1963). *Solid State Physics*. Macmillan, London.
6. Deguchi, S. (1980). *Astrophysical J.*, **236**, 567.
7. Donn, B. and Nuth, J. A. (1985). *Astrophysical J.*, **288**, 187.
8. Donn, B., Wickramasinghe, N. C., Hudson, J. P. and Stecher, T. P. (1968). *Astrophysical J.*, **153**, 451.
9. Dorschner, J. and Henning, T (1995). *Astronomy & Astrophysics Rev.*, **6**, 271.
10. Draine, B. T. (1979). *Astrophys. Space Sci.*, **65**, 311.
11. Draine, B. T. and Salpeter, E. E. (1977). *J. Chem. Phys.*, **67**, 2230.
12. Duschl, W. J., Gail, H.-P. and Tscharnuter, W. M. (1996). *Astronomy & Astrophysics*, **312**, 624.
13. Evans, R. C. (1966). *An Introduction to Crystal Chemistry*, 2nd edn, Cambridge University Press, Cambridge.
14. Feder, J., Russel, K. C., Lothe, J. and Pound, G. M. (1966). *Adv. Phys.*, **15**, 111.
15. Fegley Jr., B. and Prinn, R. G. (1989). In *The Formation and Evolution of Planetary Systems*, eds. H. A. Weaver and L. Danly, Cambridge University Press, Cambridge, p.171.
16. Gail, H.-P. and Sedlmayr, E. (1984). *Astronomy & Astrophysics*, **132**, 163.
17. Gail, H.-P. and Sedlmayr, E. (1986). *Astronomy & Astrophysics*, **166**, 225.
18. Gail, H.-P. and Sedlmayr, E. (1987). In *Physical Processes in Interstellar Clouds*, eds. G. Morfill and M. Scholer, Reidel, Dordrecht, p.275.
19. Gail, H.-P. and Sedlmayr, E. (1988). *Astronomy & Astrophysics*, **206**, 153.
20. Gail, H.-P., Keller, R. and Sedlmayr, E. (1984). *Astronomy & Astrophysics*, **133**, 320.
21. Goebel, J. H., Bregman, J. D. and Witteborn, F. C. (1994). *Astrophysical J.*, **430**, 317.
22. Grevesse, N. and Noels, A. (1993). In *Origin and Evolution of the Elements*, eds. N. Prantzos, E. Vangioni-Flam and M. Cassé, Cambridge University Press, Cambridge, p.14.
23. Habing, H. J. (1996). *Astronomy & Astrophysics Rev.* **7**, 97.
24. Henning, Th., Begemann, B., Mutschke, H. and Dorschner, J. (1995). *Astronomy & Astrophysics Suppl.*, **112**, 143.
25. Iben Jr., I. (1991). *Astrophysical J. Suppl.*, **76**, 55.

26. John, M. (1995). Bildung von Eisen-Clustern in kühlen Sternwinden. Diploma thesis, Technical University Berlin.
27. John, M. and Sedlmayr, E. (1998). Formation of iron clusters in cool stellar winds (in preparation).
28. Köhler, T. M., Gail, H.-P. and Sedlmayr, E. (1997). *Astronomy & Astrophysics*, **320**, 553.
29. Kozasa, T. and Hasegawa, H. (1987). *Prog. Theor. Phys.*, **77**(6), 1402.
30. Kuskov, O. L. and Galimzyanov, R. F. (1986). In *Chemistry and Physics of Terrestrial Planets*, ed. S. K. Saxena, Springer, New York, p.310.
31. Lauretta, D. S., Kremser, D. T. and Fegley Jr., B. (1996). *Icarus*, **122**, 288.
32. Lenzuni, P., Gail, H.-P. and Henning, Th. (1995). *Astrophysical J.*, **447**, 848.
33. Lian, L., Su, C.-X. and Armentrout, P. (1992). *J. Chem. Phys.*, **97**(6), 4072.
34. Lide, D. R. (1995). *Handbook of Chemistry and Physics*, 76th edn, CRC Press, Boca Raton.
35. Loup, C., Forveille, T., Omont, A. and Paul, J. F. (1993). *Astronomy & Astrophysics Suppl.*, **99**, 291.
36. Martin, T. P. (1983). *Phys. Rep.*, **95**, 167.
37. Nuth, J. A. and Donn, B. (1981). *Astrophysical J.*, **247**, 925.
38. Nuth, J. A. and Donn, B. (1982a). *Astrophysical J.*, **257**, L103.
39. Nuth, J. A. and Donn, B. (1982b). *J. Chem. Phys.*, **77**(5), 2639.
40. Nuth III, J. A. and Hecht, J. H. (1990). *Astrophys. Space Sci.*, **163**, 79.
41. Nuth III, J. A., Donn, B. and Nelson, R. (1986). *Astrophysical J.*, **310**, L83.
42. Pollack, J. B., Hollenbach, D., Beckwith, S., Simonelli, D. P., Roush, T. and Fong, W. (1994). *Astrophysical J.*, **412**, 615.
43. Rowan-Robinson, M. and Harris, S. (1983). *Monthly Notices Roy. Astr. Soc.*, **202**, 767.
44. Salpeter, E. E. (1974). *Astrophysical J.*, **193**, 579.
45. Sedlmayr, E. and Dominik, C. (1995). *Space Science Rev.*, **73**, 211.
46. Sharp, C. M. and Hübner, W. F. (1990). *Astrophysical J. Suppl*, **72**, 417.
47. Shim, I. and Gingerich, K. (1982). *J. Chem. Phys.*, **77**, 2490.
48. Smith, V. V. and Lambert, L. L. (1990). *Astrophysical J. Suppl.*, **72**, 387.
49. Sylvester, R. J., Barlow, M. J. and Skinner, C. J. (1994). *Monthly Notices Roy. Astr. Soc.*, **266**, 640.
50. Turkdogan, E. T., McKewan, W. M. and Zwell, L. (1965). *J. Phys. Chem.*, **69**, 327.
51. Waters, L. B. E. M. and 36 others (1996). *Astronomy & Astrophysics*, **315**, L361.
52. Yamamoto, T. and Hasegawa, H. (1977). *Progr. Theor. Phys.*, **58**, 816.

14
Models of Circumstellar Masers

David Field
School of Chemistry, University of Bristol and Institute of Physics and Astronomy, University of Aarhus

1 Introduction

Maser emission involving SiO, H_2O and OH has been observed in the direction of a large number of AGB stars. These stars have the characteristic that they lose mass with rates between 10^{-8} solar masses per year to as much as 10^{-4} solar masses per year. Maser emission may occur in the circumstellar envelope (CSE) of AGB stars and is believed to be closely connected with the phenomenon of mass loss. Maser emission may be used as a sensitive probe of the number density, temperature and velocity fields in the CSE. In regions close to the photosphere SiO masers, to tens of stellar radii distant H_2O masers and to hundreds of stellar radii distant OH masers provide information about physical properties. The evolution and development of the CSE may be studied in depth by combining models of masers with hydrodynamic pulsation models of the circumstellar envelope (CSE) of a M-Mira (Bowen 1988, 1989; Pijpers and Hearn 1989; Bowen and Wilsson 1991; Fleischer *et al.* 1991, 1992; Humphreys *et al.* 1996 (H96); Bessell *et al.* 1996; Hoefner *et al.* 1996 and references therein). The present chapter outlines how this may be carried out for M-Miras, the largest class of AGB stars. We start with a brief review of observational data.

2 Observations of SiO, H_2O and OH Masers in Late-type Stars

A good review of older observational data and other aspects of circumstellar masers may be found in Cohen (1989). Further details may be found in Elitzur (1992). Some important features mentioned in those references are briefly summarized here, with additional information from more recent observations. For brevity, mention of the polarization of maser emission is excluded. The picture that the reader should have in mind is of hot post-shocked gas close to the M-Mira photosphere harbouring SiO masers, with cooler less dense gas at larger radii harbouring H_2O masers and at still greater radius cold, dilute gas supporting OH maser emission.

Before embarking on a description of observations, one should note that the specific intensity I_ν of maser emission (in $W\,m^{-2}Hz^{-1}sr^{-1}$) is often expressed in terms of a brightness temperature T_b defined by

$$T_b = \frac{c^2}{2k\nu^2} I_\nu \qquad (2.1)$$

Thus if some column of masing gas yields a factor of gain of G, and the maser amplifies some effective background temperature T_{bkgnd}, the maser brightness temperature is simply G multiplied by T_{bkgnd}. Masers do not in general radiate isotropically but are beamed through both geometrical and saturation effects. A maser therefore has an observed brightness temperature which is in inverse proportion to the solid angle into which the maser beams as it leaves the region in which it underwent amplification.

Starting in the most quiescent region, at the greatest radial distance from the photosphere of the host M-Mira, OH masers form at 1612 MHz, a line arising from a relatively weak ("satellite") lambda-doublet transition in the ground rotational state of OH. Considerably more than 1000 maser sources at 1612 MHz have been observed in the direction of late-type stars. The emission lineshape often exhibits a spectrum with two peaks separated by 20 to 50 km s^{-1}. This lineshape suggests that the active gas may be in the form of a thin expanding shell, whose expansion velocity is half of the peak separation. This has been elegantly confirmed by radio-interferometric observations with Multi-Element Radio- Linked Interferometry Network (MERLIN) initially by Booth et al. (1981). The shell of masing gas is typically situated 10^{14} to 10^{15} m from the stellar photosphere. The shells of masing gas which these and similar observations have revealed are however incomplete and maser emission is not homogeneously distributed within a shell. This inhomogeneity is consistent with high resolution single dish observations of sharp structure within emission peaks, with components as narrow as 0.1 km s^{-1}. In this connection, Very Long Baseline Interferometry (VLBI) resolves out most 1612 MHz emission, but there remain a number of hot spots of brightness temperature 10^9 to 10^{10} K, that is, more than an order of magnitude greater than bright more extended emission. 1612 MHz maser emission shows a systematic correlation of temporal variability with stellar emission, with a maser amplitude variation through a factor of $\sim$2 and a phase lag of some weeks. This provides excellent confirmation of models of OH 1612 MHz emission based on a population inversion driven through a far-infrared pumping mechanism, as described in detail in Elitzur (1992). About 30 percent of nearby Miras also show weaker emission in the lambda doublet "main line" transitions at 1667 or 1665 MHz. These masers show larger and less regular temporal variations through the stellar cycle than the 1612 MHz masers.

H_2O masers are a very common feature of the CSEs of late-type stars, with an estimated 75 percent of Miras in the solar neighbourhood showing maser action (Bowers and Hagen 1984). Maser lineshapes may either have a centrally concentrated narrow profile or may show a twin-peaked profile. H_2O maser emission tends to have a mean velocity at the stellar velocity. The velocity width of the spectrum is smaller than that of OH 1612 MHz maser emission.

Radio-interferometry shows that H_2O masers tend to form in thick irregular clumpy shells at distances of between 10^{11} and 10^{12} m from the stellar photosphere. In supergiants, H_2O masers may however be found at a very much greater radius. For example in NML Cyg, MERLIN maps show that a shell of water masers lies between 1 to 5×10^{13} m from the photosphere (Richards et al. 1996). Figure 1 shows H_2O maser emission at 22 GHz from circumstellar envelope of the supergiant S Persei.

H_2O maser hotspots detected with VLBI comprise in general rather more of the total emission than do 1612 MHz hotspots and can reach brightness temperatures of several times 10^{11} K. In keeping with their location in a more active region closer to the photosphere, H_2O masers show much more rapid and erratic time variability than OH 1612 MHz masers. Maser intensity however roughly follows the stellar cycle, with superposed short-term variations. Changes in output through the cycle may be typically an order of magnitude, with different behaviour from cycle to cycle. Phase lags are difficult to determine but appear to be considerably greater than the light travel time from the host star to the H_2O maser zone, in contrast to OH 1612 MHz masers. Thus collisional as well as radiative pumping is likely to be involved in the creation of population inversions in H_2O. The three most extensively studied transitions involving H_2O masers are at 22 GHz (6_{16}–5_{23}), which is by far the most common line observed, 321 GHz (10_{29}–9_{36}) and 325 GHz (5_{15}–4_{22}).

Recent VLBI data show that SiO masers inhabit a most inhospitable hot dense shocked zone at about one stellar radius from the photosphere. Maser spots form an incomplete ring around the host star (Diamond et al. 1994; Humphreys et al. 1996) and have complex structures down to sub-milliarcsecond scales. SiO masers probe the region in which mass loss is initiated and are for this reason potentially the most interesting of all masers in the CSE. Masers have been detected in $v = 0, 1, 2, 3, 4$ vibrational states involving rotational transitions up to $J = 7$–6 (Gray et al. 1995) and $J = 8$–7 (Humphreys et al. 1997). SiO masers have been observed in several hundred sources and extensive surveys may be found in Engels and Heske (1989) and Jewell et al. (1991). A number of features characteristic of SiO masers may be identified, as described in detail in Doel et al. (1995) and Humphreys et al. (1996, 1997a). SiO masers are centred closely on the stellar velocity and the velocity extent, full-width zero-height, is $\sim$15 km s^{-1}. A number of trends regarding the relative intensities and the lineshapes of the 23 known different rotational transitions may be discerned. Remarkably, the flux in the ground vibrational state, $v = 0$, tends to be low and the strongest lines appear in $v = 1$ and $v = 2$. The flux in $v = 1$ $J = 2$–1 (at 86 GHz) is greater than that in $J = 1$–0 (at 43 GHz), which is itself greater than that in $v = 2$ $J = 1$–0. The flux in $v = 2$ $J = 2$–1 tends to be very low. Velocities and velocity profiles of low and high J lines tend to be different. Studies at 43 and 86 GHz $v = 1$ and 2 (see Martinez et al. (1988) and references therein) show that SiO masers undergo intervals of high and low brightness within a stellar optical cycle, roughly following the light curve of the host star

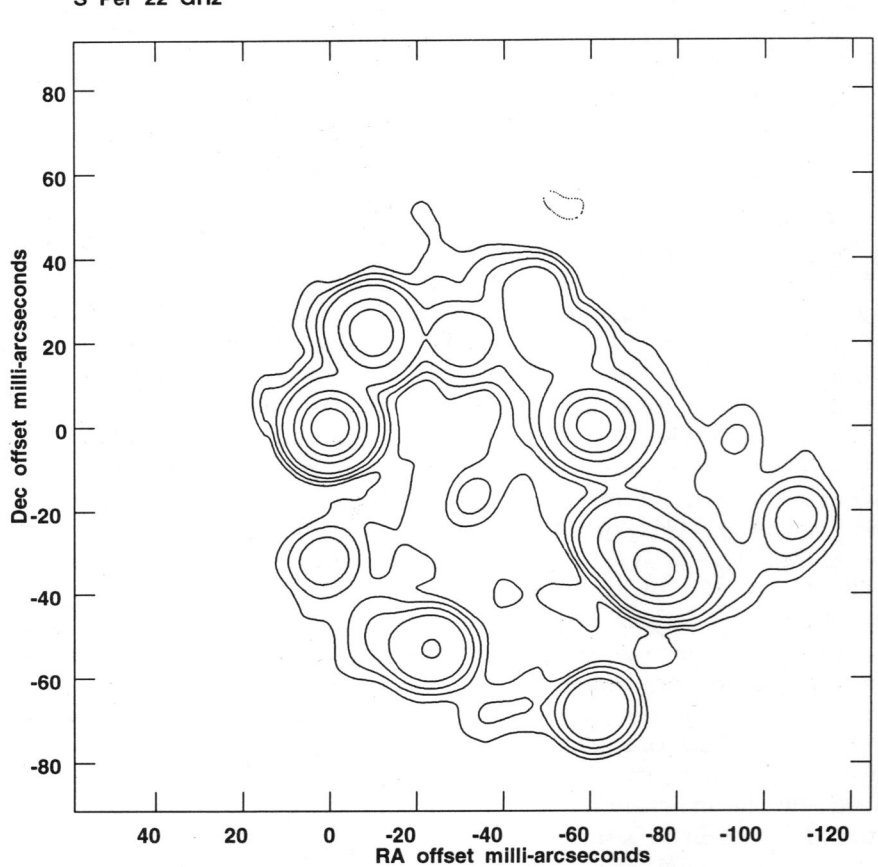

FIG. 1. Water maser emission at 22 GHz from the circumstellar envelope of S Persei. Observations were performed with MERLIN in March 1994 with an effective beamsize of 8 milliarcseconds. The outflow velocity is 9 to 16 km s^{-1} and if S Persei is assumed to be at a distance of 2.3 kpc the maximum diameter of the water maser shell is 5.2×10^{13} m. The hotspots reach brightness temperatures of 10^{11}–10^{12} K.

with a phase lag of ∼0.2. The 43 GHz line remains at more than half of its maximum brightness for 0.3 to 0.4 of a period. The contrast between high and low brightness within a stellar period is very different for different sources, ranging from more than 100 for example in o-Ceti (Mira itself) to a factor of 3 to 4 in R Leo. The maximum–minimum contrast also varies from cycle to cycle. In fact the maser emission in one cycle never resembles that in another. As for H_2O masers but still more so here, the phase lag between the stellar output and the response of the SiO masers is very much greater than the light travel time from the photosphere to the maser zone. Again this suggests a pumping mechanism in which collisional rather than radiative events dominate. Rapid variations in maser brightness are superposed on the general tendency of maser brightness to follow the optical (and infrared) stellar emission. Marked changes in lineshape occur over periods of days, or indeed within 24 hours (Pijpers et al. 1994). Variations on the very shortest timescales may be due to short wavelength sound waves generated through convection. However, changes on a timescale of 0.05 to 0.1 cycle period, that is, 15 to 30 days in o-Ceti, may well arise from rapidly changing conditions associated with shock waves passing through the masing gas, as we discuss in Section 3.

Following this very brief and far from inclusive tour of observational data, we turn now to the physical conditions in the CSE and the hydrodynamic pulsation model of the time-varying CSE. In the remainder of the chapter, we concentrate exclusively on SiO and H_2O masers.

3 A Hydrodynamic Pulsation Model for a Mira Variable

Bowen (1988, 1989), Bowen and Willson (1991) and Humphreys et al. (1996) described in detail the essential features of a model which may be used to generate the physical conditions in the CSE of a M-type Mira variable (see also Chapters 11 and 12). The spherically symmetrical atmosphere is periodically driven by a sinusoidally varying radial force, buried deep within the star. Forces on the gas in the CSE include the gravity of the star, convective forces, pressure gradients and radiation pressure on dust and molecules (Joergensen and Johnson 1992). The quantity of dust present in any zone depends on the radiation temperature (T_{rad}), to which the dust grains are strongly coupled. In the model of Bowen, there is no dust for $T_{rad} = 2000$ K and a maximum amount below 1000 K. The conditions of temperature and density at which dust forms and the rate at which it forms in the CSE remain among the largest sources of uncertainty in CSE models. The outer boundary of the model resides at 40–50 R_* where the outflow of material has become a steady supersonic wind with a velocity exceeding the escape velocity. Since the phase lag associated with the passage of the disturbance from the inner boundary into the upper photosphere is not known, it is unfortunately not possible at present to relate the model phase to the observed optical phase of an M-Mira.

If we consider a parcel of gas in the CSE close to the photosphere, this gas initially experiences a brief but very strong outward acceleration, and thereafter decelerates, subsequently reversing direction. After a period of infall it is then driven outwards again and this cycle is repeated with successively weaker impulses as the material is driven increasingly further from the photosphere. Material thus performs oscillatory motion, wending its way farther from the photosphere, resulting ultimately in mass loss from the star. If some arbitrary initial radial position is chosen within the CSE, the parcel of gas defined in the co-moving Lagrangian frame experiences a continuously changing set of number densities, temperatures and velocity fields. Some of these changes are both spatially and temporally abrupt, as one would expect in a shocked zone. The hydrodynamic pulsation model yields the number density, temperature and velocity profiles as functions of radial distance and of stellar model phase, thereby providing an important fraction of the conditions that are required to calculate the rotational and rovibrational populations of SiO and H_2O in the CSE, as described in Section 4. The next step is to see how physical conditions generated by the hydrodynamic pulsation model can be used to calculate molecular level populations.

4 Maser Models

4.1 Local Influences on Molecular Populations

Local influences include (i) the total number density, kinetic temperature and velocity field in the medium (ii) the concentration of SiO or H_2O molecules (iii) all radiative events due to spontaneous emission, stimulated emission and absorption with regard both to line radiation and any continuum radiation fields present in the medium and (iv) rotationally and rovibrationally inelastic collisions between SiO or H_2O molecules and other species.

Point (i) of the preceding paragraph is covered in the output of the hydrodynamic pulsation model (Humphreys et al. 1996). With regard to point (ii), the local concentrations of SiO and H_2O are unknown. Studies such as those described here can help to constrain the number densities that observations imply. Detailed chemical studies of the CSEs of M-Miras are presently lacking and should be undertaken, especially for regions close to the photosphere.

With regard to spectroscopic data for the maser molecules, the energy levels of SiO consist of vibrational levels separated by 1700 to 1800 K in energy, and each vibrational level supports a stack of rotational levels with a rotational constant of ~ 21.8 GHz (Tipping and Chackerian 1981; Lovas et al. 1981; Dunham 1932). Einstein A-coefficients for rotational and rovibrational transitions were taken from Tipping and Chackerian (1981). For accurate results, it is necessary to include energy levels for the first 40 rotational states ($0 \leq J \leq 39$) of each of the 5 lowest-lying vibrational states ($0 \leq v \leq 4$). Hence our model contains 200 states in all.

Spectroscopic data for H_2O are given in Flaud et al. (1976). Between 100 and 180 rotational levels are used, in the ground vibrational state, for both the ortho and para states of H_2O. The highest J and K_a quantum numbers involved are $J = 18$ and $K_a = 18$ for ortho H_2O and $J = 18$ and $K_a = 17$ for para H_2O. There are 888 allowed para transitions, and 894 allowed ortho transitions when 180 levels of H_2O are used. The Einstein A-values for these transitions were calculated using an algorithm described in Bayley (1985). Hyperfine splitting is not included. Only the ground vibrational state is involved in modelling H_2O masers.

Molecules are subjected not only to a line radiation field generated by the molecules themselves, but also to a continuum field. This has a contribution due to the host star and a contribution due to dust. The contribution due to the star may be approximated by a black-body function reduced by a geometrical dilution factor. In the case of SiO, it is assumed that significant quantities of dust are not found in regions occupied by SiO masers. In the supergiant VX Sgr (Greenhill et al. 1995) both SiO maser and dust emissions were detected at high spatial resolution, showing that the dust and the maser zone are separated by several stellar radii. The dust may also be patchy, with a covering factor significantly less than unity (Doel et al. 1995). The dust radiation field is therefore spatially diluted. The field can be represented in a crude manner by a black-body function at a single dust temperature modified by a wavelength-dependent factor. The dust temperature used in the black-body function is an independent parameter of the model. SiO masers are essentially collisionally pumped (Miyoshi et al. 1994; Doel et al. 1995) and results in Doel et al. (1995) show that the only significant property of the dust radiation field (or any other external field) is that it should be weak. For the majority of powerful H_2O masers, the requisite condition is again that the dust radiation field be weak.

Processes involving inelastic collisions between masing molecules and hydrogen and helium can be important, and a great deal of effort has been expended in the last two decades in order to develop methods to calculate cross sections and rate coefficients as functions of kinetic temperature for events involving rotational and rovibrational energy transfer. Despite all this detailed work there remains a very significant lack of accurate data for modelling purposes. Rovibrationally inelastic collisions in addition to purely rotationally inelastic collisions must be included in SiO maser models, since a number of different vibrational states are involved. Rovibrationally inelastic collision cross sections are typically two orders of magnitude smaller than those for rotationally inelastic collisions. They are also considerably more difficult to calculate accurately. Data for the rotationally and rovibrationally inelastic rate coefficients for SiO have been taken from Bieniek and Green (1983a,b). These data refer to collisions of SiO with He over a range of temperature 1000–3000 K and involve only a restricted range of SiO energy states. The data of Bieniek and Green therefore require extension to include collisionally induced transitions between all states in the model, for both higher rotational and vibrational states (Doel et al. 1995). Helium as a collision partner only crudely represents H_2 in $J = 0$ and still more poorly H_2 in rotationally excited states; note that at the kinetic temperatures of interest, $J = 1, 2, \ldots$ are heavily populated. In addition, a significant fraction of hydrogen may be in the form of H atoms where SiO masers are found. The H/H_2 ratio is unknown. Rate coefficients for collisions with H atoms, constituting a potentially reactive system, are likely to be markedly different from those for He or H_2. This compounded problem, of a poor knowledge of inelastic rate coefficients and of the nature of the collision partner, constitutes a significant source of uncertainty in the maser models reported here since, as noted above, SiO maser inversions are essentially collisionally pumped.

For H_2O, only purely rotationally inelastic collisional events are included. Rate coefficients for collisions involving the first 45 levels, up to level 7_{70} for ortho- and 7_{71} for para-H_2O are taken from Green et al.(1993). As in the case of SiO, these rate coefficients are for inelastic collisions with He. For collisional rate coefficients, for which either level lies above 7_{70} or 7_{71}, rate coefficients have been estimated using the parametrization adopted in Neufeld and Melnick (1991).

The local ingredients which influence the molecular populations have now been assembled. The molecular populations are governed by a set of coupled rate equations, called the kinetic master equations. These equations may be written

$$n_i \sum_{j \neq i}^{N} (R_{ij} + c_{ij}) - \sum_{j \neq i}^{N} n_j (R_{ji} + c_{ji}) = 0 \qquad (4.1)$$

where n_i and n_j are populations per sub-level of levels i and j. R_{kl} represents radiative events, transporting populations from level k to level l, c_{kl} represents collisional events and N is the total number of energy levels involved in the model. The values of R_{kl} are given by $R_{kl} = B_{kl} \langle J_{kl} \rangle + A_{kl}$ for $k > l$ and $R_{kl} = B_{kl} \langle J_{kl} \rangle$ for $k < l$. $\langle J_{kl} \rangle$ is given by

$$\langle J_{kl} \rangle = \int_0^{4\pi} \frac{d\Omega}{4\pi} \int_0^{\infty} d\nu \phi_\nu^\Omega I_\nu^\Omega \qquad (4.2)$$

where ϕ_ν^Ω is the normalized absorption and emission profile at frequency ν in direction Ω and I_ν^Ω is the intensity at frequency ν in direction Ω. Dropping superscripts and subscripts, the problem of calculating the populations of H_2O or SiO energy levels reduces to solving the set of steady-state master equations coupled to the equation of radiative transfer:

$$\frac{dI}{d\tau} = I - S \qquad (4.3)$$

where τ is the optical depth and S is the source function in any line. The source function is defined as the emission coefficient divided by the absorption coefficient for the transition of interest (Mihalas 1978). The solution of the set of coupled equations, eqns (4.1) and (4.3), yields the required molecular populations. The essential problem in finding accurate solutions is that the ambient line and continuum radiation fields depend on the molecular level populations, which in turn determine the line and continuum fields, at frequencies of the latter in the vicinity of spectral lines. This introduces the topic of the next section, which concerns the non-local problem of finding a self-consistent set of populations and line and continuum radiation fields.

4.2 Radiation Fields and Molecular Level Populations

The method most generally used for molecular systems to achieve a solution of the kinetic master equations (eqns (4.1)) coupled to the equation of radiative transfer (eqn (4.3)) involves the Large Velocity Gradient (LVG) approximation. This was originally developed to deal with radiative transfer in fast stellar winds, for example in Wolf–Rayet stars, but has been applied very freely to molecular systems, on occasion without much regard for a sound physical justification for its use. In a mass of gas supporting a velocity gradient, photons resonant with any molecular transition will pass out of resonance with that transition elsewhere in the medium, due to the Doppler shift associated with the bulk velocity field in the medium. Thus resonant photons interact only locally with molecules in the presence of a large velocity gradient. The attraction of the LVG approximation is that it allows the formulation of analytic expressions for the photon escape probability in terms of the bulk velocity field in the gas.

We have studied both SiO and H_2O masers using the LVG approximation. The LVG approximation is relatively rapid in yielding solutions of the master equations compared for example with the exact Accelerated Lambda Iteration (ALI) methods described in Randell et al. (1995) and Jones et al. (1994). Speed of computation is an important aspect in the context of the study of the CSE. Many radial positions within the CSE need to be sampled to obtain a representative measure of the maser output that may be associated with any phase of the host star pulsation. A preliminary study for SiO (Doel et al. 1995) used the LVG approximation in its very simplest form, with a logarithmic velocity gradient $d\ln v/d\ln r = 1.0$, implying that the velocity gradient is spherically symmetrical at every point in the medium. The photon escape probability is then given by $\beta = [1 - \exp(-\tau_s)]/\tau_s$, where τ_s is referred to as the Sobolev optical depth and

$$\tau_s = \frac{c^3}{8\pi\nu_{ul}^3}A_{ul}\frac{[n_u - n_l g_u/g_l]}{dv/dr} \qquad (4.4)$$

where dv/dr is the velocity gradient and n_u and n_l are populations per sub-level of the upper and lower energy levels, and g_u and g_l are the statistical weights of levels in the transition of frequency ν_{ul} and with Einstein A-value A_{ul}. SiO population inversions were found to form readily in gas at 1500 K with number densities around 5×10^9 cm^{-3}. A number of important qualitative characteristics of SiO masers, mentioned in Section 2, were reproduced using this model when maser propagation was included (see Section 4.3). Doel et al. (1995) showed that SiO maser inversions formed strongly under conditions which hydrodynamic pulsation models of M-Miras predicted for regions close to the photosphere. This was therefore a useful precursor to the work specifically directed to M-Mira CSEs, which is described in more detail in Section 5.

In a separate study on H_2O masers, the use of exact rather than LVG methods was adopted in order to obtain self-consistent populations and radiation fields (Yates et al. 1997). The accelerated lambda iteration method (ALI), widely used in stellar atmosphere models and applied in earlier work to analyse OH absorption data and OH megamaser emission (Jones et al. 1994; Randell et al. 1995), was used to obtain level populations of H_2O molecules over a range of conditions which included those found in the CSEs of M-Miras. Solutions of the coupled eqns (4.1) and (4.3) were obtained with ALI applied in a highly modified version of the MULTI program based upon the work of Scharmer and Carlsson (1985). For reviews and further details of the operator perturbation methods involved, see Scharmer and Carlsson (1985), Kalkofen (1987), Rybicki and Hummer (1991) and references therein. A description of the operation of the code may be found in Jones et al. (1994) and Randell et al. (1995).

There are very marked advantages in using ALI over LVG methods. The solutions are precise, though recall that the equations themselves are not exact. There is no requirement that the medium contain supersonic velocity shifts. In addition, structure may be built into the model: number density, temperature, velocity and other parameters may be varied at will as functions of position in the medium subject to geometrical constraints of the model. The ability to build structure into the medium promises to be important in future work using hydrodynamic pulsation model data as input into maser models.

4.3 Maser Radiative Transfer

Maser radiation is partially coherent and thus the manner in which maser radiation interacts with molecular populations is distinct from that in which gaussian light interacts with molecular populations. This becomes important as maser radiation begins to modify significantly population inversions in its passage through the amplifying gas, that is, as maser beams become sufficiently intense to cause saturation. There is considerable observational and theoretical evidence that SiO and H_2O masers are indeed saturated in the CSE of M-Miras; see for example Doel et al. (1995). When masers propagate into the saturated regime, the effects of saturation couple to the kinetic scheme which formed the unsaturated inversions. Level populations are modified, even when these levels are not directly involved in a maser inversion. Masers involving different transitions are therefore either directly or indirectly coupled and compete for inversions present in the medium, leading to the phenomenon of "competitive gain" (Field 1985; Field and Gray 1988). For example, SiO in any one vibrational state has a simple stack of rotational levels supporting a set of population inversions and there develops a remarkable cycle of pushing and pulling between the inversions as saturated masers propagate through the medium Doel et al. (1995). Observations indicate that masers fill only a very small proportion of the total volume of the maser zone (e.g. Greenhill et al. 1995) and the assumption is made in our models that saturating masers are sufficiently spatially confined that they do not affect the general pumping cycle elsewhere in the maser zone.

Maser saturation and coupling of the maser radiation to the kinetic master equations are treated with a semiclassical formalism, in which the radiation is treated according to Maxwell's equations but the response of the molecular ensemble is calculated using quantum mechanical density matrix theory (Field and Richardson 1984; Field 1985; Field and Gray 1988). In our models, all effects of saturation and competitive gain are included in the propagation of a maser beam through the gas containing population inversions. The masing zone is assumed to amplify a black-body background at the appropriate wavelength at the local kinetic temperature of the maser zone for SiO masers and a background field generated by dust in the CSE for H_2O masers. Calculations yield the factor of amplification of this background as a function of frequency within the maser line, supplying in turn maser lineshapes and maps of maser emission as outlined in Section 5.

5 SiO and H_2O Masers in the Model Circumstellar Envelope

The theoretical apparatus outlined in Sections 4.1, 4.2 and 4.3 has been used in conjunction with data for the physical conditions in the CSE, generated by the hydrodynamic pulsation model outlined in Section 3, to synthesize SiO and H_2O maser maps and spectra. Both the SiO and H_2O maser calculations reported in this section were performed using the LVG approximation. For both the SiO and H_2O maser calculations it is necessary to use a somewhat more sophisticated means of calculating the photon escape probability than that mentioned in Section 4.2 in connection with Doel et al. (1995). Radiative transfer in the CSE takes place in a radially expanding or contracting medium with a velocity gradient having both radial and tangential components. The photon escape probability is a function of the direction of the ray within the medium and must therefore be averaged over all directions. General expressions for the angle-averaged photon escape probability in such a system, using the relevant local values of the velocity gradient, are given in Humphreys et al. (1996) and are used here. In addition the velocity field is not monotonic and strictly the theory of Rybicki and Hummer (1978) should be used, since photons may encounter more than one common velocity surface in their passage through the CSE. It turned out to be computationally too expensive to include this; the additional errors incurred should not be severe, as discussed in Humphreys et al. (1996).

The synthetic map for $v = 1$ $J = 1\text{--}0$ 43 GHz SiO masers at model phase $= 0.1$ is shown in Fig. 2, in which maser emission has been summed over all velocity channels between -10 and $+10$ km s^{-1}. This map was obtained using 1500 values of radial distance as sites of maser action, where values of radial distance were chosen at random between an inner boundary of the stellar radius $(1.7 \times 10^{11}\text{m})$ and 5 × the stellar radius. The contours represent the degree of amplification associated with maser action. The lowest contour corresponds to a factor of $\sim$40 and the highest to a factor of $\sim$620. There are some very sharp spikes of amplification up to a maximum of $\sim$2200 which have been smoothed out in the interpolation routine used to produce Fig. 2. The ring of masers seen in Fig. 2 forms because the presence of strong radial velocity fields tends to favour tangential amplification of SiO masers. A synthetic map in Humphreys et al. (1996), for a model phase of zero, showed very similar structure, again in the form of a ring of masers around the host star at a distance of $\sim$0.8 stellar radii from the photosphere. Calculated maps closely resemble observational maps obtained at the Very Long Baseline Array (VLBA) by Diamond and co-workers,

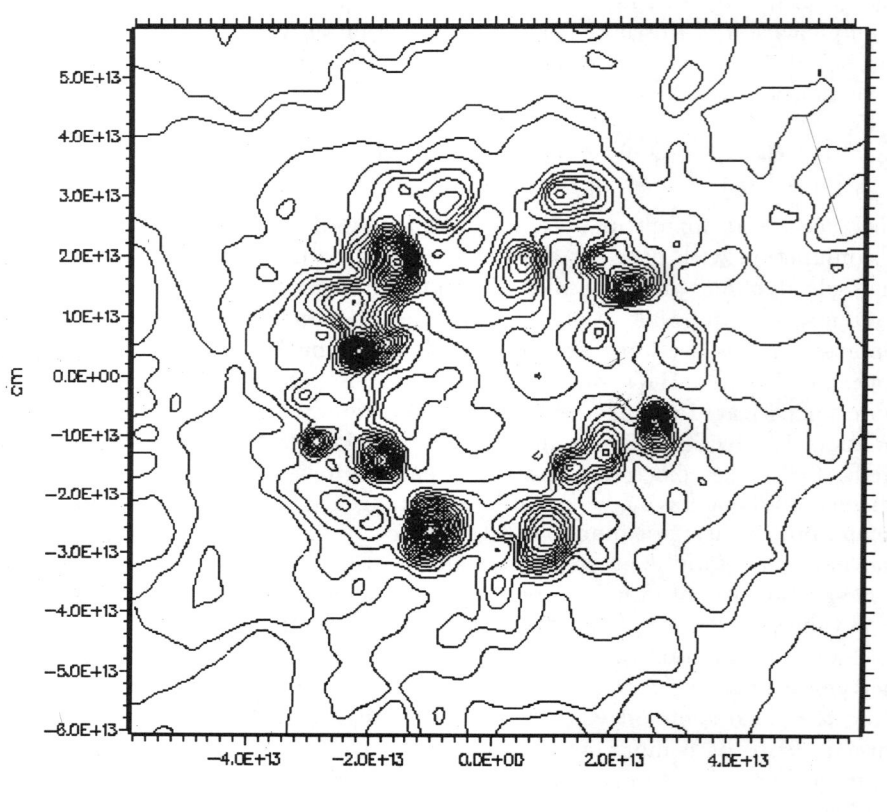

FIG. 2. A synthetic image of the $v = 1$ $J = 1\text{-}0$ 43 GHz SiO maser emission in the CSE of the model M-Mira, at model phase = 0.1. The lowest contour corresponds to a factor of $\sim$40 and the highest to a factor of $\sim$620. Emission is integrated from -10 km s^{-1} to $+10$ km s^{-1} about the stellar velocity. Further details may be found in Section 5.

for the M-Mira TX Cam (Humphreys et al. 1996). In particular the diameter of the ring of maser spots closely matches that observed. In Fig. 2 for phase = 0.1, the masers lie at distance of $\sim$0.5 stellar radii from the photosphere, a little closer to the photosphere than for phase zero. The choice of 1500 random positions of possible maser sites, both here and in Humphreys et al. (1996), stemmed from the finding that this number reproduces the appearance of the VLBA observations in terms of the number of bright maser spots around the host star, that is, about 10 to 15 such spots.

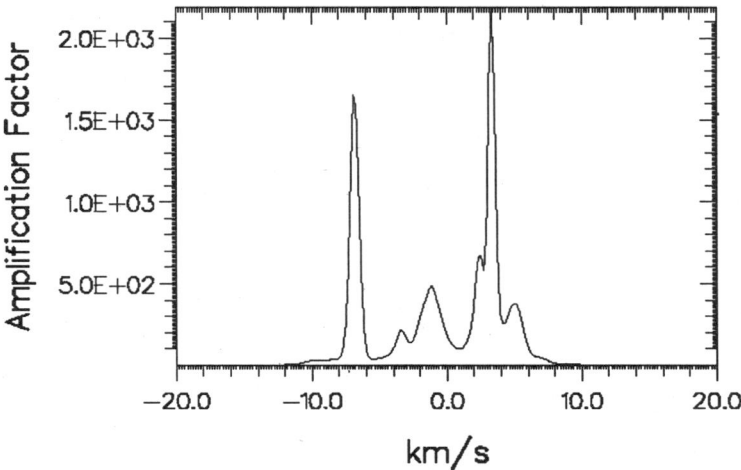

FIG. 3. The calculated spectrum of the $v = 1$ $J = 1$–0 43 GHz SiO maser emission corresponding to the 43 GHz maser map shown in Fig. 2. The stellar velocity is represented by 0.0 km s^{-1}.

The SiO maser lineshape corresponding to the map in Fig. 2 is shown in Fig. 3. The line shows the multiple and irregular structure and width characteristic of SiO 43 GHz maser emission. At this phase the emission turns out to be weaker at the stellar velocity (0.0 km s^{-1}) than at red and blue shifted velocities. The overall emission is in fact rather weak at a model phase of 0.1. Calculations performed at 20 different phases throughout the stellar cycle (Humphreys et al. 1997) show that strong emission occurs at the stellar velocity, as observed.

We now turn to results for H$_2$O masers. A synthetic map for 22 GHz H$_2$O masers at model phase = 0.1 is shown in Fig. 4, summed over all velocity channels between -10 and $+10$ km s^{-1}. Again 1500 radial positions for potential maser sites were chosen at random, in this case lying between the stellar radius and 30 times the stellar radius.

Figure 4 shows that masers are found disposed around the host star forming irregular clumpy structures at distances between $\sim 10^{13}$ and $\sim 10^{14}$ cm. The contours in Fig. 4 represent maser emissivity and cover a range of 2.5×10^{-13} ergs s^{-1} cm^{-2} Hz^{-1} sr^{-1} to 2.2×10^{-9} ergs s^{-1} cm^{-2} Hz^{-1} sr^{-1}. Masers are represented in terms of emissivity rather than amplification factor, as in Fig. 2 for SiO, because of the large range of radial distance over which H$_2$O maser emission is computed. The amplified background dust emission varies in brightness significantly over this range of radial distance. As in the case of SiO, there are some sharp spikes of emission up to a maximum of $\sim 6 \times 10^{-9}$ ergs

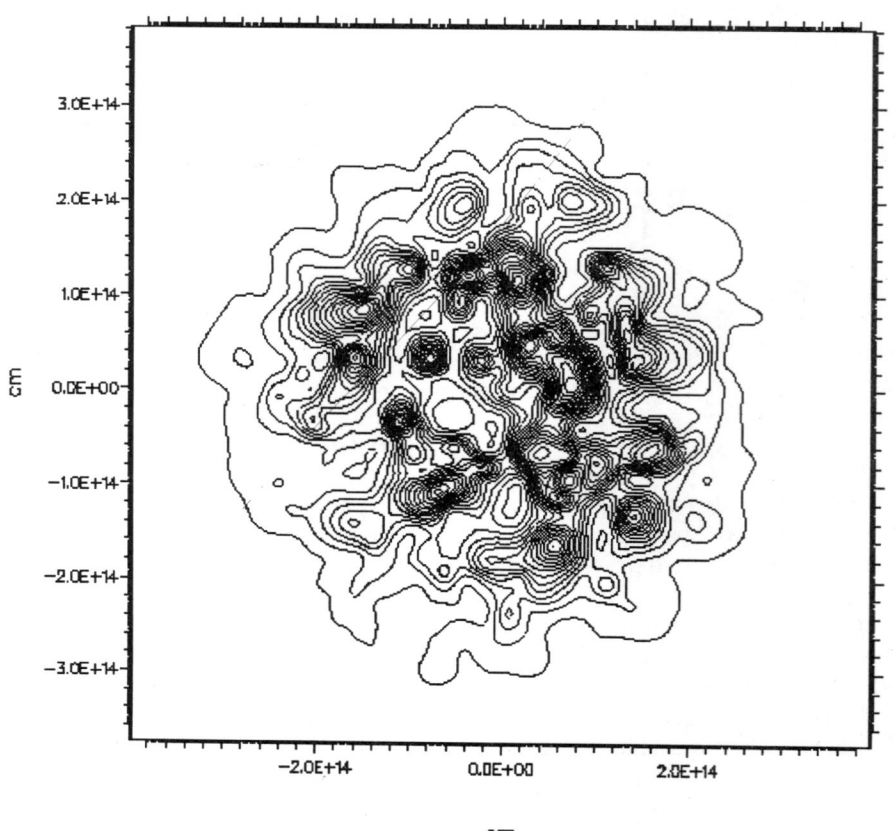

FIG. 4. A synthetic image of the 22 GHz H_2O maser emission in the CSE of the model M-Mira, at model phase = 0.1. The contours represent maser emissivity and cover a range of 2.5×10^{-13} ergs s^{-1} cm^{-2} Hz^{-1} sr^{-1} to 2.2×10^{-9} ergs s^{-1} cm^{-2} Hz^{-1} sr^{-1}

s^{-1} cm^{-2} Hz^{-1} sr^{-1} which have been smoothed out in the interpolation routine used to produce Fig. 4. A qualitative comparison with observations may be made with reference to the H_2O maser map for the supergiant S Persei shown in Fig. 1. The similarity of structure between the simulation in Fig. 4 and Fig. 1 is apparent, with S Persei showing a thick roughly circular tangentially amplifying maser zone. The scale of the maser shell in Fig. 1 is ∼50 times greater than in Fig. 4, in keeping with the relative scale of M-Miras and supergiants. The H_2O maser lineshape corresponding to the map in Fig. 4 is shown in Fig. 5. The line shows the multiple and irregular structure and width characteristic of some observed H_2O 22 GHz maser emission, for example in U Her, R Crt or RX Boo (Yates and Cohen 1996; Yates et al. 1995). Calculations are currently in progress to cover the full stellar pulsation cycle for water masers.

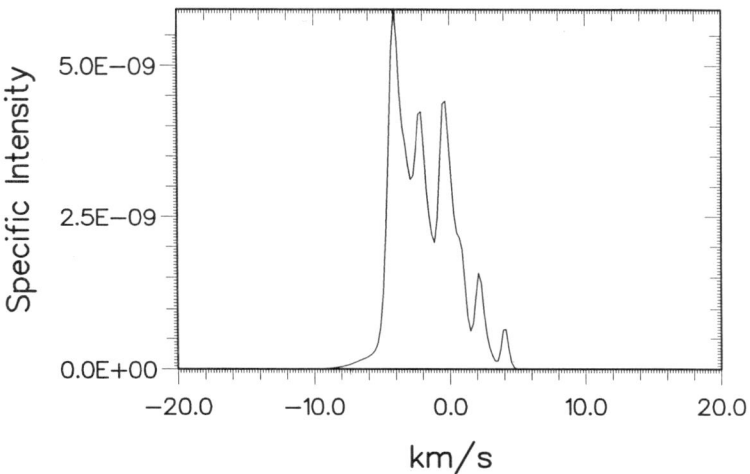

FIG. 5. The calculated spectrum of the 22 GHz H_2O maser emission corresponding to the 22 GHz maser map shown in Fig. 4. The stellar velocity is represented by 0.0 km s^{-1}.

6 Concluding Comments

The most significant conclusion which may be drawn at this stage is that current models of circumstellar maser emission coupled with hydrodynamic pulsation models perform remarkably well in reproducing the available observational data. This is particularly true of the maser morphology for SiO masers, and also looks very promising for H_2O masers. Conditions of number density, temperature and velocity field predicted from hydrodynamic pulsation models place masers in the correct position with respect to the stellar photosphere. These results show that significant progress has been made in understanding the mechanism of mass loss and of the formation of the CSEs of AGB stars.

The following suggestions are made for future work. Theoretical models are required showing the relationship between the model phase and the optical phase in order that calculated and observed phase lags between the stellar and maser emission may be compared. The hydrodynamic pulsation model should in future incorporate some chemical modelling both of SiO and H_2O formation and of the

H/H$_2$ ratio, the latter in the region close to the photosphere. In addition, a time-dependent description of silicate, that is, olivine dust formation, should be introduced, along the lines on which carbon-based dust formation has been introduced into hydrodynamic pulsation models of carbon-rich stars (Hoefner et al. 1996; Sedlmayr and Dominik 1995). The present SiO maser model suffers from a poor knowledge of the rate coefficients for collisional energy transfer; the computation of rate coefficients for energy transfer between H$_2$ and SiO and H and SiO should therefore be a high priority. Accelerated lambda iteration (ALI) methods should be used to replace LVG, since as noted in Section 4.3, ALI methods have the important characteristic that they can incorporate the velocity, temperature and number density structure of the CSE explicitly into the calculation of the populations of molecular energy levels. ALI methods require development for a spherical geometry for molecular systems. One should look forward to a global model of the CSE, incorporating ALI SiO, H$_2$O, and OH maser models, providing results over distances from the photosphere spanning three orders of magnitude. The outstanding observational requirements are for a set of 43 GHz SiO maser VLBI images disposed at various phases within a stellar period and medium baseline (MERLIN) H$_2$O maser images at 22 GHz, gathered at the same epoch, for a suitable long period Mira variable. OH images at 1612 MHz and in the main lines at 1665 and 1667 MHz would also be a valuable addition.

7 Acknowledgements

The work described here has been performed in collaboration with a number of researchers at the University of Bristol, Jodrell Bank, Iowa State University, the National Radioastronomy Observatory and elsewhere. A major input has been through the work of Malcolm Gray (theory and observations), Robert Doel (theory), Elizabeth Humphreys (theory and observations), Jeremy Yates (theory and observations), Kevin Jones (theory) and Robin Walker (theory) at the University of Bristol, School of Chemistry and Department of Physics. The input from George Bowen, Iowa State University, has been vital to the project. The maps of SiO masers made by Phil Diamond and co-workers using the VLBA have also been essential. I should also like to thank Jim Cohen (Jodrell Bank) and Frank Pijpers (Institute of Physics and Astronomy, University of Aarhus) for valuable discussions. Thanks are also due to the Directors and Staff at Jodrell Bank and the James Clerk Maxwell Telescope for the use of their facilities. I should like to acknowledge Anita Richards, Jim Cohen and Richard Hills for allowing me to use their 22 GHz H$_2$O maser MERLIN image of S Persei. The Collaborative Computational Project no. 7 (CCP7), funded by PPARC, should be gratefully acknowledged for providing the original copy of the MULTI code, and for useful advice in its implementation.

Bibliography

1. Bayley M, (1985), PhD thesis, University of Bristol.

2. Bedijn, P. J. (1987) *Astron. Astrophys.*, **186**, 136.
3. Bessell, M. S., Scholz, M. and Wood, P. R. (1996) *Astron. Astrophys.*, **307**, 481.
4. Bieniek, R. J. and Green, S. (1983a) *Ap.J.*, **265**, L29.
5. Bieniek, R. J. and Green, S. (1983b) *Ap.J.*, **270**, L101.
6. Booth, R. S., Kus, A. J., Norris, R. P. and Porter, N. D. (1981) *Nature*, **290**, 382.
7. Bowen, G. (1988) *Ap.J.*, **329**, 299.
8. Bowen, G. (1989) in *NATO Advanced Workshop on "Numerical Modelling of Nonlinear Stellar Pulsations. Problems and Prospects"*. Les Arcs, France.
9. Bowen, G.and Willson, L. A. (1991) *Ap.J.*, **375**, L53.
10. Bowers, P. F. and Hagen, W. (1984) *Ap.J.*, **285**, 637.
11. Cohen, R. J. (1989) *Rep. Prog. Phys.*, **52**, 881.
12. Diamond, P. J., Kemball, A. J., Junor, W., Zensus, A., Benson, J. and Dhawan, V. (1994) *Ap.J.*, **430**, L61.
13. Doel, R. C., Gray, M. D., Humphreys, E. M. L., Braithwaite, M. F. and Field, D. (1995) *Astron. Astrophys.*, **302**, 797.
14. Dunham, J. L. (1932) *Phys.Rev.*, **41**, 721.
15. Elitzur, M. 1992, *Astronomical Masers*, Kluwer Academic Publications, Dordrecht.
16. Engels, D. and Heske, A. (1989) *Astron. Astrophys.Suppl.*, **81**, 323.
17. Field, D. (1985) *Mon.Not.R.Astr.Soc.*, **217**, 1.
18. Field, D. and Richardson, I. M. (1984) *Mon.Not.R.Astr.Soc.*, **211**, 799.
19. Field, D. and Gray, M. D. (1988) *Mon.Not.R.Astr.Soc.*, **234**, 353.
20. Flaud, J. M., Camy-Peyret, C. and Maillard, J. P. (1976) *J.Mol.Phys.*, **32**, 499.
21. Fleischer, A. J., Gauger, A. and Sedlmayr, E. (1991) *Astron. Astrophys.*, **242**, L1.
22. Fleischer, A. J., Gauger, A. and Sedlmayr, E. (1992) *Astron. Astrophys.*, **266**, 321.
23. Gray, M. D., Ivison, R. J., Yates, J. A., Humphreys, E. M. L., Hall, P. J. and Field, D. (1995) *Mon.Not.R.Astr.Soc.*, **277**, L67.
24. Green, S., Maluendes, S. and McLean, A. D. (1993) *Ap.J.Suppl.*, **85**, 181.
25. Greenhill, L. J., Colomer, F., Moran, J. M., Backer, D. C., Danchi, W. C. and Bester, M. (1995) *Ap.J.*, **449**, 365.
26. Hoefner, S., Fleischer, A. J., Gauger, A., Feuchtinger, M. U., Dorfi, E. A., Winters, J. M. and Seldmayr, E. (1996) *Astron. Astrophys.*, **314**, 204.
27. Humphreys, E. M. L., Gray, M. D., Yates, J. A., Field, D., Bowen, G. and Diamond, P. J. (1996) *Mon.Not.R.Astr.Soc.*, **282**, 1359.
28. Humphreys, E. M. L., Yates, J. A., Gray, M. D. and Field, D. (1997) *Mon.Not.R.Astr.Soc.*, **287**, 663.

29. Jewell, P. R., Snyder, L. E., Walmsley, C. M., Wilson, T. L. and Gensheimer, P. D. (1991) *Astron. Astrophys.*, **242**, 211.
30. Joergensen, U. G. and Johnson, H. R. (1992) *Astron. Astrophys.*, **265**, 168.
31. Jones, K. N., Field, D., Gray, M. D. and Walker, R. N. F. (1994) *Astron. Astrophys.*, **288**, 581.
32. Kalkofen, W. (1987) *Numerical Radiative Transfer*, Cambridge University Press, Cambridge.
33. Lovas, F. J., Maki, A. G. and Olsen, W. B. (1981) *J.Mol.Spectr.*, **87**, 449.
34. Martinez, A., Bujarrabal, V. and Alcolea, J. (1988) *Astron. Astrophys.Suppl.*, **74**, 273.
35. Mihalas, D. (1978) *Stellar Atmospheres*, WH Freeman, San Francisco.
36. Miyoshi, M., Matsumoto, K., Kameno, S., Takaba, H. and Takahiro, I. (1994) *Nature*, **371**, 395.
37. Neufeld, D. A. and Melnick, G. J. (1991) *Ap.J.*, **368**, 215.
38. Pijpers, F. P. and Hearn, A. G. (1989) *Astron. Astrophys.*, **209**, 198.
39. Pijpers, F. P., Pardo, J. R. and Bujarrabal, V. (1994) *Astron. Astrophys.*, **286**, 501.
40. Randell, J., Field, D., Jones, K. N., Yates, J. A. and Gray, M. D. (1995) *Astron. Astrophys.*, **300**, 659.
41. Richards, A. M. S., Yates, J. A. and Cohen, R. J. (1996) *Mon.Not.R.Astr.Soc.*, **282**, 665.
42. Rybicki, G. B. and Hummer, D. G. (1978) *Ap.J.*, **219**, 654.
43. Rybicki, G. B. and Hummer, D. G. (1991) *Astron. Astrophys.*, **245**, 171.
44. Scharmer, G. B. and Carlsson, M. (1985) *J. Comp. Phys.*, **59**, 56.
45. Sedlmayr, E.and Dominik, S. (1995) *Astro.Sp.Sci.*, **73**, 211.
46. Tipping, R. H. and Chackerian, C. (1981) *J.Mol.Spectr.*, **88**, 352.
47. Yates, J. A. and Cohen, R. J. (1996) *Mon.Not.R.Astr.Soc.*, **278**, 655.
48. Yates, J. A., Cohen, R. J. and Hills, R. E. (1995) *Mon.Not.R.Astr.Soc.*, **273**, 529.
49. Yates, J. A., Field, D. and Gray, M. D. (1997) *Mon.Not.R.Astr.Soc.*, **285**, 303.

15
Molecular Synthesis in the External Envelopes of AGB Stars

T. J. Millar
Department of Physics, UMIST

1 Introduction

Observations of molecules in the circumstellar envelopes (CSEs) of late-type asymptotic giant branch (AGB) stars provide a natural laboratory in which to apply and test astrochemical models. Such regions, particularly C-rich (i.e. C/O > 1) CSEs, are rich in molecular species, have a known, usually spherically symmetric, geometry, are bathed in the interstellar radiation field, and undergo a simple expansion at constant velocity in the outer CSE, so that single-dish observations are able to give some information about the spatial locations of molecules from the line profiles (Olofsson 1997). In addition, for an increasing number of molecules, interferometer measurements are available and give detailed maps of the molecular distributions in the CSE (Bieging and Tafalla 1993; Gensheimer et al. 1995; Lucas et al. 1995; Guélin et al. 1996). Observations and models of the CSEs can be used as an input into studies of the later stages of stellar evolution when the AGB stars evolve through the proto-planetary nebula phase to planetary nebulae, a subject discussed in Chapter 16. Table 1 shows a list of molecules observed to date in CSEs.

For a spherically symmetric outflow at a constant expansion speed, v, the number density of H_2 molecules, $n(r)$, the outward radial column density, $N(r)$, the outward radial extinction at 100 nm of the external radiation field, $A(r)$, and the kinetic temperature, $T(r)$, are given by

$$
\begin{aligned}
n(r) &= 10^5 \dot{M}_{-5}/r_{16}^2 v_{15} \text{ cm}^{-3} \\
N(r) &= 10^{21} \dot{M}_{-5}/r_{16} v_{15} \text{ cm}^{-2} \\
A(r) &= 5.4 \dot{M}_{-5}/r_{16} v_{15} \text{ mag} \\
T(r) &= 100 r_{16}^{-0.7} \text{ K}
\end{aligned}
$$

where $\dot{M}_{-5}$ is the mass-loss rate measured in units of $10^{-5} M_\odot$ yr^{-1}, r_{16} is the radial distance in units of 10^{16} cm, and v_{15} is the expansion speed in units of 15 km s^{-1}. The temperature at r_{16} depends on $\dot{M}$, such that winds having lower mass-loss rates have higher temperatures, and is typically 100 K for $\dot{M} = 10^{-5} M_\odot$ yr^{-1} and 300 K for $\dot{M} = 10^{-7} M_\odot$ yr^{-1} (Kastner 1992; Millar and Olofsson 1993).

Table 1 Molecules detected in CSEs

Carbon-rich CSEs							
H_2	C_2H	$l\text{-}C_3H$	C_4H	C_5H	C_6H	C_7H	C_8H
CO	C_2S	C_3S					
CN	HCN	C_3N	HC_3N		HC_5N		HC_7N
CP	HNC	H_2CN	HC_2NC	CH_3CN			HC_9N
CS	MgCN	NH_3	$c\text{-}C_3H_2$				
SiC	SiC_2		SiC_4				
SiN	C_3		C_5				
SiO			H_2C_3	H_2C_4			
SiS		C_2H_2		C_2H_4			
NaCl	NaCN						
AlCl	MgNC		CH_4				
AlF			SiH_4				
KCl							
Oxygen-rich CSEs							
CO	H_2O	H_2CO					
CN	HCN	NH_3					
OH	HNC						
SO	SO_2						
CS	OCS						
SiO	H_2S						
SiS							

Parent molecules can be formed in LTE or in reactions behind pulsating shocks just inside the dust formation zone or in dust–gas interactions (Willacy and Cherchneff 1998) and, through a dust-driven outflow, eventually reach a zone in which the density is relatively large but in which the extinction is low. Parent molecules are photodissociated and photoionized and daughter products can undergo reaction to form further species. In this respect, chemical modelling of CSEs is similar to that in photodissociation regions (PDRs), although the UV radiation fields are much weaker in CSEs, and in cometary comae where the parents arise from the evaporation of nuclear ices by solar radiation.

In the following section we discuss the chemistry around C-rich CSEs. Section 3 concerns the O-rich CSEs whilst Section 4 gives a brief summary of future directions. A recent discussion of CSE chemistry, including that in the inner envelope, has been given by Glassgold (1996).

2 Carbon-rich CSEs

Around 50 molecular species have been detected in the CSEs around C-rich AGB stars (see Table 1), although about one half of them have been detected only in

IRC+10216, a nearby star with a high mass loss rate. Most of the detections have been at millimetre wavelengths and include the cyanopolyyne chains, $HC_{2n+1}N$, $n = 1$–4, the hydrocarbon chain radicals, C_nH, $n = 2$–8, and other related chains including C_2S, C_3S, C_4Si, H_2C_3 and H_2C_4. The methyl group is detected in only one molecule, CH_3CN, while because of their extremely large electric dipole moments, a number of relatively low abundance halide species, NaCl, KCl, AlCl and AlF, have been observed in IRC+10216. Metal cyanides, NaCN and MgCN, as well as magnesium isocyanide, MgNC, and several diatomic molecules containing silicon, SiO, SiS, SiC and SiN, have also been observed. In addition to these radio detections, a number of molecules which possess no allowed radio transitions, C_3, C_5, CH_4, SiH_4, C_2H_2 and C_2H_4, have been detected via infrared absorption measurements.

Abundances are difficult to determine from the measurements because of radially varying conditions in the CSEs; millimetre observations generally result in beam- or source-averaged column densities while the infrared observations give a line-of-sight column density and often probe different regions of the CSE than do the millimetre emission lines. Olofsson (1997) has discussed the derivation of column densities and presents typical values. As he has stressed, abundance *ratios* are more accurately determined and clearly show the effects of chemical processing. For example, the abundance ratios $AlCl:KCl:NaCl = 1:0.01:0.01$ whereas the elemental abundance ratios are $Al:K:Na = 1:21:0.6$.

For C-rich objects, the abundances of parent molecules are well determined in general, although some likely parents, for example N_2, have not been observed. In addition to H_2 and CO, which both self-shield and are photodissociated only at radii which are too large to be important chemically, parents include C_2H_2 and HCN, which provide the bulk of the reactive carbon and nitrogen, N_2, NH_3, SiH_4 and SiS. Cosmic ray ionization and cosmic-ray-induced photodissociation also play a part in the destruction of parents. The former is important only for H_2 and He which produce H_3^+ and He^+ and can help drive an ion–molecule chemistry. However, this process is only effective deep in the envelope, in regions not penetrated by the interstellar UV radiation field. For H_3^+, which is destroyed in proton transfer reactions with abundant neutrals, one can show that the fractional abundance at radius r, $x(H_3^+, r)$, is given by

$$x(H_3^+, r) = \zeta/[kn(r)x(CO, r)].$$

Since ζ, k and $x(CO, r)$ are constants, $x(H_3^+, r) \propto 1/n(r) \propto r^2$ for an assumed constant mass-loss rate. Thus the fractional abundance, and hence ionization, is small deep in the envelope and becomes large only in regions in which UV photons provide a much greater degree of ionization, primarily through the photoionization of C_2H_2. Ionization can also be produced by positrons formed in the β-decay of radioactive ^{26}Al (Glassgold 1994) but the ionization rate calculated is similar to that inferred for low-energy cosmic rays and is probably less than the standard interstellar value for ζ, given the non-detection of HCO^+ in IRC+10216 (Glassgold 1994). Early chemical models of C-rich CSEs were published by Glassgold *et al.* (1986), Nejad and Millar (1987) and Glassgold *et al.* (1987). In these, hydrocarbons build up as a result of the photodissociation and photoionization of acetylene, C_2H_2,

$$C_2H_2 + h\nu \longrightarrow C_2H + H \tag{2.1}$$

$$C_2H_2 + h\nu \longrightarrow C_2H_2^+ + e \tag{2.2}$$

followed by ion–neutral reactions, such as

$$C_2H_2^+ + C_2H_2 \longrightarrow C_4H_2^+ + H_2 \tag{2.3}$$

$$C_2H_2^+ + C_2H_2 \longrightarrow C_4H_3^+ + H \tag{2.4}$$

and dissociative recombination with electrons,

$$C_4H_2^+ + e \longrightarrow C_4H + H \tag{2.5}$$

$$C_4H_3^+ + e \longrightarrow C_4H_2 + H \tag{2.6}$$

and by neutral–neutral reactions, such as

$$C_2H + C_2H_2 \longrightarrow C_4H_2 + H. \tag{2.7}$$

These early schemes were based on models of interstellar chemistry, specifically derived for the dark dust cloud TMC-1, and gave reasonably good agreement with observations of the smaller hydrocarbons, particularly the abundance and shell size of C_2H. However, they had trouble in building enough of the larger molecules since to do so requires a series of ion–neutral reactions taking place before dissociative recombination occurs. Since this series of reactions takes a finite time, in which abundances are diluted by the expansion flow, two problems arise. Firstly, the collision times increase as the radius increases and become longer than the expansion time, and secondly, the ionization fraction also increases with radius, thereby making dissociative recombination rather than chain growth more likely.

Howe and Millar (1990) recognized the potential of neutral–neutral reactions in synthesizing large molecules and used the reaction

$$CN + C_2H_2 \longrightarrow HC_3N + H \tag{2.8}$$

to reproduce the observed beam-averaged column density of HC_3N. They suggested that similar reactions might build the larger cyanopolyynes:

$$CN + C_4H_2 \longrightarrow HC_5N + H \tag{2.9}$$

$$CN + C_6H_2 \longrightarrow HC_7N + H. \tag{2.10}$$

Because of the rapid loss of neutrals through photodissociation, efficient schemes based on such chemistry require that the reactions should (i) be fast at low temperatures (<50 K), and (ii) preferably involve direct daughter products reacting with parent neutrals. While reaction (2.8) has been studied in the laboratory and fulfils these two requirements, reactions (2.9) and (2.10), which have not been investigated experimentally, do not.

Cherchneff et al. (1993) suggested that a variant of reaction (2.8) could produce the large cyanopolyynes:

$$C_3N + C_2H_2 \longrightarrow HC_5N + H \qquad (2.11)$$
$$C_5N + C_2H_2 \longrightarrow HC_7N + H. \qquad (2.12)$$

These reactions, although not yet studied in the laboratory, should be exothermic. However, C_3N and C_5N are not direct daughter species and the column densities derived, on the assumption of rapid rate coefficients, for HC_5N and HC_7N, are less than those observed, although those of C_3N and HC_3N agree well (Cherchneff et al. 1993).

Because of these problems, Cherchneff and Glassgold (1993) and Millar and Herbst (1994) extended the neutral chemistry to include processes based on those neutral–neutral reactions measured to be extremely rapid at low temperatures. These reactions include (2.8), which has a rate coefficient of the form (2.9) $10^{-10}(T/300)^{-0.53}$ cm^3 s^{-1} measured down to 25 K by Sims et al. (1993). On this basis, Millar and Herbst (1994) adopted rate coefficients of $2.6 \times 10^{-11}(T/300)^{-1}$ cm^3 s^{-1} for reactions of the form

$$CN + C_{2n}H_2 \longrightarrow HC_{2n+1}N + H \qquad (2.13)$$

for $n = 2$–4.

The other major primary daughter of C_2H_2 is C_2H. Reaction (2.7) forms diacetylene and has a measured rate coefficient of $\sim 5 \times 10^{-11}$ cm^3 s^{-1} at 300 K. Polyacetylenes may be synthesized via a generalization of reaction (2.7)

$$C_2H + C_nH_2 \longrightarrow C_{n+2}H_2 + H. \qquad (2.14)$$

The reactions of C atoms with hydrocarbons are now known to be fast at room temperature (Haider and Husain 1993; Clary et al. 1994) and proceed as

$$C + C_nH_m \longrightarrow C_{n+1}H_{m-1} + H \qquad (2.15)$$

with rate coefficients proportional to $n^{1/3}$ (Clary et al. 1994).

Millar and Herbst (1994) included these latter sets of reactions and also reactions involving atomic nitrogen

$$N + C_nH \longrightarrow C_nN + H \qquad (2.16)$$

as well as

$$C_{2n}H + HCN \longrightarrow HC_{2n+1}N + H \qquad (2.17)$$
$$C_2H + HC_{2n+1}N \longrightarrow HC_{2n+3}N + H \qquad (2.18)$$

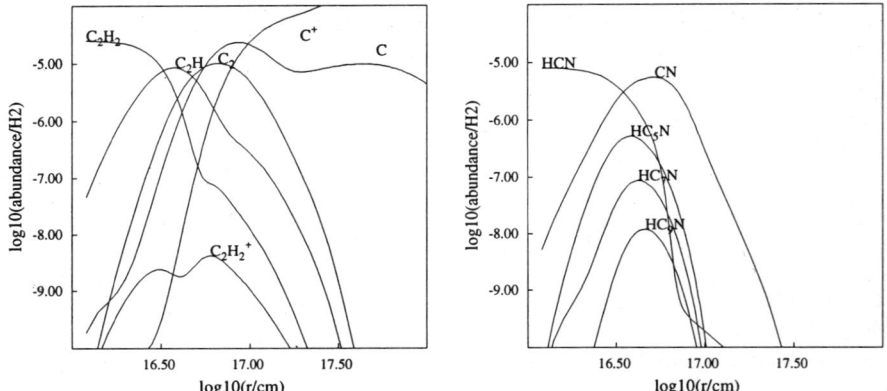

FIG. 1. Molecular radial distributions, relative to H_2, in IRC+10216 for an input abundance of 2.5×10^{-5} for C_2H_2 (Millar and Herbst 1994).

and considered all reactions up to and including the formation of HC_9N. The results of model calculations for IRC+10216 by Millar and Herbst (1994) are shown in Fig. 1, which presents radial distributions of the fractional abundances of several hydrocarbon molecules for various assumptions about the abundance of injected C_2H_2 and with the intensity of the incident UV field taken to be the average interstellar UV field. The calculations also allow one to predict column densities and peak abundance radii for comparison with observations. In general, the agreement between theory and observation is remarkably good—see, for example, Fig. 2 from Guélin et al. (1997), which shows this for the hydrocarbon chains C_nH, $n = 2$–8—and gives credence to the overall description of the chemistry. However, detailed discrepancies do remain. The Millar and Herbst model predicts that C_2H, C_3H and C_4H peak at different distances from the star whereas the observations do not show this (Guélin et al. 1993). However, the differences are small and are sensitive to the choice of the incident radiation field and, in particular, the photodissociation rates, which affect the sizes of the distributions, are not well determined. It is important to check such measurements carefully.

In an alternative formation route to large molecules they result from the breakdown of even larger species released from grains or grain mantles (Jura and Kroto 1990). While such a mechanism can take place in shocks, and hence may be much more important in the protoplanetary nebula (PPN) phase (see Chapter 16), it is possible that in CSEs the ambient UV field may photodesorb large molecules from grains and break these into smaller species. Thus, in the case of grain chemistry, one would not expect to find the correlation between chain size and radial peaks suggested by the gas-phase chemistry. Indeed, the opposite correlation might occur. Large species, e.g. HC_9N, could be desorbed

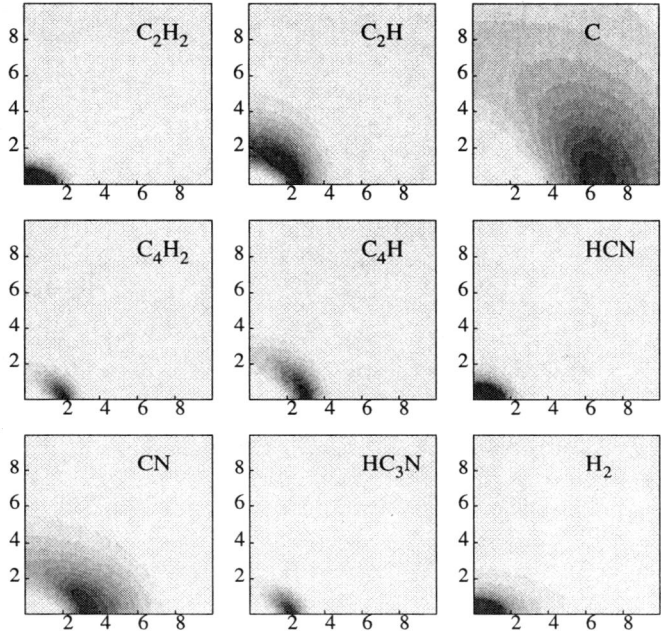

FIG. 2. Number density distributions for selected species for velocity-dependent mass loss, with an anisotropy factor $a = 0.5$. Shades of grey represent a 10% change in the number densities (see Howe and Millar 1996 for details).

and broken down progressively by UV photons so that molecular shells would be larger for smaller species.

To date, almost all of the complex chemical kinetic models of the external regions of CSEs have been studied as cases of spherical symmetry. However, recent interferometric observations of IRC+10216 have shown evidence for slightly asymmetric distributions (Guélin et al. 1993; Lucas et al. 1995; Gensheimer et al. 1995). Howe and Millar (1996) have generalized the work of Jura (1983), who derived effective UV optical depths out of an oblate spheroidal distribution for positions in polar directions, i.e. along the symmetry axis, $\theta = 0$, and in the plane, $\theta = \pi/2$, to an arbitrary outflow angle. They adopted the chemical scheme of Millar and Herbst (1994) and used either direction-dependent mass-loss or velocity-dependent variations to generate the oblate distribution. Axial symmetry was assumed so that there was no dependence on azimuthal angle and column densities were calculated for 45 runs at 2 degree intervals. Figure 2, from Howe and Millar (1996), shows number density distributions for a variety of molecules chosen to represent typical parents (C_2H_2, HCN), photoproducts (C_2H, CN, C) and products of chemical reactions (C_4H_2, C_4H, HC_3N) for a model with velocity-dependent anisotropy and an anisotropy factor,

Table 2 Ratios of peak values plane/axis for model results from Howe and Millar (1996).

Species	A	B	C
CN	1.1	2.4	1.4
HC_3N	1.5	6.3	1.6
HC_5N	2.3	20.9	1.9
HC_7N	3.4	78.3	2.2
C_2H	1.0	1.1	1.2
C_4H	1.3	3.9	1.5
C_6H	2.1	17.9	1.8
SiC_2	2.4	16.9	1.9

Column A: ratios of peak *number densities* for angle dependent mass loss rate model with anisotropy factor $a = 0.5$; Column B: ratios of peak *number densities* for angle dependent velocity model with anisotropy factor $a = 0.5$ as shown in Fig. 2; Column C: ratios of peak *velocity channel columns* for angle dependent velocity model with anisotropy factor $a = 0.9$.

a, of 0.5—a spherical distribution has $a = 1$. In general, the distribution of parent species is less anisotropic than that of H_2, which traces the mass density. This occurs because, in this figure, which corresponds to Column B in Table 2 the underlying asymmetry is caused by velocity as in polar directions the larger velocity implies that the scale length against photodissociation, $\sim v/\beta$, where β is the photodissociation rate, is larger than in the plane. Thus, in polar directions, chemistry amongst the daughters occurs further from the star in regions of higher interstellar UV flux so that they are more rapidly destroyed and polar holes arise. These holes are particularly noticeable in the longer carbon chains which form less efficiently as their progenitors are photodissociated. Table 2 gives the predicted abundance ratios for polar/plane directions for a number of species and model calculations (Howe and Millar 1996).

Although the general anisotropies observed are reproduced in the model, observations of SiS and SiC_2 cannot be reproduced (Lucas *et al.* 1995; Gensheimer *et al.* 1995). Indeed, SiS appears to be elongated in the polar direction in contrast to other molecules. This elongation could be due to the efficiency of incorporation of silicon into dust grains differing with direction.

3 Oxygen-rich CSEs

Oxygen-rich CSEs are known to contain a number of molecules including CO, H_2O, OH, SiO, the sulphur-bearing species H_2S, SO, SO_2, SiS, and carbon-bearing molecules CN, CS, HCN, HNC, HCO^+, OCS and H_2CO (Olofsson 1997). The primary parent is H_2O whose daughter OH determines much of the chemical composition in the outer CSE. Since OH is very reactive, diatomic oxides can be very abundant. Unlike the case of C-rich CSEs, where the photoionization of parent C_2H_2 can drive a significant ion–neutral chemistry, the chemistry in O-rich CSEs is essentially based on neutral–neutral reactions since the primary photodissociation product of H_2O is OH. It has been thought, based on the results of LTE calculations, that all carbon should be tied up in CO and therefore not available to take part in the chemistry. However, the detection of a number of carbon-bearing molecular species shows that it is necessary to provide a source of reactive carbon in the outer CSE.

Nejad and Millar (1988) suggested that this might be CH_4 and found reasonable agreement with the observations of carbon-bearing molecules, but only if the fractional abundance of CH_4 injected into the outer envelope was large, $\sim 3 \times 10^{-5}$. Such a large abundance could, perhaps, arise through production on active grain surfaces in the inner envelope which converts atomic carbon to methane via hydrogenation (Nejad and Millar 1988) or through the effects of UV and shock chemistry (Nercessian et al. 1989) although detailed calculations remain to be done.

Lindqvist et al. (1992) detected H_2CO in the bipolar PPN OH231.8+4.2 and since in the interstellar medium H_2CO is formed primarily via

$$O + CH_3 \longrightarrow H_2CO + H \tag{3.1}$$

it is possible that injection of CH_4 could also drive this reaction in CSEs. Millar and Olofsson (1993) made detailed calculations for conditions typical of four sources, IK Tau (NML Tau) and TX Cam, which are Mira variables, the supergiant VY CMa and OH231.8+4.2. They calculated intensities expected for four transitions of H_2CO and concluded that the 3_{12}–2_{11} transition at 225.7 GHz should be detectable in IK Tau and VY CMa. The presence of CH_3 would also lead to thioformaldehyde, H_2CS, and H_2CN via

$$S + CH_3 \longrightarrow H_2CS + H \tag{3.2}$$
$$N + CH_3 \longrightarrow H_2CN + H \tag{3.3}$$

but predicted intensities fall far below detectability limits, partly because S is readily photoionized and, in the reactive zone, the N abundance is low, so that $x(O) \sim 4x(N) \sim 10x(S)$.

Charnley et al. (1995) have extended the O-rich chemistry to hydrocarbons and, based on the inclusion of CH_4, predict large abundances of CH_3OH, C_2H and C_2, which form via

$$CH_3^+ + H_2O \longrightarrow CH_3OH_2^+ + h\nu \tag{3.4}$$

and reactions such as

$$CH_3^+ + CH_4 \longrightarrow C_2H_5^+ + H_2 \tag{3.5}$$
$$C^+ + CH_4 \longrightarrow C_2H_2^+ + H_2 \tag{3.6}$$
$$C^+ + CH_4 \longrightarrow C_2H_3^+ + H \tag{3.7}$$
$$C + CH_3^+ \longrightarrow C_2H^+ + H_2 \tag{3.8}$$

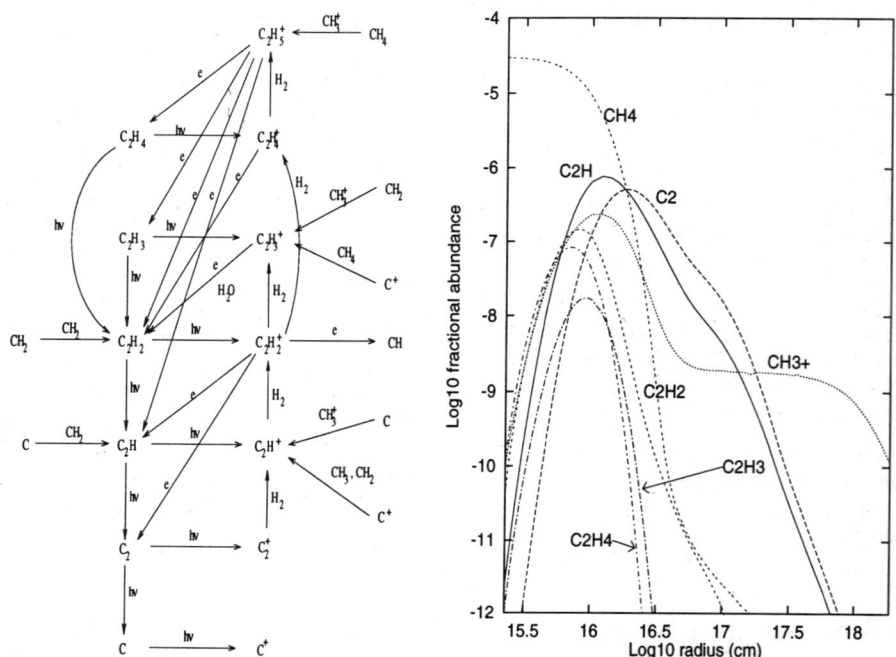

FIG. 3. The left hand panel shows how species containing two carbon atoms build up from injected methane. The right hand panel shows the resulting radial distributions relative to H_2 (Willacy and Millar 1997).

followed by dissociative recombination. Figure 3, taken from Willacy and Millar (1997), shows the basic carbon chemistry in the outer CSE and the radial distributions of several carbon-bearing molecules.

Cyanide species are detected in many O-rich CSEs. The HNC/HCN abundance ratios derived from observations are ∼1/3 and are very similar to those predicted by the models. These high ratios result from the interconversion of HCN and HNC via $HCNH^+$ (Nejad and Millar 1988). The ratio is much larger than that in C-rich CSEs because in these objects, HCN is a parent molecule and always has a large column density. Olofsson et al. (1991) detected CN in TX Cam and more recent observations by Bachiller et al. (1997) show that CN/HCN ∼0.45 in C-rich CSEs and ∼0.04 in O-rich CSEs. The former is in good agreement with the photochemical models but the latter is much smaller than predicted. Since CN is a direct photodissociation product of HCN in both C-rich and O-rich envelopes, and its calculated abundance in the former agrees with observation, it appears that there is an additional destruction mechanism for CN operating in O-rich CSEs. The destruction process must be a factor of 10–100 faster than photodissociation and involve an abundant O-bearing species since CN agrees well in the C-rich CSEs. Obvious candidates are O_2 and O atoms. The former reaction has been measured to be rapid down to very low temperatures and indeed has a rate coefficient inversely proportional to temperature (Sims et al. 1992). O_2 can be produced in the outer envelope by

$$O + OH \longrightarrow O_2 + H \quad (3.9)$$

where O and OH are the photodissociation products of H_2O. In this case, the O_2 abundance never becomes large enough in the region of the CN abundance peak to dominate CN destruction. Another possibility is that a large abundance of O_2 is injected into the outer envelope as the result of LTE or high density chemistry in the inner CSE. Note that an additional destruction process acting within the photodissociation radius of CN is consistent with the fact that the rotation temperatures derived for CN are larger in O-rich CSEs than in C-rich CSEs (Bachiller et al. 1997) indicating that CN exists closer to the star in the former case. A further possibility is that the HCN (1–0) transition is a weak maser in O-rich CSEs and that its abundance has been overestimated as a result (Bachiller et al. 1997).

Willacy and Millar (1997) discussed the chemistry of other elements, including S, Si, P and Cl, in models for specific objects, IK Tau, TX Cam and OH231.8+4.2. Parent sulphur is in the form of SiS and H_2S and their photodissociation leads to SH, which reacts with O, and to S, which reacts with OH, to form SO:

$$SH + O \longrightarrow SO + H \quad (3.10)$$
$$S + OH \longrightarrow SO + H \quad (3.11)$$

much of which is converted into SO_2:

$$SO + OH \longrightarrow SO_2 + H. \quad (3.12)$$

The CS molecule is also formed efficiently via

$$S^+ + CH \longrightarrow CS^+ + H \quad (3.13)$$
$$CS^+ + H_2 \longrightarrow HCS^+ + H \quad (3.14)$$
$$HCS^+ + e \longrightarrow CS + H \quad (3.15)$$
$$SO + C \longrightarrow CS + O \quad (3.16)$$
$$SH + C \longrightarrow CS + H. \quad (3.17)$$

Another interesting species which forms is SO^+:

$$S^+ + OH \longrightarrow SO^+ + H. \quad (3.18)$$

Since SO^+ does not react with H_2 it may be abundant in O-rich CSEs. This ion has been detected in interstellar clouds (Turner 1994) and its detection in O-rich

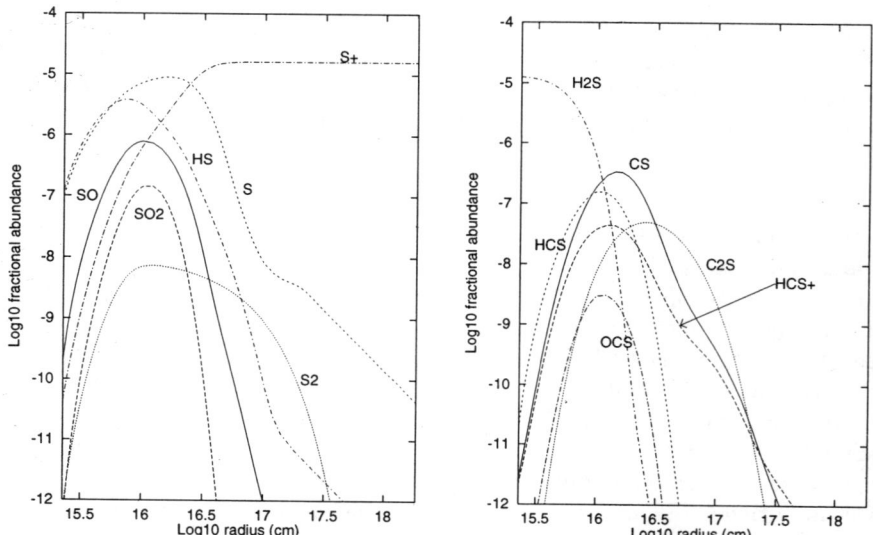

FIG. 4. Radial distributions, relative to H_2, for S-bearing species in a model for the star TX Cam (Willacy and Millar 1997). In this model, H_2S is the parent species and its photodissociation leads to the variety of molecules shown here.

CSEs, which will be difficult because of their small mass, would be an important test of these models.

The radial distributions of several S-bearing molecules, calculated with parameters appropriate for TX Cam, are shown in Fig. 4. Note that the model for OH231.8 may not be very realistic since the source contains a high-velocity outflow which is not included in the model. However, the agreement between theory and observation for this object is surprisingly good—except for SO_2, for which the model abundance is a factor of ~ 40 too small. Indeed, the observed fractional abundance of SO_2, 1.3×10^{-5} (Morris et al. 1987; Omont et al. 1993), is about 80% of the cosmic abundance of sulphur. Such an abundance might be realized in the high-velocity flow or in a shocked region close to the star (Jackson and Rieu 1988).

If a source of CH_4 is needed in the outer envelope of O-rich CSEs—as seems to be implied by the above discussion—one can ask whether it is possible to determine a limit on its abundance through observation. The indirect method is to search for species which result from the presence of CH_4. Charnley and Latter (1997) have found upper limits to CH_3OH and C_2H in several CSEs and modelled three in detail including TX Cam and IK Tau. They calculated that, with the abundance of CH_4 required, 3×10^{-5}, to match the abundance of HCN, there should be up to an order of magnitude more CH_3OH and C_2H than allowed by their upper limits, suggesting that CH_4 is not present in the CSE. However,

Willacy and Millar (1997) found no conflict between calculated column densities and observed upper limits using the *same* CH_4 input abundance. The difference in the conclusions of these two papers has not yet been explained.

A direct test for CH_4 is to search for it in absorption in the infrared. Markwick and Millar (1997) have begun an *Infrared Space Observatory (ISO)* programme to look for the ν_3 and ν_4 bands of CH_4 at 3.25 and 7.75 µm, respectively, in a number of evolved O-rich stars. To date, spectra have been taken of IRC+10240 and R Cas and show no evidence of CH_4. Interestingly, the S-type star, W Aql, shows OH absorption bands at 3.2–3.4 µm as well as clear evidence of absorption in the R-branch transitions of the $v = 0$–1 band of HCl (see Fig. 5).

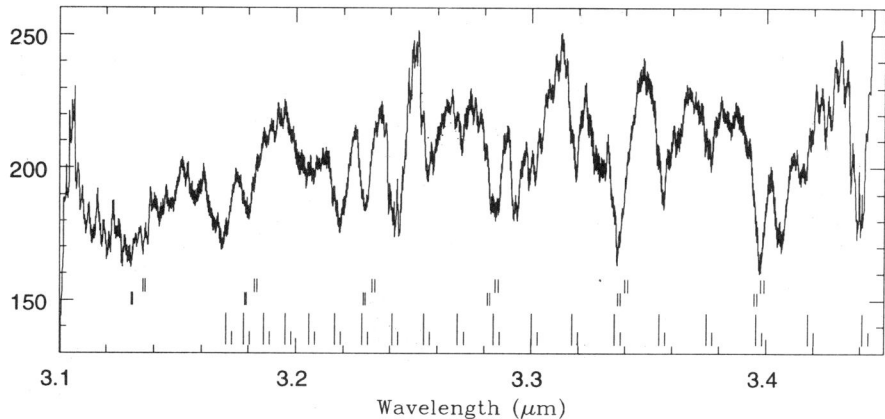

FIG. 5. 3 µm ISO spectrum of W Aql showing absorption due to OH and HCl. Tick marks denote the positions of transitions from the $v = 0\ ^2\Pi_{3/2}$ to $v = 1\ ^2\Pi_{3/2}$ (upper set) and $v = 0\ ^2\Pi_{1/2}$ to $v = 1\ ^2\Pi_{1/2}$ (middle set) of OH and the $v = 0$ to $v = 1$ R-branch transitions in $H^{35}Cl$ and $H^{37}Cl$ (lower set).

4 Future Directions

The ultimate aims of chemical kinetic modelling of CSEs are twofold. Firstly, it can be used to test specific mechanisms of molecular synthesis thought to be important in interstellar clouds. In recent years, there have been arguments raised which suggest that interstellar chemistry is an ill-defined subject, partly due to uncertainties in rate coefficients, partly due to the difficulty of including physical processes such as turbulence, small-scale, possibly fractal, structure and radiative transfer in a clumpy medium, partly due to the possibility of multiple solutions to the kinetic equations and partly due to the uncertainty concerning grain surface chemistry. However, CSEs are in the main very simple regions whose chemistry appears to be dominated by similar gas phase processes thought to occur in interstellar clouds. If essentially the same model can be applied successfully to both these types of objects, then it would indicate that other effects are second-order in nature. However, one should note that the difficulty in comparing theoretical and observational abundances imply that the theorists will need to calculate line profiles from their radial distributions to more accurately compare to observation. The second aim is to use these models to probe the underlying elemental abundances of the star and the means by which this material ends up in dust and molecules in the outer CSE and the nature of the dust particles ejected into the interstellar medium. To achieve this requires a much more holistic approach to the modelling. That is, it needs to involve high-density, high-temperature chemical equilibrium calculations for parent molecule formation, plus dust nucleation, the interaction of gas and dust in the inner envelope, and the subsequent photochemistry, probably incorporating photodesorption from grains, in the outer envelope. Such a calculation needs to span about ten orders of magnitude in density, needs to include both two- and three-body interactions and needs to deal with the heterogeneous nucleation of solid particles. The latter is particularly difficult but determines the composition and abundance of material left in the gas phase.

Observational studies will drive much of the theoretical work. In particular, the infrared observations of the gas and dust becoming available from *ISO* and the new generation of millimetre-wave array telescopes currently being planned will revolutionize our view of CSEs and, in particular, the relationship between AGB stars and PPNe and PNe. As a result, one can expect research on the chemical tracers of this evolution to be an exciting topic for the future.

Acknowledgements

I am grateful to my collaborators, D. A. Howe, A. J. Markwick and K. Willacy for many helpful discussions. Astrophysics at UMIST is supported by a grant from PPARC.

Bibliography

1. Bachiller, R., Fuente, A., Bujarrabal, V., Colomer, F., Loup, C., Omont, A. and de Jong, T. (1997). *A&A*, **319**, 235.
2. Bieging, J. H. and Tafalla, M. (1993). *AJ*, **105**, 576.
3. Charnley, S. B. and Latter, W. B. (1997). *MNRAS*, **287**, 538.
4. Charnley, S. B., Tielens, A. G. G. M. and Kress, M. E. (1995). *MNRAS*, **274**, L53.
5. Cherchneff, I. and Glassgold, A. E. (1993). *ApJ*, **419**, L41.
6. Cherchneff, I., Glassgold, A. E. and Mamon, G. A. (1993). *ApJ*, **410**, 188.
7. Clary, D. C., Haider, N., Husain, D. and Kabir, M. (1994). *ApJ*, **422**, 416.
8. Gensheimer, P. D., Likkel, L. and Snyder, L. E. (1995). *ApJ*, **439**, 445.
9. Glassgold, A. E. (1994). *ApJ*, **438**, L111.
10. Glassgold, A. E. (1996). *Ann. Rev. Astron. Astrophys*, **34**, 241.

11. Glassgold, A. E., Lucas, R. and Omont, A. (1986). *A&A*, **157**, 35.
12. Glassgold, A. E., Mamon, G., Omont, A. and Lucas, R. (1987). *A&A*, **180**, 183.
13. Guélin, M., Lucas, R. and Cernicharo, J. (1993). *A&A*, **280**, L19.
14. Guélin, M., Lucas, R. and Neri, R. (1996). In IAU Symp. 170, *CO: 25 Years of Millimetre-Wave Spectroscopy*, eds. W. B. Latter, S. J. E. Radford, P. R. Jewell, J. G. Magnum and J. Bally. Kluwer, Dordrecht, p.359.
15. Guélin, M., Cernicharo, J., Travers, M. J., McCarthy, M. C., Gottlieb, C. A., Thaddeus, P., Ohishi, M., Saito, S. and Yamamoto, S.. (1997). *A&A*, **317**, L1.
16. Haider, N. and Husain, D. (1993). *J. Photochem. Photobiol.*, **A70**, 119.
17. Howe, D. A. and Millar, T. J. (1990). *MNRAS*, **244**, 444.
18. Howe, D. A. and Millar, T. J. (1996). *MNRAS*, **282**, L21.
19. Jackson, J. M. and Rieu, N.-Q. (1988). *ApJ*, **335**, L83.
20. Jura, M. (1983). *ApJ*, **275**, 683.
21. Jura, M. and Kroto, H. (1990). *ApJ*, **351**, 222.
22. Kastner, J. (1992). *ApJ*, **401**, 337.
23. Lindqvist, M., Olofsson, H., Winnberg, A. and Nyman, L.-Å. (1992). *A&A*, **263**, 183.
24. Lucas, R., Guélin, M., Kahane, C., Audinos, P. and Cernicharo, J. (1995). *Ap&SS*, **224**, 293.
25. Markwick, A. J. and Millar, T. J. (1997). **251**, 255.
26. Millar, T. J. and Herbst, E. (1994). *A&A*, **288**, 561.
27. Millar, T. J. and Olofsson, H. (1993). *MNRAS*, **262**, L55.
28. Morris, M., Guilloteau, S., Lucas, R. and Omont, A. (1987). *ApJ*, **321**, 888.
29. Nejad, L. A. M. and Millar, T. J. (1987). *A&A*, **183**, 279.
30. Nejad, L. A. M. and Millar, T. J. (1988). *MNRAS*, **230**, 79.
31. Nercessian, E., Guilloteau, S., Omont, A. and Benayoun, J. J. (1989). *A&A*, **210**, 255.
32. Olofsson, H. (1997). In IAU Symp. 178, *Molecules in Astrophysics: Probes and Processes*, ed. E. F. van Dishoeck. Kluwer, Dordrecht, p.457.
33. Olofsson, H., Lindqvist, M., Nyman, L.-Å., Winnberg, A. and Rieu, N.-Q. (1991). *A&A*, **245**, 611.
34. Omont, A., Lucas, R., Morris, M. and Guilloteau, S. (1993). *A&A*, **267**, 490.
35. Sims, I. R., Queffelec, J.-L., Defrance, A., Rebrion-Rowe, C., Travers, D., Rowe, B. R. and Smith, I. W. M. (1992). *J. Chem. Phys.*, **97**, 8798.
36. Sims, I. R., Queffelec, J.-L., Travers, D., Rowe, B. R., Herbert, L. B., Karthaüser, J. and Smith, I. W. M. (1993). *Chem. Phys. Lett.* **211**, 461.
37. Turner, B. E. (1994). *ApJ*, **430**, 727.

38. Willacy, K. and Cherchneff, I. (1998). *A&A.* **330**, 676.
39. Willacy, K. and Millar, T. J. (1997). *A&A.* **324**, 237.

16
The Chemistry of Planetary Nebula Formation

D. A. Howe
Department of Physics, UMIST

and

D. A. Williams
Department of Physics and Astronomy, University College London

1 Introduction

The high mass loss rates of red giant stars during the AGB (asymptotic giant branch) phase of stellar evolution obviously cannot continue for very long. For example, the central star of IRC+10216 is losing roughly 10^{-4} $M_\odot \text{yr}^{-1}$ (solar masses per year), and contains only a few $M_\odot$. The chemistry of matter ejected during the post AGB (PAGB) evolution of such stars is the subject of this chapter. The matter will continue to drift from its parent star, even after the end of the high mass loss rate phase. The changes in physical properties of this material (density, opacity, temperature) occur on timescales of the order of hundreds of years, or less—even shorter than those that characterize the AGB phase. In addition, the remnant star will continue to evolve quite rapidly, perhaps on comparable timescales. The interaction of the ejected envelope with the remnant star thus provides a fascinating field of study, both for the theoretician and the observer. The attraction of this field is the challenge of following a dynamic, evolving system, displaying widely differing physical conditions that change during the course of a human lifetime. However, the short evolutionary timescale implies that few such objects will be observable at any one time.

The fate of the gas ejected during the AGB phase is crucially dependent on the timescales over which various aspects of PAGB evolution occur, especially the expansion of the ejected gas and dust itself and the heating of the remnant star. For example, the remnant star (consisting mostly of the core of the original star) will heat up, on a timescale which varies from one object to the next (cf. Chapter 11). If heating occurs less than a few thousand years after the end of the AGB phase, the ejected envelope will be close enough (within about 0.1 parsecs) to the hot remnant to be ionized (at least in part), forming a "planetary nebula" (PN, plural PNe). A possible example of a star undergoing fairly rapid stellar heating is HD56126, which was assigned to spectral class G5 in 1919

(HD catalogue), but was classified as F5 in 1965 (Nassau et al. 1965). On the other hand, if the stellar heating takes too long, the envelope will merge with the interstellar medium, and be too diffuse and distant to form a PN. The physical evolution of the circumstellar gas around a star, from the AGB to the PN phase and beyond, is illustrated schematically in Fig. 1, in a simple, spherically symmetric case, starting with the AGB star surrounded by the dense, slow wind. Eventually, the period of high mass loss rate terminates and the wind forms a shell that detaches from the star. The star continues to evolve, as illustrated schematically in Fig. 2. As it does so, the star heats up considerably and its radiation field hardens to become capable of ionizing hydrogen. The star generates a fast (typically 1000–3000 km s^{-1}) low density wind with a mass loss rate of around 10^{-10}–10^{-7} $M_\odot \text{yr}^{-1}$. This fast wind may continue during the PN phase, dying away when the star eventually cools towards the white dwarf state. The rate of stellar evolution in this phase is a very sensitive function of the mass of the stellar remnant. Figure 11 in Chapter 11 shows evolutionary tracks in the Hertzsprung–Russell diagram for stars of different masses. For example, at a time 3000 years after detachment of the slow wind, a star of 0.546 $M_\odot$ is still relatively cool (6000 K), while a slightly more massive star of 0.625 $M_\odot$ has attained a much higher temperature (10^5 K).

If the star evolves sufficiently quickly to this hot phase, then the fast wind impinges on the slow wind, and according to the so-called interacting stellar winds model (Kahn 1983; Kwok 1983) sets up a two shock structure. This is illustrated schematically in Fig. 3. The fast wind from the evolved hot star is slowed by a stationary inward facing shock. Behind this shock is a zone (A) of very hot gas which expands into the slow wind driving an outward-facing shock into it. The shocked slow wind may be photoionized in part by the stellar radiation (zone B), while the outer part (zone C) may be mainly neutral. These zones are surrounded by unshocked slow wind (D). The structure illustrated in Figs. 1 and 3 represents an idealized young PN or PPN ("Pre-Planetary Nebula" or (confusingly) "Protoplanetary Nebula", plural: PPNe). As the structure continues to expand, the density and temperature fall; this is the fully fledged PN phase. Eventually the envelope drifts beyond a point where the PN is visible, while the central star cools and fades to a white dwarf.

Figure 1 also illustrates the alternative evolutionary mode, in which the central star heats up relatively slowly. The slow wind may then have moved so far from the star that it has become too diffuse to form a PN.

The kinematic timescale of the gas around a PAGB star is likely to be shorter than or comparable to that of the stellar evolution. For example, a nebula of radius 0.1 pc expanding at 20 km s^{-1} would have a kinematic age of about 5000 years. An important factor for PN chemistry is the dust opacity of the AGB remnant. Consider for simplicity a detached AGB envelope, drifting at constant speed. Assuming a constant mass loss rate during the preceding AGB phase, the density between the inner edge (the most recently ejected material) and the outer edge (of material ejected at the start of the high mass loss phase) follows

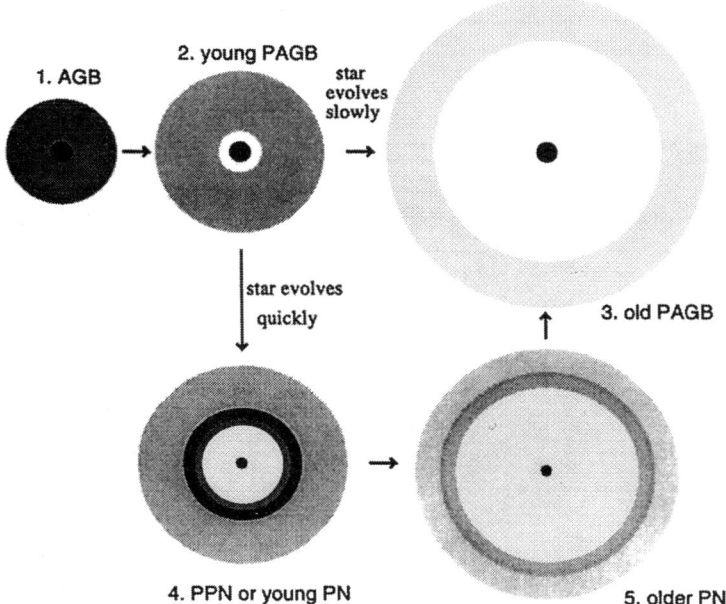

FIG. 1. Simplified diagram of the evolution of a star and its ejected mass after the AGB stage (not to scale).
1. Here the AGB red giant star (black disc) is still emitting a dense wind (dark grey).
2. The dense wind has ceased, leaving a detached shell of AGB ejecta.
The central star will shrink, and heat up, no longer containing enough matter to support a red giant structure. The effect of this on the ejecta depends upon the speed with which this occurs:
3. If the star evolves very slowly, the AGB ejecta (pale grey) has drifted too far from it, and is too diffuse to produce a planetary nebula (Section 2.1).
4. If the star heats up quickly, the fast, low density wind it now produces (pale grey), sweeps up the AGB material, forming an outward moving shock (solid circle). Outside this shock, there is unshocked AGB wind. Immediately inside the shock is shocked, neutral AGB wind; while inside this the wind has been ionized by the hard UV flux from the central star (small black disc). At early times, the opacity of the AGB remnant may still obscure the central star in the visible and UV, allowing large abundances of molecules left over from the AGB stage to be present. The high densities in the shocked AGB gas may allow a rich chemistry to occur, very different from that during the AGB.
5. In an older planetary nebula, more of the slow AGB wind has been swept up, but the density and opacity of the swept-up gas is lower, perhaps reducing the variety of molecular species able to form. As shown by the arrow, this gas too will eventually drift beyond the point where the planetary nebula is visible (picture 3). Meanwhile the central star cools and fades into a white dwarf.

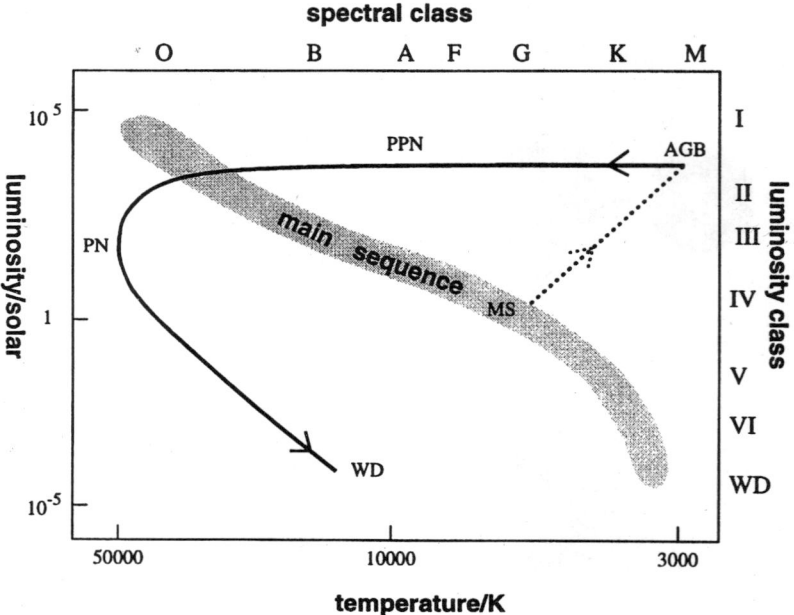

FIG. 2. Simplified Hertzsprung–Russell diagram, illustrating stellar evolution from the asymptotic giant branch (AGB), through pre-planetary nebula (PPN) and planetary nebula (PN) stages, to the white dwarf (WD) stage. The star shown was of a mass similar to that of the Sun, when on the main sequence (MS). The path from MS to AGB is shown dotted, as its details have been omitted.

Once the AGB wind is terminated, the star heats up, remaining at fairly constant luminosity, reaching temperatures high enough to ionize large amounts of hydrogen. If the star evolves fast enough, the remnant AGB wind material will still be nearby, and a PN is formed. After this, the star cools and fades to become a WD. The rate of evolution of the star is critically dependent on the mass of the star after the AGB. In some cases, possibly including a star like our Sun, it is thought that the evolution may be so slow that the remnant AGB material is too far away to produce a PN by the time the star has heated up sufficiently to form one.

an inverse square law with distance, owing to geometrical dilution, and is given as follows:

$$\rho(r) = \frac{\dot{M}}{4\pi v r^2}$$

where $\dot{M}$ is the stellar mass loss rate, v is the speed of the wind and r is the distance from the star. Assuming uniform dust abundance and optical properties,

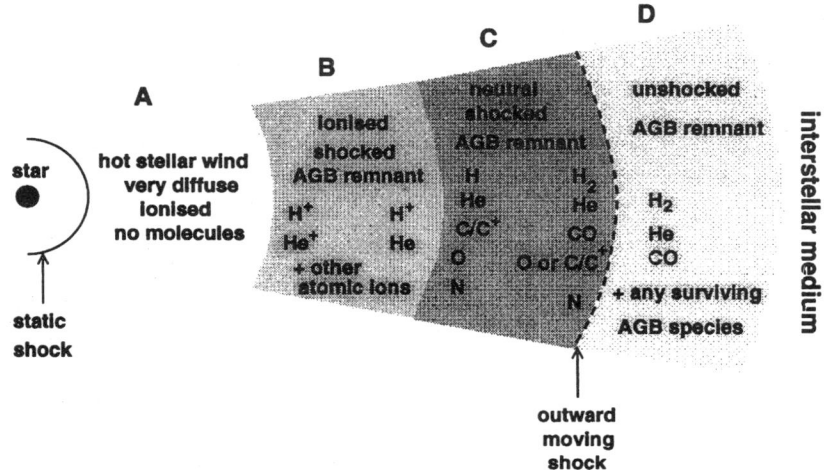

FIG. 3. Simplified cross section through a sector of gas in a PPN or young PN, according to the interacting stellar winds (ISW) model, showing the major regions, and their main chemical constituents (not to scale). In the more simple forms of the ISW model, this structure is spherically symmetric about the central star (as shown in Fig.1, picture 4); in most real cases, however, a bipolar symmetry is thought to be produced quite quickly, in which the more dense gas described here is concentrated in an approximately toroidal ring. The central star is on the left, producing a fast wind ($\sim$2000 km s^{-1}), which is thought to undergo a stationary, inward-facing shock (solid line) which arrests the fast wind and heats it, forming a hot, very diffuse ionized "bubble" (region A). This bubble expands, compressing the inner face of the remnant AGB high density, slow wind, and progressively sweeping it up, at supersonic speeds (relative to the AGB gas). This forms an outward moving shock front at the boundary of regions C and D, with compressed AGB remnant gas behind it (to the left). The inner part of the compressed AGB gas (region B) is ionized by the strong UV photon flux of the central star. Region B contains mainly ionized hydrogen, and ionized or atomic helium, along with other atoms and atomic ions, some ionized several degrees. Region C consists of largely neutral gas, whose inner parts contain mostly atomic neutrals, though carbon and some other minor elements will be significantly ionized, owing to their ionization potentials being below that of hydrogen. Further from the star, it is likely that H$_2$ and CO abundances will increase, due to self shielding. If the shock is fast, there may also be a region just inside the C/D boundary, which is hot enough to dissociate these molecules collisionally. Other elements will mostly be atomic, as the extinction of the shocked gas (regions B and C) is likely to be low, except perhaps in clumps. Where CO is most abundant, free carbon will still be available in oxygen-rich nebulae, and free oxygen in carbon-rich nebulae. Outside the shockfront, the unshocked AGB remnant (region D) will consist mostly of H$_2$, He and CO. The survival of other molecular species formed during the preceding AGB phase will only be possible either at very early times, when the extinction of region C (with regard to stellar UV photons from within) and region D (with regard to interstellar UV photons from without) is still large, or in dense clumps, where high extinction is maintained locally. The outer edge of region D represents material ejected at the start of the AGB phase, beyond which lies the interstellar medium.

the optical depth τ in a radial direction, between the inner edge at r_1 and some point further out at r_2 will be proportional to the integral of $\rho(r)$ between these points:

$$\tau \propto \left(\frac{1}{r_1} - \frac{1}{r_2}\right)$$

For $r_2 \gg r_1$, τ becomes roughly inversely proportional to r_1. Taking an example with $\dot{M} = 5\times10^{-5}$ $M_\odot$ yr^{-1}, $v = 15$ km s^{-1}, and with standard interstellar dust abundance and properties, the optical depth of UV radiation at 100 nm wavelength will be about 25 when $r_1 = 4.7\times10^{14}$ cm, 100 years after mass loss has ceased, effectively blocking the stellar radiation field, and allowing AGB remnant molecules to remain, even if the star is already hot. However, the optical depth will have fallen to only $\tau = 5$ at 500 years. The effect of an outward moving shock driven by the fast wind, sweeping the material up from the inside, as predicted by the interacting stellar winds model, would reduce the optical depth still more rapidly. This rapid fall in opacity will allow photolysis of most molecules, apart perhaps from those, such as H_2 and CO, which can shield themselves, those which are resistant to photolysis, such as PAH (polycyclic aromatic hydrocarbons) and fullerene species (whose presence in AGB outflows and PNe is not yet firmly established), or those contained in dense condensations (for which the above analysis does not apply). Any other species observed after this stage are expected to have formed, or have been released into the gas phase, after the end of the AGB, as a direct result of PN formation. If the evolution of the remnant star is rapid, the early stages of PN formation provide particularly fertile ground for studies in molecular astrophysics, as these ejecta have high densities and large optical depths, and are subjected to high UV photon fluxes and to violent dynamic processes. In more evolved PNe, the low densities and optical depths mean that the chemistry is likely to be simpler, except perhaps in localized knots or clumps.

An increasing variety of molecular species is being observed in PPNe and PNe, though theoretical modelling of this chemistry is still in its preliminary stages. As planetary nebulae represent the fate perhaps of the majority of stars, this subject deserves more attention.

In Section 2 we describe the chemistry likely to occur in the ejected envelopes of PAGB stars. In Section 2.1 we discuss briefly what is likely to happen if the star remains cool for a relatively long time (more than a few thousand years) after the termination of the AGB wind. However, the main interest in PAGB evolution, from the point of view of molecular astrophysics, is when the central star heats up while the ejected envelope is still nearby, and in Section 2.2 we pay particular attention to this, taking in turn the regions of differing physical conditions depicted in Fig. 3. In Section 3 we suggest the directions future work might take.

2 Chemistry around Post AGB Stars

2.1 PAGB Envelopes with Cool Remnant Stars

Though most stars more massive than about 1 $M_\odot$ are thought to go through an AGB phase, losing their outer layers in the form of a wind and becoming white dwarfs, there is evidence, both observational and theoretical, which suggests that some of these may not pass through a PN stage during the AGB–white dwarf transition. Though the evolution of such stars is expected to share the same qualitative behaviour (heating at constant luminosity, followed by fading and cooling) as that shown in Fig. 2, theoretical models of late stellar evolution suggest that the time this takes is a very sensitive function of the remnant stellar mass (see Chapter 11). For example, it is suggested that RV Tauri stars are PAGB objects, some of which will never form a PN (Jura 1986; Bujarrabal et al. 1988).

Further support for the concept that some PAGB stars fail to become PNe comes from studies of the so-called "High latitude supergiants" (HLSG stars). Some stars originally identified (from their spectra) as bright "supergiants" are found at high Galactic latitudes (Bidelman 1950). This means that if they are true supergiants, they must lie well out of the Galactic plane—where such stars are not expected to be. Though some of these stars may have formed in the Galactic plane, then drifted out, only a few have speeds sufficient to have moved that far since formation (since very bright stars have short lives). Therefore it seems likely that many are not supergiants at all.

A significant group of HLSGs are now thought to be of lower luminosity than true supergiants. This means that they are closer to us than supergiants of the same apparent magnitude, and that they lie closer to the Galactic plane. It also means that these objects, and others like them, represent an aspect of the evolution of a different type of star. An important feature often shown by HLSGs, is the presence of an infrared excess, i.e. their spectra show infrared (IR) intensities higher than expected from the star alone. This is attributed to the presence of circumstellar dust, and has led to the identification of these objects as low mass PAGB stars, the dust being part of the slow, massive wind produced during the preceding AGB phase. If such a star heats up before the wind has drifted far away, the wind will be ionized, and become a PN. However, some HLSGs, identified as post-AGB stars from their elemental abundances, show little or no IR excess, and it is likely that these objects have lost their winds to the interstellar medium, and will never form PNe (Waters et al. 1993).

The relatively slow evolution of the central stars of objects like these means that the chemistry in their envelopes shortly after the termination of the AGB wind should be similar to the photochemistry in AGB winds (see Chapter 15), except that, as the detached shell of ejected matter drifts away from the central star, there will be a reduction from the inside, on a timescale of about a thousand years, of the AGB molecular abundance profiles. Photochemistry in AGB winds is extremely well understood and accurately modelled, with theory often successfully guiding and explaining observations. Molecular abundances in AGB winds consist predominantly of "parent" and "daughter" molecules, though grain surface processes may also contribute (see Keady and Ridgway 1993). Parent molecules are formed in or near the atmosphere of the star, and have approximately inverse square abundance distributions, due to geometrical dilution of the outflowing wind. These distributions are truncated by photolysis at a distance governed by the mass loss rate, shielding properties of the dust, the speed of the wind, and the strength of the ambient interstellar UV flux. Daughter species are those formed from the parents by photochemistry in regions of significant penetration of UV photons. The daughter species are abundant in hollow shell distributions, bounded on the inside by their formation, and on the outside by their own photolysis, as the photon flux becomes stronger, and fall of density slows reaction rates. After the termination of copious mass loss, parent species will begin to fall in abundance, followed by the daughters. This is due to the inner edge of the detached wind passing out through the AGB abundance profiles, and no longer being replenished. For example, the molecule SiO often appears in the inner regions of oxygen-rich AGB winds but is depleted in objects which have suitable elemental abundances but whose remnant AGB winds are detached (see e.g. Bujarrabal et al. 1988; Bujarrabal et al. 1994).

2.2 Chemistry of Pre-planetary Nebulae and Planetary Nebulae

Much work is being done on PN formation, both observationally and theoretically (see e.g. the review by Iben 1995). The starting point for PN formation is widely accepted to be the high mass loss rate phase of the AGB, which is where we begin this discussion. In terms of the dominant chemical processes, the ejected gas surrounding AGB stars with high mass loss rates can be divided into two zones. Very close to the star ($\sim 10^{14}$ cm) the flow is far from steady, and includes shocks (due to stellar pulsation) and infall of material. Particular problems here include the formation of the dust in these dynamically turbulent, warm (~ 1000 K) regions (see Chapters 12 and 13) and the molecular makeup of the remaining gas (which goes to form the "parent" molecules referred to in Section 2.1). Beyond this, the flow is thought to become more uniform, and as the material moves further out, penetration by interstellar UV photons initiates photochemistry in cool (~ 50 K) regions (see Chapter 15). The chemistry during and after PN formation is also essentially photon dominated, but complicated by several factors:

(1) As there is no longer anything approximating a steady outflow of dense gas, no lasting steady state chemical profiles can be built up.
(2) The central stellar UV source is evolving, perhaps on timescales comparable to those in the ejected material, and may produce UV fluxes several orders of magnitude larger than in the interstellar medium.
(3) Though AGB winds are fairly spherically symmetric, PNe usually have some sort of bipolar morphology.

(4) The shocks which are thought to occur at the interface between the central star's hot, low density wind and the remnant AGB high density wind, involve a wide range of temperature, density and perhaps photon flux. This same interface is expected to be dynamically unstable, producing distortions and mixing, which will again affect the physical conditions in which any chemistry will occur.

(5) The dust grains produced during the AGB phase may play an active role, for example by colliding together and sputtering off large molecules.

(6) Finally, dense clumps are thought to have been left over from the AGB wind. These may provide an environment in which rich chemistry can continue well into the PN stage, long after most molecules have been photodissociated by the harsh stellar UV flux.

These changes in physical conditions radically alter chemical abundances as compared to the AGB.

There is no generally agreed distinction between the PPN and PN phases. Some authors define high mass loss AGB objects such as IRC+10216 as PPNe, and consider the PN stage as being reached as soon as there is a region of significant ionization caused by the central star. By this definition, CRL618 would be classed as a PN, albeit a young PN. We will define the start of the PPN phase to coincide with the termination of the high mass loss rate phase of the AGB (but only for cases which will turn into PNe). Transition to PN will be assumed to coincide with the formation of an ionized inner region, or with the dusty AGB remnant material falling in opacity (so that the ionized nebula is clearly visible in the optical, and most remnant AGB molecules are photolysed), whichever occurs last.

In spite of their rarity, several likely candidates for PPNe have now been identified, and some of these are receiving intense observational attention. Probably the most studied of these are the nebulae CRL2688 (the "Egg Nebula") and CRL618. The central star of CRL618 is hotter ($\sim 25\,000$ K) than that of CRL2688 (~ 7000 K), probably because the former has evolved further toward the PN stage (indeed, it has already developed a compact ionized region). Both of these objects are carbon-rich, and it is possible that they represent the fate of carbon-rich AGB outflows, such as IRC+10216. In our discussion of chemistry in the PPN stage, we confine ourselves to the case where the star becomes hot very early, within about 200 years of the end of the AGB, so that an ionized nebula is formed within a dense, opaque (at least in some directions) AGB remnant.

The observations of PPN candidates provide ample evidence of the violent processes involved in PN formation. For example, fast moving molecular gas is detected, and shocks deduced (e.g. Morris et al. 1987; Neri et al. 1992; Martín-Pintado et al. 1993; Meaburn 1995; Trammell and Goodrich 1996; Riera et al. 1996). The morphologies of many PPNe, like those of many PNe, frequently involve bipolar structure at optical wavelengths, corresponding to highly ionized gas, in a roughly toroidal region of denser, neutral material, also showing clumpiness and other small scale structure (e.g. Healy and Huggins 1990; Bachiller et al. 1989; Martín-Pintado et al. 1995).

An increasing number of molecular species is being detected, and in these observations, the transition from AGB to PN is traced through the accompanying changes in molecular abundances and distributions. The most abundant molecules are H_2 and CO, which can often survive right up to fully fledged PN formation, due to their robustness against photodissociation (see Section 2.2.2). Though CO is easily detected when present, by means of emission in its rotational spectrum, the H_2 in many astrophysical regions (including AGB winds) is inferred rather than detected, owing to its lack of a dipole moment. However, in hot gas, such as is found in PNe, detectable 2 μm emission via weak quadrupole transitions is possible, and the detection of such emission in PNe appears correlated to a bipolar morphology, the H_2 being in the toroidal region of neutral remnant AGB gas (Kastner et al. 1996). Other molecules characteristic of the AGB are found either to be absent, or to have different abundances, indicating their complete or partial destruction, or formation, in the new conditions pertaining. As might be expected, objects thought to be early in the AGB–PN transition show more similarity to the AGB than older PNe. For example, the long carbon chain molecule HC_9N is observed in the AGB outflow IRC+10216 as well as in the PPN CRL2688 (Truong-Bach et al. 1993). The formation of this species probably occurs during the AGB and its presence in CRL2688 is thought to be due to survival, via a combination of a still fairly low stellar UV output and efficient shielding by dust. The presence of a hotter central star in CRL618, appears to have resulted in a more marked departure from the molecular composition of IRC+10216. The early emergence of a stellar UV flux is expected to enhance the abundance of direct photolysis products of the AGB "parent" species relative to their precursors, at least while this flux remains at moderate levels, or where the remnant AGB gas is still fairly opaque. A possible early example of this is the object HD56126, whose central star has a temperature of about 6500 K. Bakker et al. (1996) have observed C_2 and CN in this object, which they attribute to photolysis of the AGB parental species C_2H_2 (via C_2H) and HCN respectively.

In older objects, or those with less massive AGB remnants, molecular abundances are lower, even CO often being depleted, in spite of its relative robustness (Bachiller et al. 1988). By contrast, NH_3, though a common species in the AGB, appears to have an enhanced abundance in parts of the PPN CRL618 (Martín-Pintado et al. 1993). From the spectra, it is seen that the regions of enhanced NH_3 are moving much faster than those where its abundance is more typical of the AGB. The apparent formation of this molecule after the AGB phase is therefore thought to be due to shock waves heating and accelerating the remnant AGB gas and speeding up the reactions by which nitrogen is hydrogenated to form NH_3 (see section 2.2.2). Additional support for this scenario is found in the filamentary structures observed in NH_3 by Martín-Pintado et al. (1995). Such structures are typical of a thin, compressed concentration of material, such as occurs behind a shock front, the intensity being enhanced when viewed parallel to the front, where large column densities can be built up. Other molecules thought to be formed behind fast shocks include, for example, the HCN, HC_3N and CO observed in the fast flowing gas (up to 200 km s^{-1}) in CRL618 (Neri et al. 1992).

Some species thought to be formed after departure from the AGB are not at all common in AGB outflows, and their occurrence may provide a powerful diagnostic of the evolutionary status of the objects in which they are found. Important examples of this are HNC and HCO^+. A suppressed HCN/HNC ratio may be an important indicator of the onset of PN formation in carbon-rich nebulae (Bujarrabal et al. 1988; Cox et al. 1992; Sahai et al. 1994). For example, the HCN/HNC ratio in CRL618 is suppressed relative to the less evolved objects CRL2688 and IRC+10216, which is likely to reflect higher ionization (Morris et al. 1987; Bujarrabal et al. 1988; Bujarrabal et al. 1994). HCO^+ is also prominent in CRL618 as well as the young PN, NGC7027. Another species, whose chemistry may be closely linked to that of HCO^+, is CO^+, which has been observed in NGC7027 (Latter et al. 1993). In addition to the information gained from the relative abundances of different molecules, the detection of different species provides information on the density of the regions in which they are found, as different spectral lines require different gas densities ("critical densities") in order to maintain an excited population. For example, observations of HCN, HNC, HC_3N, NH_3 and HC_7N in the PPN CRL2688 show an equatorial disc (Bieging and N-Q-Rieu 1988, N-Q-Rieu et al. 1986, N-Q-Rieu and Bieging 1990), while the isotopomer ^{13}CO of CO shows little departure from symmetry on scales larger than about 6×10^{16} cm (Yamamura et al. 1995). This is probably due to the lower critical density of ^{13}CO as compared to the other species, which are more sensitive to relatively small density variations.

For theoretical chemical modelling, PN formation presents a unique set of challenges. In the ionized part of the nebula, the conditions will be similar to those in HII regions (ionized gas close to hot stars, as in star forming regions). The conditions in the neutral gas in PPNe and many PNe will be similar to those in other photon dominated regions (PDR), such as exist in interstellar gas close to HII regions. Sternberg and Dalgarno (1995) have given a very detailed description of the chemical processes likely to occur in the mainly neutral gas; see Chapter 9.

The molecular abundances resulting from these processes in PPNe and PNe will depend on details of the density and temperature structure, the strength and spectrum of the photon field, the elemental abundances (especially the C/O ratio), as well as which (if any) molecular species survive from the AGB phase. The effect of the outward moving shockfront must also be considered.

As widely differing conditions of temperature, density, ionization and dissociation exist simultaneously and change over time, chemical modelling has understandably concentrated on limited regions and phases of evolution.

2.2.1 Chemistry in the Highly Ionized Gas

We now consider the chemistry occurring in region B of Fig. 3. This region is hot (with temperatures up to about 10^4 K) and is exposed to a very high UV flux from the hot, luminous central star, whose spectrum includes a large proportion of photons capable of ionizing hydrogen and helium. These are very harsh conditions for molecular species to exist in, as the high UV flux and free electron abundance will make destruction very rapid, through, for example photodissociation or dissociative recombination. Throughout much of the ionized region, formation will be inefficient, as most species present are positively charged ions, which will not react, owing to electrostatic repulsion. Molecule formation may however occur near the interface with region C, where neutral and ionized species are predicted to exist together.

The formation of molecular hydrogen in the interface region is likely to differ from that in cooler gas. The normal *interstellar* H_2 formation route, via grain surface catalysis, was invoked in some of the calculations of Cecchi-Pestellini and Dalgarno (1993) (those where dust was assumed present), but not by Black (1978), who assumed that the temperature of the grain surfaces was sufficiently high that H atoms striking a grain would not be bound to the surface. In any case, the gas phase sequences:

$$H + e \longrightarrow H^- + h\nu \qquad (2.1)$$

$$H^- + H \longrightarrow H_2 + e \qquad (2.2)$$

and

$$H + H^+ \longrightarrow H_2^+ + h\nu \qquad (2.3)$$

$$H_2^+ + H \longrightarrow H_2 + H^+ \qquad (2.4)$$

dominate H_2 formation (see Dalgarno and Lepp 1985). Black (1978) uses (2.1) and (2.2), Cecchi-Pestellini and Dalgarno (1993) also invoke reactions (2.3) and (2.4).

Another molecule predicted in this region is HeH^+, whose chemistry is described in detail by Roberge and Dalgarno (1982) and Zygelman and Dalgarno (1990). It is likely that the main formation mechanism in nebulae with hot central stars (such as NGC7027) is the radiative association:

$$He^+ + H \longrightarrow HeH^+ + h\nu \qquad (2.5)$$

Cecchi-Pestellini and Dalgarno (1993) obtained substantial column densities of HeH^+, $\sim 10^{12}$ cm^{-2} for stellar temperatures above 50 000 K. Black (1978), who did not include reaction (2.5), found that much smaller column densities (4.7×10^9 cm^{-2}) were formed.

HeH^+ is destroyed by photodissociation and proton transfer to H_2:

$$HeH^+ + h\nu \longrightarrow He + H^+ \qquad (2.6)$$

$$HeH^+ + H_2 \longrightarrow H_3^+ + He \qquad (2.7)$$

and by dissociative recombination:

$$HeH^+ + e \longrightarrow He + H \tag{2.8}$$

and proton transfer to atomic hydrogen:

$$HeH^+ + H \longrightarrow H_2^+ + He \tag{2.9}$$

According to Cecchi-Pestellini and Dalgarno (1993), photodissociation will dominate close to the star, dissociative recombination further out, and proton transfer near the edge of the ionized gas (where most HeH^+ is predicted to arise).

Though HeH^+ is predicted to form efficiently, and be observable (for example in NGC7027) a search for this molecule by Moorhead et al. (1988) failed to detect it. Cecchi-Pestellini and Dalgarno (1993) suggested that this may be due to the telescope beam being smaller than the nebula and missing the regions of maximum HeH^+ emission.

In addition to H_2 and HeH^+, Black (1978) modelled the chemistries of the simple species OH and CH^+. OH is suggested to form in two ways; either by hydrogen atom abstraction by atomic oxygen:

$$O + H_2 \longrightarrow OH + H \tag{2.10}$$

or by reactions of atomic oxygen and atomic hydrogen, via an excited state:

$$O + H \longrightarrow OH^* \longrightarrow OH + h\nu \tag{2.11}$$

Destruction is thought to be by photodissociation or charge exchange with ionized hydrogen:

$$OH + h\nu \longrightarrow O + H \tag{2.12}$$

$$OH + H^+ \longrightarrow OH^+ + H \tag{2.13}$$

CH^+ is thought to form in a similar way, with the carbon ion replacing the oxygen atom:

$$C^+ + H_2 \longrightarrow CH^+ + H \tag{2.14}$$

$$C^+ + H \longrightarrow CH^+ + h\nu \tag{2.15}$$

It is destroyed by electrons, hydrogen atoms and photons:

$$CH^+ + e \longrightarrow C + H \tag{2.16}$$

$$CH^+ + H \longrightarrow C^+ + H_2 \tag{2.17}$$

$$CH^+ + h\nu \longrightarrow C^+ + H \tag{2.18}$$

The column density of OH thus predicted by Black (1978) is very small (1.5×10^9 cm^{-2}), but it is suggested that (as with HeH^+) this might be detectable in NGC7027, at infrared wavelengths, in absorption against the nebula.

Black (1983) modelled the chemistry of H_2^+ and CO^+. In his scenario an ionization front is moving rapidly into neutral gas rich in H_2 and CO, producing the H_2^+ and CO^+ ions:

$$CO + h\nu \longrightarrow CO^+ \quad (2.19)$$

$$H_2 + h\nu \longrightarrow H_2^+ \quad (2.20)$$

The survival time of these species against dissociative recombination:

$$CO^+ + e \longrightarrow C + O \quad (2.21)$$

$$H_2^+ + e \longrightarrow 2H \quad (2.22)$$

is very short, resulting in a very thin region of H_2^+ and CO^+ immediately behind the ionization front. The predicted column density of CO^+ (7×10^{11} cm^{-2}) is within a factor of a few of the value (3.9×10^{12} cm^{-2}) derived from the observations of NGC7027 by Latter et al. (1993). However, it is not certain that direct ionization of CO is the most important way of forming this species in the specific conditions in the nebula. For example, according to the simplest form of the interacting stellar winds model, the ionization front moves only slowly through the shocked, compressed neutral gas, after reaching its initial Strömgren radius. If, in such a case, the ionization front is preceded by dissociation fronts of H_2 and CO within the largely neutral component, then this would reduce the importance of the formation mechanism suggested by Black (1983). However, other processes, such as instabilities and turbulent mixing, may introduce some CO into the ionized gas. Nevertheless, neutral PDR processes seem to show more potential (see Section 2.2.2).

In this highly ionized region and its interface with the neutral gas, molecule formation is evidently very inefficient, and probably involves mostly diatomic species. As pointed out by Black (1978), larger molecules, such as H_2O and H_3^+ are unlikely to form observable abundances, as they require reactions between two molecules. A much more suitable environment is the mainly neutral gas.

2.2.2 Chemistry in the Neutral Gas

We now turn to regions C and D in Fig. 3. The diatomic molecules H_2 and CO are the most abundant in PPNe and PNe, as they are in AGB winds (and many other astrophysical regions). In PPNe and PNe they occur in neutral gas, either in the main nebula, or in clumps. The robustness of these species in such harsh conditions is partly due to the fact that they are photodissociated through line rather than continuum absorption, so that self-shielding can occur. If these species are being formed (as in interstellar clouds), or physically fed into a region (as in AGB winds), these molecules, on being dissociated, will each remove one dissociating photon from the radiation passing through, and H_2 and CO in regions further from the source of UV will be shielded, allowing them to exist where other molecules are destroyed (by UV photons lying outside the lines). In a PN or a PPN, CO and H_2 can only be maintained by reformation, as mass ejection by the central star has largely ceased. By contrast, in AGB winds, these species are being produced by the star, and only need to be shielded from the relatively weak interstellar UV field, so they are able to exist well out into regions where the density and temperature are too low for much chemistry to occur. The locking up of C and O in the form of CO thus places limits on AGB chemistry, with oxygen-rich objects forming few other carbon containing molecules, and carbon-rich objects (e.g. IRC+10216) forming few oxygen containing ones. However, during PN formation, large UV fluxes are likely to impinge on dense, hot gas, and the products of any destruction of CO and H_2 would enter into the chemistry in an important way, as long as significant amounts of gas are inadequately self shielded (if only a small amount of gas suffers dissociation of CO and H_2, the products of any subsequent chemistry will not be abundant enough to be observed).

Howe et al. (1992) calculated the abundances of various molecules in the neutral remnant AGB gas during the PPN phase, including the effect of the passage of an outward moving shock. The opacity of the dust component falls rapidly, as the inner edge of the AGB remnant moves away from the central star (see Section 1). Once the star becomes hot, and emits a strong UV flux, the remnant AGB molecules are thus predicted to be photolysed very quickly (in about 100 years or so). The shock in the model of Howe et al. (1992) expands at 25 km s^{-1}, while the AGB wind speed is 15 km s^{-1}. The resulting 10 km s^{-1} shock heats and compresses the gas, sweeping it into a progressively more massive shell. This heating initiates carbon and (to a lesser extent) nitrogen hydrogenation reactions, but cooling of the postshock gas (by H_2 quadrupole transitions) is rapid, resulting in immediate postshock peaks in the abundances of these molecules. However, though the cooling reduces the magnitude of many temperature sensitive reaction rate coefficients, the fractional molecular abundances in the cooled postshock gas are still higher than in the preshock gas, as the cooling is accompanied by further compression, enhancing collision rates.

Any molecules existing during or after the PPN phase must either be remnants of the AGB stage or be created in the harsh conditions of PN formation. Most AGB remnant molecules can only survive for significant times in regions well shielded by dust, either in clumps and/or in the very early stages, when the AGB dust opacity is still high. For example, HC_5N has been detected in CRL618 (Bujarrabal et al. 1988), and is almost certainly a remnant of the AGB precursor chemistry, while much of the ammonia (NH_3) in this same object is thought to be the result of formation in dense, shocked clumpy gas, since the beginning of transition to PN (Martín-Pintado et al. 1993). Though the conditions used in the model of Howe et al. (1992) do not permit the formation of significant amounts of ammonia, they do cause hydrogenation to be the main source of molecules (other than H_2 and CO) in the shocked neutral gas. The formation of molecules in significant abundances must be carried out in spite of the intense stellar UV flux, which is much stronger than in most of the ISM (for example, typical unshielded photodissociation survival timescales may be of the order of an hour or so, instead of a hundred years, as in most of the interstellar medium).

Hydrogenation of atom X under these conditions proceeds typically through abstraction of hydrogen atoms from H_2 molecules, by ions and/or neutral species:

$$XH_n^+ + H_2 \longrightarrow XH_{n+1}^+ + H \tag{2.23}$$

$$XH_n + H_2 \longrightarrow XH_{n+1} + H \tag{2.24}$$

(where $n = 0, 1, 2, \ldots$)

The former have to compete with the destructive effects of photodissociation and/or dissociative recombination, while the latter must compete mainly with photodissociation. Though photoionization of neutral molecules often occurs, this tends merely to change the chemistry from process (2.24) to process (2.23), which may even be more efficient, as reactions of the form (2.24) usually have endothermicities of the order of several thousand kelvin. A major hurdle in hydrogenation via processes (2.23) and (2.24) is often the first step. For example, the reaction:

$$C + H_2 \longrightarrow CH + H \tag{2.25}$$

has an energy barrier of about 1.4×10^4 K. The reaction:

$$C^+ + H_2 \longrightarrow CH^+ + H \tag{2.26}$$

has a much lower barrier of 4640 K, and will be more important, owing to the high rate of photoionization of atomic carbon. However, if high densities are achieved, for example in gas which has cooled since the passage of the shock, the radiative association:

$$C^+ + H_2 \longrightarrow CH_2^+ + h\nu \tag{2.27}$$

may be more important, despite its low rate coefficient. Preliminary results from ISO indicating the detection of CH^+ in NGC7027 (Cernicharo et al. 1997) suggest that carbon hydrogenation chemistry of this kind is occurring.

The failure of the model of Howe et al. (1992) to produce ammonia is due to two features of nitrogen chemistry, combined with the parameters of their model. The ionization potential of atomic nitrogen is higher than that of hydrogen, meaning that in the neutral regions considered in their model, all photons able to ionize nitrogen were assumed absorbed in the ionized part of the nebula. In addition, the reaction:

$$N + H_2 \longrightarrow NH + H \tag{2.28}$$

has an endothermicity of 10 700 K, making it fairly slow even in the gas immediately behind their 10 km s^{-1} shock, whose temperature would be below 4000 K. Shock speeds of double this or more are quite feasible, easily overcoming the endothermicities of nitrogen hydrogenation, and would be expected to reproduce the required NH_3 abundances (the postshock temperature is proportional to the square of the shock speed).

Howe et al. (1992) made some simplifying assumptions, important among which are that negligible amounts of CO and H_2 are photolysed by the stellar

UV flux, and that the central star "switches on" instantaneously (i.e. it suddenly achieves a temperature of 3×10^4 K, sufficient to produce a large UV flux, while the ejected AGB wind is still very close by). The assumption of negligible dissociation of CO and H_2 by stellar UV would require that any dissociation occur in a negligibly thin "skin" of dense gas, where re-formation is rapid enough to maintain an abundance sufficient to shield the gas further from the star. This assumption was based on the facts that CO is widely observed in the neutral haloes of gas surrounding the bright, ionized inner parts of PNe (see e.g. Huggins and Healy 1989), and that the CO dissociation model of Bachiller et al. (1988) predicts survival of significant amounts of CO over timescales (up to ~300 yr after the star becomes hot) similar to those considered in the model of Howe et al. (1992). Dissociation of CO would be expected to release C and O, allowing enhanced abundances of carbon bearing species, and introducing oxygen chemistry in carbon-rich objects.

The possible dissociation of H_2, whether by photons, or by rapid chemistry (for instance H atom abstraction in reactions with ions) raises the question of its re-formation mechanism. In cold gas, H_2 is formed predominantly by grain surface catalysis. This process requires either cool grains (no more than about 20 K) or the existence of deep potential wells on the grain surface. It is not clear whether re-formation of H_2 will be efficient under the violent physical conditions encountered in PPNe and PNe. As H_2 is a potentially important coolant in the shocked AGB gas, and temperature being a significant factor in the chemistry, influencing reaction rate coefficients, as well as density, one can see how important it could be to include the dissociation and self shielding of H_2 (as well as CO) in chemical models of PPNe and PNe.

For more evolved objects, such as NGC7027, Gouldsworthy and Flower (1993) adapted a model for the edge of a PDR (Abgrall et al. 1992) in order to estimate the abundances of molecules in the neutral halo of a PN. They predicted high column densities of several polyatomic species, though their preliminary work has not been followed up or described in detail.

As mentioned in the previous subsection, the CO^+ observed in NGC7027 (Latter et al. 1993) seems more likely to form in the neutral gas, rather than the ionized region. For example, column densities comparable to those observed were obtained in the PDR models of Burton et al. (1990) and Sternberg and Dalgarno (1995). Formation of CO^+ as well as HCO^+ in the neutral gas of PPNe and PNe probably occurs as part of the CO formation process, in regions where large amounts of free carbon and oxygen are maintained by CO photodissociation. In this mechanism, the formation of OH occurs by atomic oxygen reacting with molecular hydrogen, either endothermically:

$$O + H_2 \longrightarrow OH + H \qquad (2.29)$$

or with H_2 which has been vibrationally excited by UV photons:

$$O + H_2^* \longrightarrow OH + H \qquad (2.30)$$

The dominant form of carbon in these regions will be C^+, owing to the ionization potential of atomic carbon being lower than that of H, He, or H_2. Therefore the following reaction is likely to form CO^+:

$$OH + C^+ \longrightarrow CO^+ + H \tag{2.31}$$

Destruction of CO^+ occurs by reaction with H and H_2, the latter being a major source of HCO^+:

$$CO^+ + H \longrightarrow CO + H^+ \tag{2.32}$$

$$CO^+ + H_2 \longrightarrow HCO^+ + H \tag{2.33}$$

The HCO^+ ion is likely to be destroyed by photodissociation (back to CO^+) or dissociative recombination (forming CO):

$$HCO^+ + h\nu \longrightarrow CO^+ + H \tag{2.34}$$

$$HCO^+ + e \longrightarrow CO + H \tag{2.35}$$

the CO being destroyed by photodissociation. The intimate connection between the chemistry of CO^+ and HCO^+ suggests that they, and perhaps OH, may be observed together.

About 1000 years after the end of the AGB mass loss, a PN becomes more diffuse, with the exception of small knots or globules (which we consider below). This, in spite of the dilution of the stellar UV flux (due to the nebula being further from its source, and possible fading of the central star), means that formation of significant amounts of polyatomic species may depend either on the existence of dense condensations (see subsection 2.2.3), or on the dissociation of the CO molecule, thus liberating potential reactants, previously locked in the form of CO.

2.2.3 Chemistry in Neutral "Cometary Globules"

The Helix Nebula (NGC7293) is seen to have many condensations within the ionized gas. These are neutral, and show "tails" directed away from the central star, and are hence called "cometary globules". Their origin has been variously attributed to Rayleigh–Taylor instabilities (Capriotti 1973); very dense condensations present in the AGB, which survive PN formation (Dyson et al. 1989); and even remnant comet-like objects—albeit much more massive than those in our Solar System (Gussie 1995). They are thought to have a number density up to about 10^6 hydrogen nuclei cm^{-3} and a kinetic temperature of about 20 K; each has a mass of of up to $\sim 10^{-5}$ $M_\odot$. They are observable via continuum absorption of photons emitted by O^{++} in the highly ionized part of the nebula, indicating that each globule contains dust creating about 0.5 magnitudes of extinction at visual wavelengths (Meaburn et al. 1992; Huggins et al. 1992). For assumed dust optical properties similar to those in the interstellar medium, the high density and optical depth, coupled with the star being both further away and of much lower luminosity than in a younger PN, permit CO and H_2 to be almost fully associated over much of the volume of the clump (Howe et al. 1994). On the assumption that these are clumps which have existed since the AGB phase and have survived the sweeping out of the less dense gas during PN formation, Howe et al. (1992) showed that large molecules characteristic of the AGB could exist in such a clump well into PN formation. Howe et al. (1994) considered even later epochs, when the clump has fallen in opacity sufficiently for any molecules other than CO and H_2 to be the result of chemical processes in the PN stage. As photodissociation timescales are typically of the order of years in such an environment, and thus much smaller than those for any significant morphological changes, steady state chemical models are suitable. The effect of self shielding by CO and H_2 can be treated by sequential solution for the chemical abundances at positions in the globule that are progressively further from the star. The previously calculated abundances of CO and H_2 for shallower depths can be used to estimate the optical depths in the CO and H_2 dissociation bands at the next spatial position (e.g. Wagenblast 1992). As the chemistry of carbon is much richer than that of oxygen, the chemistry in these and similar globules in other nebulae (e.g. see Meaburn and Lopez 1993) is likely to be richer in carbon-rich PNe, where carbon remains available after most oxygen has been incorporated into CO. However, even if high abundances are achieved, the detectability of molecules is likely to present a problem, since the mass of the globules is relatively low. For a carbon-rich globule, Howe et al. (1994) derived potentially observable amounts of CN and C_2H. If the globule is oxygen-rich, these species are only abundant in a thin skin facing the star, where significant amounts of available carbon are maintained through photodissociation by stellar UV. A carbon-rich composition for the Helix Nebula has been suggested by Young et al. (1997), who have detected neutral atomic carbon with a large abundance relative to CO (as in the carbon-rich model of Howe et al. 1994). Recent observations (e.g. Bachiller et al. 1997) of several species in various PNe (including the Helix), suggest that they occur in dense clumps with physical conditions similar to the cometary globules studied by Howe et al. (1994). In particular the CN abundance is roughly compatible with the carbon-rich case of Howe et al. (1994), though HNC and HCO^+ are more abundant than predicted, and may indicate either a difference in physical conditions between the cometary globules and the clumps observed by Young et al. (1997), or the omission of some important physical or chemical processes in the model of Howe et al. (1994).

3 Future Developments

Observationally, the continuing advances in telescope design and detector sensitivity will no doubt allow the identification of yet more molecular species, especially in young PNe and PPNe, and improve our knowledge of their dynamics and morphology. Current theoretical models indicate the importance of searching for HeH^+ in NGC7027, as well as in PPNe such as CRL618, and for simple molecules, especially CN and C_2H, in the cometary globules of the Helix Nebula (NGC7293). The conditions in neutral gas facing a hot star in young objects can cause CO to be dissociated, either by UV photons or fast shocks, and may allow oxygen-bearing species (such as OH and H_2O) to occur in carbon-rich nebulae, and carbon based species (such as CH and CH_2) to occur in oxygen-rich nebulae. Ratios of, for example, HCN/HNC and CN/HCN should be monitored frequently (over timescales of 10 years or so) in PPNe, as they appear to be sensitive diagnostics of the early stages of the AGB–PN transition. Laboratory measurements of reaction rate coefficients will continue to be of major importance for the theoretical chemical models of PNe and PPNe, as for other branches of theoretical molecular astrophysics. Much of the modelling of PNe and PPNe so far performed is of a preliminary nature. Simplifying assumptions used include static and/or symmetric and/or constant density morphologies, full association of CO and H_2, and instantaneous stellar heating. Other simplifications involve the assumption of a sudden cessation of the dense AGB wind, producing a detached shell of gas, moving uniformly out from the star, until the fast wind begins, equally suddenly. Though detached winds are observed in PAGB objects (see e.g. Olofsson et al. 1990), the transition to a fast, low density wind is undoubtedly more complex. The fast molecular flows, of the order of 100 km s^{-1}, observed in PPNe may indicate a wind which is increasing in speed while the mass loss rate falls. Future developments will no doubt produce models more applicable to the special physical and morphological features of PNe and PPNe, especially regarding the interaction of the AGB and fast winds, including the effects of instabilities, which may develop over timescales of several hundred years, while a rich chemistry may still be occurring (see e.g. Breitschwerdt and Kahn 1990). Of particular importance for future models (at least of the neutral gas), is the question of the formation mechanism of H_2. This molecule, where it exists, plays a very important role, frequently being the first reaction partner encountered by newly formed ions and radicals. H_2 may also exert an important influence on the gas temperature, for example in the shocked neutral gas in PPNe, as emission via quadrupole transitions can be an important source of cooling for hot (above $\sim$1000 K) gas (Hartquist et al. 1980). Also, when H_2 is formed on grain surfaces, it may be ejected with significant kinetic energy to heat the gas, when distributed by collisions. In some models, formation of H_2 on grain surfaces is assumed, as in the interstellar medium (e.g. Howe et al. 1992; Howe et al. 1994; Cecchi-Pestellini and Dalgarno 1993); in others, the grains, being close to an intense source of photons, are assumed to be too warm for this to occur (Black 1978). This is likely to be crucial for models of the neutral gas, where H_2 is important chemically, as well as thermally. In their model of the ionized gas, Cecchi-Pestellini and Dalgarno (1993) obtained HeH$^+$ abundances which differed only slightly with and without the presence of dust, whereas in models of the neutral gas, the H_2 "dissociation front" will represent a boundary between different types of chemistry.

Like H_2, CO self shields, and is also partly shielded by H_2 owing to the overlapping of spectral lines. It will therefore have its own dissociation front, at least at early times and/or in condensations, where the higher density assists its re-formation. It would be of interest to extend the theoretical modelling of the PPN stage by incorporating the effects of both the expanding shockfront and the dissociation of CO and H_2. The PPN model of Howe et al. (1992) relies on a high carbon/oxygen ratio in order to make carbon available for formation of hydrogenated species such as CH_3^+. Dissociation of CO in the presence of abundant H_2 would be expected to enhance these predicted abundances, by liberating more carbon, as well as liberating atomic oxygen, possibly allowing formation of oxygen bearing species such as OH and H_2O. Preliminary work by Howe et al. (in preparation), though so far using a very simplified chemical reaction network, suggests that a treatment of CO and H_2 dissociation and self shielding could dramatically enrich the chemistry of the neutral gas in a PPN, compared with the simpler model of Howe et al. (1992).

The role of dust should be investigated in more detail. So far, it has been assumed that dust grains remain largely unaffected by the new physical conditions encountered after the AGB, their functions being to serve either as a source of heating (via photoelectric ejection of electrons), of continuum shielding from stellar and interstellar UV photons, or as formation sites of H_2 from atomic hydrogen. It has been suggested, for example (Jura and Kroto 1990), that differential acceleration of grains of different sizes may result in grain–grain collisions of sufficient energy to shatter them, releasing large molecules such as HC_7N, which has been observed in the PPN CRL2688 (N-Q-Rieu et al. 1986). In addition, a detection of the PAH molecule chrysene ($C_{18}H_{12}$) has been reported (Justtanont et al. 1996), and it is likely that this species, if indeed present, has also formed from the disruption of dust grains in the violent conditions in young PAGB objects. The nature of any products of dust erosion and shattering is of interest in connection with AGB dust formation models as well as the replenishment of interstellar dust.

The chemistry in condensations should be extended through the study of clumps of lower density than those considered by Howe et al. (1992, 1994)—which were assumed to be affected little by the dynamics of PN formation. Dynamics may also be important; for example, the NH_3 observed in a fast molecular flow in CRL618 (Martín-Pintado et al. 1993) has a fractional abundance higher than in this object's AGB remnant, indicating that NH_3 is being formed during the PPN phase. Martín-Pintado et al. (1993) suggested that this molecule is formed in shocked clumps. The PPN model of Howe et al. (1992) predicted negligible NH_3, probably due to the assumption of a fairly slow shock. Therefore, work on the chemistry of material behind faster shocks would be of relevance to these data.

Bibliography

1. Abgrall, H., LeBourlot, J., Pineau des Forêts, G., Roueff, E., Flower, D. R. and Heck, L. (1992). *A&A*, **253**, 525
2. Bachiller, R., Gómez-Gonzalez, J., Bujarrabal, V. and Martín-Pintado, J. (1988). *A&A*, **196**, L5

3. Bachiller, R., Planesas, P., Martín-Pintado, J., Bujarrabal, V. and Tafalla, M. (1989). *A&A*, **210**, 366
4. Bachiller, R., Forveille, T., Huggins, P. J. and Cox, P. (1997). *A&A*, **324**, 1123.
5. Bakker, E. J., Waters, L. B. F. M., Lamers, H. J. G. L. M., Trams, N. R. and Van der Wolf, F. L. A. (1996). *A&A*, **310**, 893
6. Bidelman, W. P. (1950). *Ap.J.*, **113**, 304
7. Bieging, J. H. and N-Q-Rieu (1988). *Ap.J.*, **324**, 516
8. Black, J. H. (1978). *Ap.J.*, **222**, 125
9. Black, J. H. (1983). In *Planetary Nebulae* IAU symp.103 p. 91 ed. D. R. Flower. Reidel, Dordrecht
10. Breitschwerdt, D. and Kahn, F. D. (1990). *MNRAS*, **244**, 521
11. Bujarrabal, V., Gómez-Gonzalez, J., Bachiller, R. and Martín-Pintado, J. (1988). *A&A*, **204**, 242
12. Bujarrabal, V., Fuente, A. and Omont, A. (1994). *A&A*, **285**, 247
13. Burton, M. G., Hollenbach, D. J. and Tielens, A. G. G. M. (1990). *Ap.J.*, **365**, 620
14. Capriotti, E. R. (1973). *Ap.J.*, **179**, 495
15. Cecchi-Pestellini, C. and Dalgarno, A. (1993). *Ap.J.*, **413**, 611
16. Cernicharo, J., Liu, X.-W., González-Alfonso, E., Cox, P., Barlow, M, J., Lim, T. and Swinyard, B. M. (1997). *Ap.J.*, **483**, L65
17. Cox, P., Omont, A., Huggins, P. J., Bachiller, R. and Forveille, T. (1992). *A&A*, **266**, 420
18. Dalgarno, A. and Lepp, S. (1985). In *"Astrochemistry"* ed. S. P. Tarafdar and M. P. Varshni. Reidel, Dordrecht
19. Dyson, J. E., Hartquist, T. W., Pettini, M. and Smith, L. J. (1989). *MNRAS*, **241**, 625
20. Gouldsworthy, S. N. and Flower, D. R. (1993). In *"Planetary Nebulae"* IAU symp.155 p.221 ed. R. Weinberger and A. Acker. Kluwer, Dordrecht
21. Gussie, G. (1995). *Pub.Astron.Soc.Aus.*, **12**, 170
22. Hartquist, T. W., Oppenheimer, M. and Dalgarno, A. (1980). *Ap.J.*, **236**, 182
23. Healy, A. P. and Huggins, P. J. (1990). *A. J.*, **100**, 511
24. Howe, D. A., Millar, T. J. and Williams, D. A. (1992). *MNRAS*, **255**, 217
25. Howe, D. A., Hartquist, T. W. and Williams, D. A. (1994). *MNRAS*, **271**, 811
26. Huggins, P. J., Bachiller, R., Cox, P. and Forveille, T. (1992). *Ap.J.*, **401**, L43
27. Huggins, P. J. and Healy, A. P. (1989). *Ap.J.*, **346**, 201
28. Iben, I. (1995). *Phys.Reports*, **250**, 2
29. Jura, M. (1986). *Ap.J.*, **309**, 732

30. Jura, M. and Kroto, H. (1990). *Ap.J.*, **351**, 222
31. Justtanont, K., Barlow, M. J., Skinner, C. J., Roche, P. F., Aitken, D. K. and Smith, C. H. (1996). *A&A*, **309**, 612
32. Kahn, F. D. (1983). In *"Planetary nebulae"*, IAU symp. 103 p.305 ed. D. R. Flower. Reidel, Dordrecht
33. Kastner, J. H., Weintraub, D. A., Gatley, I., Merrill, K. M. and Probst, R. G. (1996). *Ap.J.*, **462**, 777
34. Keady, J. J. and Ridgway, S. T. (1993). *Ap.J.*, **406**, 199
35. Kwok, S. (1983). In *"Planetary nebulae"*, IAU symp.103 p.293 ed. D. R. Flower. Reidel, Dordrecht
36. Latter, W. B., Walker, C. K. and Maloney, P. R. (1993). *Ap.J.*, **419**, L97
37. Martín-Pintado, J., Gaume, R., Bachiller, R. and Johnston, K. (1993). *Ap.J.*, **419**, 725
38. Martín-Pintado, J., Gaume, R. A., Johnston, K. J. and Bachiller, R. (1995). *Ap.J.*, **446**, 687
39. Meaburn, J. (1995). *Ap&SS*, **233**, 1
40. Meaburn, J. and Lopez, J. A. (1993). *MNRAS*, **263**, 890
41. Meaburn, J., Walsh, J. R., Clegg, R. E. S., Walton, N. A., Taylor, D. and Berry, D. S. (1992). *MNRAS*, **255**, 177
42. Moorhead, J. M., Lowe, R. P., Maillard, J. P., Wahalu, W. H. and Bernath, P. F. (1988). *Ap.J.*, **326**, 899
43. Morris, M., Guilloteau, S., Lucas, R. and Omont, A. (1987). *Ap.J.*, **321**, 888
44. Nassau, J. J., Stephenson, C. B. and MacConnell, D. J. (1965). *Luminous stars in the northern Milky Way VI*, Hamburg Sternwarte and Warner and Swasey Observatory, Hamburg
45. Neri, R., García-Burillo, S., Guélin, M., Guilloteau, S. and Lucas, R. (1992). *A&A*, **262**, 544
46. Olofsson, H., Carlström, V., Eriksson, K., Gustafsson, B. and Willson, L. A. (1990). *A&A*, **230**, L13
47. Riera, A., Garcia-Lario, P., Manchado, A., Pottasch, S. R. and Raga, A. C. (1996). *A&A*, **302**, 137
48. N-Q-Rieu and Bieging, J. H. (1990). *Ap.J.*, **359**, 131
49. N-Q-Rieu, Winnberg, A. and Bujarrabal, V. (1986). *A&A*, **165**, 204
50. Roberge, W. and Dalgarno, A. (1982). *Ap.J.*, **255**, 489
51. Sahai, R., Wootten, A., Schwarz, H. E. and Wild, W. (1994). *Ap.J.*, **428**, 237
52. Sternberg, A. and Dalgarno, A. (1995). *Ap.JS.*, **99**, 565
53. Trammell, S. R. and Goodrich, R. W (1996). *Ap.J.*, **468**, L107
54. Truong-Bach, Graham, D. and N-Q-Rieu (1993). *A&A*, **277**, 133

55. Wagenblast, R. (1992). *MNRAS*, **259**, 1 155
56. Waters, L. B. F. M., Waelkens, C. and Trams, N. R. (1993). In *Mass loss in the AGB and beyond*, p.298 ed. H. E. Schwartz. European Southern Observatory Conference and Workshop Proceedings No. 46, Garching.
57. Yamamura, I., Onaka, T., Kamijo, F., Deguchi, S., and Ukita, N. (1995). *Ap.J.*, **439**, L13
58. Young, K., Cox, P., Huggins, P. J., Forveille, T. and Bachiller, R. (1997). *Ap.J.*, **482**, L101
59. Zygelman, B. and Dalgarno, A. (1990). *Ap.J.*, **36**, 2, 239

17
Dust Formation in the Environments of Hot Stars

Holger Beck and Erwin Sedlmayr
Institut für Astronomie und Astrophysik, Technische Universität Berlin

1 Introduction

Dust is a ubiquitous component of the interstellar medium (see Chapter 4 and 5) and plays a vital and indispensable role in the overall cycle of cosmic material processing. During stellar evolution, from stellar birth up to the late stages, the solid phase is intimately interwoven with the gas component and the radiation field (see Chapters 11–13). The processes of nucleation and grain growth are well known to occur in the outflows of cool late-type stars.

Surprisingly hot stellar environments have also been identified to be places of efficient grain production. Although their central objects are characterized by stellar temperatures of about 7 000 K (RCB-stars and novae) up to about 25 000 K (Wolf–Rayet and Be-stars), thus supplying a very intensive radiation field interacting with the surrounding material of their shells, nevertheless, dust is still produced. In contrast to cool late-type giants where the physical modelling can be based on the observationally confirmed facts of a steady and nearly spherically symmetric outflow, these assumptions do not hold in the case of hot shells. Unlike the dust driven winds, where the gas temperature and matter density monotonically decrease in a slowly expanding ($v_\infty \approx$ 30–40 km/s) envelope, the outflows of hot stars consist of fast winds (up to 2500 km/s) and non-homogeneous environments.

Thermodynamic conditions favouring the formation of solid structures are generated as a local phenomenon which results from a local density increase and/or temperature decrease. Basically, the situation of primary grain formation can be described in the scenario of a cooling flow, where initially hot matter expands, subsequently cools down, and finally is diluted up to the point where the gas temperature is well below the condensation temperature of the favoured monomer species which can nucleate and form small stable clusters. By further addition of suitable molecules the clusters will grow to homogeneous or heterogeneous specimens (Gail *et al.* 1984; Gail and Sedlmayr 1988).

Many cosmic sources are contributing to the solid phase in the ISM, but dust is created primarily in the vicinity of stars, and the explosive events of novae and supernovae (see Chapters 18 and 19). Although these objects significantly differ in their observational appearances as well as in their stellar masses and temperatures, they have one thing in common: solid particles are formed in their environments. Thus the phase transition of gaseous material to the solid state must be accepted as a very general phenomenon independent from a specific object. It can be viewed as a physical process taking place under certain conditions of pressure and temperature which can be supplied by a variety of different astrophysical situations. Exemplifying this idea, we concentrate in this review on the description of the dust formation in the hot environments of Wolf–Rayet (WR) and R Coronae Borealis (RCB) stars. In this connection "hot" winds are seen in contrast to the dusty outflows of cool AGB-giants, with stellar temperatures below 3500 K.

2 Observational Status of Hot Winds

2.1 Wolf–Rayet Stars

Wolf–Rayet stars are massive, evolved and hot stellar objects at the end of their nuclear burning phase (see Chapter 11), and they have a formative influence on the interstellar environment by their large input of mass and momentum (van der Hucht et al. 1988; van der Hucht 1995; Cohen 1991). WR stars are the sources of hot and dense stellar winds and the emission lines originating from the outflows provide direct information about their physical properties (for typical WR parameters see Tables 1 and 3). Two different emission contributions can be distinguished in their prominent infrared excess: the part emerging from free–free radiation and those photons due to heated circumstellar dust. Analysis is undertaken by combining spectrophotometry between 3300 Å and 11 100 Å, with infrared photometry between 1.6 µm and 11.3 µm (Nussbaumer et al. 1982; Eenens and Williams 1992; van der Hucht et al. 1996). Whereas WN-type stars show pure free-free emission, the IR radiation of some WC stars could be unequivocally attributed to dust particles. There is clear evidence for a correlation between dust formation and the spectral subtype of the WR-star (Cohen et al. 1991). In 18 of 21 WC9 stars emission from heated dust has been found, but only 4 of 9 WC8 and 3 of 13 WC7 stars show this feature. Thus single WC stars earlier than WC8 generally do not show any evidence of dust formation. Exceptions only exist for long period binaries with eccentric orbits. Therefore, an IR excess due to solid particles only develops in cool and carbon enhanced WR environments.

Observations of the *persistent* infrared emission by solid particles are found in about 20% of the WCL stars (WC "late" subtypes: WC7–WC10), which indicates that dust is formed continuously, permanently replacing the older grains, which have been swept further out by the fast wind. Those fossil dust particles outside the inner wind region couple to the outflowing gas on short time scales (ca. 1 day) and are therefore blown away from the star (Williams et al. 1987). This material is now part of some type of ring nebulae (Chu 1991) surrounding some of the WR-stars (Lozinskaia et al. 1986; Cassinelli et al. 1991). The dust clouds are identified by their characteristic 60 µm IRAS images showing a significantly higher colour temperature than the background, which is due to heating of the dust from the central star (Smith 1995). The typical size of a surrounding WR dust cloud has been derived from speckle observations of WR104 at 2.2 µm, 3.8 µm and 4.8 µm (Allen et al. 1981; Dyck et al. 1984) to a value of 0.04 arcsec ($d = 1.9$ kpc).

By contrast, some candidates of type WC4–WC7 show *episodic* dust events. The conditions for grain formation are sustained only for a limited time and then, as it is indicated by the fading of the IR emission, they become dust free for a certain period before showing distinct events of new dust formation. This time-dependent behaviour has been observed in the cases WR137 (Williams et al. 1985) and WR140 (Williams et al. 1978; Hackwell et al. 1979), the most prominent example for episodic dust formation. By analysis of the IR data and its spectroscopic variation this object seems to have a period of 7.94 yr (Annuk et al. 1991; van der Hucht et al. 1991) and a brightening of the IR-luminosity by $\Delta L \approx 2.5$ mag. Thus dust formation takes place at a distance of about 2300 R_*.

The re-occurrence of this phenomenon may be caused by the interaction of stellar winds in a binary system, undergoing a period of compression during periastron. Time-dependent optical depths and phase-dependent orbital modulation due to an eccentric orbit affects the infrared appearance and leads to a substantial variability. Additionally, colliding winds in WR binary systems cause strong shocks emitting X-rays. Weak X-ray emission is a common phenomenon especially for binary WR+OB stars with luminosities of the order of $L_X/L_{bol} \approx 10^{-7}$ (van der Hucht 1992; Rawley 1993; Willis et al. 1995).

Further observational indicators of the interaction of winds in binaries containing WRs are the large and variable non-thermal radio emissions (Dougherty et al. 1996; White and Becker 1995), tracing particle acceleration in the magnetic field of the shock region, where the fast winds of the WR-star ($v \approx 2500$ km/s) and its accompanying O-star ($v \approx 2000$ km/s) collide. Therefore, it may be concluded that episodic dust formation results from an increased gas density in the wind collision zone leading to enhanced infrared emission (see Table 1 for data on episodic dust formation). Therefore, it seems to be very likely that all episodic WR dust producers are long-period binaries with large eccentricities. There is also some evidence for the supposition that persistent WR dust sources are part of colliding wind binaries, whose separations, luminosities and wind parameters fall in a specified range (Williams 1995).

The diagnostic of spectrometric and polarimetric data (Brown and Richardson 1995) clearly demonstrates the inhomogeneous structure of the WR winds (Robert et al. 1989; Brown et al. 1995). Outflowing mass is located in so called "blobs". It is still under investigation whether these blobs arise from non-uniform mass loss or as a result of turbulent instabilities. Poe et al. (1989) developed a two-component wind theory with a fast polar and a slow equatorial wind. In this scenario rapid rotation as a result of WR evolution (Cassinelli 1991, 1992) is an indispensable ingredient. Radio observations of White and Becker (1995) point to the fact that the WR wind is strongly enhanced in the equatorial plane.

Table 1 Typical parameters for persistent and episodic WR dust makers (Williams et al. 1987). R_{dust} is the inner radius of the dust shell, ρ is the mass density at the base of the dust shell and M_{dust} is the total mass of the dust shell.

	WR star	type	R_{dust}/R_*	T_{dust} [K]	ρ [g/cm^3]	M_{dust} [M$_\odot$]
persistent	WR 53	WC8	380	1600	$4.7 \cdot 10^{-21}$	$9.2 \cdot 10^{-8}$
	WR117	WC8	580	1390	$1.1 \cdot 10^{-21}$	$2.3 \cdot 10^{-8}$
	WR 76	WC9	310	1320	$4.2 \cdot 10^{-19}$	$1.5 \cdot 10^{-5}$
	WR104	WC9	470	1110	$1.6 \cdot 10^{-19}$	$6.7 \cdot 10^{-5}$
	WR118	WC10	260	1410	$4.9 \cdot 10^{-19}$	$1.1 \cdot 10^{-5}$
episodic	WR 137	WC7	6130	780	$9.0 \cdot 10^{-24}$	$3.8 \cdot 10^{-8}$
	WR 140	WC7	1490	1460	$2.6 \cdot 10^{-22}$	$1.4 \cdot 10^{-8}$
	WR 48a	WC8	1305	1000	$1.8 \cdot 10^{-19}$	$1.4 \cdot 10^{-5}$

Therefore, most of the mass loss is confined to a plane implying an obvious similarity to the two-component outflows of Be-stars, to which this theoretical approach can also be applied (Cassinelli 1990; Chen et al. 1992).

The mass loss rates of WR stars have been reviewed by Prinja et al. (1990) and Willis (1991) and have typical values of $\dot{M} \approx (2\text{--}10) \cdot 10^{-5} M_\odot \text{yr}^{-1}$. Those mass loss rates are deduced from the electron densities and their values depend on the uncertain state of ionization. To date there exists no apparent difference in the mass loss rates of single and binary WR-stars.

Based on the estimations of Gehrz (1989) and Tielens (1990), Cohen (1991) derived the contribution of WR stars to the galactic dust component, utilizing the galactic model of Wainscoat et al. (1992). Despite the fact that approximately 85% of WC9 and about 50% of the WC8 stars show evidence for heated circumstellar dust, the contribution of WC stars to the galactic dust production is only small (ca. 1%). However, massive WR stars provide a substantial part of the heavy element enrichment in the interstellar medium (Cohen 1991) during their lifetime of ca. $5 \cdot 10^5$ yr. This results in a typical mass loss of $M \approx 1 M_\odot$ per star. Early analysis of IR-data (Gehrz and Hackwell 1974; Dyck et al. 1984) first led to the assumption of a graphitic nature of WR dust. The alternative suggestion of Lewis and Ney (1979), that Fe or FeC (cohenite) are possible condensates around WR stars, could not be confirmed by later observations. Further studies of Williams et al. (1987) showed that the infrared emission spectra from WR140 can be best fitted with the assumption of a power law for the mass absorption coefficient like that of amorphous carbon ($\kappa \propto \lambda^{-1}$). It seems to be certain that pure graphite can be ruled out because neither in the persistent dust-maker WR104 nor in the episodic case of WR140 has the 11.52 µm resonance feature been observed (Glasse et al. 1986; Williams, 1995), which would undoubtedly allow for an unambiguous identification.

The temperature of the heated circumstellar amorphous carbon dust has been deduced to lie between 900 K and 1250 K (Allen et al. 1972; Cohen et al. 1975).

For WR104 and WR112A a broad emission feature exists (IRAS spectra at 7.7 µm). This emission is attributed to a carbonaceous carrier with small domains of aromatic structure (Cohen et al. 1989). However, the common 11.3 µm feature indicating PAHs is absent. The nature of the solid phase is in agreement with the observed enhanced abundances of carbon in WC star shells with $[(C + O)/He] \geq 0.03$ (Smith and Maeder 1991).

The dominant observed carbon ion is C^{2+} (Smith and Hummer 1988). No emission of molecular carbon, suspected of being an important precursor of carbon dust formation, has been detected in WR spectra. Since the quantity of neutral seed particles needed to start the nucleation process is only a small part of the amount of dust subsequently formed, the density of nuclei is possibly too low for them to be detectable.

2.2 R Coronae Borealis Stars

Another class of objects showing spectacular behaviour connected with solid particle formation is that of the R Coronae Borealis stars (RCB). About 40 objects have been identified to date (Clayton 1996; Milone 1990). These low-mass but high luminosity stars, mostly with effective temperatures of $T_{\text{eff}} \approx 5000\text{--}7000$ K, are characterized by non-predictable decline events accompanied by brightness decreases of up to 8 magnitudes (for RCB parameters see Table 3). During an event optically thick dust clouds are produced which re-radiate about 10% to 50% of the stellar energy in the infrared spectral region. Spectroscopic observations (Alexander et al. 1972; Cottrell et al. 1990; Rao and Lambert 1993a,b; Clayton et al. 1995a) have been made for several declines of the three brightest RCBs - RCrB, RY Sag, V854 Cen. During the phase of brightness decrease, typically lasting for some weeks, an emission spectrum appears simultaneously with the dust cloud, showing lines of neutral and single ionized metals. In some cases lines of C_2 and CN bands have also been observed (Whitney et al. 1992; Benson et al. 1994).

Recent investigations to detect the presence of fullerenes in the atmospheres of RCBs (Clayton et al. 1995b; Jeffery 1995b) have not been successful. Already Buss et al. (1993) have pointed out that, from the low resolution of the IR spectroscopic data, definite spectral identification of characteristic dust features cannot be obtained. The extinction curve has been analyzed by Hecht (1991) pointing to amorphous carbon as the most likely form of the dust. The grain sizes derived by the extinction spectrum during declines range from 0.005 µm to 0.06 µm (Hecht et al. 1984).

RCB stars normally show elemental abundances differing significantly from the solar mixture. Helium is by far the most abundant element, hydrogen is under-abundant by a factor of 10^2 to 10^5, whereas carbon is overabundant by a factor 10 ($[C/He] \approx 0.03$ and $[C/Fe] \approx 1$ (Lambert 1986; Rao and Lambert 1994).

Comparison of different wavelength bands of infrared photometry (Feast 1990) shows that dust forms only over a small solid angle over the surface of the

RCBs, not producing a complete shell, but establishing a massive cloud. This observational clue is supported by the fact that there is no obvious correlation between the IR brightness and the frequency of decline events, implying that dust is only produced in the line of sight of the observer in some random cases (Feast 1986).

Pulsations are now identified in all RCB stars by photometry over long periods (Lawson et al. 1994; Lawson and Kilkenny 1996). Long term spectroscopic and polarimetric measurements of RCrB (Clayton et al. 1995a) near maximum light show large C_2 and CN band variations correlated with pulsational phase. Additionally, a coincidence of pulsational phase and the onset of decline (RY Sgr, V854 Cen) strongly indicates a connection between the RCB star atmosphere itself and the process of dust formation (Clayton et al. 1995b). These facts imply that pulsation induced shocks could lead to a substantial local density increase of volume elements of gas which, first heated but then effectively cooling down, thus intersect favourable regions for grain formation in the p–T plane (Woitke et al. 1996).

2.3 General Aspects

WR stars and RCB stars are hot stellar objects surrounded by a rather hostile environments for chemistry and dust formation (see Table 2), compared to those around late-type stars. However, dust formation associated with these objects seems to be a quite natural phenomenon. In some cases the dust even determines

Table 2 Different astrophysical situations and their relevant physical parameters and time scales for dust formation: $n_{<H>}$ total density of hydrogen in the gas, t_{typ} is the hydrodynamical time scale of the system and t_{cond} is a typical condensation time scale of carbon dust. For comparison the values for the hot winds of classical novae have been added (Sedlmayr and Dominik 1995).

object	$T[K]$	$\log n_{<H>}[\text{cm}^{-3}]$	$\log t_{\text{typ}}[s]$	$\log t^C_{\text{cond}}[s]$
R Coronae Borealis stars	7000	7–9	≈ 6–7	≈ 9
Wolf–Rayet stars	< 25000	6–8	≈ 8	≈ 9
Classical novae	≈ 10000	≈ 8–10	≈ 6–7	≈ 7–8

the structures of the stellar winds (Sedlmayr and Dominik 1995; Gail and Sedlmayr 1987a, 1985). For the majority of the RCBs a photospheric temperature of about 7000 K seems to be appropriate. However, in some peculiar cases like DY Cen, MV Sgr and V348 Sgr, the temperature is substantially higher, up to 20 000 K (Drilling et al. 1984; Jeffery and Heber 1993; Jeffery 1995a), comparable with the WCL stars. The elemental abundances for both classes of objects vary, but the large carbon overabundance, in combination with an absence of hydrogen, is a common feature. The derived elemental composition of the atmospheres is sensitive to the radiative transfer used to model the spectrum. Depending on the intrinsic parameters and the quality of the models, large

Table 3 Parameters characterizing WR and RCB stars. For comparison the values of classical novae have been added. Notes: ^{a} mass loss only in connection with dust. ^{b} velocity seen in spectral lines during decline phases. * values are approximately valid for maximum light.

	WR	RCB	Novae
system class.	single/double WCL	only single F–G supergiant	only double A–F*
L_* [$L_\odot$]	$10^{4.5}$–$10^{5.9}$	$\approx 10^4$	10^5
T_* [K]	< 25.000	5.000–7.000	7.000–10.000*
R_* [$R_\odot$]	2–5	$\approx$ 60–80	70*
M_* [$M_\odot$]	30–50	$\approx$ 1	$\leq$ 1
$\dot{M}$ [$M_\odot$/yr]	$\approx 10^{-5}$	$(10^{-7})^a$	10^{-4}–10^{-5}
v_{wind} [km/s]	1000–2000	$(50–200)^b$	500–3000

differences in the deduced abundances can occur, as can be seen for RCBs in Table 4. Switching from the model grid of Schönberner (1975) to the Uppsala models (Gustafsson and Asplund 1996), which include updated opacity data, demanded a revised input ratio of [C/He] = 0.1 resulting in a change of the abundances by −1 dex.

Nevertheless, in all those RCB environments with carbon abundances ϵ_C ranging from log ϵ_C = 7.8 (MV Sgr) to log ϵ_C = 10.6 (V348 Sgr) and hydrogen abundances ϵ_H from log $\epsilon_H \leq$ 4.1 (XX Cam) to log $\epsilon_H \leq$ 10.8 (V348 Sgr) (Jurcsik 1996) solid particles are effectively formed. Therefore, one may conclude that a phase transition can take place despite the specific elemental composition of the gas. This conclusion also applies to WR stars, where the abundances also vary widely (Willis 1996). One important difference between WR and RCB stars is

Table 4 Abundances of WR and RCB star. For RCB stars the abundances were taken from Rao and Lambert (1994) (RCB-a) and Rao and Lambert (1996) (RCB-b) which are both valid for the majority of the RCBs (for details see text). However there are substantial deviations from these values for about four stars (V854Cen, VCra, V3795 Sgr, VZ Sgr). The RCB and the solar abundances (Grevesse and Noels 1993) are given as logarithmic numbers. For the WR stars the given abundance ratios are taken from Willis (1996).

	H	He	C	N	O	Si	Fe
RCB a	$H/He \ll 1$	11.51	9.8	9.7	9.1	8.0	7.5
RCB b	$H/He \ll 1$	11.54	8.91	8.64	8.20	7.15	6.55
sun	12.00	10.77	8.55	7.97	8.87	7.51	7.55

	H/He	C/He	N/He	O/C
WR	≈ 0	$\approx 0.1 \ldots 0.7$	$\approx 10^{-3}$	≈ 0.2

the type of mass loss. In WR stars a strong radiation driven wind exists, and mass loss is a dramatic effect which can eject well over half of the initial stellar mass (Moffat and Robert 1994). The determination of a more or less reliable value for $\dot{M}$ depends on the underlying theoretical models (e.g. Hamann 1995), which always include assumptions (e.g. a homogeneous atmosphere) to make the problem tractable. In fact, the outflowing gas is far from homogeneously distributed, but shows strong deviations from spherical symmetry (Moffat et al. 1994b; Lepine 1994). Strong shocks induced by the radiative instability of the wind (Antokhin 1995) along with supersonic turbulence (Moffat et al. 1994a) is assumed to be a likely cause of the observed clumpy structure of the winds.

Mass loss in RCBs does not arise from a stellar wind driven by some intrinsic mechanism. In these objects mass loss is a phenomenon periodically initiated by the formation of dust in a clump of limited extension. This characteristic event shapes the course of the lightcurve and may be closely connected to the characteristic 40 day stellar pulsation (Lawson et al. 1992). WR stars, however, mostly vary erratically (Bratschi and Blecha 1996) on a shorter time scale, which might be responsible for the stellar wind acceleration.

3 Ionization Structure and Chemistry

A basic requirement for the applicability of dust formation models is that carbon remains predominantly neutral (cf. Chapters 12, 18 and 24). Thus, a decisive factor for the possibility of carbon nucleation is the degree of ionization in the outflowing gas. For the abundances in WC stars and under the assumption of a stationary wind, the radius of the carbon Strömgren sphere can be determined. Considering the local balance of ionization and recombination in a pure carbon mixture, one finds that the radius R_C of carbon ionization is given by (Sedlmayr and Gass 1991)

$$R_C/R_* = \left[1 - \left\{2\int(\pi F/h\nu)d\nu\right\}/\alpha n^2 R_*\right]^{-1} \quad (3.1)$$

Attempting to have a more realistic estimate of R_C, one has to modify the influence of the radiation field by the inclusion of a reasonable optical depth factor $\exp(-\tau)$. Assuming a value of $\tau = 5$, one obtains finite values of R_C for $T_* < 25\,000$ K and $n < 10^{11}$ cm^{-3} (Sedlmayr and Gass 1991).

A more sophisticated treatment accounting for the non-steady nature of the wind must be used for the time development of the degree of ionization, as has been done for nova shells (Beck et al. 1995, 1990). Hydrodynamic models also involving the non-equilibrium ionization evolution under the influence of evaporation from dense clumps have been constructed by Arthur et al. (1996). It is found that the observed correlation between the velocities of ultraviolet absorption features and the ionization potentials of the absorbing ions can be qualitatively reproduced by such a model. Although at first glance all of the considered objects (WR, RCB) seem not to have physical conditions favouring effective molecule formation, the observations of solid particles tells us that complex chemical processes must occur. We shall restrict the analysis of the chemistry to one main case. In RCB stars and dust-forming WR stars of the classes WC8-WC9 in which carbon is overabundant, hydrogen is underabundant (Willis 1996; Rao and Lambert 1996; Nugis 1994). Thus, the chemistry around one can be characterized by a pure carbon chemistry.

Although the specific abundance ratios in RCB and WCL stars are varying, with $[(C + O)/He]$ from 0.03 to 1 for WCL $\rightarrow$ WCE (Smith and Maeder 1991) and $[C/He] \approx 0.1$ to 0.5 (Nugis 1994), $[C/O] \approx 1.0$ to 10 for RCBs (Goeres 1992), the outcome of the condensation process is well known to be some kind of pure carbon solid.

In this environment the CO molecule will be the first to form, thus inhibiting an oxygen chemistry from developing. N_2 is the next molecule to form by virtue of its high binding energy. Like CO, N_2 will not take part in further nucleation processes. A typical formation sequence of chemical species according to their abundances then would be (Goeres 1996):

$$\text{He - } (C_n, CO) \text{ - } N_2 \text{ - } \quad (CN, C_2N_2, C_2N, HCN, CNO, C_2H, CH)$$
$$(CS, SiS, SiC_2, Si_2C, SiO)$$

Thus, in RCB stars the chemical situation is characterized by a gaseous mixture dominated by pure carbon with some contributions of CO and N_2 and only minor traces of other molecules, in an inert helium buffer gas.

In WR stars the overall picture differs due to the fact that the radiation field is of substantially higher intensity in the UV; therefore helium is ionized from the outer atmosphere (Koesterke and Hamann 1995) up to the wind zone (Eenens et al. 1991; Niedzielski 1994). However, if helium is ionized, undoubtedly carbon will be too. The presence of carbon ions in the condensation zone of the wind would significantly influence dust formation which is commonly assumed to occur through neutral species (cf. Chapter 12). Up to now the nucleation of carbon clusters on the basis of ionized precursors has not been investigated. From the observations of emission lines in WR star envelopes (Moffat et al. 1994b) it is evident that the outflows of hot stars are dynamically unstable due to the inherent turbulence. This effect also occurs in the winds of Be-stars and novae but is assumed to be most pronounced in WR stars. A direct consequence will be the formation of local high density material distributions, which may be observed as clumps (Moffat and Robert 1994; Chen et al. 1995). Although the degree of density increase in such clumps is still uncertain (Brown and Richardson 1995; Moffat 1995), such non-homogeneous material may offer the opportunity for shielding effects and more favourable conditions for nucleation and growth of grains which, once formed, will amplify the formation processes.

4 Dust Formation

As discussed in Section 2.1, dust formation in Wolf–Rayet stars is a strong function of spectral subtype, e.g. of the stellar temperature, as the UV-flux differs by approximately one order of magnitude between 17 000 K and 30 000 K. The ability for clusters to survive is a sensitive function of the UV photon flux.

The ability of grains to form is in principle determined by the temperature and the density of the ambient gas, although a significant degree of ionization would complicate the process.

The density differences between dust forming and non-dust forming cases appear to be relatively small and therefore only a small increase in the local density, coupled to a decrease in the local temperature, thus amplifying this effect, must be sufficient to overcome the apparently small barriers between dust forming and non-dust forming situations. For dust particles to be observable, the particle density in the wind has to be in excess of $3 \cdot 10^5$ cm^{-3} (Williams et al. 1987), which corresponds to $\rho > 8 \cdot 10^{-18}$ g cm^{-3}. For dust particles to form a minimum density of 10^8 cm^{-3} is necessary.

4.1 Chemical Models

Assuming radiative equilibrium of a dust grain, one can determine a minimum distance where dust grains can survive without sublimation. For a typical WC star this distance is a few 10^{17} cm ($\approx 10^4 R_*$), where the particle density under the assumption of spherical symmetry and a uniform wind has dropped to 10^6 cm^{-3}. This low density seems to be insufficient for the formation of dust particles.

For carbon rich situations the problem of dust condensation has been investigated by Keller (1987), Gail and Sedlmayr (1987b, 1988), Frenklach and Feigelson (1989), Cherchneff et al. (1993, 1991) and Goeres (1993). However, these chemical models cannot be easily applied to WC environments, as they do not include the consequences of the specific abundance distribution in hot winds for the chemical pathways to dust. A situation where the gas is carbon rich and hydrogen deficient has been intensively investigated in the laboratory since fullerene synthesis has been detected (Krätschmer et al. 1990). Although the scaling of the chemical results from the laboratory to the astrophysical conditions of WR and RCB stars incorporates some uncertainties, it is possible to describe a plausible dust formation process. The nucleation chemistry of the carbon species may proceed along the following steps (Goeres 1996):

> A. Linear chains (unbranched) → B. Monocyclic rings →
> C. Polycyclic aromatic carbon clusters (PACs).

A. At high temperatures short chains completely dominate the chemical processes. These small carbon molecules are linear chains of polyyne type (alternating single and triple C–C bonds) for even numbers and cumulene type (C–C double bond) for odd numbers of carbon atoms. For longer chains ($n \geq 7$) the cumulene becomes more favourable (Raghavachari and Binkley 1987). The largest cluster with a linear ground state is C_9 (Heath and Sakally 1990, 1991), but there is experimental evidence for linear chains up to $n \approx 20$ (Broyer et al. 1993).

B. If the temperature in the gas decreases, the chains successively grow and with increasing chain length the energy of the first vibrational mode rapidly decreases. Also the energy gain of saturating the radical sites by ring closure exceeds the energy loss by bending the chain to a ring. Therefore, the chains close up to monocyclic ring structures, which are less reactive than chains as it requires the opening of a π-bond, which are more preferred the more the temperature decreases.

C. The third step in the formation sequence is quite controversial. The favourite possibility is based on the idea that the reaction enthalpy of a ring–ring addition enables the cluster to overcome the activation energy barrier towards the PACs (Goeres and Sedlmayr 1991, 1993). An alternative path involves the subsequent growth of monocyclic rings by C_2 addition to sizes beyond $n = 60$, a large curved framework containing hexagons as well as pentagons and which has the tendency to close to form a fullerene stabilized by thermal C_2 dissociation (McElvaney et al. 1993). Goeres (1996) has pointed out that there is a profound difference between these two model pathways. The effectiveness of fullerene production would be significantly higher in the second case, whereas the first path only allows for the production of larger fullerenes ($n > 70$). However, the observations of the Wolf–Rayet stars WR104 and WR112 (Cohen et al. 1989) detecting the 7.7 µm C–C stretching feature supports the PAC hypothesis, whereas no fullerenes have been observed at all in WC or RCB winds.

A comparable dust formation route is described by Cherchneff and Tielens (1995). Their model, based on the theoretical arguments of Tielens (1990), Frenklach and Feigelson (1989), and Cherchneff et al. (1993), also favours a chemical picture where small, spherical carbon clusters are the basic units of soot particles, and involving monocyclic ring condensation in an intermediate phase, it also ends in PAC production. Furthermore they argue that it is unlikely that the growing curved PACs will close perfectly but that they will form spiral structures, which can serve as condensation nuclei in the formation process of amorphous carbon clusters. The classical approach involving a pathway $C_1 \longrightarrow C_2 \longrightarrow C_3 \longrightarrow C_4 \longrightarrow \ldots \longrightarrow C_n$ and essentially based on the addition of neutral carbon monomers seems to be in question or maybe not applicable. As inferred from the strength of the CII and CIII lines, the abundance of neutral carbon, suspected to be an essential precursor molecule of carbon dust formation, is very low (10^{-6}, CI lines have not been found (Torres and Conti 1984)). This is in contrast to normal C stars as well as RCB stars and it might be due to the fact that the lifetime of the precursor molecules is too short to build up a reasonable optical depth.

If one assumes the existence of grains, further growth by heterogeneous nucleation of ions may be possible (Czyzak et al. 1982). It requires the presence of negatively charged grains, where the actual grain charge is determined by a competition between the processes of grain–electron collisions and the photoelectric effect (Feuerbacher et al. 1973; Spitzer 1978). Critical parameters therefore are the UV-field and the local electron density. A small increase in the density of the electrons may push the grain charge to become negative and as a result would create ideal conditions for the growth of particles by the accretion of ions. However, a serious problem remains for the nucleation phase, where no negative charge for small clusters can be expected.

4.2 Condensation in Clumpy Media

A first approach to clarify the chemical situation in WR winds has been undertaken by Cherchneff and Tielens (1995). In their chemical kinetic model consisting of 13 carbon species and 40 reactions using standard WC9 star parameters, they assumed a steady spherically symmetric outflow, a blackbody radiation field and a simple adiabatic cooling of the gas starting from $T_{gas} = 4000$ K. From their modelling some important results come to light: a) the carbon chains were predominantly present in ionic form, but b) the concentrations even of C_2^+ and C_3^+ are quite low due to high net rates of dissociative recombination, even though the CI concentration is a factor 100 higher than observed. These disadvantageous chemical conditions lead to a very small formation yield, e.g. the number of carbon atoms bound in chains compared to the number of initially available carbon atoms is $\approx 10^{-15}$ for carbon dust precursors. The bottleneck reactions are the formation reactions of C_2 and C_3, which do not proceed effectively. This result has to be compared with the formation yield derived by Williams et al. (1987) of 10^{-3}–10^{-4}, which is necessary to comply with the observations.

However, the gas outflow in WR and also in RCBs is far from being spherically symmetric. An artificial increase of the gas density by a factor of 10^4 to simulate the existence of clumps and to stimulate the dust formation therein results in a higher concentration of neutral carbon. Generally the clump model shows a dominance of neutral carbon species, which were mainly formed via radiative associations. Destruction routes of the carbon chains are mainly via He^+ and O attacks and with lower efficiency by photodissociation. The largely enhanced formation yield now is 10^{-6}. We conclude that while dust formation in WC stars is impossible under the assumptions of a spherically symmetric and homogeneous outflow, a large density increase may promote carbon dust formation.

4.3 Shock Induced Condensation

Dust formation in WR stars not only takes place in the vicinity of single stars, but is also closely connected with the occurrence of strong shocks in colliding winds of binary WR+O systems. In the wind collision zone a gas element is first heated, compressed and afterwards re-expands resulting in very effective cooling (Usov 1991, 1995). This mechanism generally shows similarities with those assumed to be applicable for RCB stars. Woitke et al. (1996) have developed a detailed model for shock induced condensation. Chemistry, nucleation and the thermal balance were studied for fixed fluid elements, which are periodically hit by shock waves caused by stellar pulsation. Non-LTE radiative heating and cooling processes in the shocked gas have been accurately modelled including free–free, bound–free and atomic line transitions, as well as rotational and vib-rotational contributions of molecules.

The gas, heated and compressed by a shock, radiates away its excess internal energy and re-expands afterwards. The phase of re-expansion is dominated by adiabatic cooling, which may cause gas temperatures substantially lower than in radiative equilibrium, if the particle density of the gas is tuned to a

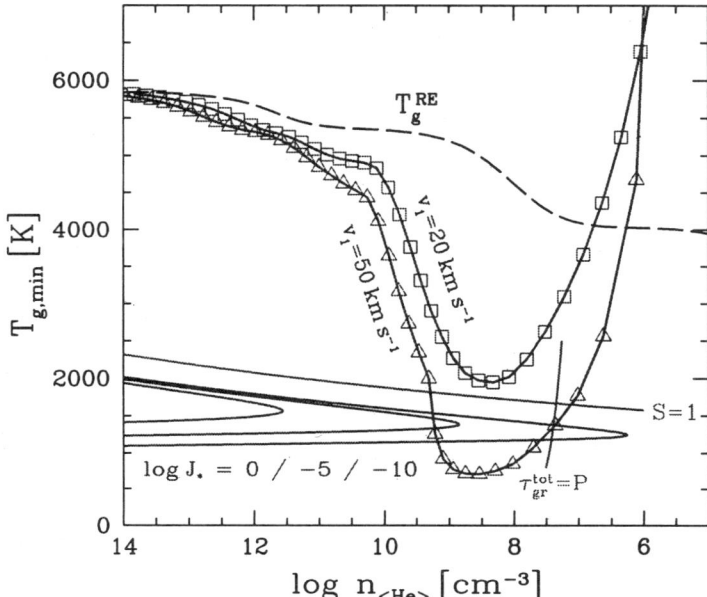

FIG. 1. Dust formation in shocked gas (Woitke et al. 1996). The figure shows the minimum gas temperature $T_{g,min}$ (full lines and points) occurring in a periodically shocked gas element for two different shock velocities as indicated, and the radiative equilibrium gas temperature T_g^{RE}. The condensation regime is depicted by contour lines of the logarithm of the classical nucleation rate J_* [cm^{-3} s^{-1}]. As the growth is limited by the pulsation period **the** maximum growth time is indicated by τ_{gr}^{tot}.

specific, but not too narrow range of $10^{7.5}$–10^{10} cm^{-3}. Favourable preconditions for carbon nucleation may occur, if a mechanism is present, which pushes the gas volume out of its equilibrium state. Temporarily a fluid element with an appropriate density can cool down below the condensation temperature of dust resulting in high supersaturation ratios. A large nucleation rate J_* (see Fig. 1), e.g. a high number of newly formed seed particles (per volume and per second), will evolve, leading to a sufficient number of small clusters, which can subsequently grow to macroscopic particle sizes. The specific range of favourable conditions of a shocked gas element in a RCB atmosphere is indicated by the intersection of the condensation regime (below $S = 1$ line) and the slope of the minimum gas temperature available during one pulsational period.

Generally this mechanism only causes the gas element to depart from its equilibrium state. Thus it can be expected that it is not restricted to pulsational shocks, but whenever a strong shock hits an atmospheric element with the appropriate density shock induced, condensation may take place.

The inherent instability of radiation driven winds in WR stars also can lead to shock waves which permeate the WR wind, and the density enhanced shock regions might be identified with the inhomogeneities (blobs) (Moffat and Robert 1994; Antokhin 1995). Recent observations of WR 104 (Crowther 1997), the brightest WR star in the IR due to its huge circumstellar dust shell and of the occasional eclipses of the dusty WC9 star WR121 (Veen et al. 1997) can both be interpreted in terms of the temporary condensation of an optically thick dust shell in the line of sight, which is subsequently dispersed by the radiation pressure. All this points to the fact that there is a close similarity between the physical processes related to dust formation in RCBs as well as in WR stars.

5 Summary

Although the amount of dust formed in the vicinity of hot stellar objects is considerably less than in the expanding shells of obviously more favourable systems like cool late-type stars, the detection of solid particles shows that even in these apparently hostile environments a rich chemistry may arise. Present modelling of thermodynamics, chemistry and dust formation are satisfactory for cool, homogeneous winds but are not applicable to the conditions prevailing in the shells of WR and RCB stars. The observational evidence points to three types of phenomena, which are of key importance for the dust formation process: these are density inhomogeneities, shocks and temperature fluctuations. All of these phenomena lead to one fundamental consequence: they generate favourable conditions for the onset of dust nucleation and growth.

The full understanding of the physical processes involved is a challenge for at least the next generation of model calculations.

Bibliography

1. Alexander, J. B., Andrews, P. J., Catchpole, R. M., Feast, M. W., Lloyd Evans, T., Menzies, J. W., Wisse, P. N. J. and Wisse, M. (1972) *Mon. Not. R. Astron. Soc.*, **158**, 305.
2. Allen, D. A., Swings, J. P. and Harvey, P. M. (1972) *Astron. & Astrophys.*, **20**, 333–336.
3. Allen, D. A., Barton, J. R. and Wallace, P. T. (1981) *Mon. Not. R. Astron. Soc.*, **196**, 797–800.
4. Annuk, K. (1991). In *Wolf-Rayet Stars and interrelations with other massive stars in galaxies*, eds. K. A. van der Hucht and B. Hidayat. Kluwer Academic Publishers, Dordrecht, p.245.
5. Antokhin, I. I. (1995). In *Wolf-Rayet stars: binaries, colliding winds, evolution*, eds. K. A. van der Hucht and P. M. Williams. Kluwer Academic Publishers, Dordrecht, pp.87–95.
6. Arthur, S. J., Henney, W. J. and Dyson, J. E. (1996). *Astron. & Astrophys.*, **313**, 897–908.

7. Beck, H., Gail, H.-P., Gass, H. and Sedlmayr, E. (1990). In *Physics of Classical Novae*, eds. A. Cassatella and R. Viotti. Reidel, Dordrecht, p.299.
8. Beck, H. K. B., Hauschildt, P. H., Gail, H.-P. and Sedlmayr, E. (1995). *Astron. & Astrophys.*, **294**, 195–205.
9. Benson, P. J., Clayton, G. C., Garnavich, P. and Szkody, P. (1994). *Astron. J.*, **108**, 247–250.
10. Bratschi, P. and Blecha, A. (1996). *Astron. & Astrophys.*, **313**, 537–544.
11. Brown, J. C. and Richardson, L. L. (1995). In *Wolf–Rayet stars: binaries, colliding winds, evolution*, eds. K. van der Hucht and P. M. Williams. Kluwer Academic Publishers, Dordrecht, pp.186–190.
12. Brown, J. C., Richardson, L. L., Antokhin, I., Robert, C., Moffat, A. F. J. and St.-Louis, N. (1995). *Astron. & Astrophys.*, **295**, 725–735.
13. Broyer, M., Goeres, A., Pellarin, M., Sedlmayr, E., Vialle, J. L. and Wöste, L. (1993). *Chem. Phys. Lett.*, **198**, 128–134.
14. Buss, R. H., Tielens, A. G. G. M., Cohen, M., Werner, M. W., Bregman, J. D. and Witteborn, F. C. (1993). *Astrophys. J.*, **415**, 250.
15. Cassinelli, J. P. (1990). In *Angular momentum and mass loss for hot stars*, eds. L. A. Willson and R. Stalio, Kluwer Academic Publishers, Dordrecht, pp.135–144.
16. Cassinelli, J. P. (1991). In *Wolf–Rayet Stars and interrelations with other massive stars in galaxies*, eds. K. A. van der Hucht and B. Hidayat. Kluwer Academic Publishers, Dordrecht, pp.289–307.
17. Cassinelli, J. P. (1992). In *Non–isotropic and variable outflows from stars*, eds. L. Drissen, C. Leitherer, and A. Nota. ASP Conference Ser., p.134.
18. Cassinelli, J. P., Mathis, J. S., van der Hucht, K. A., Prusti, T. and Wesselius, P. R. (1991). In *Wolf–Rayet Stars and interrelations with other massive stars in galaxies*, eds. K. A. van der Hucht and B. Hidayat. Kluwer Academic Publishers, Dordrecht, p.421.
19. Chen, H., Marlborough, J. M. and Waters, L. B. F. M. (1992). *Astrophys. J.*, **384**, 605.
20. Chen, Y., Wang, Z.-R. and Qu, Q.-Y. (1995). *Astrophys. J.*, **438**, 950–956.
21. Cherchneff I. and Tielens, A. G. G. M. (1995). In *Wolf–Rayet stars: binaries, colliding winds, evolution*, eds. K. van der Hucht and P. M. Williams. Kluwer Academic Publishers, Dordrecht, pp.346–354.
22. Cherchneff, I., Barker, J. R. and Tielens, A. G. G. M. (1991). *Astrophys. J.*, **377**, 541.
23. Cherchneff, I., Barker, J. R. and Tielens, A. G. G. M. (1993). *Astrophys. J.*, **413**, 445.
24. Chu, Y.-H. (1991). In *Wolf–Rayet stars and interrelations with other massive stars in galaxies*, eds. K. A. van der Hucht and B. Hidayat. Kluwer Academic Publishers, Dordrecht, p.349.

25. Clayton, G. C. (1996). *Publ. Astron. Soc. Pac.*, **108**, 225–241.
26. Clayton, G. C., Whitney, B. A., Meade, M. R., Babler, B., Bjorkman, K. S. and Nordsieck, K. H. (1995a). *Publ. Astron. Soc. Pac.*, **107**, 416.
27. Clayton, G. C., Kelly, D. M., Lacy, J. H., Little-Marenin, I., Feldman, P. A. and Bernath, P. F. (1995b). *Astronom. J.*, **109**(5), 2096ff.
28. Cohen, M. (1991). In *Wolf–Rayet stars and interrelations with other massive stars in galaxies*, eds. K. A. van der Hucht and B. Hidayat. Kluwer Academic Publishers, Dordrecht, pp.323-333; Discussion, p.334.
29. Cohen, M., Barlow, M. J. and Kuhi, L. V. (1975). *Astron. & Astrophys.*, **40**, 291–302.
30. Cohen, M., Tielens, A. G. G. M. and Bregman, J. D. (1989). *Astrophys. J.*, **344**, L13–L16.
31. Cohen, M., van der Hucht, K. A., Williams, P. M. and Thé, P. S. (1991). *Astrophys. J.*, **378**, 302–306.
32. Cottrell, P., Lawson, W. A. and Buchhorn, M. (1990). *Mon. Not. R. Astron. Soc.*, **244**, 149.
33. Crowther, P. A. (1997). *Mon. Not. R. Astron. Soc.*, **290**, L59.
34. Czyzak, S. J., Hirth, J. P. and Tabak, R. G. (1982). *Vistas in Astronomy*, **25**, 337–382.
35. Dougherty, S. M., Williams, P. M., van der Hucht, K. A., Bode, M. F. and Davis, R. J. (1996). *Mon. Not. R. Astron. Soc.*, **280**, 963–970.
36. Drilling, J. S., Schönberner, D., Heber, U. and Lynas-Gray, A. E. (1984). *Astron. & Astrophys.*, **278**, 224.
37. Dyck, H. M., Simon, T. and Woolstencroft, R. D. (1984). *Astrophys. J.*, **277**, 675–677.
38. Eenens, P. R. J. and Williams, P. M. (1992). *Mon. Not. R. Astron. Soc.*, **255**, 227–236.
39. Eenens, P. R. J., Williams, P. M. and Wade, R. (1991). *Mon. Not. R. Astron. Soc.*, **252**, 300–312.
40. Feast, M. W. (1986). In *Hydrogen Deficient Stars*, eds. C. S. Jeffery and U. Heber. ASP Conf. Ser. 96, p.151.
41. Feast, M. W. (1990). In *Confrontation between stellar pulsation and evolution*, eds. C. Cacciari and G. Clementini. ASP Conference Ser. 11, p.539.
42. Feuerbacher, B., Willis, R. F. and Fitton, B. (1973). *Astrophys. J.*, **181**, 101–113.
43. Frenklach, M. and Feigelson, E. D. (1989). *Astrophys. J.*, **341**, 372–384.
44. Gail, H.-P. and Sedlmayr, E. (1985). *Astron. & Astrophys.*, **148**, 183–190.
45. Gail, H.-P. and Sedlmayr, E. (1987a). In *Physical Processes in Interstellar Clouds*, eds. G. E. Morfill and M. Scholer. Reidel, Dordrecht, pp.275-303.
46. Gail, H.-P. and Sedlmayr, E. (1987b). *Astron. & Astrophys.*, **171**, 197–204.

47. Gail, H.-P. and Sedlmayr, E. (1988). *Astron. & Astrophys.*, **206**, 153–168.
48. Gail, H.-P., Keller, R. and Sedlmayr, E. (1984). *Astron. & Astrophys.*, **133**, 320–332.
49. Gehrz, R. D. (1989). In *Interstellar Dust*, eds. L. J. Allamondola and A. G. G. M. Tielens, Kluwer Academic Publishers, Dordrecht, p.445.
50. Gehrz, R. D. and Hackwell, J. A. (1974). *Astrophys. J.*, **194**, 619–622.
51. Glasse, A. C. H., Towlson, W. A., Aitken, D. K. and Roche, P. F. (1986). *Mon. Not. R. Astron. Soc.*, **220**, 185–188.
52. Goeres, A. (1992). *Staubbildung in den Hüllen von Kohlenstoffsternen: R Coronae Borealis.* PhD thesis, TU Berlin, Germany.
53. Goeres, A. (1993). *Rev. in Mod. Astron.*, **6**, 165–178.
54. Goeres, A. (1996). In *Hydrogen-Deficient Stars*, eds. C. S. Jeffery and U. Heber. ASP Conf. Ser. 96, pp.69–81.
55. Goeres A. and Sedlmayr, E. (1991). *Chemic. Phys. Lett.*, **184**(4), 311–317.
56. Goeres A. and Sedlmayr, E. (1993). *Fullerenes in Science and Technology*, **1**(4), 563–570.
57. Grevesse, N. and Noels, A. (1993). In *Origin and Evolution of the elements*, eds. N. Prantzas, Vangioni-Flam, and M. Casse. Cambridge University Press, Cambridge, p.14.
58. Gustafsson, B. and Asplund, M. (1996). In *Hydrogen-Deficient Stars*, eds. C. S. Jeffery and U. Heber. ASP Conf. Ser. 96, p.27.
59. Hackwell, J. A., Gehrz, R. D. and Grasdalen, G. L. (1979). *Astrophys. J.*, **234**, 133–139.
60. Hamann, W.-R. (1995). In *Wolf–Rayet stars: binaries, colliding wind, evolution*, eds. K. van der Hucht and P. M. Williams. Kluwer Academic Publishers, Dordrecht, pp.105–115.
61. Heath, J. R. and Sakally, R. J. (1990). *J. Chem. Phys.*, **93**, 8392.
62. Heath, J. R. and Sakally, R. J. (1991). *J. Chem. Phys.*, **94**, 3271.
63. Hecht, J. H. (1991). *Astrophys. J.*, **367**, 635.
64. Hecht, J. H., Holm, A. V., Donn, B. and Wu, C. C. (1984). *Astrophys. J.*, **280**, 228.
65. Jeffery, C. S. (1995a). *Astron. & Astrophys.*, **297**, 779.
66. Jeffery, C. S. (1995b). *Astron. & Astrophys.*, **299**, 135.
67. Jeffery C. S. and Heber, U. (1993). *Astron. & Astrophys.*, **270**, 167.
68. Jurcsik, J. (1996). *Acta Astronomica*, **46**, 325–333.
69. Keller, R. (1987). In *Polycyclic Aromatic Hydrocarbons and Astrophysics*, eds. A. Léger, L. d'Hendecourt and N. Boccara. ASI Series, Reidel, Dordrecht, pp.387–397.
70. Koesterke, L. and Hamann, W.-R. (1995). *Astron. & Astrophys.*, **299**, 503ff.

71. Krätschmer, W., Fostiropoulos, K. and Huffman, D. (1990). *Nature*, **347**, 354.
72. Lambert, D. L. (1986). In *Hydrogen Deficient stars*, eds. C. S. Jeffery and U. Heber. ASP Conf. Ser. 96, p.129.
73. Lawson, W. A. and Kilkenny, D. (1996). In *Hydrogen Deficient Stars*, eds. C. S. Jeffery and U. Heber. ASP Conf. Ser. 96, pp.349–361.
74. Lawson, W. A., Cottrell, P. L., Gilmore, A. C. and Kilmartin, P. M. (1992). *Mon. Not. R. Astron. Soc.*, **256**, 393f.
75. Lawson, W. A., Cottrell, P. L., Kilkenny, D., Gilmore, D., Kilmartin, P. M., Marang, P. M., Roberts, G. and van Wyk, F. (1994). *Mon. Not. R. Astron. Soc.*, **271**, 919.
76. Lepine, S. (1994). *Astrophys. Space Sci.*, **221**, 371–382.
77. Lewis, J. S. and Ney, E. P. (1979). *Astrophys. J.*, **234**, 154.
78. Lozinskaia, T. A., Sitnik, T. G. and Lomovskii, A. I. (1986). *Astrophys. Space Sci.*, **121**, 357–385.
79. McElvaney, S. W., Ross, M. M., Goroff, N. S. and Diederich, F. (1993). *Science*, **259**, 1594.
80. Milone, L. A. (1990). *Astrophys. Space Sci.*, **172**, 263.
81. Moffat, A. F. J. (1995). In *Wolf-Rayet stars: binaries, colliding wind, evolution*, eds. K. van der Hucht and P. M. William. Kluwer Academic Publishers, Dordrecht, pp.213–222.
82. Moffat, A. F. J. and Robert, C. (1994). *Astrophys. J.*, **421**, 310–313.
83. Moffat, A., Lepine, S., Robert, C. and Henriksen, R. N. (1994a). *Bull. CFH*, **31**, 14–18.
84. Moffat, A. F. J., Lepine, S., Henriksen, R. N. and Robert, C. (1994b). *Astron. & Astrophys. Suppl. Ser.*, **216**, 55–65.
85. Niedzielski, A. (1994). *Astron. & Astrophys.*, **282**, 529–534.
86. Nugis, T. (1994). In *Wolf-Rayet stars: binaries, colliding winds, evolution*, eds. K. van der Hucht and P. M. Williams. Kluwer Academic Publishers, Dordrecht, pp.371–372.
87. Nussbaumer, H., Schmutz, W., Smith, L. J. and Willis, A. J. (1982). *Astron. & Astrophys. Suppl. Ser.*, **47**, 257–294.
88. Poe, C. H., Friend, D. B. and Cassinelli, J. P. (1989). *Astrophys. J.*, **337**, 888.
89. Prinja, R. K., Barlow, M. J. and Howarth, I. D. (1990). *Astrophys. J.*, **361**, 607.
90. Raghavachari, R. and Binkley, J. S. (1987). *J. Chem. Phys.*, **87**, 2191.
91. Rao, N. K. and Lambert, D. L. (1993a). *Astronom. J.*, **105**, 1915.
92. Rao, N. K. and Lambert, D. L. (1993b). *Publ. Astron. Soc. Pac.*, **105**, 574.
93. Rao, N. K. and Lambert, D. L. (1994). *J. Astrophys. & Astron.*, **15**, 47.

94. Rao, N. K. and Lambert, D. L. (1996). In *Hydrogen deficient stars*, eds. C. S. Jeffery and U. Heber. ASP Conf. Ser. 96, p.43.
95. Rawley, G. L. (1993). In *Final Report, 2 Nov. 1992–1 Nov. 1993 Applied Research Corp., Landover, MD.*, volume 2.
96. Robert, C., Drissen, L. and Moffat, A. F. J. (1989). In *Physics of luminous blue variables*, eds. K. Davidson, A. F. J. Moffat, and H. J. G. L. M. Lamers. Kluwer Academic Publishers, Dordrecht, 299–300.
97. Schönberner, D. (1975). *Astron. & Astrophys.*, **44**, 381.
98. Sedlmayr, E. and Dominik, C. (1995). *Space Sci. Rev.*, **73**, 211–272.
99. Sedlmayr E., and Gass, H. (1991). In *Wolf–Rayet stars and interrelations with other massive stars in galaxies*, eds. K. A. van der Hucht and B. Hidayat. Kluwer Academic Publishers, Dordrecht, p.335.
100. Smith, L. J. (1995). in *Wolf–Rayet stars: binaries, colliding winds, evolution*, eds. K. van der Hucht and P. M. Williams. Kluwer Academic Publishers, Dordrecht, p.25.
101. Smith, L. F. and Hummer, D. G. (1988). *Mon. Not. R. Astron. Soc.*, **230**, 511.
102. Smith, L. F. and Maeder, A. (1991). *Astron. & Astrophys.*, **241**, 77.
103. Spitzer Jr., L. (1978). *Physical Processes in the Interstellar Medium*. Wiley, New York.
104. Tielens, A. G. G. M. (1990). In *Carbon in the Galaxy: Studies from Earth and Space*, eds. J. C. Tarter, S. Chang, and D. J. DeFrees. NASA, Ames Research Center, Moffett Field, pp.59–111.
105. Torres, A. V. and Conti, P. S. (1984). *Astrophys. J.*, **280**, 181–188.
106. Usov, V. V. (1991). *Mon. Not. R. Astron. Soc.*, **252**, 49–52.
107. Usov, V. V. (1995). In *Wolf–Rayet stars: binaries, colliding winds, evolution*, eds. K. van der Hucht and P. M. Williams. Kluwer Academic Publishers, Dordrecht, p.495.
108. van der Hucht, K. A. (1992). *Astron. & Astrophys. Review*, **4**, 123–159.
109. van der Hucht, K. (1995). In *Wolf–Rayet stars: binaries, colliding winds, evolution*, eds. K. van der Hucht and P. M. Williams. Kluwer Academic Publishers, Dordrecht
110. van der Hucht, K. A., Hidayat, B., Admiranto, A. G., Supelli, K. R. and Doom, C. (1988). *Astron. & Astrophys.*, **199**, 217–234.
111. van der Hucht, K. A., Williams, P. M., Pollock, A. M. T., Hidayat, B., McCain, C. F., Spoelstra, T. A. T. and Wamsteker, W. M. (1991). In *Wolf–Rayet Stars and interrelations with other massive stars in galaxies*, eds. K. A. van der Hucht and B. Hidayat. Kluwer Academic Publishers, Dordrecht, p.251.
112. van der Hucht, K. A., Morris, P. W., Williams, P. W., Setia Gunawan, D. Y. A., Beintema, D. A., Boxhoorn, D. R., de Graauw, T., Heras, A.,

Kester, D. J. M., Lahuis, F., Leech, K. J., Roelfsema, P. R., Salama, A., Valentijn, E. A. and Vandenbussche, B. (1996). *Astron. & Astrophys. Letters*, **315**, L193–L196.

113. Veen, P. M., van Genderen, A. M., van der Hucht, K. A., Li, A. and Sterken, C., (1997). *Astron. & Astrophys.*, **329**, 199.

114. Wainscoat, R. J., Cohen, M., Volk, K., Walker, H. J. and Schwartz, D. E. (1992). *Astrophys. J. Suppl. Ser.*, **83**, 111–146.

115. White, R. L. and Becker, R. H. (1995). *Astrophys. J.*, **451**, 352ff.

116. Whitney, B. A., Clayton, G. C., Schulte-Ladbeck, C. and Meade, M. R. (1992). *Astronom. J.*, **103**, 1652.

117. Williams, P. M. (1995). In *Wolf–Rayet stars: binaries, colliding winds, evolution*, eds. K. van der Hucht and P. M. Williams. Kluwer Academic Publishers, Dordrecht, pp.335–345.

118. Williams, P. M., Beattie, D. H., Lee, T. J., Stewart, J. M. and Antonopoulou, E. (1978). *Mon. Not. R. Astron. Soc.*, **185**, 467–472.

119. Williams, P. M., Longmore, A. J., van der Hucht, K. A., Talavera, A., Wamsteker, W. M., Abbott, D. C. and Telesco, C. M. (1985). *Mon. Not. R. Astron. Soc.*, **215**, 23P–29P.

120. Williams, P. M., van der Hucht, K. A. and Thé, P. S. (1987). *Astron. & Astrophys.*, **182**, 91.

121. Willis, A. J. (1991). In *Wolf–Rayet Stars and interrelations with other massive stars in galaxies*, eds. K. A. van der Hucht and B. Hidayat. Kluwer Academic Publishers, Dordrecht, pp.265–280.

122. Willis, A. J. (1996). *Astrophys. Space Sci.*, **237**(1), 145.

123. Willis, A. J., Schild, H. and Stevens, I. R. (1995). *Astron. & Astrophys.*, **298**, 549ff.

124. Woitke, P., Goeres, A. and Sedlmayr, E. (1996). *Astron. & Astrophys.*, **313**, 217–228.

PART V

Novae and Supernovae

18
Dust Formation in Novae

Jonathan Rawlings
Department of Physics and Astronomy, University College London

1 Introduction

Dust is a ubiquitous and vital component of the interstellar medium. Its formation in the hot winds ejected during the outburst of novae is a well-established phenomenon. However, the mechanisms by which dust is formed and the criteria which determine the efficiency of that process are very poorly understood. Central to these problems is the fact that the chemistry of dust grain nucleation and growth exists on the periphery of our understanding of interstellar chemical processes (cf. Chapters 12, 13). It bridges the microscopic world of molecules and chemical kinetics and the macroscopic world of bulk grain solid-state chemistry and thermodynamics. The intermediate regime is complex and largely unknown. To illustrate this point we should bear in mind that even a relatively simple organic intermediate based on a few tens of carbon atoms may have many millions of isomers. It is quite clear that for the most general case we have neither adequate data, nor the requisite computational power to be able to model the chemical kinetics of these chemical intermediates. In this chapter we investigate the mechanisms that lead to dust formation in the ejecta of novae, and how the apparently hostile environment of a nova still manages to be conducive to molecular processes. The harshness of the environment has both beneficial and complicating effects on the chemical modelling. Although novae may seem to be a somewhat specialized test bed for dust grain nucleation theory, the general conclusions that can be drawn have relevance to all studies of astrophysical dust formation.

2 Novae as Laboratories of Dust Formation

The formation of dust in nova winds is a well-observed phenomenon, both at infra-red (e.g. Gehrz 1990) and ultraviolet (e.g. Shore et al. 1994) wavelengths. As far back as 1937 it was proposed by McLaughlin (1937) that the behaviour of the light curve of nova DQ Her (1936) could be explained by the rapid formation of an optically thick dust shell. For some time it has been clear that carbon is the main dust component in most novae, although more recently the presence of silicates (or SiO_2), carriers of the so-called "unidentified infra-red" (UIR) features, and silicon carbide have become apparent (e.g. Gehrz 1990).

Bearing in mind the the difficulties involved in trying to model the process of dust formation, novae offer several advantages over other astrophysical "laboratories" for grain growth: novae are one-off eruptive events, resulting in the ejection of shell(s) of gaseous material from which dust condenses. The chemical and physical developments of the ejecta leading up to and through dust nucleation and growth can therefore be studied time-dependently on timescales ranging between days and years. In this respect novae are very different to cool stellar winds (such as the well-studied IRC+10216) in which dust formation is a *continuous* process and no information relating to the kinetics of grain formation can be inferred directly (cf. Chapter 12). By contrast, in novae we can identify the molecular precursors to dust formation and the likely nature of the nucleation sites, we can monitor the growth and subsequent development of the dust grains and, by comparing the gas-phase abundances before and after dust formation, we can establish the bulk chemical composition of the dust.

There are of course other stellar sources that create dust in episodic outbursts, rather than continuously, most notably Wolf–Rayet stars and R Coronae Borealis stars. These bear some resemblance to novae. For example, in the case of the latter, C_2 Swan bands are seen in emission and CN is commonly detected (e.g. Benson et al., 1994). The presence of these species is consistent with the dust nucleation scheme in novae discussed below, although for different reasons—the winds from R Coronae Borealis stars are very hydrogen-deficient (helium is the most abundant element) and the underlying source is a cool F or G type supergiant (with photospheric temperatures of $T_\star \sim 5000\text{--}7000$ K, cf $T_\star \gtrsim 10\,000$ K for novae). In the case of Wolf–Rayet stars, dust formation is apparently most effectively triggered by shock induced condensation in colliding winds, which are similarly hydrogen-poor. Nevertheless, both classes of object lack the singular nature and the apparent relative geometric simplicity of novae (in that dust formation occurs extensively, covering the nova sky).

The theoretical modelling of dust formation in novae is simplified somewhat by the presence of an intense ultraviolet radiation field, which has the effect of suppressing, by photodissociation, most of the larger molecular species. The formation pathways for the grain precursors must therefore involve simple reactions in which the building blocks for large molecule growth are small, stable molecules. Complex reactions between larger hydrocarbons are not important in the context of novae ejecta.

There, are however, some major drawbacks associated with studies of dust formation in novae. Observationally, novae are targets of opportunity; a nova event will occur without warning. Since the timescales for the chemical and physical evolution are short (of the order of a few days), data relating to the early stages of the evolution are sparse. The advent of high quality space-borne telescopes, such as the HST and ISO, coupled with a co-ordinated ground-based programme, is helping to alleviate this problem.

The theoretical problems are even more daunting: the presence of an intense UV radiation field and the extremely high temperatures and densities that are prevalent in the ejectae of novae imply that the chemical networks applicable to "interstellar" chemistry are entirely inappropriate. In particular, the various chemical species are likely to be highly excited, both electronically and, in the case of molecular species, vibrationally and rotationally. Unfortunately, there is very little chemical kinetic data available for molecular species in excited states, so that in most cases it has been necessary to extrapolate or infer the rate coefficients. The chemistry of the ejecta is at all times very much photon dominated. This also poses major problems; the photodissociation cross sections for even the smaller molecular species are highly uncertain. The data for larger species (such as the intermediary hydrocarbons en route to dust formation) are almost entirely conjectural. Moreover, the photophysics of molecules and radicals in excited states are very different to those of the ground state species. A major complication arises from the fact that the nova radiation field is itself poorly determined—partly because of the inadequate data coverage of novae at relevant wavelengths, and partly because the radiation transport in novae is complex and not well defined. We may therefore conclude that, whilst novae give us a unique opportunity to elucidate the process of dust formation, the uncertainties in the observations and the theory imply that, at present, these studies are still in their infancy.

In the discussions that follow we assume that the dust itself is *homogeneous*, that is to say it is formed from essentially one monomer type and has a uniform composition. Mixed-molecule or core-mantle grain types are not included in our discussion. However, as we shall discover below, the formation of the nucleation sites is likely to proceed as the the result of a mixed homogeneous/heterogeneous chemistry. Our discussions will also be solely concerned with the microscopic, molecular domain and the problems involved with the formation of nucleation sites. Once nucleation sites have formed, the proto-grains can be regarded as three-dimensional entities of sufficient size for reactive sites to be always available. As a result, the rates of most growth processes cease to be governed by microscopic chemical kinetics and simply scale as the surface area of the grain.

3 The Physics of Classical Novae

In common with most "cataclysmic variables" (CVs), novae are believed to originate from close binary systems, each system consisting of a red star, overflowing its Roche lobe through the inner Lagrangian point and accreting hydrogen-rich material onto the primary, a degenerate white dwarf ($\sim 1L_\odot$, $1M_\odot$) via an accretion disc. Novae are divided into three broad classes; dwarf, recurrent and "classical" novae. Of these, classical novae are the most dramatic and by far the most prolific dust producers. The rest of this chapter is therefore solely concerned with the chemistry and dust-formation properties of classical novae. Obviously, the accretion of a hydrogen rich material onto a degenerate object is potentially very explosive, especially as theory suggests that the accreted material mixes with core material from the white dwarf. A classical nova is believed to originate from a thermonuclear runaway (TNR) occurring at the base of the accreted layer.

Table 1 Typical physical characteristics of a classical nova

Outburst energy	= 10^{45} ergs
Maximum outburst luminosity	= 10^{38}–10^{39} erg s^{-1}
Mass loss	= 10^{-5} – 10^{-4} M$_\odot$
Bolometric luminosity	= 2×10^4 L$_\odot$ (constant)
Inter-outburst period	= 10^4-10^5 years
Ejecta velocity	= 300–10 000 km s^{-1}
Ejecta number density	= 5×10^{11} cm^{-3} (at $t = 5$ days)
Composition:-	Metals (e.g. C, N, O) 10–1000 × cosmic

The subsequent eruption results in the ejection of a series of winds, or shells, of matter at high velocities. The bulk of the material is carried in a wind (the "principal" ejecta) travelling at 300–1300 km s^{-1}. At later times a less massive secondary system (the "diffuse enhanced" wind) is detected travelling at velocities that are approximately twice those of the principal ejecta. At much later times very high velocity, high excitation/ionization winds are also detected (the so-called "Orion" stage of the nova evolution). Whilst the bulk of the wind is driven by shock and pressure ejection, the closeness of the luminosity of novae to the Eddington limit luminosity for a solar mass star (where the radiation pressure just balances gravity) suggests that the winds may, in part, be super-Eddington radiation pressure driven. The characteristics of a classical nova are summarized in Table 1.

3.1 The Evolution of a Nova

There are approximately 5–15 galactic novae per year, and of these about a third produce optically thick dust shells. The evolution of these objects is briefly described below.

A nova is characterized by an initial expansion phase leading to a visual maximum within a few days after the outburst (defined to be the zero point in time, $t = 0$, in the reference frame for the nova). During this period the nova will brighten (from a very faint progenitor) by several magnitudes. In the process of a nova eruption some 10^{-4}–$10^{-5} M_\odot$ of material are ejected. TNR models require significant overabundances of C, N, O, and other elements (by a factor of 10–100 or more over cosmic values), with very strong enrichments of isotopic variants (e.g. ^{13}C, ^{15}N, and ^{17}O), provided by the dredge-up of the underlying white dwarf material. Following the visual maximum, the mass loss rate falls, the initially optically thick ejecta expands and dilutes and the position of the photosphere falls back towards the surface of the white dwarf. This period is therefore characterized by a decline in the visual light curve accompanied by a corresponding rise in the UV light curve as the photospheric temperature rises from an initial value of around $\sim 10^4$K to $\sim 10^6$K. To a first approximation the photospheric temperature is given by the semi-empirical formula of Bath and Shaviv (1976):

$$T_{phot} = 15\,280 \times 10^{\Delta m_v/7.5} \text{K}$$

Table 2 Dust formation characteristics of fast and slow novae

Nova speed class	Fast	Slow
White dwarf type	He–Mg–Al	C–N–O
Dust type	Silicates, SiC	Carbon
Formation efficiency	Optically thin/none	Optically thick/copious
Examples	V1370 Aql, QU Vul	FH Ser, NQ Vul

where Δm_v is the size of the decline in magnitudes from maximum. Throughout the decline, and indeed often for many hundreds of days after the original outburst, it is not uncommon for a nova to maintain a constant total (bolometric) luminosity (e.g. Hyland and Neugebauer 1970). The rate at which the visual light curve declines and hence, by inference, the rate at which the photospheric temperature rises, defines the so-called nova "speed". Fast novae are those that decline at a rate of $\dot{m}_v \simeq 0\overset{m}{.}1$ day^{-1}, whilst slow novae decline at a rate of $\dot{m}_v \simeq 0\overset{m}{.}01$ day^{-1}.

Some 50–100 days after maximum, and particularly in the case of "moderate" or "slow" novae, the nova light curve may enter a "transition" phase where the V light curve drops on a timescale of a few days to a deep minimum (by several magnitudes) and the infra-red luminosity rises sharply at the same time. This behaviour is attributed to the rapid and efficient formation of an optically thick dust shell that covers the "sky" of the nova. Infra-red spectra often, but by no means always, show a featureless continuum with a black-body temperature of between ~600–800 K. The lack of spectral features and the temperature of the newly formed dust suggest that the main condensate is amorphous or graphitic carbon.

Some general correlations between the nova speed and the type of white dwarf and the efficiency of dust formation are summarized in Table 2. Faster novae are also brighter but are less efficient dust-producers. Many of the most well known fast novae (e.g. V1500 Cyg) did not produce any observable quantities of dust at all.

In the fast novae where dust is formed, the predominant molecular lines seen are of SiC, SiO and other silicate features sitting on a thermal dust continuum (e.g. Gehrz et al., 1984). One of the first novae in which silicate features were clearly identified was Aquilae 1982. This extraordinary oxygen-rich nova (with C:O < 0.17) possessed winds travelling at up to 10 000 km s^{-1} and showed a very clear feature at a wavelength of 9.7 µm which is well matched by laboratory "amorphous olivine smokes" (Roche et al., 1984). The absence of the 20 µm O–Si–O bend feature is puzzling, but may simply be due to an excitation effect.

Recent observations indicate that the dust formation characteristics of novae may be more complex than this general categorization suggests. A notable example is V842 Cen (1986) (Gehrz 1990) in which at least three different dust/molecular types seem to have formed in the ejecta; carbon, hydrocarbon, and silicates.

The dust formation phase is followed by a period when the newly formed dust shell expands, dilutes and becomes optically thin. This "recovery" phase lasts some tens to hundreds of days and is characterized by a rise in the visual luminosity and increased ionization levels. Ultimately, the nova will fade back into obscurity, in the "decline" phase. On some occasions the recovery can be very quick—lasting a few days or less—as was the case with nova V842 Cen (1986), suggesting that a more dramatic process may be occurring, such as the break-up of an optically thick shell into clumps. One curious, and not fully explained, characteristic of the dust shell is that the dust temperature often stays constant, or even rises slightly, throughout the recovery phase and for some hundreds of days afterwards. Simple arguments based on the grain absorption and emissivity characteristics and the dilution of the radiation field would suggest that the dust temperature should fall. This "isothermal phase" is often attributed to grain destruction and the effects that this has on the optical characteristics of the dust, but the issue has not yet been fully resolved.

3.2 The Ionization Structure of Idealized Novae

Although the wind structure is too complicated to allow a complete description of the varying physical conditions we can make some generalizations. In the discussion that follows, we do not address the complicating factors of compositional variations in the ejecta and non-spherically symmetric density distributions.

The temperature and density structure of the ejecta are most unclear, but some general deductions can be made from the state of ionization. If we represent the radial dependence of the density by a simple power law:

$$n = n_0 \left(\frac{r}{r_0}\right)^{-p}$$

(where n_0 and r_0 are the ejecta density and radius, respectively, at $t = 0$) then hydrodynamic simulations of nova atmospheres indicate that the power law index (p) is small; $\sim$2–3 (Starrfield, 1992). This compares with $p \sim$ 5–13 for supernovae and implies that novae atmospheres are extended and a wide range of temperatures and densities exist along a line of sight. This is borne out by the observation of the simultaneous presence of many ionization states of individual elements (e.g. as seen in V4169 Sgr, Scott et al. 1995). Immediately after outburst, the recombination timescale for the ejecta is very short (<1 s), and the wind is predominantly neutral. As the radiation field hardens and flux is effectively redistributed from the visible into the UV an outward flowing parcel of gas is overtaken by a series of ionization fronts. Gallagher (1977) identified three distinct ionization "zones":

I: H^0, C^0, Fe^+, Mg^+, ...
II: H^0, C^+, N^0, O^0, ...
III: H^+, C^+, N^+, O^+, Fe^{++}, Mg^{++}, ...

Simple ionization models (Mitchell and Evans 1984; Rawlings 1988) for a uniform, spherically symmetric ejecta distribution irradiated by a black-body with

T_{phot} given by the Bath and Shaviv (1976) formula, suggest that the ejecta of most novae may be completely carbon ionized (zone II) within some 5–10 days of the outburst, whilst complete hydrogen ionization may take up to 100 days or more.

Hauschildt et al. (1992) developed model atmospheres for spherically expanding, line-blanketed nova photospheres which are not in local thermodynamic equilibrium (LTE). Although idealized in some respects, these models give significant improvements over the black-body assumption. Most importantly, Hauschildt et al. showed that for the cooler photospheric temperatures, as appropriate to the early stages of development of a slow nova, the FeII absorption "forest" forms a quasi-continuum at UV wavelengths, resulting in fluxes that are several orders of magnitude lower than the black-body estimate. Beck et al. (1990, 1995) used these model atmosphere calculations to re-assess the ionization and temperature structure of novae ejecta. One of the most important conclusions of their work was that, for slow and moderate novae, the ionizing radiation is completely absorbed in a thin layer, with the result that complete carbon ionization cannot occur within $t = 50$ days. The CI region actually contains a significant fraction of C^+ and, with gas expansion dominating the cooling, can cool rapidly to $\sim$1000 K. Even for moderately fast novae ($\dot{m}_v \simeq 0\overset{m}{.}04$ day^{-1}), less than 2% of the ejecta are hydrogen ionized after 100 days. However, the observed CO excitation temperature is often substantially larger than this model might suggest. In NQ Vul, T_{ex}(CO) $\sim$ 3500±750 K at $t = 19$ days (Ferland et al., 1979), whereas in V705 Cas (1993) the ^{12}CO first overtone bands are strong, with bands up to $v = 6 \rightarrow 4$ present, even before visual maximum is achieved. The deduced excitation temperature for V705 Cas is high; T_{ex}(CO) $\sim$ 4500 ± 300 K at $t = 26.5$ days (although it may be as low as 2000 K if the CO emission is very optically thick), but the emission features had vanished by $t = 46$ days, perhaps indicative of efficient cooling of the ejecta. Indeed we may even speculate that the CO itself is an important ejecta coolant (Evans et al. 1996). The CO:C ratio in this nova was of the order of 10^{-4}, consistent with the chemical models described below.

4 The Early Chemical Evolution of Novae

It is against this background of changing temperature, density and ionization level that we must consider the four basic stages of chemical evolution:

- pre-dust formation epoch chemistry;
- the formation of (microscopic) nucleation sites;
- the growth of (macroscopic) dust grains;
- the subsequent evolution of the dust and gas as the ejecta expands and disperses.

As each of these stages is chemically dependent on the preceding stage, it is important that we understand the chemical processes that are occurring in the pre-dust nucleation epoch. The physical conditions in the ejecta of novae are very complex and time-dependent. Over the period of interest ($t = 0$–100 days) the density and temperature of the winds vary over a huge range ($T \sim 10\,000$–500 K, $n \sim 10^{13}$–10^7cm^{-3}) so that a variety of chemical processes can operate. Moreover, the ejecta are subject to a harsh, intense and strongly time-dependent radiation field. Throughout most of the period of interest the chemistry of novae ejecta is very strongly photon-dominated.

To date, only carbon-rich novae have been studied in any detail. There are good observational and theoretical reasons why this should be the case; observationally, carbon is the dominant form of dust in most novae, especially those which produce optically thick dust shells. Theoretically, carbon chemistry has been well-studied in the laboratory and the terrestrial analogues of the kinetics of soot formation are reasonably well understood. The same cannot be said of the chemistry of refractory grains (see Chapter 13).

The chemical modelling has concentrated on two epochs of the nova evolution: (i) pseudo-equilibrium models of the pre-dust formation epoch (Rawlings 1988), and (ii) non-LTE kinetic models of the dust nucleation epoch (e.g. Rawlings and Williams 1989).

The first molecular species to be detected in the ejecta of a nova was CN, as seen in the prototype of dust-forming novae—DQ Her (1935), (Wilson and Merrill 1935). Some fifty years later, H_2 (in the $v'' = 1 \to 0$ S(1) transition at 2.122 μm) was detected in the remnant of this old nova (Evans, 1991). However, the clearest indicators of chemical processes taking place in novae were the broadband photometric excesses seen in the infra-red M- and K-bands (the so-called "5 μm excess") and attributed to the CO $v'' = 1 \to 0$ transition at 4.8 μm (and its first overtone at 2.4 μm). These features are usually seen very early in the evolutionary development of novae. In the case of NQ Vul (1976), for example, the 5 μm excess was clearly apparent by $t = 19$ days, some 20 days before the nova went into transition (Ferland et al. 1979). This suggests that CO may be some sort of molecular precursor or necessary intermediate for the dust formation process. Indeed, for NQ Vul, Rawlings (1988) deduced a CO column density of 10^{17}–10^{18}cm^{-2} corresponding to some 1–10% (depending on the ejecta configuration) of all of the available carbon in the ejecta being locked up in CO. On the other hand, not all novae that exhibit a strong 5 μm excess go on to form dust. For example, the bright, fast, dustless nova V1500 Cyg demonstrated a very strong 5 μm excess (Shenavrin et al. 1977). CO is also seen at millimetre wavelengths in old novae (Hessman 1989).

In addition to CO, CN and H_2, in recent years there have been detections of SiO, SiO_2, SiC and polycyclic aromatic hydrocarbon (PAH) features (e.g. Gehrz et al. 1986; Bode et al. 1984; Greenhouse et al. 1990). However, other than the PAH features, there have been no detections of intermediate to large-sized molecules.

The intensity and spectrum of the radiation field are directly coupled to the chemistry and the state of the ionization through the various continuum and line opacities. Most significantly, the carbon continuum, which extends up to a wavelength, λ, of ~ 1100 Å, shields several molecular species against the dissociative UV flux. This is particularly important in the case of H_2 which dissociates following absorption in the Lyman and Werner bands ($\lambda = 912$–1100 Å). H_2 can self-shield, providing its column density is greater than about 10^{19}–10^{20} cm^{-2}. In addition, CO is protected against photodissociation in a CI region, and most of the CO absorption bands that lead to dissociation are blocked by the H_2 Lyman and Werner bands if they are very optically thick (Mamon et al. 1987). The situation is complicated further by the high densisties and temperatures. In such situations the H_2 and CO may be vibrationally excited. The oscillator stengths for Lyman and Werner transitions out of the $v'' = 1$ state of the ground Rydberg state ($X^1\Sigma_g^+$) of H_2 are typically some 5–10 times larger than for the ground ($v'' = 0$) vibrational state (Allison and Dalgarno 1970). Moreover, the wavelengths for the transitions are different and the self-shielding problem takes on an additional degree of complexity. It is thus apparent that very complicated combinations of self/mutual shielding processes exist in novae ejecta. These factors imply that H_2 formation is predicted to be *very* much more propitious in a CI region (zone I) than it is in a CII region (zone II). In the extreme case of carbon and hydrogen ionization (an HII region, zone III) all molecular chemistry will be suppressed.

4.1 The Chemistry of H_2 and CO

Rawlings (1988) established that, as with interstellar chemistry, the efficient formation of H_2 is a vital prerequisite for a vigorous nova chemistry to take place. The abundances of both simple molecules, like CO, and more complex hydrocarbons are highly sensitive to the H_2 fractional abundance in the ejecta. In the absence of dust a variety of gas-phase H_2 formation pathways are possible. In the early stages (within a few days of the outburst) the ejecta are dense and neutral. In tnese circumstances (when the density is $\gtrsim 10^{11}$ cm^{-3}) the dominant formation channel is three-body H_2 formation:

$$H + H + H \longrightarrow H_2 + H$$

In the neutral region the carbon continuum effectively shields the H_2 against photodissociation and the main loss channel for H_2 is collisional dissociation:

$$H_2 + H \longrightarrow H + H + H$$

Both these reactions are strongly dependent on the density. The formation rate scales as the cube of the density. Collisional dissociation operates by populating the excited vibrational states of H_2 faster than they can decay, so it is effectively a many-body rather than a bimolecular reaction. Consequently the rate is extremely sensitive to both the temperature and the density and the physical conditions of the inner ejecta effectively control the H_2 fractional abundance. Generally, if the temperature is above about 3500 K then it is not possible to create sufficient H_2 (with a fractional abundance of $\sim 10^{-4}$) for the ejecta to be self-shielding. However, if the ejecta are sufficiently cool then a substantial fraction of H_2 can build up, until the density drops and the H_2 abundance remains "frozen in" to the expanding ejecta.

Throughout most of the nova evolution, and certainly after about 3 days, three-body H_2 formation ceases to be important and the main formation channel is via the negative ion, H^-:

$$H + e^- \longrightarrow H^- + h\nu$$

followed by

$$H^- + H \longrightarrow H_2 + e^-$$

The importance of this reaction in dust-free circumstellar environments is well known (e.g. Rawlings et al. 1988; Glassgold, Mamon and Huggins 1989). The electrons are provided by the ionization of low ionization potential metals such as sodium, magnesium and silicon, so that formation efficiency is intimately coupled to the metallicity of the nova ejecta. The H^- intermediate is susceptible to photodetachment by the radiation field, especially at near-infra-red wavelengths:

$$H^- + h\nu \longrightarrow H + e^-$$

but the H^- pathway dominates the H_2 formation throughout most of the rest of the evolution of the ejecta. As the ejecta's carbon becomes ionized the electron abundance obviously rises and this channel is enhanced. However, this effect is greatly outweighed by the increase in the UV Lyman flux which results from the increased transparency of the carbon ionization continuum. This leads to the rapid photodissociation of H_2:

$$H_2 + h\nu \longrightarrow H + H$$

Other H_2 formation channels are possible, such as the route involving the positive ion H_2^+:

$$H + H^+ \longrightarrow H_2^+ + h\nu$$

$$H_2^+ + H \longrightarrow H_2 + H^+$$

The H_2^+ formation reaction is endothermic and, again, the H_2^+ intermediate is susceptible to photodissociation (at UV wavelengths, and particularly so if, as expected, the H_2^+ is vibrationally excited) as well as to dissociative recombination (which is also very sensitive to the vibrational state of the H_2^+). The strong radiation field and high excitation temperatures of novae ejectae imply that this channel is not significant in novae. The H_2^+ can also be formed by reactions involving excited hydrogen (Rawlings, Drew and Barlow 1993):

$$H(n=2) + H \longrightarrow H_2^+ + h\nu$$

or H_2 can be formed by direct radiative association (Latter and Black 1991):

$$H(n=2) + H \longrightarrow H_2 + h\nu$$

Of these two reactions the associative ionization is the more efficient at producing H_2, despite the 1.1 eV endothermicity of the reaction, but extreme departures from LTE are required for either of these channels to be significant (such as are provided in extreme cases of Lyman α trapping). Recent work (Pontefract and Rawlings 1998) suggests that these reactions are not significant in novae ejecta.

A model of the early time chemistry (Rawlings 1988) which was limited to the chemistry of hydrogen, carbon and oxygen, confirmed that large molecules cannot survive for long in the nova environment. In all cases the chemistry, which is dominated by neutral–neutral, negative ion, dissociative recombination and photodissociation reactions, is fast (with timescales of the order of seconds, so that the chemistry is effectively in steady state throughout most of the early stages), fairly inefficient and the ejecta are primarily atomic. Moreover, the models confirmed that simple molecules such as H_2 and CO can have appreciable abundances *only* in the CI region (ionization zone I).

In the CI region the abundances are *very* sensitive to the temperature (primarily because the H_2 abundance is limited by the temperature-sensitive collisional dissociation reaction). It is possible for the CO abundance to reach saturation levels only if $T \lesssim 5000$ K. Temperatures of less than 3500 K are required for H_2 to be optically thick. The CO formation route is typical for hot circumstellar environments:

$$O + H_2 \longrightarrow OH + H$$

$$O^- + H \longrightarrow OH + e^-$$

$$C + H_2 \longrightarrow CH + H$$

$$C^- + H \longrightarrow CH + e^-$$

followed by

$$OH + C \longrightarrow CO + H$$

$$CH + O \longrightarrow CO + H$$

In the absence of H_2, OH and CH can also be formed by the very slow direct radiative association of O and C with H atoms. Additionally, CO can form by direct radiative association:

$$C + O \longrightarrow CO + h\nu$$

The main CO loss routes are photodissociation and collisional dissociation by hydrogen atoms. Photodissociation dominates to such an extent that it is quite clear that efficient H_2 production and/or the shielding of CO from the dissociative UV flux is essential. In the CII region, the unshielded Lyman flux suppresses CO and H_2 so that neither species can build up appreciable column densities to allow self-shielding to occur.

The observed 5 μm excess persists for times that are much longer than the simple ionization models of Mitchell and Evans (1984) predict for complete carbon ionization of the ejecta. Since CO formation can only occur in a region in which carbon is neutral, this lends support to the more detailed models of Beck et al. (1990, 1995) which predict much longer timescales for complete carbon ionization. Alternatively, inhomogeneities in the ejecta could significantly delay the progress of the ionization front.

5 The Dust Nucleation Epoch

The chemical conditions in the dust forming epoch are very far from LTE so that a microscopic, kinetic approach to the formation of nucleation sites is necessary. Once the nucleation sites have formed, the (macroscopic) growth of dust grains is assumed to be rapid and efficient. This latter stage of the evolution has been well studied and was first described by the pioneering work of Clayton and Wickramasinghe (1976). The essential criterion for grain growth is that the ejecta temperature must lie below some condensation temperature, so that the partial pressure of carbon gas is greater than the vapour pressure of the solid-state condensate (such as graphite of amorphous carbon). In these conditions and *providing* nucleation sites are present, dust grains will grow until the monomer is depleted, or ejecta expansion freezes out the grain population. However, in all cases it is essential to form the nucleation sites beforehand.

In most nucleation schemes it is usual to consider the kinetics of molecular growth until one approaches a "critical cluster size", $n_{crit.}$. Molecular species which are larger than $n_{crit.}$ become more stable as they grow and hence runaway growth is possible. For carbon, this critical cluster size is often associated with the conversion of linear carbon chains (which are very stable for 8 or 9 carbon atoms) to monocyclic rings (such as benzene, C_6H_6) which are then able to grow into larger PAHs, graphitic planar structures etc. (cf. Chapter 13). Whilst the concept of a critical cluster size may be very useful in the context of novae, we should not assume that the details of the chemistry bear much similarity to the nucleation schemes that are applicable in other astrophysical environments. In the atmospheres of cool stars the most dominant carbon-bearing molecule, after CO, is acetylene (C_2H_2). Heterogeneous nucleation schemes for dust nucleation in these objects are then concerned with the conversion of C_2H_2 to dust, either through simple carbon growth reactions:

$$C_n + C_2H_2 \longrightarrow C_{n+2} + H_2$$

(cf. Chapter 13), or by reactions leading to the formation of the C_3H_3 radical, which can react with itself to form phenyl, and then through a series of C_2H_2 additions and H-extractions form more complex multi-ring structures (cf. Chapter 12). However in the case of novae, C_2H_2, which has dissociation channels extending up to 1550 Å (Suto and Lee, 1984), is strongly suppressed by the radiation field which ensures a low level of hydrogenation for all molecular species. Indeed any comparison with terrestrial hydrocarbon pyrolysis (e.g. as applied to carbon star winds by Frenklach and Feigelson, 1989) is entirely inappropriate to nova winds for these reasons. A completely different nucleation scheme must be sought. The fact that dust formation in novae *is* so efficient, even in such harsh conditions, suggests that there may be other highly efficient dust nucleation pathways in cool stellar winds whose significance have not yet been appreciated.

At the dust formation epoch in novae, the ejecta temperature is typically 1000–2000 K whilst the the photospheric temperature is $\gtrsim 10$–$20\,000$ K. The intensity of the radiation field implies chemical timescales of the order of seconds, even in the CI region several tens of days after the outburst. This is therefore the timescale on which the formation of nucleation sites must occur.

When considering the possible nucleation mechanisms we discover that the choice is very limited. In the ionized carbon region (zone II), coulombic effects severely inhibit any basic chemistry/grain growth pathways, so, in the absence of H_2, the only available chemical routes involve slow, simple radiative association reactions, such as

$$C^+ + H \longrightarrow CH^+ + h\nu$$

In the neutral carbon zone the conditions are much more conducive to the formation of nucleation sites. Firstly there may be significant quantities of H_2 present which can drive complex hydrocarbon chemistries. At the temperatures and densities appropriate to the nucleation epoch, the H_2 fractional abundance can rise to 10^{-4}–10^{-2} and, as already discussed, the CO abundance can rise to saturation. This has an important and well-known implication: if O>C then *all* of the chemically active carbon will be abstracted into the relatively inert form of CO and oxygen chemistry (and hence the formation of silicates etc.) will dominate the "active" chemistry. Conversely, if C>O then the oxygen will be taken out of the chemistry and (hydro-)carbon chemistry will dominate (and carbon dust formation will ensue). Since hydrocarbons would be attacked and destroyed ("burnt") by free oxygen atoms, the efficient formation of CO is an important step in the chemistry leading to the formation of hydrocarbon nucleation sites. In the earlier work of Rawlings and Williams (1989) this fact was emphasized and the efficient formation of CO was seen as an essential prerequisite for carbon dust formation. However, this conclusion is now being challenged by observations of novae such as V842 Cen which indicate that silicate and carbon dust types may be formed concurrently and co-extensively in the ejecta. The possibility of the co-extensive formation of multiple dust types has yet to be addressed in any detail.

Returning to the possible nucleation mechanisms, we find that homogeneous nucleation, such as

$$C_n + C \longrightarrow C_{n+1} + h\nu$$

is too slow to be significant, whilst more "exotic" pathways, for example involving iron complexes, such as $[Fe_2(CO)_9]^{2+}$, still involve slow chemical pathways and intermediates that are susceptible to the UV radiation field.

Most of the species that may be considered as potential monomers for grain growth are suppressed by the radiation field so that only the smallest species (such as C and C_2) are effective "building blocks" in the chemistry. C_2 has two pre-dissociation bands of interest; the main one, corresponding to $3^1\Pi_u$-$X^1\Sigma_g^+$, is completely shielded by the carbon continuum (Pouilly et al. 1983), so that C_2 is a relatively stable molecule. The fact that dust nucleation involves small monomers only, such as C_2, and that larger intermediates are continually being photodissociated, has an important implication for the viability of dust formation. Unlike the situation in cool, weakly irradiated stellar atmospheres, where the chemical pathways essentially consist of a *web* of cross-linked reaction networks, the nucleation chemistry in novae must be based on *chains* of chemical reactions. Thus, the efficiency with which larger nucleation sites are formed is *critically* dependent on the stability and abundances of the smaller intermediates. This may help to explain why some novae produce very optically thick dust shells, whereas others fail to produce any dust at all.

C_2, like H_2, is a homonuclear molecule and the chemistry of its formation bears some similarities to that of H_2: the direct radiative association of carbon atoms is very slow ($k \sim 10^{-17} \mathrm{cm}^3 \mathrm{s}^{-1}$), whilst the negative ion analogy of the H_2 formation route is relatively efficient:

$$C + e^- \longrightarrow C^- + h\nu$$

$$C^- + C \longrightarrow C_2 + e^-$$

Other formation routes include

$$C + CH \longrightarrow C_2 + H$$

$$C + CN \longrightarrow C_2 + N$$

$$C + CO \longrightarrow C_2 + O$$

although the last two reactions are strongly endothermic.

Rawlings and Williams (1989) investigated a simple heterogeneous hydrocarbon chemistry for the formation of nucleation sites. Their model is based on an extended hydrocarbon chemistry. In their scheme, a typical reaction pathway for the complex hydrocarbons may proceed as follows:

$$C_2 + (H^+, H_2^+, He^+, N^+, O^+) \longrightarrow C_2^+ + \ldots$$

$$C_2^+ + H_2 \longrightarrow C_2H^+ + H$$

$$C_2H^+ + H_2 \longrightarrow C_2H_2^+ + H$$

Then carbon addition can occur by

$$C_m H_n^+ + C \longrightarrow \begin{cases} C_{m+1}H_{n-1}^+ + H \\ C_{m+1}H_{n-2}^+ + H_2 \end{cases}$$

and

$$C_m H_n^+ + C_2 \longrightarrow \begin{cases} C_{m+2}H_n^+ + h\nu \\ C_{m+2}H_{n-1}^+ + H \\ C_{m+2}H_{n-2}^+ + H_2 \end{cases}$$

Further hydrogenation may also occur

$$C_mH_n^+ + H_2 \longrightarrow C_mH_{n+1}^+ + H$$

and the neutrals form by dissociative recombination

$$C_mH_n^+ + e^- \longrightarrow \begin{cases} C_mH_{n-1} + H \\ C_mH_{n-2} + H_2 \end{cases}$$

Other reaction types also contribute to the growth of large molecules. In this fashion, long-chain hydrocarbons can quickly be built up. There is very little laboratory data relating to the rate coefficients for many of these reactions (which, in any case, do not distinguish between different isomers, for the reasons stated above) so these rates were extrapolated from trends in known reactions of similar types. The radiation field ensures that the hydrogenation of these intermediate species is quite low (as is borne out by laboratory data for the UV irradiation of hydrocarbons, e.g. Wild and Koidl 1987) and the chemistry was limited to species up to and including C_8H_6. This termination was somewhat arbitrary but was justified on the grounds that larger molecules probably become more stable against photodissociation as their size increases (cf. Gail and Sedlmayr 1987). C_8-bearing molecules were therefore assumed by Rawlings and Williams to represent the critical cluster size and act as a sort of kinetic "bottleneck" to the chemistry.

The results from these models show that the abundances of the large molecules initially rise, then fall again on a timescale of a few days. This can be understood by the fact that the (pre-existing) H_2 drives the chemistry and is "absorbed" into the larger molecules which are subject to photodissociation (neutrals) and dissociative recombination (ions). It therefore follows that the chemistry is extremely sensitive to both the ionization level (through dissociative recombinations) and the initial H_2 abundance. This sensitivity can be very large; a change in the H_2 abundance from 10^{-4} to 5×10^{-2} results in the larger molecules being enhanced by a factor of between 8 and 12 orders of magnitude. The fractional abundance of nucleation sites (as represented by the largest molecules) required to produce a dust shell that is optically thick is not very large: $\sim 10^{-14}$–10^{-15} (Rawlings and Williams 1989), and can be produced providing the H_2 abundance is $>10^{-4}$ and the fractional ionization is $<10^{-3}$. This in turn requires that the gas temperature is low (1000–1500 K) so as to suppress the collisional dissociation reactions which inhibit H_2 formation. These studies would seem to suggest that the nucleation centres must grow into stable macroscopic grains within a day of their formation and that the nova dust is characterized by a small population of large grains, rather than vice versa.

There are obviously several important criteria for carbon dust formation to be viable. Firstly, at least part of the ejecta must remain carbon neutral until it reaches the "condensation radius" at which the formation of macroscopic grains is thermodynamically feasible. Secondly, the C:O ratio must be >1 and the CO abundance must be close to saturation, and thirdly, the ionization level and H_2 abundance (which is sensitively related to the temperature and density structure of the ejecta) must be in the ranges as described above. We have already noted that "slow" novae tend to be better at producing dust and a direct correlation between the nova speed (and hence the development of the radiation field) and dust formation is clearly apparent from the first point above. However, we should be wary of attempting to correlate the dust-forming capabilities of novae with a single nova parameter, such as the speed class. The third point above, for example, depends on the metallicity of the nova and the density and temperature structures. These may, or may not, be correlated with the speed class.

6 The Evolution of Dust Grains

Once the nucleation centres have formed, the subsequent growth is well described by the macroscopic models of Clayton and Wickramasinghe (1976) and others, and will not be described further here; it suffices to say that observational constraints require that the growth is highly efficient. Thus, for example, grains grow to sizes of >0.2 μm in the ejecta of V705 Cas on a timescale that requires that the sticking probability of carbon atoms is close to unity (Shore et al. 1994).

6.1 The Effects of Increased Ejecta Ionization

We may like to conjecture how once the dust has formed the grains evolve in response to their changing environment as the ejecta expand and dilute. As we have already discussed, the formation of nucleation sites can only occur in a CI region and the initial period of growth must occur on a timescale of a day or so. As the grains grow and the dust shell becomes more optically thick we may conclude that the growth phase must also occur in a CI region. Eventually, as the ejecta disperse, the dust shell will become optically thin again and hence its ionization level will rise. The onset of carbon ionization has several important consequences (Rawlings and Evans 1998): the gas kinetic temperature will rise and the H_2 and CO will cease to be shielded against dissociation and will be converted to H, C^+ and O. The ionization front is accompanied by a shock front that leads to grain–grain collisions and perhaps the formation of free-flying PAHs. The PAHs so-formed will, however, be dissociated by the radiation field and so will not persist for more than a few days (see below). Gas-phase molecular chemistry will be suppressed and the grains will be exposed to chemical erosion by the H and O atoms. Hydrogen attack of the grains initially results in the formation of surface hydrocarbon emission features but ultimately leads to chemisputtering of the dust with molecules like CH, CH_4 and C_2H_6 as products. These gas-phase products will then of course be subject to photodissociation and ionization. The oxygen atoms will "burn" the grains with CO as the predominant product.

The arrival of the carbon ionization front also has important implications for the nitrogen chemistry. Throughout most of the evolution of the nova, nitrogen is mainly taken up in the form of the relatively inert N_2, since the dissociation bands of N_2 only extend up to $\sim$979Å. The dissociation of N_2 may then lead to nitrogen attack of the carbon grains (leading to products like H_2CN). More importantly, from the observational point of view, nitrogen insertion into the carbon lattice of grains may excite the Raman G (6.35 μm) and D (7.35 μm) bands of the solid-state hydrocarbons.

There is considerable uncertainty as to the nature of the dust condensate and the picture of carbon dust nucleation described above may be a gross over-simplification. In a recent survey of ten novae, Smith et al. (1995) found that six of the novae produced dusts with a mixed carbon/silicate composition. However, what is not clear is whether these dust types formed concurrently and co-extensively, or whether, for example, they formed in different regions of the ejecta, where differing elemental abundances may exist, at different times. This is probably true for many novae, including (for example) V838 Her (1991) in which the 10 μm silicate feature was not seen until some 38–49 days *after* the initial dust formation epoch. If, however, different dust types can form co-extensively, then many of our pre-conceptions concerning the prerequisites for dust formation to occur will have to be revised.

6.2 UIR Features in Novae

In recent years, extensive observational coverage with higher resolution and higher sensitivity instrumentation has revealed considerable complexity in the dust composition of novae. Three novae in particular have been closely studied and revealed to be rich in "molecular" emissions (Evans and Rawlings 1994; Evans et al. 1996, 1997): QV Vul (1984), V842 Cen (1986) and V705 Cas (1993).

In each case there is evidence for at least three grain types: (i) carbon, from the underlying continuum emission between 1–8 μm, (ii) silicate, as determined from the 9.7 μm emission feature, and (iii) hydrocarbons, from the "UIR" features at 3.28 μm (aromatic CH stretch), 3.41 μm (aliphatic CH stretch), 11.3 μm (out of plane aromatic CH bend) and the features at 8.2 μm and 8.7 μm (although it should be noted that these may well be solid-state features associated with carbon dust, rather than gas-phase hydrocarbons—see below). Not all "hydrocarbon" novae show all of these features and the various lines come and go at different times. In the case of V705 Cen, Evans et al. (1997) have fitted a function of the form $\nu^\beta B(\nu, T_d)$ to the dust continuum, where $B(\nu, T_d)$ is the Planck function, T_d is the dust temperature and the β-index is defined in terms of the emission efficiency, $Q(\nu)$, by $Q(\nu) \propto \nu^\beta$. It was found that $\beta \sim 1$ gives the best fit to the continuum, suggesting that the carbon is amorphous, rather than graphitic (for which $\beta \sim 2$). The 9.7 μm features is also best fit by an amorphous, rather than a crystalline, silicate.

However, it is interesting to note that the UIR features only appear some time *after* the novae have recovered from the visual minimum. Evans and Rawlings (1994) speculated as to the possible carriers of the UIR features in novae. The possibilities include: (i) stable, free-flying PAHs, (ii) transient PAHs formed, for example, as a result of grain–grain collisions in a shock (possibly associated with an ionization front), and (iii) solid-state surface features associated with C–C and C–H stretching and bending modes in Hydrogenated Amorphous Carbon (HAC). Of these, (iii) seems to be the most viable. PAHs are susceptible to photodissociation by the radiation field after recovery, so that the lifetime of even quite large PAHs like coronene ($C_{24}H_{12}$) against photodissociation is of the order of a day or less once the Lyman window has opened. Smaller PAHs would be even less stable. Moreover, PAHs are susceptible to *chemical* erosion by hydrogen and oxygen atoms on timescales of the order of seconds to hours. HAC may also exhibit UIR features; it consists of PAH-like units weakly bonded by van der Waals forces or chemical bridging groups (e.g. Duley and Williams, 1988b). The ratio of soft, polymeric sp^3 to hard, graphitic sp^2 carbon mainly depends on the radiation field and the hydrogen atom density. Hydrogen atom attack on sp^2 carbon converts it to the hydrogenated sp^3 form, whereas irradiation tends to expel the hydrogen converting sp^3 to sp^2 carbon. This is, in effect, an annealing process which results in photodarkening and a reduction of the bandgap from ~ 2.5 eV (for sp^3) to $\lesssim 0.5$ eV (for sp^2). Conversion to a more graphitic form may help the grains survive chemical erosion. As the grains will almost certainly be positively charged (due to the photoelectric effect), monomeric accretion of the C^+ ions will be suppressed. Whether or not the grains can survive these conditions remains to be determined and depends on the timescales of the destruction processes as compared to the dynamical dilution timescale for the ejecta. Simple considerations suggest the carbon dust in novae initially has a high sp^3:sp^2 ratio. As the dust shell becomes transparent to ionizing radiation, the dust is annealed to sp^2, but at later times (of the order of tens to hundreds of days, depending on the ejecta configuration) the rate of re-hydrogenation exceeds that for annealing and the sp^3:sp^2 ratio increases again. Observationally, some measure of the sp^3:sp^2 ratio is provided by the relative strength of the 3.41 µm and 3.28 µm features. In those novae where these features have been detected, and which produce optically thick dust shells, the ratio of the strengths of the 3.41 µm and 3.28 µm features is high (e.g. V705 Cas, Evans et al. 1997). This is indicative of a high sp^3:sp^2 ratio and is consistent with the simple model of Evans and Rawlings (1994).

The broad, low contrast emission feature between 5000 and 9000 Å, known as the Extended Red Emission (ERE) has been detected in Nova QV Vul (1987) more than five years after maximum (Scott et al. 1994). As discussed above, the lifetime of free-flying PAHs in the late recovery epoch is typically less than one day, so it seems more likely that the ERE is consistent with photoluminescence from high band-gap solid-state HAC in the sp^3 form, as predicted for the general interstellar case by Duley and Williams (1988a; 1990). The fact that ERE is not seen in transition, when we expect the polymeric, sp^3, form to dominate, may simply be due to the high opacity of the dust shell and the low UV flux received by the dust grains. An additional feature of this model is that it may help to explain the so-called "isothermal" phase of the dust evolution (where the dust temperature is seen to stay constant, or even rise, over an extended period of time), without recourse to grain destruction mechanisms: as the Lyman window opens up, the dust "sees" more UV and the annealing process leads to photodarkening. Both effects help to raise the grain temperature, counteracting the decline caused by geometrical dilution.

7 Future Directions and Conclusions

Until very recently, the modelling of nova chemistry has been confined to the two separate epochs as described above: the CO-formation, pre-dust period and the epoch of the formation of dust grain nucleation sites. The two models are not directly causally linked and must be regarded as more exploratory in nature than definitive. New studies (Pontefract and Rawlings 1998) aim to remedy this situation by modelling the chemistry from outburst through to the nucleation epoch in one self-consistent model. The new models incorporate a more complex and up-to-date chemistry involving the elements H, He, C, N, O and a representative low ionization potential metal, Na. The inclusion of nitrogen is particularly relevant as CN is one of the very few molecular species to be seen in novae. Preliminary results indicate that CO, CN and C_2 are all formed highly efficiently during the early stages of the nova evolution. As with previous models, the efficiency of the chemistry is found to be very strongly dependent on the ionization level. Some of the major omissions and assumptions of the early models (such as the assumed absence of oxygen in the dust-forming epoch, deviations from spherical symmetry and chemical homogeneity and the *ad hoc* assumption that the nucleation chemistry is triggered by conditions suddenly becoming chemically favourable) are being addressed in these studies. There is observational evidence for strong compositional variations in the ejecta of some novae (e.g. Duerbeck 1987). Thus, for example, in the remnant of the nova RR Pic, the C:O ratio seems to be $\ll 1$ in the equatorial ring of the ejecta while C:O $\gtrsim 1$ in the polar blobs (Evans *et al.* 1992).

For the reasons given above, theoretical modelling of silicate dust formation in novae has not yet been attempted. Nevertheless, there are some conclusions that we can draw by analogy to the carbon dust nucleation: silicates form in O-rich atmospheres, where O:C > 1, so that the C in this environment is locked up in CO. Amorphous silicates consist of distinct regions of fully connected SiO_2, bounded by tetrahedra of non-bridging O atoms, the detailed structure being determined by the number and type (e.g. Mg or Fe) of modifying cations present (e.g. Thompson *et al.* 1996, and references therein). The nucleation of silicates presumably initially involves the formation of simple silicon-bearing molecules. The simple oxides, SiO and SiO_2, are extremely stable and may act as nucleation sites themselves. However, the SiO molecule has a dissociation energy of 7.8 eV (cf. 11.1 eV for CO) and must be shielded against the radiation field. It would seem plausible that this shielding is provided by the silicon ionization continuum (IP = 8.2 eV). The ionization models of Beck *et al.* (1995) suggest that the ejecta can remain silicon neutral even at $t = 45$ d, whereas other low ionization potential metals, such as Fe, Na and Mg, will all be ionized within a few days of the outburst.

The most important criterion for the successful nucleation of dust would then, by analogy, be that the ejecta reach the appropriate condensation radius for silicate grains before they are overtaken by the silicon ionization front. The progress of the silicon ionization front is obviously determined by the silicon abundance in the ejecta. In this context it is interesting to note that of the novae whose Si abundances have been determined, and are listed in Table 1 of Gehrz (1990) those with [Si/H] > 10 (relative to the solar value) possessed a silicate dust component while those with [Si/H] < 10 did not.

In conclusion, novae provide a challenge to anyone who studies the processes of dust formation in astrophysical environments. If the subject were not observationally driven, and we simply considered theoretical nucleation mechanisms in the context of the extremely hostile environment of a nova, then we would probably come to the conclusion that dust formation is not viable! The fact that dust formation *does* occur, and does so with such efficiency, is a major challenge. It implies that nature has found a way to overcome what would seem to be major kinetic obstacles. Perhaps more significantly, it warns us that there may be many highly efficient pathways to dust formation in astrophysical environments that have either not been identified in the laboratory, or are simply not applicable to the terrestrial environment. In either case, the clear message is that while we have made some significant advances, there is still a long way to go in our understanding of the formation of interstellar dust.

Bibliography

1. Allison, A. C. and Dalgarno, A. (1970). *Atomic Data* **1**, 289.
2. Bath, G. T. and Shaviv, G. (1976). *Monthly Notices of the Royal Astronomical Society* **175**, 305.
3. Beck, H. K. B., Gail, H.-P., Gass, H. and Sedlmayr, E. (1990). *Astronomy and Astrophysics* **238**, 283.
4. Beck, H. K. B., Hauschildt, P. H., Gail, H.-P. and Sedlmayr, E. (1995). *Astronomy and Astrophysics* **294**, 195.
5. Benson, P. J., Clayton, G. C., Garnavich, P. and Szkody, P. (1994). *Astronomical Journal* **108**, 247.
6. Bode, M. F., Evans, A., Whittet, D. C. B., Aitken, D. A., Roche, P. F. and Whitmore, B. (1984). *Monthly Notices of the Royal Astronomical Society* **207**, 897.
7. Clayton, D. D. and Wickramasinghe, N. C. (1976). *Astrophysics and Space Science* **42**, 463.
8. Duerbeck, H. W. (1987). *ESO Messenger* **50**, 8.
9. Duley, W. W. and Williams, D. A. (1988a). *Monthly Notices of the Royal Astronomical Society* **230**, 1P.

10. Duley, W. W. and Williams, D. A. (1988b). *Monthly Notices of the Royal Astronomical Society* **231**, 969.
11. Duley, W. W. and Williams, D. A. (1990). *Monthly Notices of the Royal Astronomical Society* **247**, 647.
12. Evans, A. (1991). *Monthly Notices of the Royal Astronomical Society* **251**, 54P.
13. Evans, A. and Rawlings, J. M. C. (1994). *Monthly Notices of the Royal Astronomical Society* **269**, 427.
14. Evans, A., Bode, M. F., Duerbeck, H. W. and Seitter, W. C. (1992). *Monthly Notices of the Royal Astronomical Society* **258**, 7P.
15. Evans, A., Geballe, T. R., Rawlings, J. M. C. and Scott, A. D. (1996). *Monthly Notices of the Royal Astronomical Society* **282**, 1049.
16. Evans, A., Geballe, T. R., Rawlings, J. M. C., Eyres, S. P. S. and Davies, J. K. (1997). *Monthly Notices of the Royal Astronomical Society* **292**, 192.
17. Ferland, G. J., Lambert, D. L., Netzer, H., Hall, D. N. B. and Ridgeway, S. T. (1979). *Astrophysical Journal* **227**, 489.
18. Frenklach, M. and Feigelson, E. D. (1989). *Astrophysical Journal* **341**, 372.
19. Gail, H. P. and Sedlmayr, E. (1987). In: *Physical Processes in Interstellar Clouds*, eds. G. E. Morfill and M. Scholer. Reidel, Dordrecht, p.275.
20. Gallagher, J. S. (1977). *Astrophysical Journal* **82**, 209.
21. Gehrz, R. D. (1990). In: *Physics of Classical Novae*, eds. A. Cassatella and R. Viotti, Springer-Verlag, Berlin Heidelberg, 138.
22. Gehrz, R. D., Ney, E. P., Grasdalen, G. L., Hackwell, J. A. and Thronson, H. A. Jr. (1984). *Astrophysical Journal* **281**, 303.
23. Gehrz, R. D., Grasdalen, G. L., Greenhouse, M. A., Hackwell, J. A., Hayward, T. and Bentley, A. F. (1986). *Astrophysical Journal* **308**, L63.
24. Glassgold, A. E., Mamon, G. A. and Huggins, P. J. (1989). *Astrophysical Journal* **336**, L29.
25. Greenhouse, M. A., Grasdalen, G. L., Woodward, C. E., Benson, J., Gehrz, R. D., Rosenthal, E. and Strutskie, M. F. (1990). *Astrophysical Journal* **352**, 307.
26. Hauschildt, P. H., Wehrse, R., Starrfield, S. and Shaviv, G. (1992). *Astrophysical Journal* **393**, 307.
27. Hessman, R. (1989). *Monthly Notices of the Royal Astronomical Society* **239**, 759.
28. Hyland, A. R. and Neugebauer, G. (1970). *Astrophysical Journal* **160**, L177.
29. Latter, W. B. and Black, J. H. (1991). *Astrophysical Journal* **372**, 161.
30. McLaughlin, D.B (1937). *Pub. Mich. Obs.* **6**, 107.
31. Mamon, G. A., Glassgold, A. E. and Huggins, P. J. (1987). *Astrophysical Journal* **323**, 306.

32. Mitchell, R. M. and Evans, A. (1984). *Monthly Notices of the Royal Astronomical Society* **209**, 945.
33. Pontefract, M. and Rawlings, J. M. C. (1998). *Monthly Notices of the Royal Astronomical Society* In preparation.
34. Pouilly, B., Robbe, J. M., Schamps, J. and Roueff, E. (1983). *Journal of Physics B* **16**, 437.
35. Rawlings, J. M. C. (1988). *Monthly Notices of the Royal Astronomical Society* **232**, 507.
36. Rawlings, J. M. C. and Evans, A. (1998). *In preparation*
37. Rawlings, J. M. C. and Williams, D. A. (1989). *Monthly Notices of the Royal Astronomical Society* **240**, 729.
38. Rawlings, J. M. C., Williams, D. A. and Cantó, J. (1988). *Monthly Notices of the Royal Astronomical Society* **230**, 695.
39. Rawlings, J. M. C., Drew, J. E. and Barlow, M. J. (1993). *Monthly Notices of the Royal Astronomical Society* **265**, 968.
40. Roche, P. F., Aitken, D. K. and Whitmore, B. (1984). *Monthly Notices of the Royal Astronomical Society* **211**, 535.
41. Scott, A. D., Evans, A. and Rawlings, J. M. C. (1994). *Monthly Notices of the Royal Astronomical Society* **269**, L21.
42. Scott. A. D. et al. (1995). *Astronomy and Astrophysics* **296**, 439.
43. Shenavrin, V. I., Moroz, V. I. and Liberman, A. A. (1977). *Soviet Astrophysics* **21**, 358.
44. Shore, S. N., Starrfield, S., Gonzalez-Riestra, R., Hauschildt, P. H. and Sonneborn, G. (1994). *Nature* **369**, 539.
45. Smith, C. H., Aitken, D. K., Roche, P. F. and Wright, C. M. (1995). *Monthly Notices of the Royal Astronomical Society* **277**, 259.
46. Starrfield, S. (1992). In: *"Binary Stars"* (Kluwer), Eds. Y. Kondo and J. Sahade
47. Suto, M. and Lee, L. C. (1984). *Journal of Chemical Physics* **80**, 4824.
48. Thompson, S. P., Evans, A. and Jones, A. P. (1996). *Astronomy and Astrophysics* **308**, 309.
49. Wild, Ch. and Koidl, P. (1987). *Applied Physics Letters* **51**(19), 1506.
50. Wilson, O. C. and Merrill, P. W. (1935). *Publications of the astronomical Society of the Pacific* **47**, 53.

19
Supernova Chemistry

Weihong Liu
Oak Ridge National Laboratory

1 Introduction

Supernovae are violent stellar explosions and they are classified into different types according to their spectroscopic characteristics. From the point of view of stellar evolution, Type II and Type Ib/Ic supernovae are explosions of massive stars produced by core collapse following late stages of nuclear burning, whereas a Type Ia supernova results from the thermonuclear runaway burning of an accreting white dwarf in a binary system. Each explosion releases the order of 10^{51} $M_\odot$ of energy and roughly one to several $M_\odot$ of ejecta expanding at velocities of up to $\sim 10^4$ km s^{-1}. A supernova explosion synthesizes heavy elements and injects them into the interstellar medium where they take part in subsequent star formation. The explosive nucleosynthesis produces large amounts of radioactive materials which power the supernova ejecta and drive the supernova physics. The understanding of these violent cosmic phenomena is crucial to our knowledge of the evolution of the Universe.

The core-collapsed supernova 1987A is the brightest supernova to be observed since SN 1604 seen by Kepler, the first to be observed in every band of the electromagnetic spectrum and in neutrinos, and the first in which molecules have been detected. SN 1987A has given us an unprecedented opportunity to infer details of supernova nucleosynthesis and explosion dynamics. Observations of the molecular emissions have probed an exciting environment for the study of molecular astrophysics. The detection of the molecules in the supernova provided a pleasant surprise for the 60th birthday of Alex Dalgarno. To celebrate Alex's 60th birthday, Richard McCray opened a chapter on the atomic and molecular physics of SN 1987A (McCray 1990). Since then, Alex has made major contributions to our understanding of the chemistry of SN 1987A. Now, the 10th birthday of SN 1987A has not only produced another round of fireworks for Alex's 70th birthday but has also presented an ideal occasion to close this chapter. Ancient Chinese astronomers considered supernovae to contribute to the excellent health and long lives of their emperors. I would follow in this tradition to wish Alex, a kind mentor, a generous colleague and a good friend, a long future of leadership in molecular astrophysics and perpetual influence on our understanding of the Universe.

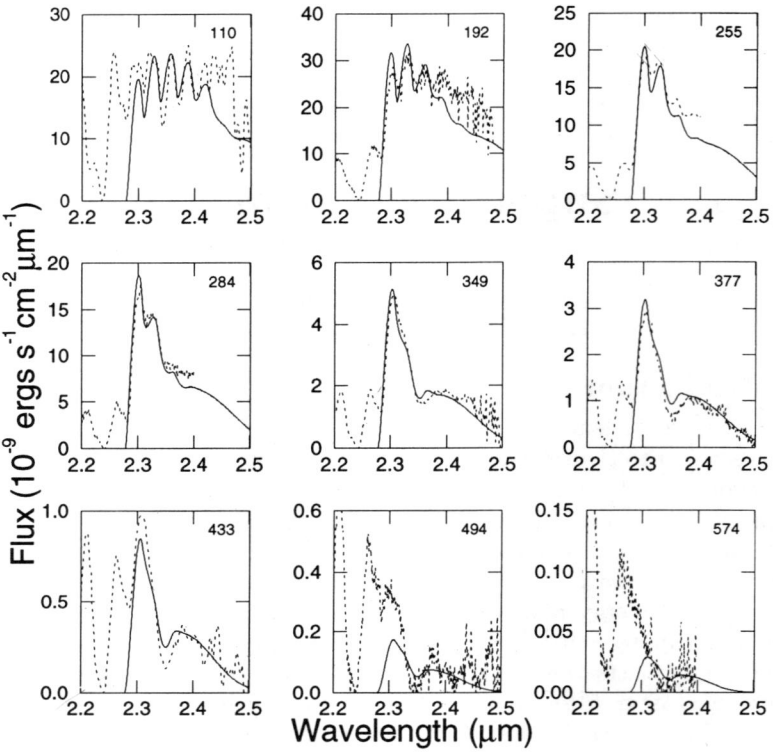

FIG. 1. Comparison of the observed (dashed curves) and calculated (solid curves) spectra of the first-overtone band of CO in SN 1987A from 110 days to 574 days. (Adapted from Liu and Dalgarno (1995).)

This chapter summarizes the studies of supernova chemistry. In Section 2, I will review the essential observations of the molecules and describe the modelling of the molecular spectra and light curves. Then, in Section 3, I will discuss supernova physics with an emphasis on the ionization and thermal structures. Finally, in Section 4, I will give a detailed account of the chemistries of Type II and Type Ia supernovae.

2 Molecular Spectra and Light Curves

The identification of carbon monoxide and silicon monoxide in the infrared spectra of SN 1987A represents the first ever detection of molecules in a supernova. The first overtone emission of CO at 2.3 µm was detected as early as 110 days after the supernova explosion (Meikle *et al.* 1989) and as late as 574 days (Meikle *et al.* 1993). This identification is supported by the presence of fundamental emission of CO at 4.6 µm which was detected from 157 days to 532 days (Bouchet

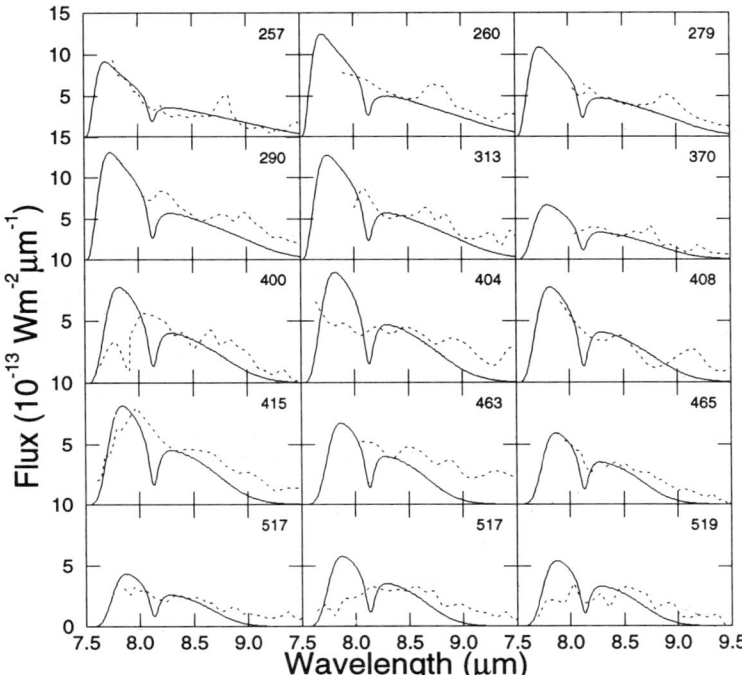

FIG. 2. Comparison of the observed (dashed curves) and calculated (solid curves) spectra of the fundamental band of SiO in SN 1987A from 257 days to 519 days. (Adapted from Liu and Dalgarno (1994).)

and Danziger 1993). The fundamental emission of SiO at 8 μm was detected after 160 days (Aitken *et al.* 1988) and it remained clearly detectable until 519 days (Bouchet *et al.* 1991). The absence of SiO emission after 519 days (Bouchet and Danziger 1993) and the onset of dust formation in the supernova around 530 days (Lucy *et al.* 1989; Danziger *et al.* 1991) indicate that the SiO molecules were depleted into dust grains for which the SiO molecules themselves may have provided the seed.

The observed spectra of CO and SiO are compared with the synthetic spectra in Fig. 1 (Liu and Dalgarno 1995) and Fig. 2 (Liu and Dalgarno 1994), respectively. In the model calculations the temperatures and masses of molecular gas were derived by fitting the observed spectra.

The rovibrational levels of the molecules are populated by electron impact excitation and de-excitation and by radiative processes. Although the rotational level populations are in thermal equilibrium at the kinetic temperature, the vibrational levels are not. The vibrational level populations have been obtained by solving a set of equations of statistical equilibrium. The CO emission is optically

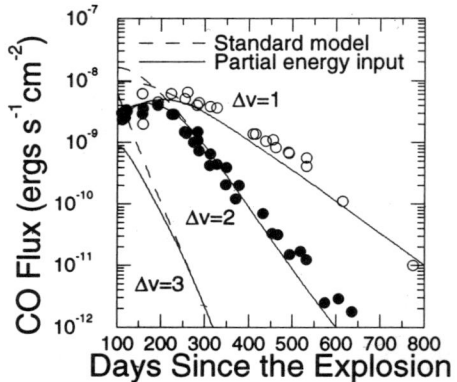

FIG. 3. The predicted (curves) and observed (filled circles) light curves of the fundamental and overtone bands of CO in SN 1987A. (Adapted from Liu and Dalgarno (1995).)

thick at early times and the vibrational level populations fall significantly out of thermal equilibrium at later times. The model reproduces the observed spectra of CO satisfactorily until 433 days. As the CO flux declines rapidly at later times, the observed fluxes at 494 and 574 days in the R branch region are overwhelmed by a strong unidentified emission feature at 2.26 μm. As a result, the predicted flux lies largely below the observed flux in the R branch region at 494 days; at 574 days there is minimal emission from CO in the R branch region and the predicted flux is only an upper limit to the observation. However, the model predicts well the fluxes in the P branch region at these later times.

On the other hand, significant differences between the predicted SiO spectra and the observed spectra exist. The fundamental emission of SiO is dominated by the $v = 1 \to 0$ band and the higher components contribute little to the total intensity because of the much lower populations. The photon fluxes in the P and R branches should be approximately equal because the intensity ratio of the R and P branches has a value near unity for any rotational quantum number and it is insensitive to the temperature. However, the observational data yield ratios for the integrated photon fluxes in the R and P branches which are significantly less than unity. The difference between the model results and the observations may be caused partly by additional emission in the P branch wavelength range from species other than SiO or absorption by other species in the R branch range.

The predicted and observed light curves of CO are compared in Fig. 3 (Liu and Dalgarno 1995). As the dominant coolant, the total CO emission is expected to decrease at the ^{56}Co-decay rate if there is a uniform energy deposition in the ejecta. At early times, the energy distribution in the ejecta is significantly non-uniform due to the large optical depth effects of the γ-rays, and the fraction of energy deposition in the iron core declines as demonstrated by the significant decrease of the iron emission intensities (Li et al. 1993). As the supernova expands and the optical depth decreases, the fraction of the decay energy deposited in the regions other than the iron core increases and the CO emission intensities increase. As the energy distribution becomes more uniform at later times, the CO light curves follow the ^{56}Co-decay rate closely. A partial energy input model (Liu and Dalgarno 1995) reproduces the observed light curves very well. Similar behaviours have been observed for many strong atomic emission lines other than those of the iron group elements including the [OI]$\lambda\lambda 6300, 6364$ (Hanuschik 1991; Li and McCray 1992), [CaII]$\lambda\lambda 7300$, and CaII$\lambda\lambda 8600$ (Hanuschik 1991).

3 Ionization and Thermal Structures

The supernova ejecta expand homologously, and the density of the oxygen core is roughly

$$n = 7 \times 10^{10} (t/100 \,\text{days})^{-3} \,\text{cm}^{-3}, \tag{3.1}$$

at time t after the supernova explosion. Ionization of the supernova in the nebular phase is driven by the γ-rays released in the decay of ^{56}Co and later of ^{57}Co and ^{44}Ti. The γ-rays lose energy in Compton scattering with the free and bound electrons and are degraded into X-rays. X-rays are preferentially absorbed in inner shell ionizations of heavy elements, producing energetic electrons. Auger processes produce multiply charged ions which recombine radiatively

$$X^{m+} + e \to X^{(m-1)+} + h\nu, \tag{3.2}$$

and by dielectronic recombination

$$X^{m+} + e \to X^{(m-1)+*} \to X^{(m-1)+} + h\nu. \tag{3.3}$$

As the most rapidly recombining ion becomes neutral, charge transfer

$$X^{m+} + Y \to X^{(m-1)+} + Y^+ \tag{3.4}$$

quickly removes the multiply charged ions because processes of this kind are almost always fast for $m > 1$. The only proposed identification of a line of a doubly charged ion for SN 1987A was based on an extremely weak spectral feature at 22.9 µm which was tentatively assigned to an emission line of Fe^{2+} (Moseley et al. 1989). However, this identification is very unlikely given the fact that the Fe^{2+} is removed quickly by charge transfer

$$\text{Fe}^{2+} + \text{Fe} \to \text{Fe}^+ + \text{Fe}^+, \tag{3.5}$$

which proceeds with a rate coefficient of about 7×10^{-9} cm^3 s^{-1} at 3000 K (Liu et al. 1998).

The ejecta are mostly singly ionized because of the low radioactive power input and the high gas density. The rates of charge transfer reactions of singly charged and neutral atoms are uncertain. Charge transfer reactions involving noble ions are likely to be slow with rate coefficients of generally less than 10^{-13} cm^3 s^{-1}. The charge transfer reaction,

$$\text{He}^+ + \text{C} \rightarrow \text{C}^+ + \text{He}, \tag{3.6}$$

proceeds slowly, with a rate coefficient of 5.2×10^{-14} cm^3 s^{-1} at 2000 K (Kimura et al. 1993). The rate coefficient for the charge transfer reaction,

$$\text{He}^+ + \text{O} \rightarrow \text{O}^+ + \text{He}, \tag{3.7}$$

is negligibly small (Kimura et al. 1994). On the other hand, charge transfer reactions such as

$$\text{C}^+ + \text{Si} \rightarrow \text{Si}^+ + \text{C}, \tag{3.8}$$

$$\text{O}^+ + \text{Si} \rightarrow \text{Si}^+ + \text{O}, \tag{3.9}$$

are likely to proceed rapidly with rate coefficients of the order of 10^{-9} cm^3 s^{-1}.

The energetic electrons created by the radioactive decay deposit their energy in the supernova by exciting and ionizing atoms and molecules and by dissociating the molecules, and as heat through Coulomb scattering with thermal electrons, and in dissociating the molecules. The fractions of energy deposition in excitation, ionization, and heating are shown in Fig. 4 (left panel) for the oxygen core of SN 1987A (Liu and Dalgarno 1995). The computational method needed to obtain these results was described in detail by Liu and Victor (1994). Shown in Fig. 4 (right panel) are the mean energies per ion pair, defined as the energies of primary electrons divided by the numbers of ion pairs produced (Liu and Dalgarno 1995). The ionization rates are given by the ratios of the rate of radioactive energy input and the mean energies per ion pair. Type II supernovae are only moderately ionized with maximum ionization fractions of the order of 10^{-2} (Liu et al. 1992) due to the low mass of radioactive materials and high gas density. On the other hand, Type Ia supernovae are characterized by highly ionized gas and higher ionization stages (Liu et al. 1998). Although a significant abundance of doubly charged ions exist in the outer regions of Type Ia supernovae, multiply charged ions are negligibly small due to rapid charge transfer between ions and neutral atoms (Liu et al. 1998).

As shown in Fig. 5 (left panel), the cooling of the CO-forming region is dominated by the vibrational emission of CO. Contributions from the forbidden transitions of atomic oxygen and carbon are unimportant. In fact, they heat rather than cool the gas as a result of non-thermal excitations by the energetic electrons. Radiative recombination, free–free emission, and adiabatic expansion make minor contributions to the cooling. The fine-structure transitions of atomic oxygen and carbon contribute very little to the cooling due to the low transition energies and large optical depths and the long cooling time scales, and they are not important before 800 days, though they may become so later as the ejecta continues to cool.

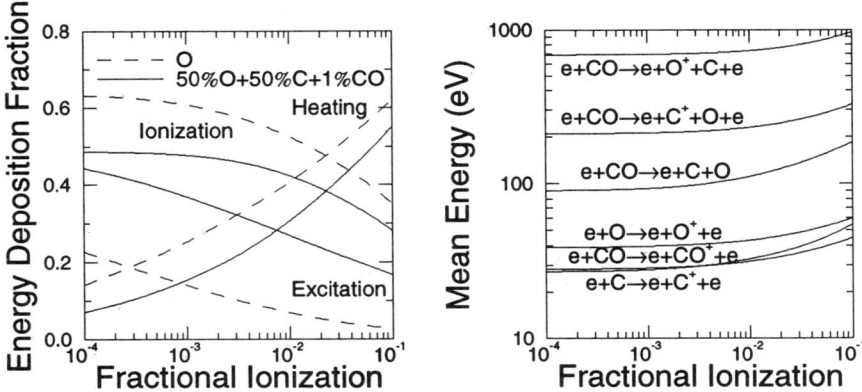

FIG. 4. Fractions of energy deposition in excitation, ionization, and heating (left panel) and mean energies per ionization or dissociation (right panel) as a function of the fractional ionization. (Adapted from Liu and Dalgarno (1995)).

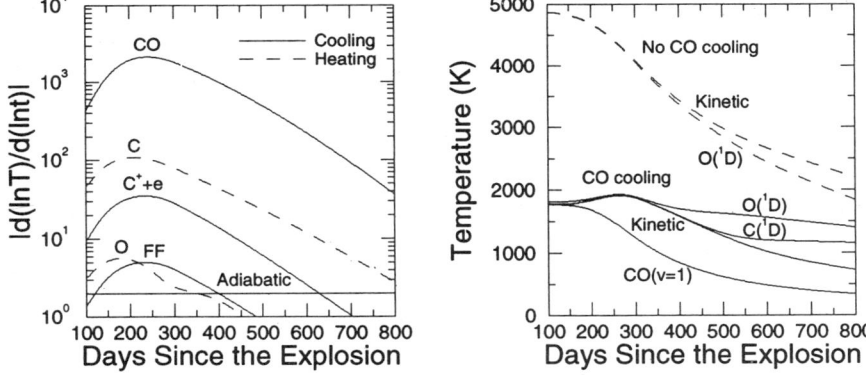

FIG. 5. Cooling efficiencies of the CO-forming region in SN 1987A (left panel) and the temperatures of the oxygen core of the supernova with and without the cooling by CO (right panel) as a function of time since the supernova explosion. (Adapted from Liu and Dalgarno (1995).)

The kinetic temperature of the CO-forming region is roughly constant at about 1800 K in the first year and drops to 700 K at 800 days (Fig. 5, right panel). The vibrational level populations of CO are in thermal equilibrium at early times and fall out of it at later times, as the result of both the decreasing electron density which prolongs the collision times and the decreasing rovibrational optical depths which shortens the effective radiative lifetimes as the supernova evolves. The ^1D populations of O and C are superthermal as the result of the non-thermal excitations by the energetic electrons. The excitation temperatures are significantly higher than the kinetic temperature after about 350 days for oxygen and about 500 days for carbon. In the SiO-forming region, cooling occurs by the electron collision-induced vibrational radiation of SiO, and the temperature is expected to be close to that of the CO-forming region.

On the other hand, the temperature in the O-emitting region is relatively high, ranging from 4800 K at 100 days to 2200 K at 800 days, in agreement with that derived from the observations of [OI]$\lambda\lambda 6300, 6364$ (Li and McCray 1992). The oxygen emission originates from the regions where carbon and silicon are absent or deficient so that neither CO nor SiO is formed. The most important cooling mechanism in these regions is the excitation and subsequent radiative decay producing [OI]$\lambda\lambda 6300, 6364$, which is less efficient than the vibrational excitations of CO or SiO in the molecule-forming region. [OI]$\lambda\lambda 6300, 6364$ emission in the supernova must occur in a hot region where neither CO nor SiO is present and the oxygen can emit efficiently in the metastable ^1D–^3P transition. The oxygen would be effectively invisible in the cold regions of the oxygen core where CO and SiO are formed.

4 Chemistry

Supernova chemistry differs very much from interstellar chemistry because the physical environment in which the molecules are formed and destroyed in a supernova differs considerably from that in an interstellar cloud. Dense cores in molecular clouds are weakly ionized by energetic cosmic rays and have fractional ionizations of the order of 10^{-7}, their gas densities are of the order of 10^5 cm^{-3}, they are cold with temperatures of the order of 20 K, and they are hydrogen-rich and contain dust grains on whose surfaces molecular hydrogen can form. On the other hand, the supernova ejecta are much more ionized by the energetic electrons produced by the radioactive decay. Type II supernovae (SNe II) are only moderately ionized with fractional ionizations of the order of 10^{-2} while Type Ia supernovae (SNe Ia) are nearly fully ionized. The ejecta are much denser with densities of the order of 10^{11} cm^{-3} for SNe II and 10^6 cm^{-3} for SNe Ia, they are hot with temperatures of several thousand kelvins, and they are stratified (with, e.g., the two shells containing primarily hydrogen and helium being distinct from that containing primarily carbon and oxygen though different layers can be mixed though (as argued below) only macroscopically rather than microscopically. Since the molecules form long before the dust particles do, the molecular formation in supernovae must be initiated by radiative processes. The chemistries of SNe II and SNe Ia are significantly different due to their very different physical conditions. Type Ib/Ic supernovae should have a chemistry similar to that of SNe II because of their similar physical conditions.

4.1 Type II Supernovae

4.1.1 *Carbon Monoxide*

The formation of CO molecules in Type II supernova 1987A is dominated by the direct radiative association

$$C + O \to CO + h\nu, \tag{4.1}$$

with a rate coefficient of 1.5×10^{-17} cm^3 s^{-1} at 2000 K (Dalgarno *et al.* 1990). Some contribution may be made by the sequence of negative ion formation

$$e + C \to C^- + h\nu, \tag{4.2}$$
$$e + O \to O^- + h\nu, \tag{4.3}$$

followed by associative detachment,

$$C^- + O \to CO + e, \tag{4.4}$$
$$O^- + C \to CO + e, \tag{4.5}$$

though its effectiveness is strongly limited by photodetachment,

$$C^- + h\nu \to C + e, \tag{4.6}$$
$$O^- + h\nu \to O + e, \tag{4.7}$$

and by mutual neutralization such as

$$C^+ + C^- \to C + C, \tag{4.8}$$
$$C^+ + O^- \to C + O. \tag{4.9}$$

C_2 and O_2 are also produced by radiative association

$$C + C \to C_2 + h\nu, \tag{4.10}$$
$$O + O \to O_2 + h\nu, \tag{4.11}$$

and by associative detachment

$$C^- + C \to C_2 + e, \tag{4.12}$$
$$O^- + O \to O_2 + e, \tag{4.13}$$

followed by the neutral reactions

$$C + O_2 \to CO + O, \tag{4.14}$$
$$O + C_2 \to CO + C, \tag{4.15}$$

to form CO. The reaction sequence

$$C^+ + O \to CO^+ + h\nu, \tag{4.16}$$
$$C^+ + O_2 \to CO^+ + O, \tag{4.17}$$
$$CO^+ + O \to CO + O^+, \tag{4.18}$$
$$CO^+ + C \to CO + C^+, \tag{4.19}$$

is unimportant partly because radiative association reactions (4.16) and (4.11) proceed slowly with rate coefficients of about 2×10^{-18} cm^3 s^{-1} (Dalgarno et al. 1990) and about 8.5×10^{-21} cm^3 s^{-1} at 2000 K (Babb and Dalgarno 1995), respectively, and mainly because of the rapid dissociative recombination

$$CO^+ + e \to C + O. \tag{4.20}$$

Without microscopic mixing (in the sense that a chemical reaction between material initially in two distinct fluid elements occurs) of helium into the CO-forming region, CO can not be destroyed by the reaction

$$He^+ + CO \to He + C^+ + O. \tag{4.21}$$

Instead, CO is mainly destroyed by impact by the energetic electrons,

$$\begin{align} e + CO &\to e + CO^+ + e, \tag{4.22} \\ &\to e + C + O, \tag{4.23} \\ &\to e + C^+ + O + e, \tag{4.24} \\ &\to e + O^+ + C + e. \tag{4.25} \end{align}$$

CO may also be destroyed by charge transfer

$$\begin{align} Ne^+ + CO &\to Ne + CO^+, \tag{4.26} \\ O^+ + CO &\to CO^+ + O, \tag{4.27} \end{align}$$

followed by dissociative recombination (4.20). However, the charge transfer reaction (4.26) is unusually slow with a rate coefficient of about $2.5 \times 10^{-14}(T/300)^2$ cm^3 s^{-1}, where T is the temperature in kelvins, and the abundances of Ne$^+$ and O$^+$ are reduced via charge transfer

$$\begin{align} Ne^+ + O &\to O^+ + Ne, \tag{4.28} \\ O^+ + C &\to C^+ + O, \tag{4.29} \end{align}$$

though reaction (4.28) is probably a slow one. Photodissociation and photoionization of CO in the supernova,

$$\begin{align} CO + h\nu &\to C + O, \tag{4.30} \\ &\to CO^+ + e, \tag{4.31} \end{align}$$

are ineffective mainly because of strong shielding of the ultraviolet photons by the neutral carbon and oxygen atoms in the CO-forming regions (Liu and Dalgarno 1996).

As shown in Fig. 6, the chemical model (Liu and Dalgarno 1995) predicts the mass of CO in SN 1987A derived from its first-overtone spectra well. It decreases from about 0.045 $M_\odot$ at 100 days quickly to about 3×10^{-3} $M_\odot$ at

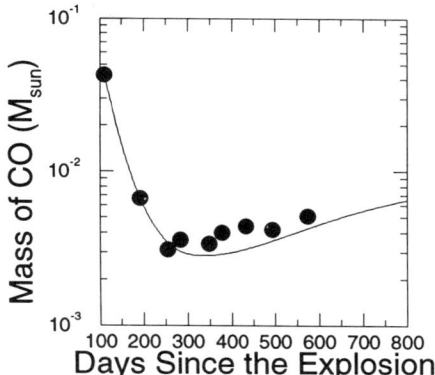

FIG. 6. Comparison of the masses of CO in SN 1987A derived from its first-overtone spectra (filled circles) and predicted by the chemical model (solid curve). (Adapted from Liu and Dalgarno (1995).)

340 days and then increases slowly to about 7×10^{-3} $M_\odot$ at 800 days. Its drastic decline at early times is caused by both the rapid destruction rate as a result of the enhanced energy deposition rate and the diminishing formation rate via radiative association (4.1) because of the decreasing density. At later times, the formation rate is proportional to the density and scales as t^{-3}, and the destruction rate decreases as $\exp(-t/111.3 \text{ days})$, the decay rate of ^{56}Co; thus, the mass of CO scales as $t^{-3} \exp(t/111.3 \text{ days})$ and reaches its minimum at 340 days, increasing relatively slowly afterwards. At later times, cooling of the supernova reduces the formation rate of CO because the rate coefficient of radiative association (4.1) decreases as the temperature decreases (Dalgarno et al. 1990). The predicted CO mass would be reduced by orders of magnitude were the CO mixed microscopically with helium because the destruction of CO by the dissociative charge transfer reaction (4.21) is fast. Thus, the CO data imply that there is no microscopic mixing of helium into the CO-forming region.

4.1.2 *Silicon Monoxide*

SiO in SN 1987A is formed predominantly by the direct radiative association

$$\text{Si} + \text{O} \rightarrow \text{SiO} + h\nu, \tag{4.32}$$

with a rate coefficient of $5.52 \times 10^{-18} \, T^{0.31 \pm 0.02}$ cm^3 s^{-1} (Andreazza et al. 1995). SiO may also be formed indirectly by the ionic radiative association

$$\text{Si}^+ + \text{O} \rightarrow \text{SiO}^+ + h\nu, \tag{4.33}$$

with a rate coefficient of $(6.22 \times 10^{-18} - 4.61 \times 10^{-22} \, T + 2.73 \times 10^{-26} \, T^2$ cm^3 s$^{-1})$ (Andreazza et al. 1995) and by the reaction

$$\text{Si}^+ + \text{O}_2 \rightarrow \text{SiO}^+ + \text{O}, \tag{4.34}$$

followed by charge transfer reactions with metals such as

$$SiO^+ + Si \rightarrow Si^+ + SiO, \qquad (4.35)$$
$$SiO^+ + Ca \rightarrow Ca^+ + SiO. \qquad (4.36)$$

However, the SiO^+ is destroyed predominantly by dissociation recombination

$$SiO^+ + e \rightarrow Si + O. \qquad (4.37)$$

SiO is formed additionally by radiative association (4.11) followed by the neutral reaction

$$Si + O_2 \rightarrow SiO + O, \qquad (4.38)$$

with a rate coefficient of 2.7×10^{-10} cm^3 s^{-1} (Husain and Norris 1978) or 9×10^{-12} cm^3 s^{-1} (Swearengen et al. 1978). The sequence initiated by radiative attachment (4.3) and

$$e + Si \rightarrow Si^- + h\nu, \qquad (4.39)$$

followed by associative detachment

$$O^- + Si \rightarrow SiO + e, \qquad (4.40)$$
$$Si^- + O \rightarrow SiO + e, \qquad (4.41)$$

is unimportant due to effective destruction of the negative ions by photodetachment (4.7) and

$$Si^- + h\nu \rightarrow Si + e, \qquad (4.42)$$

and by mutual neutralization such as

$$Si^+ + O^- \rightarrow Si + O, \qquad (4.43)$$
$$Ar^+ + Si^- \rightarrow Ar + Si. \qquad (4.44)$$

In the absence of helium, the reactions

$$He^+ + SiO \rightarrow Si^+ + O + He, \qquad (4.45)$$
$$\rightarrow O^+ + Si + He, \qquad (4.46)$$

do not occur. The reactions

$$C^+ + SiO \rightarrow SiO^+ + C, \qquad (4.47)$$
$$\rightarrow Si^+ + CO, \qquad (4.48)$$
$$\rightarrow CO^+ + Si, \qquad (4.49)$$
$$O^+ + SiO \rightarrow SiO^+ + O, \qquad (4.50)$$
$$\rightarrow Si^+ + O_2, \qquad (4.51)$$

are unimportant mechanisms for destruction of the SiO in the supernova even if the carbon co-mingles with the SiO because the carbon and oxygen are mostly neutral in the supernova due to rapid charge transfer to metals such as (3.8) and (3.9). The SiO is most likely to be destroyed by charge transfer:

$$Ar^+ + SiO \rightarrow SiO^+ + Ar, \quad (4.52)$$
$$Ne^+ + SiO \rightarrow SiO^+ + Ne, \quad (4.53)$$
$$\rightarrow Si^+ + O + Ne. \quad (4.54)$$

Silicon and argon together with sulphur and calcium are the main products of oxygen burning during the explosive nucleosynthesis, and they are intermixed in the supernova. Neon is produced in the stellar nucleosynthesis during the main sequence, and neon and oxygen are mixed microscopically in the supernova. Reactions (4.52)–(4.54) should proceed more rapidly than the corresponding reactions of CO,

$$Ar^+ + CO \rightarrow CO^+ + Ar \quad (4.55)$$

and (4.26), because SiO is less stable than CO. The rate coefficient for reaction (4.55) has been measured to be about 4×10^{-11} cm^3 s^{-1} at the centre-of-mass kinetic energy of 0.1 eV (Rebrion et al. 1989; Flesch et al. 1991). Reaction (4.26) proceeds with a rate coefficient of 2×10^{-12} cm^3 s^{-1} at 0.2 eV and the rate coefficient increases rapidly as the kinetic energy increases (Jones et al. 1981; Villinger et al. 1983). Reactions (4.52) and (4.53) may proceed as fast as reactions (4.55) and (4.26), respectively, and they are likely to be the dominant destruction mechanisms of the SiO in the supernova because the ionization fractions of argon and neon should be high due to the slow charge transfer between the noble ions and metals. The SiO can also be ionized and dissociated by the energetic electrons,

$$e + SiO \rightarrow e + SiO^+ + e, \quad (4.56)$$
$$\rightarrow e + Si + O, \quad (4.57)$$
$$\rightarrow e + Si^+ + O + e, \quad (4.58)$$
$$\rightarrow e + O^+ + Si + e. \quad (4.59)$$

However, photodissociation and photoionization of SiO,

$$SiO + h\nu \rightarrow Si + O, \quad (4.60)$$
$$\rightarrow SiO^+ + e. \quad (4.61)$$

are inefficient due to shielding of the ultraviolet photons by the neutral silicon atoms.

As shown in Fig. 7, the SiO masses in SN 1987A predicted (Liu and Dalgarno 1996) for the elemental compositions of several supernova models (Woosley et al. 1988; Thielemann et al. 1990; Nomoto et al. 1991) agree well with the SiO mass of 10^{-4}–10^{-3} $M_\odot$ derived from the SiO spectra (Liu and Dalgarno 1994). The observed SiO mass is best reproduced by the chemical model for which the composition of Woosley et al. (1988) was adopted; the observed and predicted

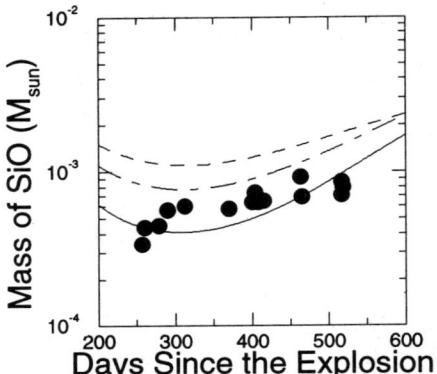

FIG. 7. Comparison of the masses of SiO in SN 1987A derived from its fundamental spectra (filled circles) and predicted by the chemical models for the elemental compositions of Woosley et al. (1988) (solid curve), Thielemann et al. (1990) (dashed curve), and Nomoto et al. (1991) (dot-dashed curve). (Adapted from Liu and Dalgarno (1996).)

SiO masses agree within a factor of 1.5. The SiO masses predicted by the chemical models with the compositions of Thielemann et al. (1990) and Nomoto et al. (1991) overestimate the observed SiO masses by factors of no more than 3.5. Similar to the case of CO, the agreement between the model and the observations in the SiO mass rules out any significant microscopic mixing of the helium in the SiO-forming region. In addition, the chemistry is robust and not very sensitive to the elemental compositions given by the supernova models.

4.1.3 Carbon Sulphide

CS molecules in SN 1987A can be formed directly by the radiative association

$$C + S \rightarrow CS + h\nu, \tag{4.62}$$

and indirectly by the ionic radiative association

$$C^+ + S \rightarrow CS^+ + h\nu, \tag{4.63}$$

followed by charge transfer

$$CS^+ + Mg \rightarrow Mg^+ + CS, \tag{4.64}$$
$$CS^+ + S \rightarrow S^+ + CS. \tag{4.65}$$

The rate coefficients at 2000 K for (4.64) and (4.65) are about 2.5×10^{-18} cm^3 s^{-1} and 1.1×10^{-18} cm^3 s^{-1}, respectively (Andreazza et al. 1995), However, CS$^+$ is predominantly lost due to dissociative recombination

$$CS^+ + e \rightarrow C + S. \tag{4.66}$$

CS may also be formed via the negative ion sequences (4.2) and

$$C^- + S \to CS + e, \tag{4.67}$$

and

$$\begin{aligned} e + S &\to S^- + h\nu, & (4.68) \\ S^- + C &\to CS + e, & (4.69) \end{aligned}$$

but their effectiveness is strongly limited by photodetachment

$$\begin{aligned} C^- + h\nu &\to C + e, & (4.70) \\ S^- + h\nu &\to S + e, & (4.71) \end{aligned}$$

and by mutual neutralization such as

$$\begin{aligned} Ne^+ + C^- &\to Ne + C, & (4.72) \\ Mg^+ + C^- &\to Mg + C, & (4.73) \\ Ne^+ + S^- &\to Ne + S, & (4.74) \\ Mg^+ + S^- &\to Mg + S. & (4.75) \end{aligned}$$

The most important source of CS in SN 1987A is provided by sequences initiated by the ionization of SiO which is formed in abundance in the supernova. The CS is formed most effectively via charge transfer (4.47), (4.52), (4.53) and

$$O^+ + SiO \to SiO^+ + O, \tag{4.76}$$

with a rate coefficient of 5.4×10^{-10} cm^3 s^{-1} (Herbst et al. 1989), and by energetic electron impact ionization (4.56) followed by the reactions

$$\begin{aligned} SiO^+ + S &\to Si^+ + SO, & (4.77) \\ C + SO &\to CS + O. & (4.78) \end{aligned}$$

Although the supply of SO through reaction (4.77) may be limited due to dissociative recombination

$$SiO^+ + e \to Si + O, \tag{4.79}$$

which is the dominant loss channel of SiO$^+$ in the supernova, alternative sources of SO may be provided directly by the radiative association

$$S + O \to SO + h\nu, \tag{4.80}$$

by the negative ion sequences (4.3) and

$$O^- + S \to SO + e, \tag{4.81}$$

and

$$e + S \rightarrow S^- + h\nu, \tag{4.82}$$
$$S^- + O \rightarrow SO + e, \tag{4.83}$$

and by the reaction

$$S + O_2 \rightarrow SO + O. \tag{4.84}$$

Thus the chemistry leading to the formation of CS is robust.

Similar to CO and SiO, the CS in SN 1987A is not destroyed by

$$He^+ + CS \rightarrow C^+ + S + He, \tag{4.85}$$
$$\rightarrow S^+ + C + He. \tag{4.86}$$

CS is destroyed predominantly by the neutral reaction

$$O + CS \rightarrow CO + S, \tag{4.87}$$

with a rate coefficient of about 2×10^{-10} cm^3 s^{-1} at 2000 K (Leen and Graff 1988). The destruction of CS through (4.87) proceeds rapidly because the oxygen is the most abundant element in the region and it is mostly neutral. The CS may also be destroyed by charge transfer reactions such as

$$O^+ + CS \rightarrow CS^+ + O, \tag{4.88}$$
$$\rightarrow S^+ + CO, \tag{4.89}$$
$$C^+ + CS \rightarrow CS^+ + C, \tag{4.90}$$

but they proceed relatively slowly because of the low abundances of the ions. The CS can be removed by the energetic electrons,

$$e + CS \rightarrow e + CS^+ + e, \tag{4.91}$$
$$\rightarrow e + C + S, \tag{4.92}$$
$$\rightarrow e + C^+ + S + e, \tag{4.93}$$
$$\rightarrow e + S^+ + C + e. \tag{4.94}$$

However, photodissociation and photoionization

$$CS + h\nu \rightarrow C + S, \tag{4.95}$$
$$\rightarrow CS^+ + e, \tag{4.96}$$

are unimportant destruction mechanisms for the CS due to the strong ultraviolet shielding (Liu and Dalgarno 1996).

The mass of CS in SN 1987A predicted (Liu 1998) for the elemental composition of Thielemann et al. (1990) is shown in Fig. 8. It is of the order of $10^{-11} M_\odot$, many orders of magnitude lower than those of CO and SiO in the

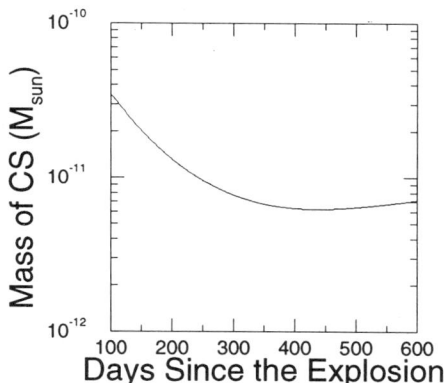

FIG. 8. Mass of CS in SN 1987A predicted by the chemical model. (Adapted from Liu (1997b).)

supernova (cf. Figs. 6 and 7). This is caused partly by the relatively low degree of mixing of carbon and sulphur in the supernova which leads to a much lower rate of formation of CS than CO and SiO and mainly by the rapid destruction of the CS by the abundant neutral oxygen through reaction (4.87). The destruction of CS proceeds orders of magnitude faster than those of CO and SiO which are through energetic electron impact (4.22)–(4.25) and charge transfer (4.52)–(4.54), respectively.

It is possible that the mass of CS required to explain the intensity of the 3.88 μm feature may not be as high as $\sim 10^{-3} M_\odot$. Since the P branch of the 2–0 vibrational transition of CS is overlapped by the broad and strong Brα line of atomic hydrogen, the $v = 2$ level may be populated more efficiently by Brα pumping than by thermal electron impact. However, many difficulties exist with this mechanism. First of all, the $v = 2$ level of the more abundant SiO can also be pumped by Brα, largely overshadowing CS pumping unless very favourable geometry exists. Secondly, it would produce a much stronger emission of SiO than observed. Finally, the SiO mass would have been largely overestimated, requiring more efficient destruction mechanisms of a very stable molecule. A definite conclusion on the identification of the 3.88 μm emission must await a detailed and quantitative study of all these issues.

4.2 Type Ia Supernovae

There is no observational evidence of molecular formation in Type Ia supernovae whose physical conditions and elemental compositions differ greatly from those of Type II supernovae such as SN 1987A. SNe Ia are characterized by the synthesis of larger masses of radioactive materials, higher expansion velocities, and lower gas densities than SNe II. The ejecta expand homologously with densities which vary between 10^5 cm^{-3} and 10^7 cm^{-3} at 100 days after the explosion depending on the ejecta shell and the explosion model. This is in contrast with the much higher density of the order of 10^{11} cm^{-3} in SN 1987A. These result in higher ionization fractions and higher ionization stages in SNe Ia than in SNe II. As a result, chemistry of SNe Ia differs significantly from that of SNe II.

As in SNe II, the formation of the CO and SiO molecules in SNe Ia is mainly initiated by the direct radiative association reactions (4.1) and (4.32), respectively. In contrast to what occurs in SNe II, significant contributions to the molecular formation in the highly ionized SNe Ia are made by the ionic radiative association reactions (4.16) and (4.33) to form the CO$^+$ and SiO$^+$ ions, followed by charge transfer (4.18), (4.19), (4.35) and (4.36). Significant sources of the molecular ions are also provided by reactions (4.17) and (4.34). However, only small fractions of the molecular ions may lead to the molecules because the ions are mostly destroyed by dissociative recombination (4.20) and (4.37). As the result of the high electron abundances, significant contributions to the molecular formation in SNe Ia may be made by radiative attachment (4.2), (4.3), and (4.39) followed by associative detachment (4.4), (4.5), (4.40) and (4.41). The effectiveness of the negative ion sequence as the source of the molecules is limited by photodetachment (4.6), (4.7) and (4.42) and by mutual neutralization including (4.8), (4.9), (4.43) and (4.44). A significant source of CO is provided by radiative association reactions (4.10) and (4.11) and associative detachment (4.12) and (4.13) followed by reactions (4.14) and (4.15). Reaction (4.38) may also be a significant source of SiO.

Unlike the molecules in SNe II, the molecules in the highly ionized SNe Ia are effectively destroyed by charge transfer (4.27), (4.50) and (4.51). An important sink for the molecules in SNe Ia is provided by charge transfer

$$C^{2+} + CO \rightarrow CO^+ + C^+, \quad (4.97)$$

$$O^{2+} + CO \rightarrow C^+ + O^+ + O, \quad (4.98)$$

$$O^{2+} + SiO \rightarrow Si^+ + O^+ + O, \quad (4.99)$$

$$Mg^{2+} + SiO \rightarrow SiO^+ + Mg^+. \quad (4.100)$$

In addition, the molecules are destroyed by impact with the energetic electrons through processes (4.22)–(4.25) and (4.56)–(4.59). But photodissociation (4.30) and (4.60) and photoionization (4.31) and (4.61) play a negligible role in the molecular destruction because of strong shielding of the ultraviolet photons.

The masses of CO and SiO in SNe Ia have been predicted to be below 10^{-9} $M_\odot$ (Liu 1997) and they are many orders of magnitude lower than those in SN 1987A (Liu and Dalgarno 1995, 1996) as a result of the very different physical conditions in the two types of supernovae. The supernova chemistry is controlled mainly by the ionization fraction and the gas density. SNe Ia are highly ionized while SNe II are only moderately ionized. The density in SNe Ia is of the order of 10^6 cm^{-3} at 100 days which is much lower than the 10^{11} cm^{-3} in SNe II. Because of these differences, the contrast between the chemistries of the two types of supernovae is remarkable. In SNe Ia, the molecular formation as initiated by radiative processes is inefficient due to the low density, and the molecular destruction as predominantly caused by charge transfer is effective due to the high ionization fraction. On the other hand, the higher density and the lower ionization fraction lead to a faster formation and a slower destruction of the molecules in SNe II. These result in much lower molecular masses in SNe Ia than in SNe II. On the other hand, the CO^+/CO mass ratio in SNe Ia is much higher (Liu 1997) than in SNe II (Lepp et al. 1990). This is mainly because the formation rate of CO^+ via the ionic radiative association (4.16) is enhanced and the formation rate of CO via the neutral radiative association (4.1) is reduced in the highly ionized SNe Ia compared with those in the primarily neutral SNe II. Because of the low molecular masses in SNe Ia, the molecular formation should have no significant effect on the physical properties of SNe Ia. This is remarkably different from the case of SN 1987A for which the thermal structure, spectra, and light curves are substantially altered due to the molecular formation (Liu and Dalgarno 1995). This conclusion is consistent with the spectral observations of SNe Ia which reveal no evidence of molecules.

Bibliography

1. Aitken, D. K., Smith, C. H., James, S. D., Roche, P. F., Hyland, A. R. and McCregor, P. J. (1988). *MNRAS*, **231**, 7P.
2. Andreazza, C. M., Singh, P. D. and Sanzovo, G. C. (1995). *Ap. J.*, **451**, 889.
3. Babb, J. F. and Dalgarno, A. (1995). *Phys. Rev. A*, **51**, 3021.
4. Bouchet, P. and Danziger, I. J. (1993). *Astr. Ap.*, **273**, 451.
5. Bouchet, P., Danziger, I. J. and Lucy, L. B. (1991). In *Supernovae*, ed. S. E. Woosley. Springer-Verlag, New York, p.49.
6. Dalgarno, A., Du, M. L. and You, J. H. (1990). *Ap. J.*, **349**, 675.
7. Danziger, I. J., Lucy, L. B., Bouchet, P. and Gouiffes, C. (1991). In *Supernovae*, ed. S. E. Woosley. Springer-Verlag, New York, p.69.
8. Flesch, G. D., Nourbakhsh, S. and Ng, C. Y. (1991). *J. Chem. Phys.*, **95**, 3381.
9. Hanuschik, R. W. (1991). In *Proc. ESO/EIPC Workshop: SN 1987A and Other Supernovae*, ed. I. J. Danziger and K. Kjär. ESO, Garching, p.237.
10. Herbst, E., Millar, T. J., Wlodek, S. and Bohme, D. K. (1989). *Astr. Ap.*, **222**, 205.
11. Husain, D. and Norris, P. E. (1978). *J. Chem. Soc. Faraday Trans. II*, **71**, 525.
12. Jones, T. T., Villinger, J., Lister, D. G., Tichy, M., Birkinshaw, K. and Twiddy, N. D. (1981). *J. Phys. B*, **14**, 2719.
13. Kimura, M., Dalgarno, A., Chantranupong, L., Li, Y., Hirsch, G and Buenker, R. J. (1993). *Ap. J.*, **417**, 812.

14. Kimura, M., Gu, J. P., Liebermann, H. P., Li, Y., Hirsch, G. and Buenker, R. J. (1994), *Phys. Rev. A*, **50**, 4854.
15. Leen, T. M. and Graff, M. M. (1988). *Ap. J.*, **325**, 411.
16. Lepp, S., Dalgarno, A. and McCray, R. (1990). *Ap. J.*, **358**, 262.
17. Li, H. and McCray, R. (1992). *Ap. J.*, **387**, 309.
18. Li, H., McCray, R. and Sunyaev, R. A. (1993). *Ap. J.*, **419**, 824.
19. Liu, W. (1997). *Ap. J.*, **479**, 907.
20. Liu, W. (1998). *Ap. J.*, **496**, 967.
21. Liu, W. and Dalgarno, A. (1994). *Ap. J.*, **428**, 769.
22. Liu, W. and Dalgarno, A. (1995). *Ap. J.*, **454**, 472.
23. Liu, W. and Dalgarno, A. (1996). *Ap. J.*, **471**, 480.
24. Liu, W. and Victor, G. A. (1994). *Ap. J.*, **435**, 909.
25. Liu, W., Dalgarno, A. and Lepp, S. (1992). *Ap. J.*, **396**, 679.
26. Liu, W., Jeffery, D. J. and Schultz, D. R. (1998). *Ap. J.*, **494**, 812.
27. Lucy, L. B., Danziger, I. J., Gouiffes, C. and Bouchet, P. (1989). In *Proc. IAU. Coll. 120, Structure and Dynamics of the Interstellar Medium*, ed. G. Tenorio-Tagle, M. Moles and J. Melnick. Springer-Verlag, Berlin, 164.
28. McCray, R. (1990). In *Molecular Astrophysics: A Volume Honouring Alexander Dalgarno*, ed. T. W. Hartquist. Cambridge Univ. Press, Cambridge, 439.
29. Meikle, W. P. S., Allen, D. A., Spyromilio, J. and Varani, G.-F. (1989). *MNRAS*, **238**, 193.
30. Meikle, W. P. S., Spyromilio, J., Allen, D. A., Varani, G.-F. and Cumming, R. J. (1993). *MNRAS*, **261**, 535.
31. Moseley, S. H., Dwek, E., Silverberg, R. F., Glaccum, W. J., Graham, J. R. and Loewenstein, R. F. (1989). *Ap. J.*, **347**, 1119.
32. Nomoto, K., Shigeyama, T., Kumagai, S. and Yamaoka, H. (1991). In *Supernovae*, ed. S. E. Woosley. Springer-Verlag, New York, p.176.
33. Rebrion, C., Rowe, B. R. and Marquette, J. B. (1989). *J. Chem. Phys.*, **91**, 6142.
34. Swearengen, P. M., Davies, S. J. and Niemczyk, T. N. (1978). *Chem. Phys. Lett.*, **55**, 274.
35. Thielemann, F.-K., Hashimoto, M. and Nomoto, K. (1990). *Ap. J.*, **349**, 222.
36. Villinger, H., Futrell, J. H., Richter, R., Saxer, A., Niccolini, St. and Lindinger, W. (1983). *Int. J. Mass Spectrom. Ion Phys.*, **47**, 175.
37. Woosley, S. E., Pinto, P. A. and Weaver, T. A. (1988). *Proc. Astron. Soc. Australia*, **7**, 355.
38. Woosley, S. E., Pinto, P. A. and Hartmann, D. (1989). *Ap. J.*, **346**, 395.

PART VI

Starburst Galaxies and Active Galactic Nuclei

20
Molecular Gas, Starbursts and Active Galactic Nuclei

Roy S. Booth
Onsala Space Observatory

and

Susanne Aalto
Onsala Space Observatory

1 Introduction

Active centres of galaxies range from extremely compact active galactic nuclei (AGNs) powered by mass accretion on to black holes to extended starbursts of more modest power. In some extreme cases, e.g. Arp 220, the star-powered luminosity may rival that of a compact AGN. Interest in the connection between the starburst phenomenon and the central AGN has been stimulated by the fact that some galaxies with central AGNs, e.g. NGC 1068 and Centaurus A, contain significant amounts of circumnuclear gas. Molecular gas fuels the star-forming activity and responds physically and chemically to the ultraviolet radiation and supernovae produced in a young population of massive stars. Molecular clouds collide dissipatively and thus sink readily towards the centres of galactic gravitational potentials and may fuel the black holes. The high velocity resolution together with the modest angular resolution achievable in millimetre wave emission line measurements have made molecular emissions useful reddening-free tracers of the central kinematics of galaxies. Indeed, more recent high velocity, high angular resolution VLBI observations of molecular masers have revealed a very close connection between molecules and accretion discs surrounding AGN black holes.

In the following, we review the possible evolutionary relationship between starbursts and nuclear activity in galaxies. We then concentrate on the observations of molecular gas (and in some cases, atomic gas) and how they may be used to investigate the relationship. Starting with the properties of prototypical starburst galaxies, we proceed through the ultraluminous infrared galaxies discovered in the IRAS survey, through radio galaxies and even high red-shift quasi-stellar objects (QSOs), using molecular observations as a link.

While black holes are the likely drivers of the strong radio galaxies, the presence of one in a conventional starburst galaxy often remains elusive.

2 The Types of Objects

Active galaxies have been classified into several different types which we briefly characterize below.

2.1 Starburst Galaxies

Starburst galaxies, as the name implies, are galaxies undergoing vigorous bursts of star formation in their nuclei; the star formation takes place at a rate far higher than the average during a galactic lifetime. They have been described by Weedman (1983). Starburst nuclei are detected optically by the presence of strong emission lines and blue continuum colours produced by hot stars. They also exhibit excessive infrared radiation attributed to dust heated by hot stars (e.g. Rieke and Lebofsky 1978) and radio emission from supernova remnants arising from massive stars (e.g. Bierman and Fricke 1997; Condon 1980). Indeed, there is a strong correlation between the thermal far infrared emission and the non-thermal radio emission of galaxies extending over three orders of magnitude (Dickey and Salpeter 1984). The explanation that such nuclei are locations of brief but intense star formation is generally accepted. As long as stars with conventional mass limits are used to explain the starbursts, the total mass of massive ($\gtrsim 10$ $M_\odot$), short-lived ($\lesssim 2 \times 10^7$ yr) stars can exceed 10^8 $M_\odot$ (Rieke et al. 1980; Gehrz, Sramek and Weedman 1983). There is mounting evidence from simulations that bursts of star formation may be induced by mergers of galaxies but gas flows in barred galaxies may also trigger the bursts (Norman 1988; Elmegreen 1994). Since stars form from interstellar matter, and from molecular clouds in particular, a complete understanding of the phenomenon can only be possible through observation of the molecular content of these systems. The galaxy-to-galaxy variation in the spatial extent for starburst activity is quite large, the size of the emission region ranges from about 100 pc to several kpc (e.g. Moorwood and Olivia 1994).

2.2 Seyfert Galaxies and Mergers

What is the fate of a starburst galaxy? Can the material of the starburst sink in the gravitational potential resulting in accretion on to a central black hole causing an active or a Seyfert nucleus? Several composite Seyfert/starburst objects, e.g. NGC 1068 (Balick and Heckman 1985) and NGC 7469 (Heckman et al. 1986) are known and several authors have pursued the evolutionary hypothesis (e.g. Weedman 1983; Norman and Scoville 1988). In an important paper, Hernquist (1989) combined these and other ideas into a hierachical framework in which starburst and Seyfert activity follow sequentially, being triggered by a merger of two galaxies. He then investigated this scenario with numerical simulations. In his model, gas distributed throughout a galaxy responds strongly to the tidal field of a companion during a merger. In some cases dynamical instability will drive a large fraction of the gas into the inner regions of the galaxy where self-gravity takes over, compressing the gas and causing the starburst. Subsequent evolution may lead to the formation of a black hole and continued accretion may provide sufficient power to explain the Seyfert activity in otherwise normal galaxies. Depending on the merging galaxies, there may even be enough power to explain quasi-stellar objects. High resolution studies of molecular gas in starburst and Seyfert galaxies, and especially in the ultraluminous IRAS galaxies, go some way to support this hypothesis but, as we shall see, the situation is by no means clear.

2.3 Seyfert 1–Seyfert 2 Galaxy Unification

Seyfert galaxies are (generally) spiral galaxies with extremely intense active nuclei. They are part of a generic group first identified in 1943 by Seyfert who detected their bright nuclei in short exposure photography. Subsequent spectroscopic observations complicated the picture and led to a division of the class of Seyferts (Sy) into two categories—Sy 1 and Sy 2—and even sub-categories, e.g. Sy 1.5, etc. Compared to normal galaxies, whose spectra are generally devoid of emission lines, Sy 1 galaxies have broad emission lines (full width at zero intensity between 7000 and 20 000 km s^{-1}) that are attributed to ionized gas within 1 pc of the black hole. In addition to the broad wings in the permitted lines, they have narrower forbidden lines. Sy 2 galaxies show only narrower emission lines, believed to originate from a much larger region around the core. There appear to be no statistically significant differences in the radio luminosities of Sy 1 and Sy 2 galaxies (e.g. Ulvestad and Wilson 1989) (but see Norris and Roy 1996); however, Sy 1s are much stronger keV X-ray sources. Diameters of the nuclear sources are small, ranging from point-like to 3 kpc with a median around 0.5 kpc (Woltjer 1990).

Another class of galaxies whose spectra resemble those of Seyfert 2s are the LINERs—low luminosity galaxies with low ionization nuclear emission-line regions. However, the low ionization lines, e.g. [O I], [N II] and [S II], are stronger than in their Seyfert 2 counterparts; additionally they may be distinguished from Sy 2 galaxies by their low values of [O III]/Hβ relative to [N II]/Hα—see Peterson (1997), Osterbrock (1989).

In 1985, following much suggestive material in the literature, Antonucci and Millar (1985) drew a unifying link between the two classes of Sy galaxies. In their model the spectroscopic differences are attributed to the presence of a dusty torus of dense molecular gas surrounding the black hole. The observed properties are determined by the orientation of the torus relative to the line of sight to the nucleus. In type 1 nuclei, the axis of the torus is close to the line of sight and one observes a naked AGN directly with its associated broad line region in full view; in type 2 the orientation of the dusty torus is such that it shields the nucleus from view and only the more extended narrow line clouds are observed. Typical size scales are 1 pc for the broad line region and 100 pc for the narrow line region.

The clinching observation for this unifying idea was the detection by Antonucci and Millar (1985) of broad emission lines in the polarized spectrum of the Sy 2 galaxy, NGC 1068; they concluded that this Sy 1 spectrum was caused by reflection (scattering) of the light from the hidden nucleus into our line of sight. This result is confirmed through spectropolarimetric observations of other samples of Sy 2 galaxies (see e.g. Heisler, Lumsden and Bailey 1997).

Some of the most extreme examples of both starburst and Seyfert phenomena are the ultraluminous infrared galaxies detected in the far infrared sky survey with the IRAS satellite and we shall discuss the studies of their molecular gas in the following pages. We will also see how observations of both molecular gas and even atomic hydrogen are providing strong evidence for the dusty torus hypothesis and even for the existence of a massive black hole in the centres of Seyfert and other active galactic nuclei.

2.4 Radio Galaxies, Quasi-stellar Objects and BL Lac Objects

Seyfert galaxies are relatively low luminosity AGNs; the most luminous are the quasi-stellar objects (QSOs) with bolometric nuclear magnitudes, $M_B < -21.5 + 5 \log H_0$ ($H_0 = 100$ km s^{-1} Mpc^{-1}). Optically QSOs are faint blue objects of small diameter, sometimes surrounded by a low surface brightness halo. They have spectra similar to those of Sy 1 galaxies, with very broad lines, except that their stellar absorption features are extremely weak or even absent. Weedman (1976) has argued, given the overlapping properties of QSOs and Seyferts and that they form a continuous sequence in luminosity, that they are the same phenomenon at different distances or evolutionary states. However this is not proven.

Some 10% of QSOs are radio loud and are numbered among the most luminous radio sources; radio loud QSOs are called quasars. However, not all extragalactic radio sources are quasars, there being a class of radio galaxies associated with giant elliptical galaxies. These sources exhibit narrow optical lines, more like Sy 2 galaxies. The generic structural form of radio galaxies and quasars is that of a central, flat spectrum core (the active galaxy), straddled by two large isotropically radiating, steep spectra, lobes fed by relativistic jets which produce highly anisotropic radio emission. The jets are powered by a supermassive black hole (see, e.g., Wilson 1995).

On the small scale (0.1–100 pc) the nuclear radio emission regions are one sided jets and material is ejected at relativistic speeds. Thus the initial collimation occurs on scales <0.1 pc and relativistic beaming is important in these objects making their actual appearance strongly dependent on orientation. In a BL Lac object we believe that we are viewing the AGN along the relativistic jet. We will address orientation effects and the possibility of obscuring tori to unify the radio galaxy and radio loud QSO populations in Section 8 of this chapter.

However, we note here that Cygnus A, the prototypical narrow line radio galaxy, by far the most powerful radio source out to $z = 1$, is now known to contain a hidden broad line AGN. An exciting new observation in polarized light by

Ogle et al. (1997) using the Keck II telescope (four nights after commissioning!) has revealed broad Hα and Hβ emission lines and a blue continuum seen in scattered light, confirming earlier hints of broad Mg II lines by Antonucci, Hurt and Kinney (1994). This observation provides very strong evidence for unification of the radio galaxy and radio loud QSO populations through obscuring tori at different orientations just as the Antonucci and Millar experiment did for Sy 1 and Sy 2 galaxies.

All AGN exhibit temporal intensity variations at all wavelengths; extreme variations occur in some objects on timescales of days and even hours and such objects are called optically violent variables (OVV). BL Lac objects are similar to the optically violent variables but contain featureless spectra.

Most strong radio sources have overall sizes which are much larger that their host galaxies, but a small minority, of order 5% in flux limited samples selected at 5 GHz, are subgalactic with all their radio emission occurring within a region smaller than 1 kpc. VLBI observations show these objects have symmetric radio structures straddling an active central component. They resemble the large extended radio sources but are two to three orders of magnitude smaller. For these reasons they are called compact symmetric objects, CSOs. Multi-epoch VLBI observations by Owsianik and Conway (1998) have detected the motion of a hot spot in one of these sources allowing them to estimate that its age is a few thousand years and thus showing that CSOs are probably younger versions of classical symmetric radio galaxies.

3 Molecular Line Observations of Starburst Galaxies

Because molecular hydrogen has no permanent electric dipole moment and the lowest quadrupole rotational transitions lie in the infrared, most molecular line observations of galaxies are conducted in the $J = 1$–0 rotational line of CO, the next most abundant molecular species; [CO]/[H$_2$] = 10^{-5}–10^{-4}. The column density of molecular hydrogen is inferred from an empirical conversion constant derived for (gravitationally bound) giant molecular clouds in the Galaxy (Solomon et al. 1987). This is normally defined as $X = N(H_2)/I(CO) = 3 \times 10^{20}$ mols cm^{-2}(K km s^{-1})$^{-1}$ (for discussion, see e.g. Young and Scoville 1991; Booth 1991; references therein; subsection 5.4 of Chapter 3). For regions of massive star formation and consequent high UV radiation fields, the use of this constant may introduce errors in the calculation of the total molecular mass and therefore in the star formation rate, etc. We discuss this problem later in this chapter.

The galaxies in which CO was first detected are generally active galaxies, their central regions being strong emitters of infrared and radio continuum radiation, having optical emission lines and in some cases showing evidence of non-circular motions. Although with improved receiver sensitivity many more normal galaxies are detected, the starburst galaxies remain among the galaxies with the highest CO luminosities; a strong correlation is found between integrated CO emission and far infrared flux. (see e.g. Young and Scoville 1991). Young and Deveraux (1991) observed a sample of nearby (15 < D(Mpc) < 40) starburst galaxies in the CO (1–0) line and showed that the inferred global molecular hydrogen masses were fairly large, being in the range 10^9–10^{10} $M_\odot$ and that most of the CO was concentrated towards the central region of each galaxy. All members of their sample have disturbed morphologies.

3.1 M82 and NGC 253—Prototypical Starburst Galaxies

The two IR-bright ($\approx 3 \times 10^{10}$ $L_\odot$) galaxies M82 and NGC 253 are both classical starburst galaxies with strong radio continuum emission. As IR luminous galaxies go, M82 and NGC 253 have quite modest IR luminosities and molecular masses ($\approx 10^8$ $M_\odot$) but their status as starburst galaxies is well-established. Furthermore, their proximity ($D \approx 3$ Mpc) makes them ideal for detailed studies of the starburst phenomenon and its mechanisms since the starburst ISM can be studied at high linear resolution. From infrared photometry, spectroscopy and mapping of M82, Rieke et al. (1980) found that the starburst models could account for the energetic nuclear sources in both galaxies, although M82 may be somewhat ahead of NGC 253 in its burst development (Rieke, Lebofsky and Walker 1988).

Although M82 is part of the M81, M82, NGC 3077 group of galaxies which shows evidence of interactions with H I tidal tails extending well beyond the optical discs of the galaxies (Yun, Ho and Lo 1994) and NGC 253 does not appear to have a companion, their nuclear sources are very similar. Observations of the radio continuum of M82 (e.g. Kronberg and Sramek 1985) showed that the source breaks up into more than 30 compact sources, assumed to be young supernova remnants (SNR), immersed in a weak diffuse background, and that their combined luminosity was sufficient to explain the radio emission. Muxlow et al. (1997) confirmed this result and have found that the supernova remnants have ages of $\lesssim 10^3$ years. There is no apparent evidence for a central AGN. VLA observations of NGC 253 by Antonucci and Ulvestad (1988) show that the radio structure breaks up into point like SNRs as for M82 (see also Section 10).

High resolution observations of M82 in the lines of CO, HCN and HCO$^+$ by Lo et al. (1987), Nakai et al. (1987) and Carlstrom (1988) indicated that most of the molecular gas mass, $\approx 10^8$ $M_\odot$, is contained in a torus of 150 pc radius. A substantial fraction of high density gas (40% of the mass has $n(H_2) \gtrsim 10^4$ cm^{-3}) is inferred from HCN observations (Brouillet and Schilke 1993) and from multi-transition observations of CO (Güsten et al. 1993). There are also indications that the molecular gas in the inner region of M82 is warm with kinetic temperature, T_k, greater than 50 K (e.g. Güsten et al. 1993).

The most recent observations of the CO (1–0) emission in M82 with the Berkeley–Illinois–Maryland Association (BIMA) array with a linear resolution of 38 pc showed that most of the CO emitting gas could be located in molecular spiral arms 125 pc and 390 pc from the nucleus. The CO associated with the outer arm shows considerable velocity dispersion and has been disrupted, probably by starbursts, an interpretation supported by a close association of the radio SNRs with the CO emission peaks. The authors suggested that these observations are consistent with modelling of the infrared, optical and radio data by Rieke et al. (1993) implying that two bursts of star formation occurred 5×10^6 and 3×10^7 years ago. The fact that the outer arm is disrupted is taken to indicate that the older starburst is located in this arm and that the younger starburst took place in the inner arm recently enough that the surrounding interstellar medium has not been severely disrupted. Shen and Lo (1995) have argued that if this is the case, the starburst is propagating inwards towards the centre of the galaxy.

In contrast to M82, the molecular gas in NGC 253 (see Peng et al. 1996; references therein) shows a central molecular bar similar to the stellar bar mapped at optical and infrared wavelengths. The CO bar (Canzian, Mundy and Scoville 1988) lies at a position angle of 64° and has dimensions 1100 pc × 260 pc. CS observations by Peng et al. (1996) show that the dense gas lies within a radius of about 300 pc. They speculated that the bar channels gas into the central region of the galaxy, fuelling the starburst activity.

4 Ultraluminous Infrared Galaxies

One of the most exciting results from the IRAS survey was the discovery of galaxies with luminosities dominated by far-infrared emission (Soifer et al. 1984). These galaxies can be more than 100 times more luminous in the infrared than in the visible, making them 100–1000 times brighter than normal spirals, and contain copious amounts of molecular gas ($2-50 \times 10^9$ $M_\odot$)—up to twenty times that of the Milky Way (Sanders et al. 1986). In most of these luminous infrared galaxies the far-infrared flux appears to come from dust heated as a consequence of intense star formation. However, as the far-IR emission rises above $L_{FIR} > 10^{11}$ $L_\odot$, these objects are often powered by active nuclei in addition to starbursts. Sanders et al. (1988) have proposed that such galaxies, each of which radiates in the infrared as much energy as a QSO (i.e. $L_{FIR} > 10^{12}$ $L_\odot$ in the 8 µm–1000 µm) wavelength range, really are QSOs enshrouded by dust. Since, as we have said, the ultimate source of fuel for intense star formation and/or an active nucleus resides in the interstellar medium, studies of the atomic and molecular gas are of importance for our understanding of these systems. Their optical morphologies, especially among the very brightest infrared objects, are usually disturbed, often showing double nuclei or tidal tails (see Sanders et al. (1990) for isophotes of some of the northern objects and Melnick and Mirabel (1990) for images of the southern sample obtained with the European Southern Observatory New Technology Telescope). These are tell-tale signs of merging systems and so it has been suggested that the enormous luminosities are the result of bursts of star formation, triggered by the mergers.

In addition to the vast quantities of molecular hydrogen, luminous infrared galaxies also contain atomic hydrogen, observed through its spin–flip transition at a wavelength of 21 cm. Many also contain powerful OH masers, radiating a million times more power than galactic masers and hence called megamasers. Most bright infrared galaxies have H I emission but the OH megamaser sources show broad H I absorption lines covering a large velocity range—up to 650 km s^{-1}. The bulk atomic (H I) and molecular (CO) properties of the luminous galaxies have been measured (Mirabel and Sanders 1988; Sanders et al. 1986; Mirabel et al. 1990). OH masers have been studied by Baan, Henkel and Haschick (1987) and Martin et al. (1991) and reviewed by e.g. Henkel, Mauersberger and Baan (1991). The field of luminous infrared galaxies has recently been reviewed by Sanders and Mirabel (1996).

4.1 CO Studies

CO observations of ultraluminous infrared galaxies ($L_{\text{bol}} > 10^{12}$ L$_\odot$) have been reviewed by Sanders (1991) and by Scoville and Soifer (1991). The CO observations have been conducted by several groups using the IRAM 30 m telescope, SEST and the NRAO 12 m antenna (Sanders et al. 1989; Scoville and Soifer 1991; Mirabel et al. 1990; Solomon, Radford and Downes 1990; Barvainis, Alloin and Antonucci 1989). The detected objects range in red shift out to $z = 0.8$, the nearest being the galaxy Arp 220 with a recession velocity $cz = 5542$ km s^{-1}. Thus, the CO signal is often quite weak, with atmosphere corrected antenna temperatures of a few mK. In addition, the lines are broad (full width >500 km s^{-1}), so detection can be difficult and depends on good receiver baselines. However, as we have already said, copious amounts of molecular gas are inferred from the CO data (>10^{10} M$_\odot$) based on the standard value of the conversion constant.

This extremely high abundance of molecular gas is taken to indicate that we are observing pairs of giant spirals in the process of merging, a fact borne out by the optical data described above. The ratio $L_{\text{IR}}/M(H_2)$, or the star formation efficiency, ranges from about 20 to 200 L$_\odot$/M$_\odot$ compared to a value of 4 L$_\odot$/M$_\odot$ for normal isolated spiral galaxies; it is also larger than the average ratio found for "classical" starburst galaxies like M82. Again, the idea of mergers leading to violent bursts of star formation are strongly supported by these data.

If the infrared luminosity is taken to be a measure of the total mass of recently formed stars, the luminosity to mass ratio will be a measure of the time-scale over which interstellar matter is converted into young stars. If the initial mass function of the stars producing the infrared luminosity is similar to that for the Galaxy, the approximate order of magnitude higher $L_{\text{FIR}}/M(H_2)$ in the ultraluminous galaxies implies that the stellar cycling time is about ten times shorter than in normal galaxies, or about 10^8 years.

4.1.1 Arp 220

The prototypical ultraluminous galaxy is Arp 220. Interferometer images of the CO (1–0) emission (Scoville et al. 1991) and also of HCN and HCO$^+$ emission (Radford, Downes and Solomon 1991) show that 90% of the dynamical mass of the system and 75% of the molecular gas is confined to a central core, 600 pc in diameter (Scoville et al. 1991). A recent Caltech interferometer map of the CO (2–1) distribution at 230 GHz (0.″8 resolution) (Scoville, Yun and Bryant 1997) reveals 3 peaks in the dense molecular gas. Two peaks correspond to a double nucleus found in the near infrared by Graham et al. (1990) and in the radio continuum (Norman et al. 1985), and the third to a more extended disc like structure elongated SW–NE, similar but separated from the dust lane seen in optical images. The elongated disc feature exhibits a monotonic velocity gradient parallel to the CO intensity distribution, and the dynamical mass, for an assumed rotating disc ($= RV^2/G$, where R is the radius, G the gravitational constant and V is the rotational velocity determined from the line width) is approximately 3×10^{10} M$_\odot$ within a radius of 360 pc.

Scoville, Yun and Bryant (1997) considered the likelihood of Arp 220 being powered by a starburst alone, based on constraints provided by the measured millimetre continuum emission. They found that this is unlikely. However, recent VLBI observations reported by Lonsdale et al. in the recent IAU Colloquium 164, held at NRAO Socorro, show that the Arp 220 nuclei consist of multiple compact emission regions which they interpret as young SNRs. These point sources account for all the radio flux density in Arp 220 with $T_B > 10^6$ K. No other 18 cm emission is detected on scales from 1–30 pc. Thus, it appears that all the compact continuum emission may be explained by the SNR (although Lonsdale et al. did not rule out the presence of an AGN) and that it is not necessary to appeal to AGN activity to account for the overall radio/infrared characteristics of Arp 220.

4.1.2 Arp 299

Arp 299 is an ultraluminous ($L_{IR} \approx 8 \times 10^{11}$ L$_\odot$) merger of two galaxies, IC 694 and NGC 3690. The nuclei of the two galaxies are still well separated by 5 kpc, making Arp 299 a merger at an earlier stage than Arp 220. The CO emission of Arp 299 is bright and the inferred molecular mass from the standard CO to $M(H_2)$ conversion factor is 8.6×10^9 M$_\odot$(Solomon and Sage 1988). The CO emission is dominated by three "condensations" of high surface brightness: two on the nuclei of the two merging galaxies, and one where the discs of the two galaxies overlap. As seen in Fig. 1, extended emission surrounds the nuclei and also seemingly connects the major condensations of gas. The CO condensation centred on the nucleus of IC 694 remains unresolved with an upper limit to its radius of only 140 pc and a lower limit to the CO brightness temperature of 18 K. The two nuclei, as well as the western overlap region, currently harbour intense star formation activity (cf. Gehrz et al. 1983; Baan and Haschick 1990). Furthermore, the nucleus of IC 694 is a flat-spectrum radio source, and may contain an AGN (Gehrz et al. 1983).

High resolution OVRO observations of CO, ^{13}CO, and HCN 1–0 of Arp 299 show that the CO/^{13}CO and CO/HCN line ratios vary dramatically (Fig. 1) across the system (Aalto et al. 1997). The CO/^{13}CO ratio is unusually large, 60 ± 15, at the IC 694 nucleus, where CO emission is very strong. Elsewhere, the CO/^{13}CO line ratio is 5–20, typical of spiral galaxies (e.g. Aalto et al. 1995). For comparison, the ratio is 6–7 for the disc of the Galaxy (Polk et al. 1988). The CO/HCN line ratio also varies across Arp 299. HCN emission is bright towards

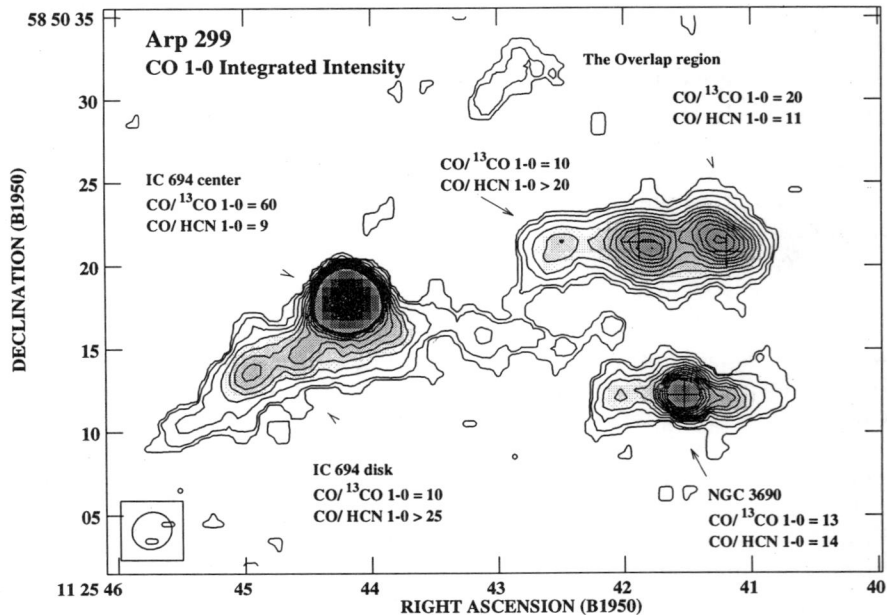

FIG. 1. The distribution of CO 1–0 emission in the merger Arp 299 (IC 694 and NGC 3690). Molecular line ratios, in terms of brightness temperature, are indicated in the figure (Aalto et al. 1997). Relative to CO, ^{13}CO 1–0 is brightest in quiescent regions of low CO surface brightness and weakest in starburst regions and the galactic nuclei. The crosses indicate peaks in the 6 cm radio continuum emission (Gehrz et al. 1983).

each galaxy's nucleus and in the extranuclear region of active star formation. At the nuclei of IC 694 and NGC 3690 the ratios are 9 ± 1 and 14 ± 3 respectively. In the western part of the overlap region it is 11 ± 3 while elsewhere in Arp 299 the ratio is >20.

The large CO/^{13}CO 1–0 intensity ratio at the nucleus of IC 694 can be attributed primarily to a low to moderate optical depth ($\tau \lesssim 1$) in the CO 1–0 line. These data support the hypothesis that unusually high CO/^{13}CO line ratios (>20) are associated with extremely compact molecular distributions in the nuclei of merging galaxies (Aalto et al. 1995). Recent OVRO high resolution CO and ^{13}CO 2–1 data (Aalto et al. in preparation) reveal that the ^{13}CO 2–1/1–0 intensity ratio is high, ≈ 3, in the nucleus of IC 694 directly indicating warm gas. A medium consisting of dense ($n = 10^4$–10^5 cm^{-3}) and warm ($T_k > 100$ K) gas will reproduce the extreme line ratios observed in the nucleus of IC 694 where the area covering factor must be at least 20%. It is interesting to note that the unusual line ratios do not occur in the interaction zone between the two galaxies, but in the nuclei of the galaxies.

5 Molecular Cloud Properties in Starburst Nuclei

The CO, ^{13}CO, and HCN results for Arp 299 displayed in Fig. 1 clearly indicate that physical conditions of molecular gas in the nuclei of starburst/active galaxies differ very much from those of molecular clouds in the disc of the Milky Way.

High kinetic temperature, $T_k \gtrsim 50$ K, seems to be a general result for molecular gas in the inner kpc of starburst galaxies (e.g. Irwin and Avery 1992; Martin and Ho 1986; Harris et al. 1991; Israel 1992; Wall et al. 1993; Solomon, Downes and Radford 1992a; Güsten 1993; Aalto et al. 1991a, b, 1995). Often, the conclusion that the molecular gas is warm is based on observations of bright emissions in high transitions of CO, ^{13}CO and HCN, on observations of NH$_3$, or on fits of data to cloud models. In some cases, high resolution aperture synthesis observations of CO yield useful lower limits to the gas kinetic temperatures. Bryant and Scoville (1996) measured, with the OVRO array, a lower limit to the CO 2-1 brightness temperature of 80 K for the ultraluminous merger Mrk 231.

In starburst nuclei, a larger fraction of the molecular mass is in a *high density* ($n \gtrsim 10^4$ cm^{-3}) phase than in Giant Molecular Clouds (GMCs) in the disc of the Milky Way. The bright emissions, relative to CO, from high density tracer molecules (such as CS and HCN) towards many starburst nuclei indicate substantial amounts of high density gas (e.g. Solomon, Radford and Downes 1990; Nguyen-Q-Rieu et al. 1989; Mauersberger et al. 1989, Mauersberger, Henkel and Sage 1990). The detection of high transition CO emission, 4-3 and 6-5 (Harris et al. 1991; Güsten et al. 1993), also requires the presence of gas at densities $n \gtrsim 10^4$ cm^{-3}. Interestingly, HCN 1-0 is particularly bright in the inner regions of the Seyfert galaxies Mrk 231 (e.g. Solomon et al. 1992; Bryant and Scoville 1997) and NGC 1068 (e.g. Sternberg, Genzel and Tacconi 1994; Helfer and Blitz 1995).

Our understanding of the CO 1-0 emission from molecular clouds in the disc of our Galaxy is that it originates in self-gravitating structures. Recent studies, however, suggest a different picture for the inner regions of starburst galaxies. Aalto et al. (1994; 1995) proposed that the lower transition emission of CO may arise in diffuse, *non-self-gravitating* structures, loosely attached to dense cores, and Radford (1993) argued that the CO emission from the nucleus of an ultraluminous galaxy may be dominated by a non-cloudy component that is bound to the potential of the galaxy.

Interestingly, it appears that normal galactic centres show similar changes in cloud properties. An increase in mean density of the H$_2$ gas may be typical of all galactic centres, and not only for molecular gas in starburst or active nuclei (e.g. Helfer and Blitz 1997). High temperatures of the molecular gas appear also to be characteristic of the non-starbursting centre of the Galaxy (e.g. Güsten 1989). Dahmen et al. (1997) found that the CO 1-0 emission from the inner region of our Galaxy is dominated by non-self-gravitating structures, analogous to the proposed non-cloudy medium of starburst galaxies. We have yet to establish how far the similarities between molecular gas in normal and starburst/active nuclei goes. Jackson et al. (1996) argued that starburst nuclei have larger masses of dense gas, and higher average gas density in the central 200 pc, than the comparatively inactive Galactic Centre. Another indication that the ISM of starburst and normal galaxies are not wholly alike comes from observations of the $\lambda = 158$ µm [C II] and $\lambda = 63$ µm [O I] lines from centres of nearby galaxies showing that the [C II] and [O I] emissions from starburst galaxies are brighter relative to that of CO than in non-starburst galaxies (Stacey et al. 1991). These far-infrared (FIR) lines are typical signatures of so-called Photon Dominated Regions (PDRs) or photo-dissociation regions. These are regions where the newborn stars interact with and impact their birth clouds. More on PDRs in galaxies can be found in Chapter 9.

Work on *molecular abundances* in starburst and Seyfert nuclei is mostly confined to CO, ^{13}CO, C^{18}O, HCN, CS and HCO$^+$, apart from in the most nearby systems. The high values (> 30) of ^{12}CO/^{13}CO 1–0 intensity ratios observed in the centres of some merging galaxies (e.g. Aalto et al. 1991a,b; Casoli, Dupraz and Combes 1992) may to some degree be caused by abundance effects. High resolution maps of the CO and ^{13}CO 1–0 emission (Aalto et al. 1997) of the merger source Arp 299 show that the region of faint ^{13}CO emission is localized to the compact nuclei of the merging galaxies—regions where the CO 1–0 emission is brightest. Since the optical depth of the CO 1–0 line is often moderate, $\tau \approx 1$, in galactic nuclei (due to high gas temperatures and/or the presence of non-self-gravitating gas), even a fairly modest decrease in the ^{13}CO abundance will become visible in the ^{12}CO/^{13}CO 1–0 intensity ratio.

Some have discussed an inflow of gas from the optical disc of the galaxy as a possible cause of a relative lack of ^{13}CO (e.g. Henkel and Mauersberger 1993). Then there are the "in situ" explanations for relatively faint ^{13}CO emission: (1) selective enrichment of ^{12}C through the early return from massive stars in a burst (Casoli, Dupraz and Combes 1992; Henkel and Mauersberger 1993) and (2) selective photo-dissociation of ^{13}CO with respect to CO due to a high density of UV photons in a starburst environment. At this stage it is unfortunately still difficult to disentangle elemental abundance effects from effects of chemistry or from excitation. It has, however, been established that the CO/^{13}CO abundance ratio for the nucleus (inner 200 pc) of IC 694 (Arp 299) is at least 60 (Aalto et al. 1997) which means that, for any of the above reasons, the abundance ratio is about a factor of two higher than that found for the centre of the Galaxy.

There have been suggestions that C^{18}O is overabundant with respect to C^{17}O in starburst nuclei (Sage, Henkel and Mauersberger 1991), caused by nucleosynthesis in massive stars. HCN has been suggested to be overabundant with respect to CO for the active nucleus of NGC 1068 (Sternberg, Genzel and Tacconi 1994), but this result has been challenged by Helfer and Blitz (1995) who argued that no abundance anomalies are necessary to explain observations. It is clear that a lot of work remains to disentangle the effects of excitation, abundance and chemistry for molecular emission.

5.1 Cloud Properties and the Conversion Factor

A long debated question concerns whether the CO-to-H_2 conversion factor, X, is universal or whether it varies significantly from galaxy to galaxy and/or with location within a galaxy. Here, there appears to be agreement that "significantly" should imply a variation greater than a factor of 2–3. In the previous section we have discussed how different the physical conditions of the nuclear starburst clouds are from those of the cool GMCs in the galactic disc, upon which the conversion factor is calibrated. The inferred molecular masses for IR luminous galaxies are very high (see Section 2), but since their CO emission in general is dominated by central starburst distributions, a significant change in the applicability of the standard conversion factor is likely to have a severe impact on the estimated masses.

The CO luminosity for a cloud of radius R, linewidth δV and peak CO brightness temperature $T_B(CO)$ is $L(CO) = T_B(CO)\delta V \pi R^2$. For a self-gravitating cloud of mass M and density ρ, $\delta V \approx \left(\frac{GM}{R}\right)^{\frac{1}{2}}$, leading to the relation

$$L(CO) \approx T_B(CO) M \left(\frac{3\pi G}{4\rho}\right)^{\frac{1}{2}}. \tag{5.1}$$

It can therefore be argued that the increase in kinetic temperature of the nuclear gas to some degree will be cancelled out by the higher mean number density of the clouds since $L(CO)$, and therefore the conversion factor X, is proportional to $\frac{T_B(CO)}{\rho^{1/2}}$ (e.g. Dickman, Snell and Schloerb 1986; Scoville and Sanders 1987).

However, this rests to some degree on the assumption that the CO emission is thermalized, but even a cloud of high mean density may have a low density outer layer that renders the observed CO emission sub-thermally excited. In this case, the coupling between $T_B(CO)$ and gas kinetic temperature, T_k, becomes poor. Therefore, the cloud's higher mean density would not be cancelled out by a larger CO brightness temperature.

A perhaps bigger problem may be the assumption that the CO emission largely arises from self-gravitating structures. Indeed, as discussed in the previous section, there is evidence that most of the CO 1–0 emission from galactic nuclei actually emerges from "non-cloudy" structures. In such structures the basic assumption that the cloud linewidth is proportional to the cloud's mass breaks down, and the L_{CO} is likely to overestimate the molecular mass. It is however possible that the conversion factor still works: assume that the CO emission originates in diffuse structures, bound to the total potential of the galactic centre; then the CO emission traces the total enclosed mass in this potential, rather than an ensemble of graviationally bound gas clouds. If this enclosed mass is mostly molecular (as argued by Downes, Solomon and Radford 1993), then the conversion factor may be roughly accurate. A molecular medium may consist of fairly low density, $n(H_2) = 10^2$–10^3 cm^{-3}, diffuse CO-emitting gas, in which is embedded dense ($n(H_2) \gtrsim 10^4$cm^{-3}), compact clumps responsible for the bright high density tracer emission. In this scenario, the bulk of the molecular mass will reside in the dense clumps (e.g. Aalto et al. 1994).

Implicit in the use of the conversion factor is also a basic assumption that the CO and the molecular hydrogen clouds are co-extensive. However, if the molecular cloud is illuminated by UV radiation, selective photo-dissociation may take place which will change this situation (e.g. van Dishoeck and Black 1988). CO may not always trace the true H_2 distribution, especially in low metallicity systems like the Magellanic Clouds where CO may already be underabundant (e.g. Booth 1990), or for objects where the UV activity is unusually high, like a starburst and/or AGN nucleus.

A considerable variation of molecular line intensity ratios within the merger Arp 299 implies equally dramatic changes of the properties of the H_2 gas, where the most extreme gas-properties are found for the two nuclei. It is clear that the prevailing physical conditions of the molecular gas within starburst and/or AGN nuclei are far from those of quiescent galactic disc clouds and that it is very likely that this fact will have an impact on the conversion factor. It is conceivable that the value of X therefore varies *within* galaxies as well as between them. This variation would trace changes in the cloud physical conditions (as implied for Arp 299) or metallicity (Arimoto, Sofue and Tsujimoto 1996). An internally varying conversion factor will naturally have consequences for our understanding of the molecular mass distribution within galaxies. In particular, we may overestimate the degree of central concentration of the molecular gas in interacting and starburst galaxies.

It is clear that observations and modelling of molecular cloud properties in starburst/AGN nuclei will have to improve further before quantitative conclusions can be drawn on the impact of cloud properties on the standard CO to H_2 mass conversion factor.

6 Molecular Emission from Nearby Radio Galaxies and QSOs

The source of the high luminosity in the ultraluminous galaxies remains uncertain despite the large observational effort. The three basic processes which have been discussed are nuclear starbursts (Rieke et al. 1985), the release of the kinetic energy of colliding galaxies (Harwit et al. 1987) and the reprocessing of radiation by dust enshrouding QSOs (Sanders, Scoville and Soifer 1988). It is quite possible that all three processes are involved: certainly the most luminous infrared galaxies show strong evidence of merging, and theoretical ideas, as discussed earlier, suggest that mergers of gas-rich galaxies can provide the fuel for starbursts and/or feed the black hole at the centre of an active galactic nucleus.

It is therefore not surprising that a number of attempts have been made to detect CO in relatively nearby QSOs and radio galaxies. CO has been detected in the UV excess QSOs Mk 1014, Pg 0838+77, Mk 876 and 1 Zw 1 (Sanders, Scoville and Soifer 1988; Barvainis, Alloin and Antonnucci 1989; Alloin et al. 1992) while Mirabel et al. (1989) have detected CO in the powerful radio galaxies

Perseus A (3C 84) and 4C 12.50 and several groups have detected molecules in the nearest radio galaxy Centaurus A. More recently CO has been detected in 3C 48, the first optically identified quasar (Scoville et al. 1993). In all cases, the mere detection of CO at red shifts >0.1 implies an equivalent mass of molecular hydrogen of the order 10^{10} $M_\odot$, suggesting that large amounts of molecular mass and therefore high rates of star formation are involved in the genesis of the central engines that power extragalactic radio sources. Perseus A and Centaurus A are relatively nearby radio galaxies (recession velocities of 5250 and 550 km s^{-1}, respectively) and have rather less molecular gas; both are interacting systems.

7 CO at High Red Shift

During a programme to measure red shifts of galaxies detected in the IRAS Faint Source Survey, Rowan-Robinson et al. (1991) detected a remarkable emission-line galaxy, IRAS 10214+4724, at a red shift of 2.286. They found this galaxy had an enormous far-infrared luminosity of 3×10^{14} times that of the Sun and stronger than the previous most luminous known object. It was not long before molecular line astronomers attempted to find molecular gas in this object and Brown and Vanden Bout (1991) detected the CO (3–2) transition red shifted to 105.2 GHz with the NRAO 12 m telescope on Kitt Peak. The detection, if not the detailed spectrum, was quickly confirmed with the IRAM 30 m telescope (Brown and Vanden Bout 1992; Solomon, Downes and Radford 1992a) and the Nobeyama 45 m instrument (Tsuboi and Nakai 1992) and other transitions were measured.

The enormity of this discovery was not the detection, in itself remarkable, but the realization that molecular line astronomy has the potential to probe the early Universe and that the epoch of first star formation might be within the grasp of our current technique.

Subsequent observations of IRAS 1014+4724 with the IRAM 30 m telescope produced detections of CO emission in the (4–3) (Brown and Vanden Bout 1992) and (6–5) transitions (Solomon, Downes and Radford 1992b). The line ratios CO (6–5)/CO (3–2) = 0.6±0.2 and CO (4–3)/CO (3–2) = 0.8±0.2 are considerably higher than overall values for the Milky Way and LVG calculations by Solomon, Downes and Radford show that the ratios are consistent with warm gas with $T_k \approx 50$ K and $n(H_2) \approx 5000$ cm^{-3}

7.1 Molecular Mass

In the case of emission at high red shift, observed quantities must be corrected for z. Solomon, Downes and Radford (1992a) have derived the expression below for the CO luminosity:

$$L_{CO} = 23.5 \Omega d^2 I_{CO} (1+z)^{-3} \qquad (7.1)$$

where Ω is the solid angle of the source convolved with the telescope beam, which is the beam solid angle if the source is much smaller than the beam. Turning this equation around, we see that for a given fixed beam size, I_{CO} does not scale as d^{-2} but as $(1+z)^3 d^{-2}$. This means that the detection of CO from distant objects, although not easy, is actually easier than had been thought.

Initial estimates of the CO line luminosity of IRAS 10214+4724 gave a value greater than 10^{11} $L_\odot$, an order of magnitude larger than for the other ultraluminous infrared galaxies discussed above, and an infrared to CO luminosity ratio twice as great. The estimated molecular mass was found to be more than 10^{11} $M_\odot$, for the standard conversion ratio. However, later observations showed that the galaxy is actually artificially brightened by gravitational lensing and these factors needed to be reduced by an order of magnitude.

7.2 The Structure of IRAS 10214+4762—a Gravitationally Lensed Object?

Images of IRAS 10214+4762 in 2.2 μm continuum emission were obtained by Matthews et al. (1994) and Graham and Liu (1995). Both show a compact (0.″7 diameter) source superposed on a weaker arc 1.″5 long. This morphology is suggestive of a gravitational lens with at least two components, i.e. the emission from the galaxy is partially focused by a massive object along our line of sight to it, giving an enhanced signal at the Earth. Depending on the precise distribution of mass in the lensing object in relation to the line of sight, arcs, rings and multiple images are known to be produced. A subsequent IRAM interferometer map of the CO (3–2) emission (Downes, Solomon and Radford 1995) has shown the source to be extended-1.″5 × 0.″9. Their image is somewhat larger than the 2.2 μm image convolved with the IRAM interferometer beam but it agrees well with the 2.2 μm arc convolved with their beam, indicating that the CO is associated with the arc.

The authors concluded that IRAS 10412+4762 is indeed a gravitationally lensed image of a galaxy. The estimated magnification is 10 times, reducing the actual infrared luminosity by an order of magnitude to 10^{13} $L_\odot$ and the molecular mass to 10^{10} $M_\odot$. These values place IRAS 10214+4762 in the same category as the ultraluminous galaxies—but still show that molecular gas exists at a red shift of 2.286.

7.3 High Red-shift QSOs

There are two more detections of CO at high z. The first is the detection in the so-called clover leaf quasar—a "classical" gravitational lensed quasar. Such a clover leaf pattern is caused by a massive body with a uniform mass distribution along the line of sight to a point background object. The CO in this quasar is at a red shift of 2.546 (Barvainis et al. 1994).

However, the most spectacular result to date is the detection of CO and dust in the QSO BR1202-0725 by Ohta et al. (1996) and Omont et al. (1996) at the almost incredible red shift of 4.69. BR1202 is one of the most distant objects in the Universe and the emission we now see left it when the Universe was only about 7% of its current age, or roughly a billion years after the Big Bang. The large mass of dust and CO in this object shows that the primordial hydrogen had already been enriched with heavy elements in galaxies at this early epoch. Interferometer maps of the CO and the dust emission from BR1202 reveal two objects, separated by a few arcseconds, which either indicate that the source is actually double, or that gravitational lensing is again at work. Nevertheless, vast amounts of molecular gas have been shown to be present at this very early epoch, showing that conditions even in the very early Universe were conducive to a huge bursts of star formation.

8 Molecular Tori Around Active Galactic Nuclei

In Section 2.3 we discussed the difference between the observational properties of Sy 1 and Sy 2 galaxies as an orientation effect and the postulated existence of a dusty torus shielding the broad line region in a Sy 2 galaxy. Such ideas have also been applied to radio galaxies and quasars: double-lobed sources and sources with single relativistic jets. For source orientations in which the radio jet is close to the sky plane, the torus will hide the central engine from our direct view giving sources which are classed as radio galaxies rather than quasars. Krolik and Begelman (1986) have proposed the existence of a torus of very dense molecular material in the central 1–10 pc with a more diffuse envelope (Pier and Krolik 1993) extending over tens of pc. Observational evidence for molecular and even atomic hydrogen tori is mounting and here we discuss some examples.

8.1 The Nearby Galaxies Centaurus A and Circinus A

The nearest radio galaxy is Centaurus A. SEST observations of the CO (2–1) lines in Centaurus A (Rydbeck et al. 1993) have shown the gas to be distributed in a molecular ring or torus with a diameter of about 200 pc (Fig. 2). This is confirmed with higher resolution CO (3–2) line observations (Rydbeck private communication). A similar result is obtained for another nearby starburst galaxy, Circinus A (Curran et al. in preparation), Circinus A is thought to contain a Sy 2 nucleus (Harnett et al. 1990). In both cases the tori have rotational velocities of around 200 km s^{-1}, indicating enclosed dynamical masses of about 10^9 M$_\odot$.

8.2 NGC 1068

This galaxy contains the prototypical Sy 2 nucleus; it also exhibits strong starburst activity and its total luminosity of 3×10^{11} L$_\odot$ is approximately equally divided between the extended starburst and the point-like (<30 pc) AGN (Telesco et al. 1984). The galaxy is rich in molecular gas with a total H$_2$ mass estimated to be 1.5×10^{10} M$_\odot$ (Scoville, Young and Lucy 1983). High resolution aperture synthesis maps of NGC 1068 have been made of the CO (1–0) emission (Planeas, Scoville and Myers 1991), HCN (e.g. Jackson et al. 1993; Tacconi et al. 1994) and ^{13}CO and C^{18}O (Papadopoulos, Seaquist and Scoville 1996). The CO is distributed in a ring-like structure, which is interpreted as the inner spiral arms of gas at 15″ radius (1.5 kpc) which originate from the ends of a stellar bar, and an unresolved source coincident with the Seyfert nucleus. These components account for about 30% of the molecular content of NGC 1068. In the HCN maps,

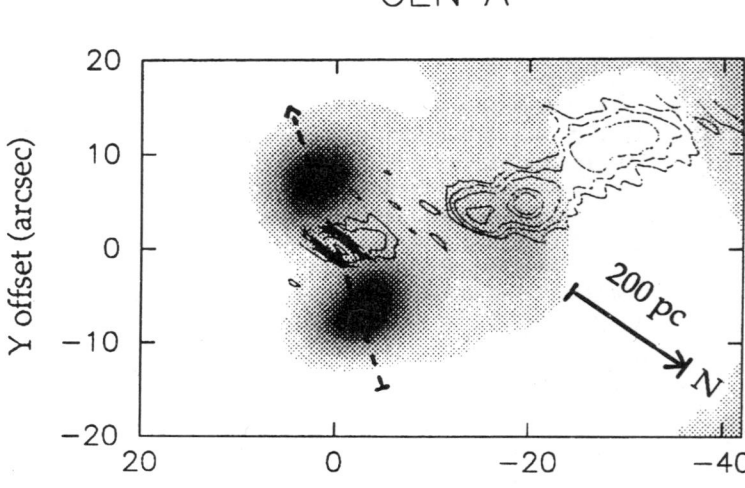

FIG. 2. Deconvolved CO, $J = 2$–1 emission intensities of highly red- and blue-shifted molecular gas in the nucleus of Cen A, superposed on a VLA image of the radio jet at 1.5 GHz.

the very compact nuclear source dominates, although clumps of gas are seen in the inner star forming spiral arm delineated by the CO emission. Tacconi et al. (1994) have also made images in the 2.12 μm H_2 $v = 1$–0 S(1) line with their near-infrared Fabry–Perot imaging spectrometer. A combination of their HCN and H_2 data indicates that the central source which has a diameter of about 300 pc contains a system of dense ($n(H_2) \approx 10^5$ cm^{-3}) warm ($T_k = 70$ K) molecular clouds. The rotation velocities obtained from the HCN map imply that the mass contained within 1″ of the nucleus is about 10^8 $M_\odot$. Tacconi et al. (1994) suggested that the thick nuclear molecular cloud is responsible for obscuring the Sy 1 nucleus in NGC 1068 in the visible, UV and perhaps even in X-rays!

8.3 Mrk 231

This ultraluminous infrared galaxy is another disturbed system known to contain a dust enshrouded Seyfert 1 nucleus (Boksenberg et al. 1977). Caltech interferometer maps of the CO (1–0) and (2–1) emission have been presented by Bryant and Scoville (1996) and the higher frequency image shows an extended source (2″ × 0.″8) with a velocity gradient along the major axis. Virtually all the single dish flux is contained within a radius of 420 pc. The authors have examined the relationship between the dynamical mass and the H_2 mass determined from the standard Galactic CO–H_2 conversion factor. They found that these could only be reconciled (molecular mass less than dynamical mass) if the molecular gas is confined to a disc which is oriented with its axis not more than 59° from the line of sight. They pointed out that this is consistent with the modest extinction measure for the Seyfert nucleus.

9 Broad Neutral Hydrogen and OH Absorption in Starbursts and Seyferts—Further Evidence for Obscuring Tori?

Many well-known starburst and Seyfert galaxies, including M82, NGC 253, NGC 660, NGC 1068, NGC 4945 and Circinus A show broad (several hundred km s^{-1}) atomic hydrogen absorption lines. Broad OH absorption lines as well as OH maser emission are also often observed (Martin 1989). Broad absorption lines are a common feature of the ultraluminous infrared galaxies. The absorption lines generally have the same velocity widths as the corresponding emission features but whereas the emission velocities are representative of the total galactic gas, the absorbing gas must lie in front of the nuclear continuum source. VLA observations of the H I absorption in Arp 220 (Baan et al. 1987) and of a sample of spiral galaxies with active nuclei (Dickey 1986) show that the absorbing gas is compact and probably lies not in the galactic disc but in circumnuclear clouds. In the case of NGC 5793, European VLBI observations reveal that the absorbing gas is barely resolved with a beam of about 30 milliarcseconds (Gardner et al. 1992).

From the studies of H I absorption in a sample of 15 active spiral galaxies (Dickey 1986) and of ultraluminous infrared galaxies by Martin et al. (1991a,b), it has been shown that statistically, there is a positive offset between the optical velocity (taken as the systemic velocity of the galaxy) and the mean absorption velocity, although some galaxies show a negative offset. This result is taken to indicate that the circumnuclear gas is generally falling into the nucleus, although it does not explain the broad velocity widths of the lines. Koribalski (1996) has suggested that the absorbing gas is in fast rotation about the nucleus—this could explain the large velocity width. In the light of observations described below, we favour the rotational hypothesis, or a mix of infall and rotation. Indeed, we would suggest that the absorbing gas may lie in the dusty torus surrounding the AGN.

9.1 Neutral Hydrogen Absorption in the Seyfert Galaxy, NGC 4151

Mundell et al. (1995) have observed the H I absorption in NGC 4151. Although the radio continuum in the nuclear region consists of five main components, the H I absorption is clearly associated with only one of them, the active nuclear component. Mundell et al. estimated a lower limit of 90 $M_\odot$ of neutral hydrogen required to produce the absorption against the 1.5×1.5 pc^2 nucleus. This gas, they argued, is in a disc or torus surrounding the nucleus although they pointed out that NGC 4151 is classified as an Sy 1 galaxy and as such should not have an obscuring torus in the unified scheme of things. They noted, however, that its radio structure is more like an Sy 2 nucleus and that it has been reported by Penston and Perez (1984) that its broad line region had almost disappeared. Pedlar et al. (1993) have suggested that we are viewing the nucleus close to the edge of its ionizing cone, which might account for its properties being intermediate between those of Sy 1 and Sy 2.

9.2 H I Absorption in CSO Radio Galaxies

Compact Symmetric Objects (CSO) are strong, compact (<1 kpc) objects with radio structures and luminosities similar to those of classical double radio sources (i.e. two lobes and a weak core), but are thousands of times smaller (Wilkinson et al. 1994). They are probably young (few times 10^3 years) radio sources which will evolve into classical double sources (Owsianik and Conway 1998). A relatively high proportion of the known CSOs show H I absorption and this interesting fact makes them important probes for the presence of circumnuclear tori. Because of their small component separation there is a high probability that the far mini-lobe, which generally contains about 50% of the total flux, will be covered by the torus.

Conway (1996, 1997) has reported VLBA observations of the H I absorption in one CSO, 4C31.04, which has a size of only 100 pc. The H I absorption in this source was discovered by Mirabel (1990); it is a double system with a broad (133 km s^{-1}) component and a narrow (<20 km s^{-1}) component. Figure 3a shows the H I absorption spectrum integrated over the whole source and 3b the H I opacity integrated over the deepest part of the broad absorption line. The opacity is largest ($\tau \approx 0.07$) and fairly uniform over the eastern lobe but in the western lobe there is a sharp change of opacity from 0.02 on the core side to effectively zero further west. Conway modelled his observational result in terms of a (>100 pc) H I disc perpendicular to the radio jet, the axis of which would be expected to be close to the sky plane, since there is no relativistic beaming. The disc completely covers the eastern lobe but only partially obscures the western lobe.(Fig. 3c).

10 Megamasers—Homing in on the Black Hole

An interesting, if poorly understood, phenomenon in molecular astrophysics is that of the interstellar masers. Several excellent reviews on cosmic masers have been written by Elitzur, culminating in his book *Astronomical Masers* (Elitzur 1991; also see Chapter 14 of this volume). Here we discuss the extragalactic masers since, in recent years, several new discoveries have been made and it has been shown that the masers may be used to determine distances to external galaxies and, more spectacularly, to weigh the massive black holes in their centres.

Several molecular species exhibit the maser phenomenon in external galaxies. The most spectacular (highest gain) are those due to water and hydroxyl (OH) but emission lines from formaldehyde (H_2CO), CH (Henkel, Mauersberger and

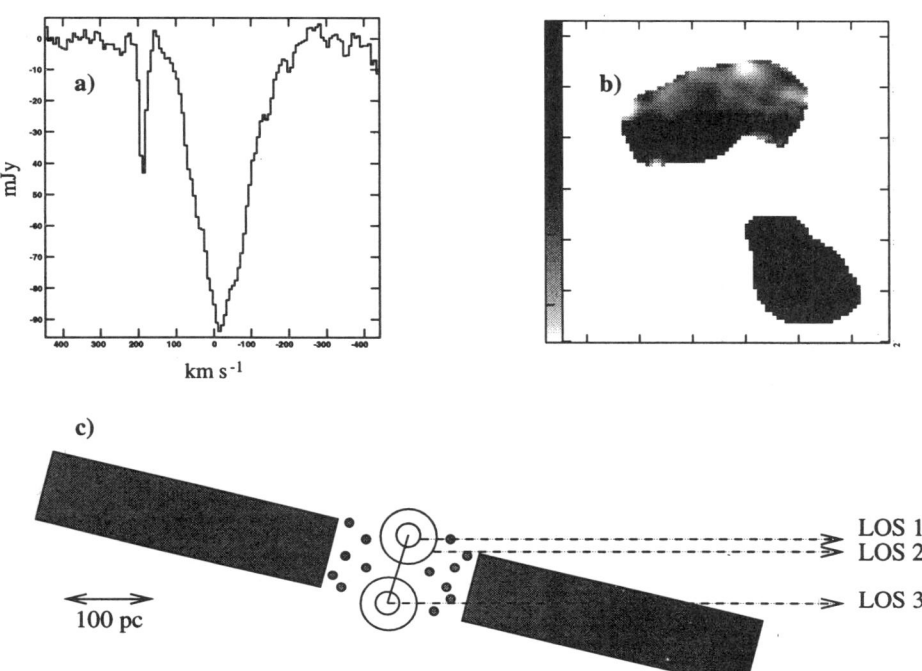

FIG. 3. H I Absorption in the nucleus of the compact (<100 pc) radio loud galaxy 4C31.04.(a) H I absorption spectrum integrated over the whole source, the peak absorption is 90 mJy which corresponds to an opacity of 0.035.(b) H I opacity averaged over the central part of the broad absorption for regions with continuum brightness greater than 40 mJy/beam. Greyscale image is plotted from −0.005 (white) to 0.040 (black), resolution is 12 by 9 mas, $PA = 5°$, tick marks are separated by 20 mas. Note image has been rotated 90° anticlockwise so that north is to the left. (c) Possible model. Grey indicates a gas disc whose axis is parallel to the radio axis; small clouds are shown evaporating off its inner edge. The regions of continuum radio emission from the lobes are shown schematically as two sets of concentric circles. Three lines of sight are shown, corresponding respectively to the regions of negligible absorption, the absorption edge and higher absorption seen in going from top to bottom in Figure b.

Baan 1991) and methyl alcohol (CH_3OH) also appear to be amplified by the maser process.

10.1 OH Megamasers

The first extragalactic OH masers were discovered in NGC 253 by Whiteoak and Gardner (1973) and a similar detection was made by Rieu *et al.* (1976) towards

M82, both in the so-called main lines of the ground state lambda doublet of OH at 1665 and 1667 MHz. In M82 the 1667 MHz line is the stronger whereas in Galactic OH masers it is the weaker. These maser features are weak narrow lines confused by broad absorption of OH against the galactic nuclei. However, it was quickly realized that they are stronger than the most luminous Galactic OH masers by factors of 10–100, and it was suggested that they are amplifying the continuum radiation from the galactic nuclei, or perhaps galactic H II regions.

Much more powerful OH masers have been detected subsequently. The prototype of this class of megamasers was discovered by Baan, Wood and Haschick (1982) towards the core of the ultraluminous infrared galaxy Arp 220 (IC 4553). Its apparent isotropic luminosity is 380 $L_\odot$ (it is convenient to assume the radiation to be isotropic although it may well be beamed), more than one million times more powerful than any Galactic OH maser, hence, leading to the designation mega! The earlier extragalactic masers are now often referred to as kilomasers. Surveys of the luminous IRAS galaxies have revealed more than 50 OH megamasers, some at relatively large distances with recession velocities, cz >10 000 km s^{-1}(Henkel, Mauersberger and Baan 1991) and some gigamasers, e.g. IRAS 20100-4156 with $cz = 38\,700$ km s^{-1} and an isotropic luminosity of 10^4 $L_\odot$ (Staveley-Smith et al. 1989).

The properties of the prototypical megamaser in Arp 220 have been studied thoroughly in order to address the general issue of the origin of megamasers (see Henkel et al. 1991). The galaxy's far-IR luminosity is $L_{\rm FIR}=10^{12}$ $L_\odot$ and its distance is 76 Mpc, if $H_0 = 75$ km s^{-1} Mpc^{-1} is assumed. The peaks of the near-IR, far-IR and radio continuum are broadly coincident, the radio structure being described as a double source with a separation of 1 arcsec (or 350 pc) between the components, each of which is barely resolved at a resolution of 0.15 arcsec (Norris 1988). Imaging VLBI observations of the OH maser have shown that the line emission is double and mimics the continuum source (Baan and Haschick 1984; Norris et al. 1985) and so they have been interpreted as low-gain images of the continuum components, amplified by intervening clouds of molecular material. The utility of this interesting new way to observe the continuum structure seemed to be confirmed by early VLBI observations of Diamond et al. (1989) which revealed OH maser components <30 milliarcsec in diameter associated with each continuum source, although they had insufficient sensitivity to detect the continuum components on these VLBI scales. However, recent global VLBI observations by Lonsdale et al. (1998) show that the OH consists of a diffuse component and several compact (pc scale) components at 1667 MHz. Four major compact emission regions are revealed with complex spatial and velocity structures and the authors speculated that they probably trace shock fronts in the dense nuclear environment and may be related to AGN activity.

It seems likely that the pump source for the masers is provided by the powerful IR radiation field and this is given credence by a rough correlation between OH and IR luminosity, $L_{\rm OH} \propto L_{\rm IR}^2$ (Martin et al. 1988; Baan 1989; Henkel and

Wilson 1990). Baan (1991) has also shown that masing seems only to occur when the spectrum between 25 and 60 μm is particularly steep, i.e. when log $(S_{25}/S_{60}) < -0.55$, where S_{25} and S_{60} are the fluxes at those wavelengths.

Since the OH megamasers are associated with starburst and Seyfert galaxies, it is interesting to speculate that they might be amplified in the dense regions of the molecular tori. Montgomery and Cohen (1992) observed the 1667 MHz OH megamaser in III Zw 35 with the MERLIN array and found the maser velocity features to be spatially aligned with position angle $-20°$ over a distance of 40 pc with a south to north velocity gradient of about 2.5 km s^{-1}pc^{-1} (Fig. 4). They interpreted this as a rotating disc of OH with an enclosed mass of 3×10^9 M$_\odot$. This exciting discovery went largely unnoticed but was in fact the first direct indication that the broad OH lines in megamaser sources are indicative of rotation about a central black hole.

10.2 Water Masers

A dramatic confirmation that this sort of model is correct has come through observations of other extragalactic masers—water masers associated with a galaxy NGC 4258, another luminous IRAS galaxy, with peculiar tidal arms and an active nucleus (see also Chapter 23).

The water maser line is at a frequency close to 22 GHz corresponding to the 6_{16}–5_{23} transition. Like OH features, it is seen in the Milky Way in association with regions of star formation and in the atmospheres of evolved stars (cf. Chapter 14). The water maser outshines OH in most cases, and in the Orion nebula the water maser may reach a total flux density of 10^6 Jy, radiating almost the total luminosity of the Sun in a line only 50 kHz wide!

To date, 15 extragalactic water masers with isotropic luminosities >10 L$_\odot$ have been found. They are referred to as megamasers by analogy with OH; their luminosities exceed that of the brightest Galactic H$_2$O maser by more than an order of magnitude. All 15 are in galaxies with active nuclei. The other extragalactic water masers are less intense and are associated with galaxies of various types from the LMC and SMC, through the normal spiral M33 to the nuclear starburst galaxy NGC 253. These masers have been thought to be associated with spiral arm H II regions, similar to those in the Galaxy.

The galaxy NGC 4258 has received much attention by the maser observers of late. Discovered by Claussen, Heilgman and Lo (1984) during a survey of bright IRAS galaxies, this maser is one of four detected in the 73 galaxies searched. It is interesting to note that already in 1986, Claussen and Lo (1986) suggested that extragalactic water masers in AGN like NGC 4258, could be associated with the putative dusty tori, put forward to explain the different visible spectral properties of Seyfert 1 and Seyfert 2 galaxies in terms of a single unified model, by Antonucci and Miller (1985). The use of the maser emission from NGC 4258 and other galaxies to probe the kinematics of the torus and derive masses of the central object, which is possibly a supermassive black hole, in each is treated in Chapter 23.

11 Summary and Discussion

The observations of molecules associated with starbursts and active galactic nuclei have provided dramatic confirmation of the existence of molecular tori/discs around black holes in the galactic centres. Violent starbursts are clearly associated with interactions and mergers but not all starburst activity is merger driven (cf. NGC 253). The idea that stellar bars have an important role to play is implied here.

The exact relationship between the starburst phenomenon and the AGN state is still not absolutely clear. In particular, the nearby classical starburst galaxies like M82 show little or no evidence of AGNs. Observations of a sample of ultraluminous IRAS galaxies (Crawford et al. 1996) with the VLA at 6 and 20 cm, while revealing radio emission in extended radio sources and confirming the correlation between far infrared and microwave luminosity (Helou, Soifer and Rowan-Robinson 1985), showed little evidence of nonthermal activity in the galactic nuclei. They claimed that even the nuclear activity in these galaxies is starburst driven.

In an observational attempt to test the orientation explanation of the Sy 1/Sy 2 differences, Norris and Roy (1996) have produced another puzzling result. They used the Parkes Tidbinbilla radio linked interferometer to observe the compact cores of a large sample of Sy 1 and Sy 2 galaxies. They argued that because the dusty tori should not obscure the radio emission, they should have detected equal numbers of Sy 1s and Sy 2s. Instead they found significantly more compact cores in Sy 2s (48% detected) than Sy 1s (27% detected). They can only explain this discrepancy if the narrow line clouds are optically thick at radio wavelengths. If this is indeed the case then the absence of detected nuclear radio emission, even in the starburst and ultraluminous infrared galaxies, could possibly be explained. Further confirmatory evidence is provided by observations of several Seyfert galaxies by Pedlar et al. (1997). They found free–free absorption in the nucleus of NGC 1068, Mk 3 and NGC 4151. They pointed out that, given the proximity of the torus to the intense UV field of the AGN, it is difficult to see how its surface can avoid being photo-ionized. They speculated that the neutral molecular torus, as well as hiding the broadline region and collimating the nuclear UV, may also be the source of ionized gas which forms the narrow line region. Could this ionized gas also hide a weak nuclear AGN?

Bibliography

1. Aalto, S., Booth, R. S., Johansson, L. E. B. and Black, J. H. (1991a), *A&A*, **247**, 291.
2. Aalto, S., Johansson, L. E. B., Booth, R. S. and Black, J. H. (1991b), *A&A*, **249**, 323.
3. Aalto, S., Booth, R. S., Black, J. H., Koribalski, B. and Wielebinski, R. (1994). *A&A*, **286**, 365.

4. Aalto, S., Booth, R. S., Black, J. H. and Johansson, L. E. B. (1995). *A&A*, **300**, 369.
5. Aalto, S., Radford, S. J. E., Scoville, N. Z. and Sargent, A. I. (1997). *ApJ*, **475**, L107.
6. Alloin, D., Barvainis, R., Gordon, M. A. and Antonucci, R. R. J. (1992). *A&A*, **265**, 429.
7. Antonucci, R. (1993). *Ann. Rev. Astron. Astrophys.*, **31**, 473.
8. Antonucci, R. J., Hurt, T., Kinney, A. (1994). *Nature*, **371**, 313.
9. Antonucci, R. and Millar, J. S. (1985). *ApJ*, **297**, 621.
10. Antonucci, R. J. and Ulvestad, J. S. 1988, *ApJ*, **330**, L97.
11. Arimoto, N., Sofue, Y. and Tsujimoto, T. (1996). *PASJ*, **48**, 275.
12. Baan, W. A., and Haschick A. D. (1984). *ApJ*, **279**, 541.
13. Baan, W. A., and Haschick A. D. (1990). *ApJ*, **364**, 65.
14. Baan, W. A., Henkel, C. and Haschick, A. (1987). *ApJ*, **320**, 154.
15. Baan, W. A., Wood, P. A. D. and Haschick, A. D. (1982). *ApJ*, **260**, L49.
16. Baan, W. A., van Gorkom, J. H., Schmelz, J. T. and Mirabel, I. F. (1987)., *ApJ*, **313**, 102.
17. Balick, B. and Heckman, T. M. (1985). *AJ*, **90**, 197.
18. Barnes, J. E. and Hernquist, L. E. (1991). *ApJ*, **370**, L65.
19. Barvainis, R., Alloin, D. and Antonucci, R. (1989). *ApJ*, **337**, L69.
20. Barvainis, R., Tacconi, L., Antonucci, R., Alloin, D. and Coleman, P. (1994). *Nature*, **371**, 586.
21. Bierman, P. and Fricke, K. (1997). *A&A*, **54**, 461.
22. Boksenberg, A., Carswell, R. F., Allen, D. A., Fosbury, R. A. E., Penston, M. V., and Sargent, W. L. W. (1977). *MNRAS*, **178**, 451.
23. Booth, R. S. (1991). In *Molecular Clouds*, eds. R. James and T. Millar, CUP, p.157
24. Brouillet, N. and Schilke, P. (1993). *A&A*, **277**, 381.
25. Brown, R. L. and Vanden Bout, P. A. (1991). *ApJ*, **397**, L11.
26. Brown, R. L. and Vanden Bout, P. A. (1992). *ApJ*, **397**, L19.
27. Bryant, P. M. and Scoville, N. Z. (1996). *ApJ*, **457**, 678.
28. Bryant, P. M. and Scoville, N. Z. (1997). *ApJ*, in press.
29. Canzian, B., Mundy, L. G. and Scoville, N. Z. (1988). *ApJ*, **333**, 157.
30. Carlstrom, J. (1988). PhD thesis, Univ. California.
31. Casoli, F., Dupraz, C. and Combes, F. (1992). *A&A*, **264**, 55.
32. Claussen, M. J., Heilgman, G. M. and Lo, K. Y. (1984). *Nature*, **310**, 298.
33. Claussen, M. J. and Lo, K. Y. (1986). *ApJ*, **308**, 592.
34. Condon, J. J. (1980). *ApJ*, **242**, 894.

35. Conway, J. E. (1997). In *High Sensitivity Radio Astronomy*, eds. N. Jackson and R. J. Davis, Cambridge University Press, Cambridge, p.153.
36. Conway, J. E. (1996). In *Extragalactic Radio Sources*, eds. R. Ekers, C. Fanti and L. Padrielli, Kluwer, Dordrecht, p.92.
37. Crawford, T., Marr, J., Partridge, B. and Strauss, M. A. (1996). *ApJ*, **460**, 225.
38. Dahmen, G., Hüttemeister, S., Wilson, T. L., Mauersberger, R. *et al.* (1998). *A&A*, **331**, 959.
39. Diamond, P. J., Norris, R. P., Baan, W. A. and Booth, R. S. (1989). *ApJ*, **340**, L49.
40. Dickey, J. M. (1986). *ApJ*, **300**, 190.
41. Dickey, J. M. and Salpeter, E. E. (1984). *ApJ*, **284**, 461.
42. Dickman, R. L., Snell, R. L. and Schloerb, F. P. (1986). *ApJ*, **309**, 326.
43. Downes, D., Solomon, P. M. and Radford, S. J. E. (1993). *ApJ*, **414**, L13.
44. Downes, D., Solomon, P. M. and Radford, S. J. E. (1995). *ApJ*, **453**, L65.
45. Downes, D., Radford, J. E., Greve, A., Thum, C., Solomon, P.M and Wink, J. E. (1992). *ApJ*, **398**, L25.
46. Elitzur, M. (1991). *Astronomical Masers*, Kluwer, Dordrecht.
47. Elmegreen, B. G. (1994). *ApJ*, **425**, L73.
48. Gardner, F. F. and Whiteoak, J. B. (1975). *MNRAS*, **173**, 77.
49. Gardner, F. F., Whiteoak, J. B., Norris, R. P. and Diamond, P. J. (1992). *MNRAS*, **259**, 296.
50. Gehrz, R. D., Sramek, R. A. and Weedman, D. W. (1983). *ApJ*, **267**, 551.
51. Graham, J. R. and Liu, M. C. (1995). *ApJ*, **449**, L29.
52. Graham, J. R., Carico, D. P., Matthews, K., Neugebauer, G., Soifer, B. T. and Wilson, T. D. (1990). *ApJ*, **354**, L5.
53. Greenhill, L. J., Henkel, C., Becker, R., Wilson, T. L. and Wouterlooot, J. G. A. (1995a) *A&A*, **304**, 21.
54. Greenhill, L. J., Jiang, D. R., Moran, J. M., Claussen, M. J. and Lo, K.-Y. (1995b) *ApJ*, **440**, 619.
55. Güsten, R. (1989). In *The Center of the Galaxy*, ed. M. Morris. Kluwer, Dordrect, p.89.
56. Güsten, R., Serabyn, E., Kasemann, C., Schinkel, A., Schneider, G., Schulz, A. and Young, K. (1993). *ApJ*, **402**, 537.
57. Harnett, J. I., Whiteoak, J. B., Reynolds, J. E., Gardner, F. F. and Tzioumis, A. (1990). *MNRAS*, **244**, 130.
58. Harris, A. I., Stutzki, J., Graf, U. U., Russell, A. P. G., Genzel, R. and Hills, R. E. (1991). *ApJ*, **382**, L75.
59. Harwit, M., Houck, J. R., Soifer, B. T. and Palumbo, G. G. C. (1987). *ApJ*, **315**, 28.

60. Heckman, T. M., Smith, E. P., Baum, S. A., van Breugel, W. J. M. and Miley, G. K., (1986). *ApJ*, **311**, 526.
61. Heisler, C. A., Lumsden, S. L. and Bailey, J. A. (1997). *Nature*, **385**, 700.
62. Helfer, T. T. and Blitz, L. (1995). *ApJ*, **450**, 90
63. Helfer, T. T. and Blitz, L. (1997). *ApJ*, **478**, 162.
64. Helou, G., Soifer, B. T. and Rowan-Robinson, M. (1985). *ApJ*, **298**, L5.
65. Henkel, C. and Mauersberger, R. (1993). *A&A*, **274**, 730.
66. Henkel, C., Mauersberger, R. and Baan, W. N. (1991). *Astron. and Astrophys. Rev.*, **3**, 47.
67. Henkel, C. and Wilson, T. L. (1990). *A&A*, **229**, 431.
68. Hernquist, L. (1989). *Nature*, **340**, 687.
69. Irwin, J. A. and Avery, L. W. (1992). *ApJ*, **388**, 328.
70. Israel, F. P. (1992). *A&A*, **265**, 487.
71. Jackson, J. M. and Paglione, T. A. D. (1993). *ApJ*, **418**, L13.
72. Jackson J. M., Heyer, M. H., Paglione, T. A. D. and Bolatto, A. D. (1996). *ApJ*, **456**, L91
73. Koribalski, B. (1996). In *Minnesota Lectures on Extragalactic HI, May 1995*, ed. E. Skillman. ASP Conference Series, Vol. 106, p.238.
74. Krolik, J. H. and Begelman, M. C. (1986). *ApJ*, **308**, L55.
75. Kronberg, P. P. and Sramek, R. A. (1985). *Science*, **227**, 28.
76. Lo, K. Y., Cheung, K. W., Masson, C. R., Philips, T. G., Scott, S. L. and Woody, D. P. (1987). *ApJ*, **312**, 574.
77. Lonsdale, C. J., Lonsdale, C. J., Diamond, P. J. and Smith, H. E. (1998). *ApJ*, **493**, L13.
78. Martin, J.-M. (1989). PhD thesis, Univ. de Paris VII.
79. Martin, J.-M., Bottinelli, L., Dennerfeld, M., Gouguenheim, L. and Le Squeren, A. M. (1988). *A&A*, **201**, L13.
80. Martin, J.-M., Bottinelli, L., Dennerfeld, M., Gouguenheim, L. and Le Squeren, A.-M. (1991a). In *Dynamics of Galaxies and their Molecular Cloud Distributions*, eds. F. Combes and F. Casoli. Kluwer, Dordrecht, p.447.
81. Martin, J.-M., Bottinelli, L., Gouguenheim, L. and Dennefeld, M. (1991b). *A&A*, **245**, 393.
82. Martin, R. N. and Ho, P. T. P. (1986). *ApJ*, **308**, L7.
83. Matthews, K., Soifer, B. T., Nelson, J., Boesgaard, H., Graham, J. R., Harrison, W., Irace, W., Jernigan, G., Larkin, J. E., Lewis, H., Lin, S., Neugebauer, G., Sirota, M., Smith, G. and Ziomkowski, C. (1994). *ApJ*, **420**, L13.
84. Mauersberger, R., Henkel, C. and Sage, L. J. (1990). *A&A*, **236**, L63.
85. Mauersberger, R., Henkel, C., Wilson, T. L. and Harju, J. (1989) *A&A*, **226**, L5.

86. Melnick, J. and Mirabel, I. F. (1990). *A&A*, **231**, L19.
87. Mirabel, I. F. (1990). *ApJ*, **352**, L37.
88. Mirabel, I. F. and Sanders, D. B. (1987). *ApJ*, **322**, 688.
89. Mirabel, I. F. and Sanders, D. B. (1988). *ApJ*, **335**, 104.
90. Mirabel, I. F., Sanders, D. B. and Kazes, I. (1989). *ApJ*, **340** L9
91. Mirabel, I. F., Booth, R. S., Garay, G., Johansson, L. E. B. and Sanders, D. B. (1990). *A&A*, **236**, 327.
92. Montgomery, A. S. and Cohen, R. J. (1992). *MNRAS*, **254**, p.23
93. Moorwood, A. F. M. and Olivia, E. (1994). *ApJ*, **429**, 602.
94. Moran, J. L., Greenhill, L. J., Herrnstein, J. R., Diamond, P. J., Miyoshi, M., Nakai, N. and Inoue, M. (1996). *Proc. Natl. Acad. Sci. USA*, **92**, 11427.
95. Mundell, C. G., Pedlar, A., Baum, S. A., O'Dea, C. P., Gallimore, G. F. and Brinks, E. (1995). *MNRAS*, **272**, 355.
96. Muxlow, T. W. B., Pedlar, A., Wills, K. A., Wilkinson, P. N. and Axon, D. J., 1997. In *High Sensitivity Radio Astronomy*, eds. N. Jackson and R. J. Davis. Cambridge University Press, p.149.
97. Myoshi, M., Moran, J. M., Herrnstein, J., Nakai, N., Diamond, P. J. and Inoue, M. (1995). *Nature*, **373**, 127.
98. Nakai, N., Inoue, M. and Miyoshi, M. (1993). *Nature*, **361**, 45.
99. Nakai, N., Hayashi, M., Handa, T., Sofue, Y., Hasegawa, T. and Sasaki, M. (1987). *PASJ*, **39**, 685.
100. Nguyen-Q-Rieu, Nakai, N. and Jackson, J. M. (1989). *ApJ*, **220**, 57.
101. Noguchi, M. (1988). *A&A*, **203**, 259
102. Norman, C. A., 1988, In *Galactic and Extragalactic Star Formation*, eds. R. E. Prudritz and M. Fich. Kluwer, Dordrecht, p.495.
103. Norman, C. A. and Scoville, N. Z. (1988). *ApJ*, **332**, 124.
104. Norris, R. P. (1988). *MNRAS*, **230**, 345.
105. Norris, R. P. and Roy, A. L. (1996). In *Extragalactic Radio Sources*, eds. R. Ekers, C. Fanti and L. Padrielli, Kluwer, Dordrecht, p.381.
106. Norris, R. P., Baan, W. A., Haschick, A. D., Diamond, P. J. and Booth, R. S. (1985). *MNRAS*, **213**, 821
107. Ogle, P. M., Cohen, M. H., Miller, J. S., Tran, H. D., Fosbury, R. A. E. and Goodrich, R. W. (1997) *ApJ*, **482**, L37
108. Ohta, K., Yamada, T., Nakanishi, K., Kohno, K., Akiyama, M. and Kawabe, R. (1996). *Nature*, **382**, 426.
109. Omont, A., Petitjean, P, Guilloteau, S., McMahon, R. G., Solomon, P. M. and Pécontal, E. (1996). *Nature*, **382**, 428.
110. Osterbrock, D. E. (1989). *Astrophysics of Gaseous Nebula*, University Science Books, Mill Valley.
111. Owsianik, I. and Conway, J. E. (1998). In preparation.

112. Papadopoulos, P. P., Seaquist, E. R. and Scoville, N. Z. (1996). ApJ, **465**, 173.
113. Pedlar, A., Hamilton, N. G. and Kukula, M. J. (1997). In *High Sensitivity Radio Astronomy*, eds. N. Jackson and R. J. Davis. Cambridge University Press, Cambridge, p.139.
114. Pedlar, A., Kukula, M., Longley, D. P. T., Muxlow, T. W. B., Axon, D. J., Baum, S., O'Dea, C. and Unger, S. W. (1993). MNRAS, **263**, 471.
115. Peng, R., Zhou, S., Whiteoak, J. B., Lo, K. Y. and Sutton, E. C. (1996). ApJ, **470**, 821.
116. Penston, M. V. and Perez, E. (1984). MNRAS, **211**, 33.
117. Peterson, B. M. (1997). *Active Galactic Nuclei*, Cambridge University Press, Cambridge.
118. Pier, E. A. and Krolik, J. H. (1993). ApJ, **418**, 673.
119. Planeas, P., Scoville, N. Z. and Myers, S. T. (1991). ApJ, **369**, 364.
120. Polk, K. S., Knapp, G. R., Stark, A. A. and Wilson, R. W. (1988). ApJ, **332**, 432.
121. Radford, S. J. E., Downes, D. and Solomon, P. M. (1991). ApJ, **368**, L15.
122. Rieke, G. H., Cutri, R. M., Black, J. H., Kailey, W. F., McAlary, C. W., Lebofsky, M. J. and Elston, R. (1985). ApJ, **290**, 116.
123. Rieke, G. H. and Lebofsky, M. J., ApJ, **220**, L37.
124. Rieke, G. H., Lebofsky, M. J., Thompson, R. I., Low, F. J. and Tokunaga, A. (1980). ApJ, **238**, 24.
125. Rieke, G. H., Lebofsky, M. J. and Walker, C. E. (1988). ApJ, **325**, 679.
126. Rieke, G. H., Luken, K., Rieke, M. J. and Tamblyn, P. (1993). ApJ, **412**, 99.
127. Rieu, N.-Q., Mebold, U., Winnberg, A., Gilbert, J. and Booth, R. S. (1976). A&A, **52**, 467.
128. Rowan-Robinson, M., Broadhurst, T., Lawrence, A., McMahon, R. G., Lonsdale, C. J., Oliver, S. J., Taylor, A. N., Hacking, P. B., Conrow, T., Saunders, W., Ellis, R. S., Efstathiou, G. P. and Condon, J. J. (1991). Nature, **351**, 719.
129. Rydbeck, G., Wiklind, T., Cameron, M., Wild, W., Eckart, A., Genzel, R. and Rothermel, H. (1993). A&A, **270**, L13.
130. Sage, L. J., Henkel, C., Mauersberger, R. (1991). A&A, **249**, 31
131. Sanders, D. B. (1991). In *Dynamics of Galaxies and their Molecular Cloud Distributions*, eds. F. Combes and F. Casoli, Kluwer, Dordrecht, p.417.
132. Sanders, D. B. and Mirabel, I. F. (1996). Ann. Rev. Astron. Astrophys, **34**, 749.
133. Sanders, D. B., Phinney, E. S., Neugebauer, G., Soifer, B. T. and Matthews, K. (1989). ApJ, **347**, 29.

134. Sanders, D. B., Sargent, A. I., Scoville, N. Z. and Philipps, T. G. (1990). In *Submillimeter Astronomy*, eds. G. Watt and A. Webster, Kluwer, Dordrecht, p.213.
135. Sanders, D. B., Scoville, N. Z. and Soifer, B. T. (1988). *Science*, **239**, 625.
136. Sanders, D. B., Scoville, N. Z., Young, J. S., Soifer, B. T., Schloerb, F. P., Rice, W. L. and Danielson, G. E. (1986). *ApJ*, **305**, L45.
137. Sanders, D. B., Soifer, B. T., Elias, J. H., Madore, B. F., Matthews, Neugebauer, G., and Scoville, N. Z. (1988). *ApJ*, **328**, L35.
138. Sargent, W. L. (1977). *MNRAS*, **178**, 451.
139. Scoville, N. Z., Padin, S., Sanders, D. B., Soifer, B. T. and Yun, M. S. (1993). *ApJ*, **415**, L75.
140. Scoville, N. Z. and Sanders, D. B. (1987). In *Interstellar processes*, eds. D. J. Hollenbach and H. A. Thronson Jr. Reidel, Dordrecht, p.21.
141. Scoville, N. Z., Sargent, A. I., Sanders, D. B. and Soifer, B. T. (1991). *ApJ*, **366**, L5.
142. Scoville, N. Z. and Soifer, B. T. (1991). In *Massive Stars in Starbursts*, eds. C. Leitherer, Walborn, N. R., Heckman, T. M., Norman, C. A. and van der Hucht, K. A., Cambridge University Press, Cambridge, p.233.
143. Scoville, N. Z., Young, J. S. and Lucy. L. B. (1983). *ApJ*, **270**, 443.
144. Scoville, N. Z., Yun, M. S., Bryant, P. M. (1997). *ApJ*, **484**, 702.
145. Shen, J. and Lo, K. Y. 1995, *ApJ*, **445**, L99.
146. Soifer, B. T. *et al.* (1984). *ApJ*, **283**, L1.
147. Soifer, B. T., Houck, J. R. and Neugebauer, G. (1987). *Ann. Rev. Astron. Astrophys.*, **25**, 187.
148. Soifer, B. T., Boehmer, L., Neugebauer, G. and Sanders, D. B. (1989). *AJ*, **98**, 766.
149. Solomon, P. M., Downes, D. and Radford, S. J. E. (1992a). *ApJ*, **398**, L29.
150. Solomon, P. M., Downes, D. and Radford, S. J. E. (1992b). *Nature*, **356**, 318.
151. Solomon, P. M., Radford, S. J. E. and Downes, D. (1990). *ApJ*, **348**, L53.
152. Solomon, P. M. and Sage, L. J. (1988). *ApJ*, **334**, 613.
153. Solomon, P. M., Rivolo, A. R., Barrett, J. and Yahil, A. (1987). *ApJ*, **319**, 730.
154. Stacey, G. J., Geis, N., Genzel, R., Lugten, J. B., Poglitsch, A., Sternberg, A. and Townes, C. H. (1991). *ApJ*, **373**, 423
155. Staveley-Smith, L., Allen, D. A., Chapman, J. M., Norris, R. P. and Whiteoak, J. B. (1989). *Nature*, **337**, 625.
156. Sternberg, A., Genzel, R. and Tacconi, L. (1994). *ApJ*, **436**, 131
157. Tacconi, L. J., Genzel, R., Blietz, M., Cameron, M., Harris, A. I. and Madden, S., 1994, *ApJ*, **426**, L77.

158. Telesco, C. M., Becklin, E. E., Wynn-Williams, C. G. and Harper, D. A. (1984). *ApJ*, **282**, 427.
159. Tsuboi, M. and Nakai, N. (1992). *PASJ*, **44**, L241.
160. Ulvestad, J. S. and Wilson, A. S. 1984, *ApJ*, **278**, 544.
161. van Dishoeck, E. F. and Black, J. H. (1988). *ApJ*, **334**, 771.
162. Wall, W. F., Jaffee, D. T., Bash, F. N., Israel, F. P., Maloney, P. R., Baas, F. (1993). *ApJ*, **414**, 98.
163. Weedman, D. (1976). *Ann. Rev. Astr. and Astrophys.*, **15**, 69.
164. Weedman, D. (1983). *ApJ*, **266**, 479.
165. Whiteoak, J. B. and Gardner, F. F. (1973). *Ap.Lett.*, **15**, 211.
166. Wilkinson, P. N., Polatidis, A. G., Readhead, A. C. S., Xu, W. and Pearson, T. J., 1994, *ApJ*, **432**, L87.
167. Wilson, A. S. (1995). In *Barred Galaxies and Circumnuclear Activity*— Nobel Symposium Nr. 98, Springer Verlag, Berlin.
168. Woltjer, L. (1990). In *Saas-Fee Advanced Course 20, Lecture Notes*, eds. T. J.-L. Courvoisier and M. Mayor. Berlin: Springer Verlag, p.1.
169. Young, J. S. and Deveraux, N. A. (1991). *ApJ*, **373**, 414.
170. Young, J. S. and Scoville, N. Z. (1991). *Ann. Rev. Astron. Astrophys*, **29**, 581.
171. Yun, M. S., Ho, P. T. P. and Lo, K. Y. (1994). *Nature*, **372**, 530.

21
Excitation and Detectability of Molecules in Active Galactic Nuclei

John H. Black
Onsala Space Observatory

1 Introduction

It is paradoxical that low-energy molecular processes might be important in the hostile, energetic environments near active galactic nuclei (AGNs). It now appears that observations of the molecular component of the central gas offer crucial insight into the physics of the AGN phenomenon. As reviewed in the preceding chapter, activity in galactic centres is displayed by nuclear starbursts on scales of 10^2 to 10^3 pc as well as the true AGNs, which are the compact, luminous sources that reside in the central 10 pc of quasi-stellar objects (QSOs), radio galaxies, and Seyfert galaxies. In this chapter we are concerned with the possible ways in which molecules can be used to investigate the inner workings of the true AGN. Observations throughout the electromagnetic spectrum now suggest that many AGN are buried inside small-scale discs or tori of gas and dust. These central gas systems reprocess the non-thermal continuum into complex emission-line spectra and shield the central source from view over much of the electromagnetic spectrum. Chapter 20 contains a brief introduction to a *unification hypothesis*, which has been proposed in order to explain the observations of Seyfert 1 and Seyfert 2 galaxies with a single model in which the spectroscopic differences reflect different orientations of the observer's line of sight to the central obscuring disc or torus (see Antonucci 1993 for a review). There is some hope that the descriptions of radio galaxies, BL Lacertae objects, radio-quiet QSOs, and radio-loud QSOs (the true quasars) can be unified similarly. A compact system of gas and dust is central to the unification hypothesis. Theory suggests that such a system will exist as small dense clouds (Krolik and Begelman 1986, 1988), which will be partly molecular if the pressure is sufficiently high (Krolik and Lepp 1989; Maloney, Hollenbach and Tielens 1996). Photoionization of gas by a non-thermal X-ray source allows atoms in a wide range of states of ionization to co-exist with molecules. Molecules can respond to heating and ionization by X-rays with unusual abundances (e.g. Lepp and Dalgarno 1996) and distinctive excitation patterns (Draine and Woods 1990; Gredel and Dalgarno 1995; Tiné et al. 1997).

Molecular hydrogen is an especially valuable probe of the central environment because it glows very brightly in infrared emission lines when excited by collisions at temperatures $T > 500$ K (cf. Chapter 8) or when exposed to intense fluxes of ultraviolet or X radiation (cf. Chapters 9 and 22). Indeed, strong H_2 emission is seen in the centres of many galaxies, both those with true AGN and those dominated by central starbursts (e.g. Koornneef and Israel 1996; Veilleux, Goodrich and Hill 1997). The pure rotational and vibration–rotation lines of H_2 are intrinsically weak, electric quadrupole transitions; therefore, these transitions are unlikely to become saturated (self-absorbed). Their intensities will be less affected by extinction than those of the visible and ultraviolet transitions of atoms. What of other molecules? Water (H_2O) shows up through its maser emission at 22.235 GHz in the centres of some galaxies. In the case of NGC 4258 (cf. Chapter 23), the H_2O maser system is demonstrably located within 0.15 pc of a low-luminosity AGN (Miyoshi et al. 1995). Molecules like OH, CO, and H_2CO in the very centres of AGN (i.e. central 10 pc) have been elusive: the failures to detect them in absorption toward radio-bright AGN may be a manifestation of very interesting excitation mechanisms that suppress the opacity in the lowest transitions (Maloney, Begelman and Rees 1994).

There have been recent advances in observing techniques. Spectroscopy with the Infrared Space Observatory (ISO) in the mid- and far-infrared has made it possible to distinguish the contributions of stellar activity and true AGN in the centres of dusty galaxies (Lutz et al. 1996). Indeed the recent mid-infrared spectrum of the obscured centre of the mildly active Circinus galaxy (ESO 97-G13 = PKS 1409–651) shows the striking juxtaposition of highly ionized atomic species and molecular hydrogen (Moorwood et al. 1996). Ground-based infrared observations continue to improve in resolution, both for spectroscopy and imaging of the H_2 and atomic emission lines. With the successful installation of the Near-Infrared Camera and Multi-Object Spectrometer (NICMOS) in the Hubble Space Telescope (HST), it will be possible to obtain high-fidelity, diffraction-limited infrared images of the centres of AGNs with ≈ 0.1 arcsec resolution (Thompson 1994). In particular, this means that the central H_2 emission can be resolved on scales of 25 pc or less in AGNs at distances out to 50 Mpc. As the resolution and sensitivity of radio interferometers (including VLBI networks) continue to improve at high frequencies, more information will become available on the conditions in and kinematics of the atomic and molecular tori in AGNs.

In the following sections, we present some elements of molecular physics that are relevant to the environment of an AGN (§2), describe how the interactions between molecules and radiation affect their detectability in such an environment (§3), and discuss a few specific examples of molecular probes of AGNs (§4).

2 Molecular Processes and Molecular Spectra

It is worth while to review some aspects of molecular physics that are likely to be relevant to the behaviour of molecules near AGNs. Table 1 summarizes some spectroscopic properties of a few astrophysically interesting molecules like H_2, H_2^+, OH, and CO.

Because H_2 is homonuclear, it has the symmetry property that associates states of aligned nuclear spins only with states of odd values of J, the rotational quantum number, and anti-aligned nuclear spins with even J in its electronic ground state. The former are called ortho states and the latter, para states. Owing to the very weak coupling of nuclear spins to electromagnetic radiation, radiative transitions between ortho and para states are strongly forbidden and completely negligible. Nor can non-reactive, inelastic collisions change the spin states. The nuclear-spin species can be interchanged by reactive processes such as collisions with H and H^+ in which an atom or nucleus changes place with a counterpart in the molecule (Dalgarno, Black and Weisheit 1973; Sun and Dalgarno 1994; Lepp, Buch and Dalgarno 1995). Hydrogen molecules will co-exist with H and H^+ in X-irradiated gas near AGNs, so that these interchange processes may well thermalize the ortho and para populations: the overall ortho/para abundance ratio might couple more directly to the kinetic temperature than the rotational excitation in either species where the excitation is partly controlled by radiative processes.

The excitation energy of the lowest rotational transition, $v = 0$–0 S(0), is $E_u/k = 510$ K (see Table 1) and the pure rotational transitions in the mid-infrared are thus thermally excited at temperatures of a few $\times 10^2$ K. The excitation energies of the vibrational transitions are $E_u/k \geq 6000$ K; therefore, these infrared lines arise in disturbed conditions (cf. Chapters 8 and 9). Thermal emission, excited by collisions, becomes important for kinetic temperatures in the range $T \approx 1000$–3000 K; at higher temperatures collisional dissociation is likely to reduce the abundance of emitting molecules significantly (Roberge and Dalgarno 1982; see Chapter 8). The lowest vibrational transition $v = 1$–0 predominates in intensity. Excited vibrational states of H_2 can also be populated by superthermal processes. Ultraviolet pumping occurs when absorption out of low-lying levels is followed by fluorescence to excited levels of the $X^1\Sigma^+$ state through the B–X, C–X, and B$'$–X electronic transitions (cf. Chapter 9). This is an efficient process yielding approximately three infrared line photons per ultraviolet absorption (Black and Dalgarno 1976), but has a limited maximum surface brightness owing to the limited depth of penetration of UV photons when the absorption lines become saturated (Black and van Dishoeck 1987; Sternberg 1988). The signature of pure fluorescent excitation is a vibrational excitation temperature that exceeds the rotational temperature: in other words, lines in highly excited bands, $v = 2$–1, 3–2, 6–4, etc., can have intensities nearly as high as those of lines in the 1–0 band. However, the situation becomes more complicated when the UV flux is high enough to heat the gas to temperatures $T > 1000$ K and the density is high enough ($n_H > 10^4$ cm^{-3}) to redistribute the excited-state populations by collisions (Sternberg and Dalgarno 1989). In this case, the line intensities from fluorescence-dominated gas can mimic those of pure thermal excitation. The description of the excitation of H_2 molecules near

AGNs is a problem of exquisite detail: the non-thermal ionizing radiation allows molecules to co-exist with atoms and atomic ions. It also produces energetic photoelectrons that lose energy step-by-step in ionization and excitation processes until they are energetically well matched to the task of exciting H_2. The destruction of H_2 produces H_2^+, which can re-form excited H_2 by charge transfer with H, even before the H_2^+ has time to radiate an infrared photon itself, on average. The excitation of H_2 in such X-irradiated gas has been investigated in detail by Gredel and Dalgarno (1995). Tiné et al. (1997) have presented H_2 line emissivities for a range of ionizing rates, gas densities, and temperatures that are relevant for AGNs (see Chapter 22 for some of the results). The emergent spectrum will be further complicated by the stratified structure of the partly ionized, X-ray-dominated nebula. If the ultraviolet radiation of the central source has not been fully attenuated before it reaches the molecular zone, then there may be an additional contribution of pure fluorescence to the H_2 emission.

The formation of H_2 is probably less efficient in many AGN environments than in the quiescent interstellar medium, where H atoms associate on surfaces of cold dust particles. Strongly irradiated dust may be more or less effective in forming H_2. At high dust temperatures, H atoms may stick and associate less efficiently, owing to increased evaporation. On the other hand, fragmentation of dust grains can increase the ratio of surface area to total dust mass, and surface damage might even increase the number of binding sites for H atoms. In the absence of dust, H_2 forms very inefficiently by gas-phase reactions from fragile intermediary species, H^- and H_2^+ (cf. Chapter 3).

From the preceding summary of molecular processes, it might seem that the abundance and excitation of H_2 are hopelessly complicated functions of the environment in which it exists. Can the emission-line spectrum of H_2 in an AGN be interpreted unambiguously? If enough different transitions can be observed in the infrared spectrum and if the distribution of brightness can be imaged with high angular resolution, then it should be possible to determine the main excitation conditions: density, temperature, and X-ray ionization rate. The relative intensities of various rotational and vibration–rotation lines of H_2 reveal the conditions under which the different excitation mechanisms compete. Since the ionization rate is related to the distance between molecular gas and the central source, the location of the molecular zone can be estimated, provided that the observed X-ray flux can be corrected for attenuation and its spectral shape determined. The total luminosity in H_2 lines must be related to the total X-irradiated surface area in the molecular zone and thus should reveal the geometry of a crucial part of the central gas distribution. Although the existing data on H_2 emission from any one AGN are inadequate for complete interpretation, it is already possible to discuss interesting limits. A few specific examples will be examined below.

Molecular hydrogen is special. It is composed of the most abundant element. Its forbidden infrared lines can be efficiently excited in gas near AGNs. These lines couple very poorly to the weakest part of the AGN continuum spectrum;

Table 1 Selected Spectroscopic Data

Molecule	Transition	$\tilde{\nu}$ (cm^{-1})	$A_{u\ell}$ (s^{-1})	g_u	g_ℓ	E_u (cm^{-1})
H$_2$	$v = $ 0–0 S(0)	354.37349	2.9E–11	5	1	354.373
	0–0 S(1)	587.03211	4.8E–10	21	9	705.519
	0–0 S(2)	814.42473	2.8E–9	9	5	1168.798
	0–0 S(3)	1034.67024	9.8E–9	33	21	1740.189
	0–0 S(7)	1720.899	2.0E–7	57	45	5001.960
	0–0 S(9)	2130.102	4.9E–7	69	57	7132.038
	1–0 S(0)	4497.8391	2.5E–7	5	1	4497.839
	1–0 S(1)	4712.9054	3.5E–7	21	9	4831.389
	1–0 S(3)	5108.4040	4.2E–7	33	21	5813.928
	1–0 Q(3)	4125.8739	2.8E–7	21	21	4831.389
	2–1 S(1)	4448.958	5.0E–7	21	9	8722.703
	6–4 Q(1)	6244.079	1.4E–6	9	9	21589.8
	2–0 S(1)	8604.2189	1.9E–7	21	9	8722.703
	B$^1\Sigma_u^+$–X$^1\Sigma_g^+$ 7–0 R(0)	98734.48	6.4E+7	3	1	98734.48
H$_2^+$	$v = $ 0–0 S(3)	510.51	3.3E–9	33	21	857.61
	0–0 S(5)	716.89	2.0E–8	45	33	1574.50
	1–0 Q(1)	2188.26	2.7E–7	9	9	2246.49
	1–0 Q(5)	2146.00	1.6E–7	33	33	3003.61
	1–0 S(1)	2461.90	2.3E–7	21	9	2520.13
	2–0 S(3)	4676.82	8.0E–8	33	21	5023.92
OH	$^2\Pi_{3/2}(J = 3/2)$ Λ-doubling	0.0554	8.6E–11	4	4	0.0554
	$^2\Pi_{3/2}(J = 5/2)$ Λ-doubling	0.2008	1.7E–9	6	6	83.9239
	$v = $ 0–0 R$_{2e}$(0.5)	61.1997	6.4E–2	4	2	187.5543
	0–0 R$_{2f}$(0.5)	61.3020	6.4E–2	4	2	187.8142
	0–0 R$_{1e}$(1.5)	83.7231	1.4E–1	6	4	83.7231
	0–0 R$_{1f}$(1.5)	83.8685	1.4E–1	6	4	83.9239
	0–0 ^{Q}R$_{21e}$(2.5)	103.8312	8.9E–3	4	6	187.5543
	0–0 ^{Q}R$_{21f}$(2.5)	103.8903	8.9E–3	4	6	187.8142
	0–0 ^{Q}R$_{21e}$(1.5)	126.3546	3.5E–2	2	4	126.3546
	0–0 ^{Q}R$_{21f}$(1.5)	126.4568	3.5E–2	2	4	126.5122
	0–0 ^{R}Q$_{21e}$(1.5)	187.4989	4.4E–2	4	4	187.5543
	0–0 ^{R}Q$_{21f}$(1.5)	187.8142	4.5E–2	4	4	187.8142
	1–0 Q$_{1e}$(1.5)	3568.4158	9.2	4	4	3568.4172
	1–0 Q$_{1f}$(1.5)	3568.5224	9.2	4	4	3568.5224
	A$^2\Sigma^+$–X$^2\Pi_i$ 0–0 P$_1$(1.5)	32440.574	9.1E+5	2	4	32440.6
CO	$v = $ 0–0 R(0) [$J = 1 \to 0$]	3.8450335	7.2E–8	3	1	3.84503
	0–0 R(1) [$J = 2 \to 1$]	7.6899199	6.9E–7	5	3	11.53495
	1–0 R(0)	2147.0816	1.2E+1	3	1	2147.08
	1–0 P(2)	2135.5466	2.4E+1	3	5	2147.08
	2–0 R(0)	4263.8372	3.5E–1	3	1	4263.84
	2–0 P(2)	4252.3022	6.9E–1	3	5	4263.84
	A$^1\Pi$–X$^1\Sigma^+$ 2–0 R(0)	67678.92	2.0E+7	6	1	67678.92

therefore, their intensities should reflect the excitation conditions rather directly. The infrared spectrum offers lines with a wide range of excitation energies. Moreover, lines arising in a common upper state should have relative intrinsic intensities that are fixed by the molecular physics; therefore, their observed intensities can be used to measure the extinction along the observer's line of sight to the emitting region. Other molecules are expected to be abundant under the same conditions. For example, CO is composed of abundant heavy elements and is even more strongly bound than H_2 (the CO dissociation energy $D_0^0 = 11.09$ eV as compared to 4.48 eV, although the ionization potential is smaller, 14.01 eV, as compared to 15.43 eV). Hydroxyl, OH, is expected to form efficiently at the elevated temperatures of X-ray heated or shocked gas by the high-temperature reaction of O with H_2. However, both CO and OH suffer additional excitation effects that may make them difficult to detect toward AGNs with strong radiofrequency continuum emission (see next section). X-ray dominated regions are also a natural refuge for exotic molecular species like HeH^+ (Cecchi-Pestellini and Dalgarno 1993; Maloney, Hollenbach and Tielens 1996). The hydrogen molecular ion, H_2^+, is a direct ionization product of H_2; however, its abundance will always be limited by its rapid destruction in reactions with H, H_2, and free electrons. As in the case of H_2, the rotational and vibration–rotation lines of H_2^+ occur only as electric quadrupole transitions with small transition probabilities (Posen, Dalgarno and Peek 1983; see Table 1). Under conditions of thermal excitation at $T = 1000$ K, one of the strongest of the H_2^+ lines, $v = 1$–0 $S(1)$ at $\nu/c = 2461.9$ cm^{-1}, has an emissivity of $\varepsilon = 1.2 \times 10^{-21}$ ergs s^{-1} molecule^{-1}, which is 4.5 times larger than the thermal emissivity of the H_2 $v = 1$–0 $S(1)$ line at the same temperature. The abundance ratio, however, will typically be quite small, $n(H_2^+)/n(H_2) \approx 10^9 \zeta/n_H$, where ζ is the ionization rate (perhaps $\leq 10^{-10}$ s^{-1} even rather close to a powerful X-ray source) and n_H is the total number density of hydrogen nuclei. By coincidence, the 2461.9 cm^{-1} line lies rather close to H I Brα.

3 The Radiation Environments Near AGNs

Although molecular rings or tori with diameters of a few hundred parsecs are common in centres of galaxies, the small-scale tori (< 10 pc) thought to exist in AGNs have been difficult to detect. Emission lines of CO have usually eluded detection (e.g. Evans et al. 1996 and references therein). One would also expect to see common molecules like CO and OH in absorption against the radio continuum emission, yet searches for the absorption lines in nuclear tori have usually been unsuccessful (Barvainis and Antonucci 1994; Conway and Blanco 1995; Braine et al. 1995; Drinkwater, Combes and Wiklind 1996). The absence of detectable molecular absorption in the putative molecular tori can have several causes: (1) the molecular abundances are low owing to the harsh conditions, (2) the gas temperature and state of excitation are high enough that the lowest transitions probe an undetectably small fraction of the molecules, and (3) the non-thermal continuum radiation of the central source enforces a high excitation temperature that suppresses the opacity in the lowest transitions (Maloney, Begelman and Rees 1994). The last of these points, radiative coupling to the non-thermal continuum, deserves further attention if we are to select the most favourable transitions for future observations.

The radiative coupling can be illustrated with reference to the spectrum of the radio source Cygnus A (3C 405). Cyg A has received considerable attention: H I has been detected in absorption (Conway and Blanco 1995) while CO, OH and H_2CO have not (Mazzarella et al. 1993; McNamara and Jaffe 1994; O'Dea et al. 1994; Barvainis and Antonucci 1994). Maloney (1996) has discussed the meaning of the observed limits on CO. The spectrum of the central source can be reconstructed in three components as follows. In the radio and infrared, the fluxes summarized by Djorgovski et al. (1991) are adopted and the corresponding intrinsic spectrum can be written in terms of the observed flux f_ν

$$I_\nu = \frac{1+z}{4\pi}\left(\frac{D_{\text{lum}}}{R}\right)^2 f_\nu \quad \text{erg s}^{-1}\text{ cm}^{-2}\text{ Hz}^{-1}\text{ sr}^{-1} \tag{3.1}$$

at a distance R from the central source, where $D_{\text{lum}} = 229(75/H_0)$ Mpc is the luminosity-distance at redshift $z = 0.0565$ for Hubble constant H_0 in km s^{-1} Mpc^{-1} and deceleration parameter $q = 1/2$. In the frequency range $\log \nu = 9$ to 14.39 Hz, this non-thermal component is

$$I_1(\nu) = 6.7 \times 10^{-10}\left(\frac{75}{H_0}\right)^2\left(\frac{10 \text{ pc}}{R}\right)^2 \nu_{\text{GHz}}^{-0.18} \quad \text{erg s}^{-1}\text{ cm}^{-2}\text{ Hz}^{-1}\text{ sr}^{-1}. \tag{3.2}$$

There is a far-infrared excess, which may arise partly outside the central compact source; this can be represented by

$$I_2(\nu) = \eta B_\nu(T_d)\left(1 - \exp(-0.068(\tilde{\nu}/100)^{1.5})\right) \quad \text{erg s}^{-1}\text{ cm}^{-2}\text{ Hz}^{-1}\text{ sr}^{-1} \tag{3.3}$$

where T_d is the thermal (dust) temperature, $B_\nu(T)$ is the Planck function, $\tilde{\nu} = \nu/c$ is the frequency in cm^{-1}, and η is a geometrical dilution factor. Finally at the highest frequencies, we extrapolate the X-ray measurement of Ueno et al. (1994) according to a ν^{-1} power law:

$$I_3(\nu) = 5.4 \times 10^{-12}\left(\frac{75}{H_0}\right)^2\left(\frac{10 \text{ pc}}{R}\right)^2\left(\frac{\nu}{\nu_0}\right)^{-1} \quad \text{erg s}^{-1}\text{ cm}^{-2}\text{ Hz}^{-1}\text{ sr}^{-1} \tag{3.4}$$

where $\nu_0 = 3.288 \times 10^{15}$ Hz is the frequency of the Lyman limit of atomic H. The combined spectrum is shown in the upper panel of Fig. 1 for two values of the distance from the central source, $R = 10$ (solid curve) and 100 pc (dashed curve). It is instructive to represent this internal intensity in terms of a radiation temperature T_{rad} defined by

$$B_\nu(T_{\text{rad}}) = I_\nu = \sum_{i=1}^{3} I_i(\nu). \tag{3.5}$$

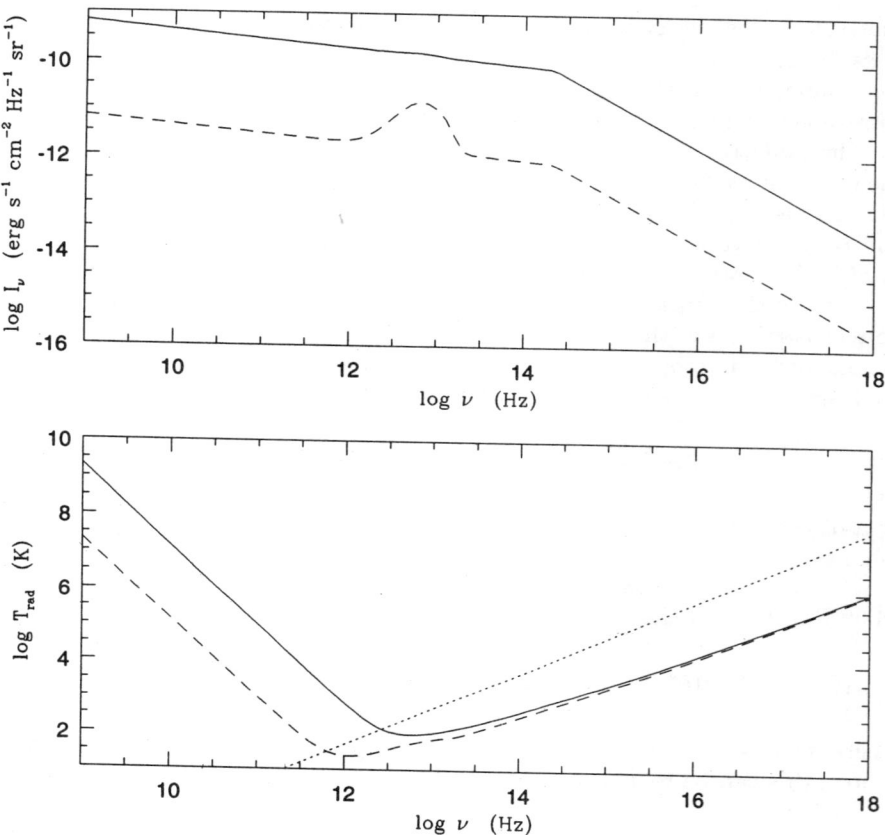

FIG. 1. The intensity of radiation at $R = 10$ pc (solid curves) and $R = 100$ pc (dashed curves) from the compact source Cygnus A. The upper panel shows the conventional intensity I_ν while the lower panel displays the corresponding radiation temperature. The dotted line in the lower panel is $h\nu/k$.

The corresponding radiation temperatures are displayed in the bottom panel of Fig. 1. The dotted line shows the photon energy $h\nu/k$ in temperature units: notice that $T_{\rm rad} > h\nu/k$ at $R = 10$ pc for all frequencies below $10^{12.5}$ Hz (wavelengths greater than 90 μm). The rate of absorption in any atomic or molecular transition can be written

$$\rho = \frac{g_u A_{u\ell}/g_\ell}{\exp(h\nu/kT_{\rm rad}) - 1} \quad {\rm s}^{-1} \qquad (3.6)$$

where $A_{u\ell}$ is the probability of spontaneous transitions between upper state u and lower state ℓ of statistical weights g_u and g_ℓ, respectively. The relative populations of any pair of states can be written in terms of an excitation temperature

$T_{\text{ex}} = T_{u\ell}$ defined in terms of the ratio of number densities

$$\frac{n_u}{n_\ell} = \frac{g_u}{g_\ell} \exp(-h\nu_{u\ell}/kT_{\text{ex}}), \tag{3.7}$$

where $h\nu_{u\ell} = E_u - E_\ell$ is the energy difference between the two states. In the limit of high densities, where the rates of collisional processes exceed the rates of radiative transitions, the excitation temperature will approach the kinetic temperature, $T_{\text{ex}} \sim T_k$. In the limit of low densities, when radiative processes dominate, $T_{\text{ex}} \sim T_{\text{rad}}$. In the galactic interstellar medium, the low-density limit will often leave the distribution of low-lying levels locked to the temperature of the cosmic background radiation, $T_{\text{ex}} \sim T_{\text{cbr}}$, where $T_{\text{cbr}} = 2.726 \pm .01$ K at redshift $z = 0$ (Mather et al. 1994).[1] In the neighbourhood of an AGN with a non-thermal spectrum, a molecule can be exposed to intense radiation with a wide range of $T_{\text{rad}}(\nu)$: its excitation will represent a complicated response to the different temperatures that characterize its rotational, vibrational, and electronic transition rates.

For the present purpose, it is useful to discuss the detectability of a molecular transition in terms of its line-centre optical depth:

$$\tau = 3.738 \times 10^{-7} \frac{A_{u\ell}}{\tilde{\nu}^3} \frac{N_\ell}{\Delta V} \frac{g_u}{g_\ell} \left(1 - \exp(-h\nu/kT_{\text{ex}})\right) \tag{3.8}$$

where N_ℓ is the lower-state column density in cm^{-2} and ΔV is the full width at half-peak of the Gaussian line shape in km s^{-1}. The optical depth is a direct measure of the strength of absorption. In the case of an emission line, the intensity produced in the same column of gas can be written

$$I_\nu = B_\nu(T_{\text{ex}})\left(1 - \exp(-\tau)\right). \tag{3.9}$$

It is clear in eqn (3.8) that the excitation enters in two ways: the fractional abundance in the lower state, N_ℓ/N_{mol}, and the correction for stimulated emission, $1 - \exp(-h\nu/kT_{\text{ex}})$, are both usually declining functions of T_{ex} for low-lying states and high values of T_{ex}.

Consider the implications for the CO molecule in its $J = 1 \to 0$ transition at $\log \nu = 11.062$ Hz, for which $A_{10} = 7.2 \times 10^{-8}$ s^{-1} (see Table 1). At $R = 10$ pc from Cyg A, $T_{\text{rad}} = 6 \times 10^4$ K at this frequency, so that $\rho = 2 \times 10^{-3}$ s^{-1}. The absorption rate in $J = 2 \leftarrow 1$ is comparable. The radiation temperature at the frequency of the $v = 1 \leftarrow 0$ vibrational fundamental, $10^{13.81}$ Hz, is $T_{\text{rad}} \approx 285$ K, which will drive radiative processes at a rate $\rho \approx 7 \times 10^{-4}$ at $R = 10$ pc from Cyg A. If the ultraviolet radiation from the central source as shown in Fig. 1 were unattenuated by dust or other CO molecules, its contribution to the radiative excitation would be even larger (for the A–X 2–0 band alone, $\rho \approx 9 \times 10^{-4}$ s^{-1}).

---

[1] It is a testable prediction of the standard Big Bang cosmology that the background radiation should vary with redshift as $T_{\text{cbr}}(z) = 2.726(1 + z)$ K (Bahcall and Wolf 1968; Ge, Bechtold and Black 1997, and references therein).

Even if the rate coefficients for collisional excitation by neutrals (H and H_2) are as large as $q_{coll} \approx 10^{-10}$ cm^3 s^{-1}, the rotational populations in CO will be thermalized only at densities $n_{coll} \approx \rho/q_{coll} > 10^7$ cm^{-3}. Thus the "low-density limit" of excitation prevails at densities that would be considered quite high in galactic molecular clouds. As pointed out by Maloney et al. (1994), the effect of radiative coupling to a nearby non-thermal radio source is to maintain a high excitation temperature in the lowest rotational levels so that their populations account for a very small fraction of the CO molecules. In addition, the large correction for stimulated emission helps to suppress the absorption lines.

Similar calculations can be done for the radiative coupling of other common molecules. At 18 cm, the wavelength of the OH Λ-doubling transitions, $T_{rad} > 5 \times 10^6$ K even at $R = 100$ pc from a source like Cyg A. At $R = 10$ pc, the pumping rate in the 18 cm transitions, $\rho \approx 1$ s^{-1}, is at least five times greater than that in the strongest pure rotational transition out of the ground state at 119 μm wavelength. Collisional processes are very slow in comparison at densities below 10^{10} cm^{-3}, so that the 18 cm absorption lines will likely be suppressed. The Λ-doubling transitions of OH X$^2\Pi_{3/2}$ $J = 5/2$ at 5 cm, however, couple to the microwave part of the spectrum at approximately the same rate as these levels are pumped in the far-infrared (principally at 85 μm), so that the excitation temperatures of these states might be significantly lower than T_{rad}. Searches for the 5 cm lines of OH might be more profitable than observations at 18 cm, if the goal is to find a molecular torus close to the central radio source. The 5 cm transitions of OH have, indeed, been observed in absorption toward the centre of the nearby, highly inclined spiral galaxy NGC 4945 (Whiteoak and Wilson 1990). This galaxy has a Seyfert 2 nucleus (Done, Madejski and Smith 1996), surrounded by a 200 pc ring of gas clouds observed in CO emission (Bergman et al. 1992; Dahlem et al. 1993). Maser emission of H_2O is concentrated within a radius ≈ 0.3 pc around the nucleus (Greenhill, Moran and Herrnstein 1997). It would be very interesting to find out whether the excited OH absorption arises in the central torus or in the 200 pc ring.

Results of some limited non-LTE excitation calculations are presented in Table 2 to illustrate the effects of radiative coupling on the optical depths of some common transitions in abundant interstellar molecules. The radiative transfer has been treated with mean escape probabilities, such that the line optical depth depends on the ratio of column density to line width, $N(\text{mol})/\Delta V$. In all cases, the molecules are exposed to the continuum spectrum at $R = 10$ pc from Cygnus A as shown in Fig. 1, and, for comparison, at $R = 300$ pc. The kinetic temperature is $T_k = 500$ K and the number density of neutral collision partners (H or H_2) is $n = 10^5$ cm^{-3}. The line-centre optical depth at distance R is τ_R and ν_{rest} is the rest frequency of each transition. In all cases tabulated, the excitation at $R = 10$ pc is dominated by the continuum source and the results are insensitive to the density and temperature. The lowest vibrational transitions have been included for CO and HCN, but the excitation of the other species includes the rotational transitions only. CO and OH would also be susceptible

Table 2 Molecular Line Optical Depths Near an AGN

Molecule	$N/\Delta V$ [cm^{-2} (km s^{-1})$^{-1}$]	ν [GHz]	τ_{10}	τ_{300}
CO	10^{16}	115.271	4.2E–4	5.2E–4
		230.538	1.6E–3	2.9E–4
		345.796	3.6E–3	1.9E–2
		461.041	7.2E–3	4.8E–2
		(λ4.580 µm)	0.91	0.18
OH	10^{15}	1.67	–7.7E–4	3.0
		6.03	8.1E–4	0.30
		13.43	1.9E–3	9.6E–4
		7.79	6.8E–4	–4.5E–4
HCN	10^{13}	88.632	4.4E–6	8.8E–3
		177.261	5.6E–5	0.29
		265.886	2.8E–4	0.90
		354.506	9.1E–4	0.96
H$_2$CO	10^{14}	4.830	–2.6E–5	–5.9E–4
		14.488	–3.3E–5	–5.8E–4
		145.603	0.010	0.22
		218.222	0.015	0.13
ortho-H$_2$O	10^{15}	22.235	4.0E–4	1.3E–7
		380.197	0.089	2.1E–3
		448.001	0.028	2.4E–5
		556.936	0.034	181.0
para-H$_2$O	10^{15}	183.310	0.18	5.7E–3
		325.153	0.015	7.8E–6

to radiative excitation in the ultraviolet, but these transitions have been ignored since the molecular region is likely to be well shielded from this part of the continuum radiation. The ultraviolet transitions of H$_2$O, H$_2$CO, and HCN are strongly predissociating and unlikely to contribute much fluorescent excitation in any case. In OH and HCN, hyperfine structure has been ignored. Ortho and para species of H$_2$CO have been treated together, including interchange reactions. Ortho and para species of H$_2$O have been treated as separate molecules.

It is possible to conclude from these examples that OH and H$_2$CO molecules close to a radio-bright AGN are not likely to be detectable through their cm-wave transitions in absorption, nor are molecules like CO and HCN likely to be observable in mm-wave absorption. The water molecule, H_2O, is predicted to be abundant in X-irradiated molecular gas (cf. Maloney, Hollenbach and Tielens 1996). Moreover, its radiative coupling is strongest to the weakest part of the continuum radiation field (in terms of T_{rad}), so that the excitation temperatures of some of its millimetre/submillimetre transitions are nearly thermal. Indeed, the radiative excitation at $R = 10$ pc maintains detectable populations in excited states that would be subthermally populated where the radiation is more dilute, e.g. at $R = 300$ pc as tabulated. Finally, the redshift of Cyg A, $z = 0.0565$, is sufficient to move the 183.310 GHz line to 173.5 GHz, out of coincidence with the corresponding atmospheric line. Searches for such non-maser transitions of H_2O might be fruitful toward flat-spectrum radio sources with measurable continuum flux at millimetre/submillimetre wavelengths.

The environment near an AGN is so extreme that radiative excitation through forbidden transitions of H_2 might have noticeable effects. The excitation of H_2 has been computed for the same conditions near Cyg A as described above. Rate coefficients for excitation of H_2 in collisions with H have been taken from Lepp, Buch and Dalgarno (1995) and Martin and Mandy (1995). A total column density $N(H_2) = 10^{21}$ cm^{-2} has been adopted. The intensities of the pure rotational transitions are the same at $R = 10$ pc and $R = 300$ pc, the strongest of these being $v = 0$–0 S(3) with $I_\nu = 2.4 \times 10^{-3}$ erg s^{-1} cm^{-2} sr^{-1}. However, the radiation temperature is high enough at $\lambda \approx 2$ μm ($T_{rad} \approx 500$ K) that IR pumping competes with collisional excitation of the $v = 1$ levels when $T_k \approx 500$ K. For example, the calculated intensity of the $v = 1$–0 S(1) line is 2.8×10^{-5} erg s^{-1} cm^{-2} sr^{-1} at $R = 10$ pc and 4.1×10^{-6} erg s^{-1} cm^{-2} sr^{-1} at $R = 300$ pc. This radiative effect will be much less important when $T_k > T_{rad}$. It is likely to be completely overwhelmed by the non-thermal collisional excitation that accompanies an X-irradiated gas (Gredel and Dalgarno 1995; Tiné et al. 1997; Chapter 22). Those effects have not been included in the simple excitation calculation described here.

Small amounts of hydrogen molecules can also be observed through their strong ultraviolet absorption lines if there is sufficient continuum flux observable from the central source. This technique has been applied successfully to the detection of H_2 in intervening absorbers toward high-redshift QSOs. At this time, H_2 has been definitely identified in only two QSO absorption systems: at an absorption redshift $z_{abs} = 2.811$ toward PKS 0528–250, which has an emission redshift $z_{em} = 2.770$ (Foltz, Chaffee and Black 1987; Lanzetta 1993), and at $z_{abs} = 1.9731$ toward Q 0013–004, for which $z_{em} = 2.0835$ (Ge and Bechtold 1997). In both cases, the H_2 is associated with a damped H I Lα absorption system. The absorber toward PKS 0528–250 is an unusual one in which $z_{abs} > z_{em}$. The difference in Doppler velocity corresponding to $z_{abs} - z_{em}$ is 785 km s^{-1}, if the special relativistic but no general relativistic corrections are taken into account. The corresponding difference in Doppler velocity for Q 0013–004 is 3878 km s^{-1}, which would place the absorbing gas 50 Mpc from the QSO for $H_0 = 75$ km s^{-1} Mpc^{-1}. Both of these molecular absorbers are

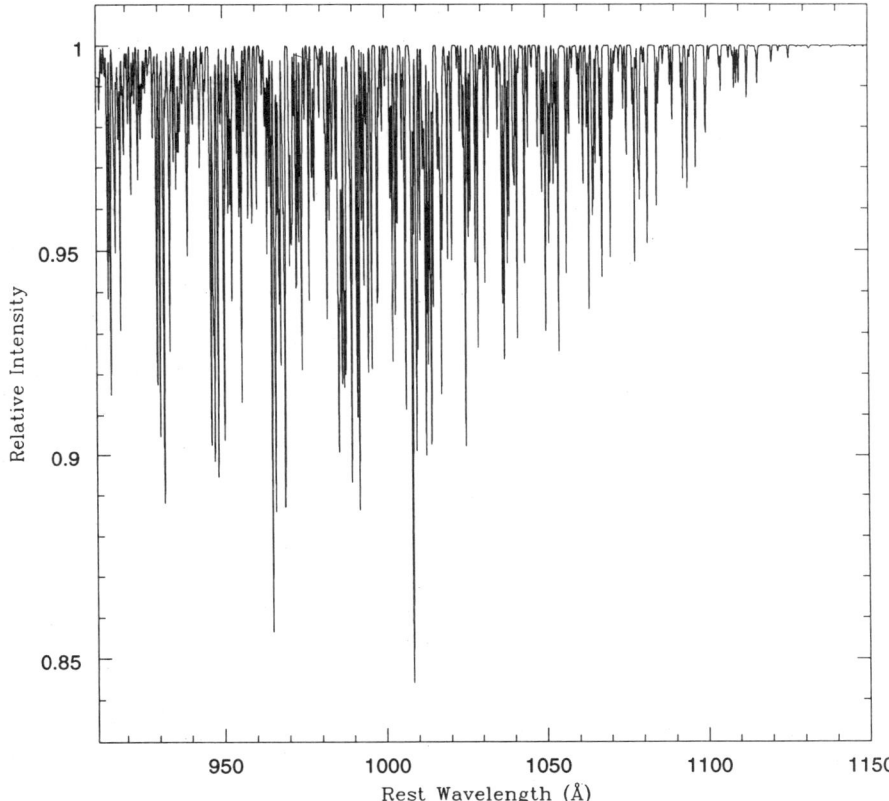

FIG. 2. An absorption spectrum of H_2 in the rest-frame ultraviolet. The molecular excitation is appropriate for a kinetic temperature $T_k = 1000$ K and a neutral density of 10^5 cm^{-3}. The column density is $N(H_2) = 10^{15}$ cm^{-2}, the intrinsic line broadening is thermal Doppler, and the resolution would correspond to 1 Å in the observer's frame at $z = 2.5$.

evidently far enough from the QSO that the excitation of H_2 resembles that in diffuse molecular clouds of the Milky Way. A molecular absorber close to the QSO would be expected to have a very high excitation as described above. The appearance of such an absorber is shown in the simulated spectrum of Fig. 2. This spectrum contains 1486 absorption lines of H_2 arising in levels $v = 0$ to 1 and $J = 0$ to 9. The practical difficulty in identifying such a spectrum is that it will coincide with the "forest" of H I Lα lines arising in the small neutral atomic components of all intervening absorbing gas. Moreover, any QSO surrounded by gas that is sufficiently dense and thick to sustain measurable H_2 abundances might be undetectably dim in the rest-frame ultraviolet owing to the associated extinction by dust. On the other hand, the periodic band structure in the H_2 spectrum should be distinguishable from a random distribution of H I Lα lines.

The observation of H I absorption in the 21 cm line toward Cygnus A can be interpreted as neutral atomic gas associated with the circumnuclear torus, even though molecular absorptions of CO, OH and H_2CO are not seen (Conway and Blanco 1995). The H I spectrum can suffer some of the extreme excitation effects described above for molecules close to such a powerful continuum source. Indeed, in these circumstances the excitation of the H I 21 cm line can no longer be treated as the simple two-level system so beloved by radio astronomers. At a distance $R = 10$ pc from Cyg A, for example, absorption and fluorescence in the $1^2S \rightarrow 2^2P$ Lyman α transition will probably transfer atoms from one hyperfine structure (hfs) level to the other at a faster rate than direct transfer by collisions (Allison and Dalgarno 1969), absorption, and stimulated emission. The Lα absorption rate at $R = 10$ pc will be at least $\rho \approx 6 \times 10^{-2}$ s^{-1} if controlled by the UV brightness of the central source, or $\rho \approx 7 \times 10^2$ s^{-1} for Lα radiation thermalized by resonance scattering at $T_{\rm rad} \approx 8000$ K (cf. Urbaniak and Wolfe 1981; Deguchi and Watson 1985). Stimulated emission in the 21 cm hfs transition itself will occur at a lower rate $\approx 4 \times 10^{-5}$ s^{-1}. The Lα pumping also permits excitation of the $2^2S_{1/2} \leftarrow 2^2P_{3/2}$ fine-structure transition at 9.9111 GHz. The population of the metastable $2^2S_{1/2}$ state will be very sensitive to the value of $T_{\rm rad}$ at the frequency of Lα. Moreover, close to a radio-loud AGN, the absorption rate out of $2^2S_{1/2}$ in the 9.911 GHz line can easily exceed the spontaneous two-photon decay rate to the ground state. Even though the 9.911 GHz line can be excited in these circumstances, it is unlikely to be observable because its radiation-damping linewidth of the order of 1 GHz exceeds the total bandwidth of exisiting spectrometers. It might even be possible to probe a component of the circumnuclear ionized gas through the hfs transition of ^{3}He$^+$ at 8.66565 GHz (because an incorrect transition probability has propagated through some of the astrophysical literature, the reader should note recent discussions by Gould 1994 and Garstang 1995).

4 A Bestiary of AGN

Recent observational developments (cf. the last few sections of Chapter 20) provide some hints of the ways in which our subject might develop in the near future. A few examples are summarized.

4.1 NGC 1068

As one of the nearest and best studied Seyfert 2 galaxies, NGC 1068 is an important test case for probes of a possible nuclear torus. Because it also has a high concentration of molecular gas and associated star-forming activity on larger scales (Papadopoulous, Seaquist and Scoville 1996 and references therein), it illustrates the common difficulties in distinguishing stellar and non-thermal sources of power. Recent infrared observations indicate that 94% of the central K-band light originates in a region of diameter of the order of 2 pc (corresponding to 0.03 arcsec: Thatte et al. 1997). This IR continuum is attributed to thermal emission of warm dust that lies close enough to the central source that it radiates at a temperature near its sublimation temperature. The central stellar cluster that supplies the remainder of the infrared flux has a size of the order of 50 pc. The presence of H_2O masers within 0.65 pc of the central source (Greenhill et al. 1996) is another indicator of a molecular torus viewed nearly edge-on (cf. Chapter 23).

4.2 NGC 4258

The high-resolution observations of H_2O maser emission (cf. Chapter 23) in the centre of NGC 4258 have wide-ranging implications for extragalactic astronomy. Miyoshi et al. (1995) demonstrated that the system of masers inhabits a thin, warped disc in Keplerian rotation. The kinematical analysis of this system implies a central mass of 3.6×10^7 $M_\odot$ within a radius less than 0.15 pc: strong evidence of a massive black hole. The accelerations of maser components can also be measured: this means that corresponding angular and linear scales can be determined directly, thus providing for the first time a geometrical measurement of distance to another galaxy. It has even been shown that the conditions (density, temperature, etc.) required for the existence of a powerful water maser are a natural consequence of the proximity of high-pressure gas, $P/k \approx 10^{11}$ cm^{-3} K, and a central X-ray source of luminosity $L_X \approx 10^{43}$ ergs s^{-1} (Neufeld, Maloney and Conger 1994; Neufeld and Maloney 1995). The warping of the disc is an important aspect of the theoretical models, which has recently received further observational support (Herrnstein, Greenhill and Moran 1996) and a theoretical basis (Maloney, Begelman and Pringle 1996).

4.3 NGC 7469

This Seyfert 1 galaxy is one of the few cases of an AGN that shows emission in the 3.3 µm feature. The 3.3 µm infrared emission band is generally attributed to UV-driven fluorescence in polycyclic aromatic hydrocarbons (PAHs) and is a prominent feature in the spectra of star-forming regions in the Milky Way and of luminous central starbursts in disturbed galaxies. The usual absence of this feature in AGN spectra may result from the destruction of PAHs by X-rays (Voit 1992). The persistence of 3.3 µm emission within the central 3" of NGC 7469 (Cutri et al. 1984) suggests that a shielded molecular region exists in the nucleus and that a significant fraction (1/3 in the K band) of the infrared continuum arises in a central starbust surrounding the AGN (Mazzarella et al. 1994). However, Miles, Houck and Hayward (1994) find that the 11.3 µm PAH emission comes from the circumnuclear ring and not from the nucleus. The infrared spectrum of NGC 7469, like that of NGC 1068, shows emission in both [Si X] and molecular hydrogen (Thompson 1996). The high-resolution (≈ 0.4 arcsec) imaging and spectroscopy of Genzel et al. (1995) shows that at least 1/3 of the H_2 infrared line flux arises within approximately 100 pc of the central source.

4.4 NGC 1275 (= Perseus A = 3C84)

This well-known Seyfert galaxy and radio source is a very interesting test case for the investigation of X-irradiated molecular gas near an AGN. It is notable for the extraordinary luminosity of emission in infrared lines of H_2. The observed CO emission in NGC 1275 is extended over ± 1.2 kpc on two sides of the nucleus, while the excited H_2 is concentrated within the central 2″, or $R \approx 340$ pc (Inoue et al. 1996). The absence of detectable absorption in CO $J = 1 \leftarrow 0$ toward the nuclear continuum source (Braine et al. 1995) may be the result of the radiative coupling discussed in §3 above. The 5.5 Jy continuum flux at 115 GHz ensures that the excitation of the $J = 0$ and 1 levels is radiatively controlled for $R < 500$ pc and densities $n(H_2) \leq 10^6$ cm^{-3}. Absorption by CO molecules within a few pc of the nucleus would be completely suppressed by the high brightness temperature of the continuum (cf. Maloney, Begelman and Rees 1994).

The observed flux in the H_2 $v = 1$–0 S(1) line, $f = 2.5 \times 10^{-14}$ ergs s^{-1} cm^{-2} (Inoue et al. 1996) corresponds to a luminosity $L = 1.5 \times 10^{40}$ ergs s^{-1} for an adopted distance $D = 70$ Mpc. The upper limit on the emitting volume is $V = 9 \times 10^{63} (D/70)^3$ cm^3. The observations suggest an excitation temperature of the H_2 emission in the range $T_{ex} \approx 1700$ (Kawara and Taniguchi 1993) to 2600 (Sams et al. 1997). If the excited molecules are fully thermalized at $T = T_{ex} = 2000$ K, then the emissivity in the $v = 1$–0 S(1) line is $\varepsilon = 4.06 \times 10^{-21}$ ergs s^{-1} molecule^{-1}. In this case, a mass of hot H_2 of the order of 6000 $M_\odot$ would be required to produce the observed emission. This is a small amount of molecular gas; however, such an estimate does not account for the much larger quantity of less excited molecular gas and atomic gas that would be expected to accompany the hot molecules. A more realistic description of such a molecular source requires detailed modelling of the physical conditions and emission mechanisms. Since NGC 1275 is a luminous, extended X-ray source, it is likely that the H_2 is exposed to heating and ionization by X-rays. From the work of Tiné et al. (1997), we can see how the emissivity of the H_2 line, $\varepsilon(n_H, T, \zeta)$ depends upon density n_H, kinetic temperature T, and X-ray ionization rate ζ (s^{-1}). Examination of the tabulated emissivities of Tiné et al. (1997) shows that at $T = 2000$ K, the observed central luminosity in the $v = 1$–0 S(1) line does not have low-density solutions because the average density over the emitting volume, $L/(\varepsilon V)$, must not exceed the local density n_H of X-irradiated gas. However, for $n_H \geq 10^3$ cm^{-3}, a wide range of solutions is possible. For example, at $n_H = 10^5$ cm^{-3} and $T = 2000$ K, the required mass of gas (including atomic hydrogen) lies in the range 4×10^4 to 1.5×10^6 $M_\odot$ for values of the ionization parameter $\zeta/n_H = 10^{-22}$ to 10^{-15} cm^3 s^{-1}. These values of gas mass are still small compared with the dynamical mass implied by the observed linewidth of the H_2 $v = 1$–0 S(1) line: $\Delta V = 250$ km s^{-1}, which corresponds to a three-dimensional dispersion of $\sigma = 180$ km s^{-1} (Inoue et al. 1996). If these gas motions are in simple virial equilibrium with the enclosed mass inside $R = 340$ pc, then $M_{dyn} \approx R\sigma^2/G \approx 3 \times 10^9$ $M_\odot$. To narrow the parameter space further will require a specific model of the distribution and spectrum of X-ray emission as well as a physically consistent description of the gas distribution (see, for example, Maloney, Hollenbach and Tielens 1996). This brief discussion of the luminous H_2 emission in the centre of NGC 1275 suggests the potential value of H_2 as a probe of molecular gas in such AGNs.

5 Summary

The investigation of molecular gas in the centres of AGNs is in its early stages. As suggested above, it is not even certain yet which molecules afford the best observational tracers of such gas. It appears that the infrared lines of H_2 are valuable for this purpose, especially if the line emission can be measured with very high angular resolution, to distinguish nuclear gas from material on larger scales. Maser emission from H_2O definitely arises in nuclear molecular gas in some AGNs. It is suggested that non-maser emission in some mm-wave transitions of H_2O might also be worth observing, particularly in galaxies for which the redshift is large enough to move the 183 GHz line out of coincidence with its atmospheric counterpart. The discussion of molecular spectra and processes hints at the large variety of molecular data that are needed for full analysis of astronomical spectra. The microscopic details of an X-irradiated molecular gas are quite complicated. Finally, we should remain mindful that the conditions near an AGN are quite unlike those that prevail in typical molecular clouds in our Galaxy. In particular, the radiation environment can have startling effects on molecular excitation.

Bibliography

1. Allison, A. C. and Dalgarno, A. (1969). *ApJ*, **158**, 423.
2. Antonucci, R. (1993). *ARA&A*, **31**, 473.
3. Bahcall, J. N. and Wolf, R. A. (1968). *ApJ*, **152**, 701.
4. Barvainis, R. and Antonucci, R. (1994). *AJ*, **107**, 1291.
5. Bergman, P., Aalto, S., Black, J. H. and Rydbeck, G. (1992). *A&A*, **265**, 403.
6. Black, J. H. and Dalgarno, A. (1976). *ApJ*, **203**, 132.
7. Black, J. H. and van Dishoeck, E. F. (1987). *ApJ*, **322**, 412.
8. Braine, J., Wyrowski, F., Radford, S. J. E., Henkel, C. and Lesch, H. (1995). *A&A*, **293**, 315.
9. Cecchi-Pestellini, C. and Dalgarno, A. (1993). *ApJ*, **413**, 611.
10. Conway, J. E. and Blanco, P. R. (1995). *ApJ*, **449**, L131.
11. Cutri, R. M., Rudy, R. J., Rieke, G. H., Tokunaga, A. T. and Willner, S. P. (1984). *ApJ*, **280**, 521.
12. Dahlem, M., Golla, G., Whiteoak, J. B., Wielebinski, R., Hüttemeister, S. and Henkel, C. (1993). *A&A*, **270**, 29.
13. Dalgarno, A., Black, J. H. and Weisheit, J. C. (1973). *Ap Letters*, **14**, 77.

14. Deguchi, S. and Watson, W. D. (1985). *ApJ*, **290**, 578.
15. Djorgovski, S., Weir, N., Matthes, K. and Graham, J. R. (1991). *ApJ*, **372**, L67.
16. Done, C., Madejski, G. M. and Smith, D. A. (1996). *ApJ*, **463**, L63.
17. Draine, B. T. and Woods, D. T. (1990). *ApJ*, **363**, 464; erratum, **387**, 732.
18. Drinkwater, M. J., Combes, F. and Wiklind, T. (1996). *A&A*, **312**, 771.
19. Evans, A. S., Sanders, D. B., Mazzarella, J. M., Solomon, P. M., Downes, D., Kramer, C. and Radford, S. J. E. (1996). *ApJ*, **457**, 658.
20. Foltz, C. B., Chaffee, F. H., Jr., and Black, J. H. (1987), *ApJ*, **324**, 267.
21. Garstang, R. H. (1995). *ApJ*, **447**, 962.
22. Ge, J. and Bechtold, J. (1997). *ApJ*, **477**, L73.
23. Ge, J., Bechtold, J. and Black, J. H. (1997). *ApJ*, **474**, 67.
24. Genzel, R., Weitzel, L., Tacconi-Garman, L. E., Blietz, M., Cameron, M., Krabbe, A., Lutz, D. and Sternberg, A. (1995). *ApJ*, **444**, 129.
25. Gould, R. J. (1994). *ApJ*, **423**, 522.
26. Gredel, R. and Dalgarno, A. (1995). *ApJ*, **446**, 852.
27. Greenhill, L. J., Moran, J. M. and Herrnstein, J. R. (1997). *ApJ*, **481**, L23.
28. Greenhill, L. J., Gwinn, C. R., Antonucci, R. and Barvainis, R. (1996). *ApJ*, **472**, L21.
29. Herrnstein, J. R., Greenhill, L. J. and Moran, J. M. (1996). *ApJ*, **468**, L17.
30. Inoue, M. Y., Kameno, S., Kawabe, R., Inoue, M., Hasegawa, T. and Tanaka, M. (1996). *AJ*, **111**, 1852.
31. Kawara, K. and Taniguchi, Y. (1993). *ApJ*, **410**, L19.
32. Koornneef, J. and Israel, F. P. (1996). *New Astronomy*, **1**, 271.
33. Krolik, J. H. and Begelman, M. C. (1986). *ApJ*, **308**, L55.
34. Krolik, J. H. and Begelman, M. C. (1988). *ApJ*, **329**, 702.
35. Krolik, J. H. and Lepp, S. (1989). *ApJ*, **347**, 179.
36. Lanzetta, K. M. (1993). In *The environment and evolution of galaxies*, eds. J. M. Shull and H. A. Thronson, Jr. Kluwer, Dordrecht, p.237.
37. Lepp, S., Buch, V. and Dalgarno, A. (1995). *ApJS*, **98**, 345.
38. Lepp, S. and Dalgarno, A. (1996). *A&A*, **306**, L21.
39. Lutz, D., Genzel, R., Sternberg, A., Netzer, H., Kunze, D., Rigopoulou, D., Sturm, E., Egami, E., Feuchtgruber, H., Moorwood, A. and de Graauw, Th. (1996). *A&A*, **315**, L137.
40. McNamara, B. R. and Jaffe, W. (1994). *A&A*, **281**, 673.
41. Maloney, P. R. (1996). In *Cygnus A—a case study of a radio galaxy*, eds. C. Carilli and D. Harris. Cambridge University Press, Cambridge, p.60.
42. Maloney, P. R., Begelman, M. C. and Pringle, J. E. (1996). *ApJ*, **472**, 582.
43. Maloney, P. R., Begelman, M. C. and Rees, M. J. (1994). *ApJ*, **432**, 606.

44. Maloney, P. R., Hollenbach, D. J. and Tielens, A. G. G. M. (1996). *ApJ*, **466**, 561.
45. Martin, P. G. and Mandy, M. E. (1995). *ApJ*, **455**, L89.
46. Mather, J. C. et al. (1994). *ApJ*, **420**, 439.
47. Mazzarella, J. M., Graham, J. R., Sanders, D. B. and Djorgovski, S. (1993). *ApJ*, **409**, 170.
48. Mazzarella, J. M., Voit, G. M., Soifer, B. T., Matthews, K., Graham, J. R., Armus, L. and Shupe, D. (1994). *AJ*, **107**, 1274.
49. Miles, J. W., Houck, J. R. and Hayward, T. R. (1994). *ApJ*, **425**, L37.
50. Miyoshi, M., Moran, J. M., Herrnstein, J. R., Greenhill, L. J., Nakai, N., Diamond, P. J. and Inoue, M. (1995). *Nature*, **373**, 127.
51. Moorwood, A. F. M., Lutz, D., Oliva, E., Marconi, A., Netzer, H., Genzel, R., Sturm, E. and de Graauw, Th. (1996). *A&A*, **315**, L109.
52. Neufeld, D. A. and Maloney, P. R. (1995). *ApJ*, **477**, L17.
53. Neufeld, D. A., Maloney, P. R. and Conger, S. (1994). *ApJ*, **436**, L127.
54. O'Dea, C. P., Baum, S. A., Maloney, P. R., Tacconi, L. J. and Sparks, W. B. (1994). *ApJ*, **422**, 467.
55. Papadopoulos, P. P., Seaquist, E. R. and Scoville, N. Z. (1996). *ApJ*, **465**, 173.
56. Posen, A. G., Dalgarno, A. and Peek, J. M. (1983). *Atomic Data Nucl. Data Tables*, **28**, 265.
57. Roberge, W. G. and Dalgarno, A. (1982). *ApJ*, **255**, 176.
58. Sams, B. J., Genzel, R., Krabbe, A., Thatte, N. and Prada, F. (1997). preprint.
59. Sternberg, A. (1988). *ApJ*, **332**, 400.
60. Sternberg, A. and Dalgarno, A. (1989). *ApJ*, **338**, 197.
61. Sun, Y. and Dalgarno, A. (1994). *ApJ*, **427**, 1053.
62. Thatte, N., Quirrenbach, A., Genzel, R., Maiolino, R. and Tecza, M. (1997). *ApJ*, **490**, 238.
63. Thompson, R. I. (1994). *Experimental Astronomy*, **3**, 93.
64. Thompson, R. I. (1996). *ApJ*, **459**, L61.
65. Tiné, S., Lepp, S., Gredel, R. and Dalgarno, A. (1997). *ApJ*, **481**, 282.
66. Ueno, S., Koyama, K., Nishida, M., Yamauchi, S. and Ward, M. J. (1994). *ApJ*, **431**, L1.
67. Urbaniak, J. J. and Wolfe, A. M. (1981). *ApJ*, **244**, 406.
68. Veilleux, S., Goodrich, R. W. and Hill, G. J. (1997). *ApJ*, **477**, 631.
69. Voit, G. M. (1992). *MNRAS*, **258**, 841.
70. Whiteoak, J. B. and Wilson, W. E. (1990). *MNRAS*, **245**, 665.

22
X-ray Dominated Regions

Stephen Lepp and Stefano Tiné
Physics Department, University of Nevada, Las Vegas

1 Introduction

X-rays are generated by many sources in the Universe including QSOs, Seyfert galaxies, shocks, supernovae, massive stars and accreting neutron stars. The X-rays heat and ionize the surrounding diffuse matter. In this chapter we will review the effects of X-rays on ambient gas. We will also discuss signatures of X-ray excited gas.

X-rays lead to emission from the gas by ionization, excitation and heating due to direct photoabsorption and by secondary ionization and excitation induced by the fast photoelectrons. Often photoabsoption will be the dominant heating mechanism, but other processes, such as dissociation in shocks, may be important; the excitation induced by fast electrons can sometimes provide a clean signature that X-ray absorption influences a region significantly.

The effects of X-rays can be wide ranging or local. Soft X-rays, those with energies of less than 1 keV, are absorbed close to the source. Absorption cross sections for hard X-rays with energy greater then 1 keV are generally very small (less than 10^{-22} cm^2 per hydrogen atom). Thus a hard X-ray source can influence an entire galaxy to some extent. When a hard X-ray is absorbed it deposits all of its energy and so a 1 keV electron may cause approximately 30 ionizations in a gas of atomic or molecular hydrogen.

X-ray illuminated regions are also often associated with regions of enhanced ultraviolet radiation. The ultraviolet is not as penetrating as X-ray radiation and it will affect only the surface of the cloud. The high ultraviolet flux associated with the X-ray source will make a region similar to a photon dominated region (PDR) on the surface of cloud, while the X-rays will penetrate the entire cloud. If the heating is dominated by X-ray energy deposition then the region is an X-ray dominated region (XDR).

We begin in Section 2 with an overview of the thermal phases that exist in a gas heated by X-rays. We discuss the chemistry of an XDR in Section 3. In Section 4 we discuss the emission from X-ray illuminated regions and in Section 5 we discuss some particular sources.

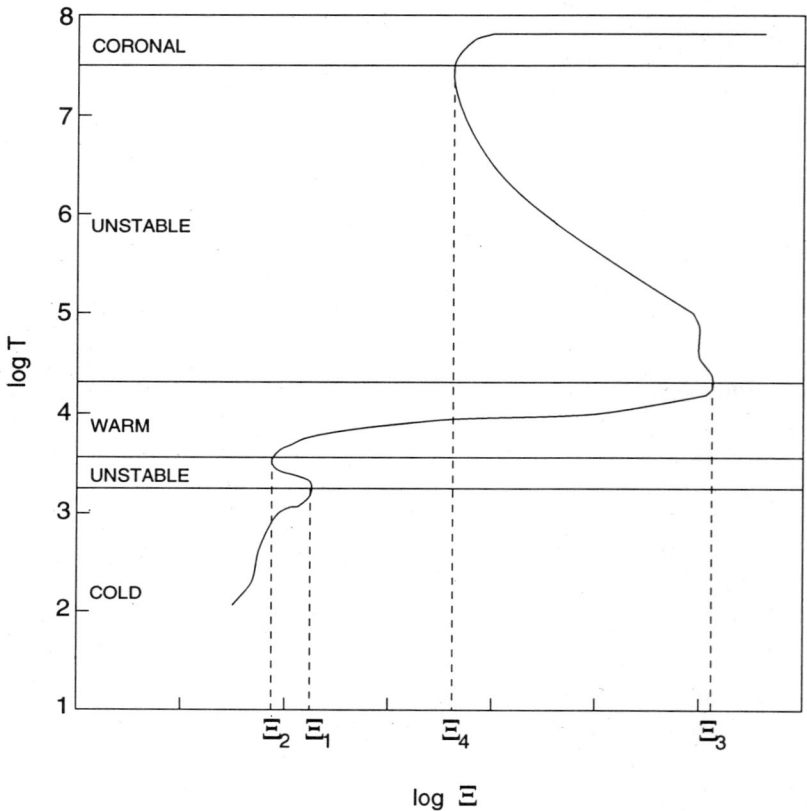

FIG. 1. A pictorial representation of the thermal phases associated with a hard X-ray source (adapted from Lepp et al. 1985). The figure shows T versus Ξ, where T is the temperature in Kelvin and Ξ is the ionizing flux divided by the pressure. The curve represents heating = cooling, with heating greater than cooling to the right of the curve and heating less than cooling to the left. Ξ_1 represents the maximum ionization parameter that sustains a cold phase. A warm phase exists for $\Xi_2 < \Xi < \Xi_3$ and a coronal phase for $\Xi > \Xi_4$. The warm and cold phases co-exist for $\Xi_2 < \Xi < \Xi_1$ and the warm and coronal phases co-exist for $\Xi_4 < \Xi < \Xi_3$.

2 Thermal Balance and Phases

The temperature of an X-ray illuminated gas depends on the flux of X-rays as well as the temperature, density and composition of the gas. There are several stable thermal phases for XDRs. These thermal phases are supported in each case by an increase in the cooling function. In order for a gas to be in stable thermal equilibrium the heating must be equal to the cooling and the (heating minus cooling) must decrease with increasing temperature for some range of temperatures. Thus, a small increase in temperature will cause cooling to be greater than heating and the gas will cool back to the equilibrium temperature; similarly a small decrease in temperature leads to heating being greater than cooling and the gas heats up. The heating rate per unit volume due to X-ray absorption is proportional to the gas density but is nearly independent of temperature, so any thermal phase must be supported by an increase in the cooling function. In pressure equilibrium the density, and therefore the X-ray heating rate per unit volume, decreases linearly with temperature and so the cooling function must be increasing faster than linearly with temperature in order for a stable phase to exist.

Figure 1 shows the thermal phases for a gas heated by hard X-rays (Lepp et al. 1985). The stable thermal phases are: a hot coronal plasma ($T \geq 10^6$ K), a warm HI or HII region ($T \simeq 10^4$ K) and a cold atomic or molecular phase ($T \leq 10^3$ K). The curve in Fig. 1 indicates conditions under which heating is equal to cooling. To the left of the curve cooling is greater than heating and to the right heating is greater than cooling. The unstable regions have heating equal to cooling but are unstable because a small increase in temperature leaves the gas in a region where heating exceeds cooling and so the temperature continues to increase. Similarly, a small decrease in temperature leads to further decreases.

This ionization structure is just an extension of the two-phase models for the interstellar medium (Field, Goldsmith and Habing 1969). The ionized/atomic phases have been investigated by several authors (e.g. Tarter, Tucker and Salpeter 1969; Kallman and McCray 1982; Wolfire et al. 1995). Kallman and McCray (1982) detailed the ionization structure and optical/ultraviolet/X-ray emission spectrum of a spherical, homogeneous cloud surrounding a compact X-ray source. Emission lines of atoms, from neutrals to highly ionized species, are expected together with a diffuse continuum produced by thermal bremsstrahlung, radiative recombination and two-photon decays of metastable levels.

In this chapter we will concern ourselves with the cold phase where the molecules occur. This phase is primarily responsible for the infrared emission from X-ray dominated regions.

3 X-ray Induced Chemistry

The first discussion of the influence of X-rays upon the chemistry was by Dalgarno (1976) who suggested doubly charged carbon as a way of making CH$^+$ near X-ray sources. The suggestion was picked up by Langer (1978) who made a detailed model. Multiply charged species are formed in X-ray ionized regions through the Auger process. In this a high energy X-ray ionizes an atom by removing one of the inner shell electrons. The atom is ionized and in an excited state, it may decay by having an electron make a transition to the ground state and at the same time emit one or more electrons.

While the multiply ionized species are certainly signatures of high energy X-rays they are not very important in the overall ionization. This is because most of the ionization is induced by the secondary electrons. A 1 keV X-ray may cause one multiple ionization and its secondary electrons then produce on the order of 30 ionizations. Higher energy X-rays have even a higher ratio of secondary to primary electrons.

Krolik and Kallman (1983) modelled the chemistry of X-ray excited regions with a particular emphasis on the Orion Star Forming Region of our Galaxy. Lepp and McCray (1983) modelled a cloud surrounding an X-ray source. More recently Lepp and Dalgarno (1996) have modelled the X-ray induced chemistry and Maloney et al. (1996) have produced detailed models of X-ray illuminated regions. In these models the X-rays affect the chemistry primarily by increasing the ionization rate and raising the temperature.

While X-rays preferentially ionize heavy elements, the secondary electrons they produce primarily ionize atomic and molecular hydrogen. Thus, the effect of X-ray ionization is much the same as increasing the cosmic-ray ionization rate. Two types of signatures that would, if detected, indicate that a region is ionized by X-rays rather then cosmic rays are emissions from multiply charged ions produced by the Auger process and the absences of emissions from molecules that are destroyed efficiently by X-ray absorption but slowly by other mechanisms.

Since the gas phase chemistry is driven by ionization, the effect of increasing ionization both increases the rate at which the chemistry is driven and increases the destruction rate of molecules. In the molecular region the chemistry is driven predominantly by ionization of molecular hydrogen to form H_2^+ which quickly reacts with another H_2 to produce H_3^+:

$$H_2 + \text{X-ray} \rightarrow H_2^+ + e, \tag{3.1}$$

$$H_2^+ + H_2 \rightarrow H_3^+ + H. \tag{3.2}$$

H_3^+ then reacts with many atoms and molecules, mostly through proton exchange reactions, driving the chemistry (cf. Chapter 4).

The ionization of molecular hydrogen along with the other H_3^+ reactions result in the removal of molecular hydrogen from the gas. Each ionization destroys approximately three hydrogen molecules. Thus, for high ionization rates the hydrogen is in atomic form rather than molecular. This changes the chemistry of the gas from one which favours the production of large molecules to one which favours simple molecules.

As an example of how the ionization driven chemistry works we consider the hydrocarbon chemistry (cf. Chapter 4). The sequence begins with H_3^+ reacting with atomic carbon:

$$H_3^+ + C \rightarrow CH^+ + H_2 \tag{3.3}$$

The CH^+ then reacts rapidly with H_2 picking up atomic hydrogens until it saturates at CH_3^+:

$$H_2 + CH^+ \rightarrow CH_2^+ + H, \tag{3.4}$$

$$H_2 + CH_2^+ \rightarrow CH_3^+ + H. \tag{3.5}$$

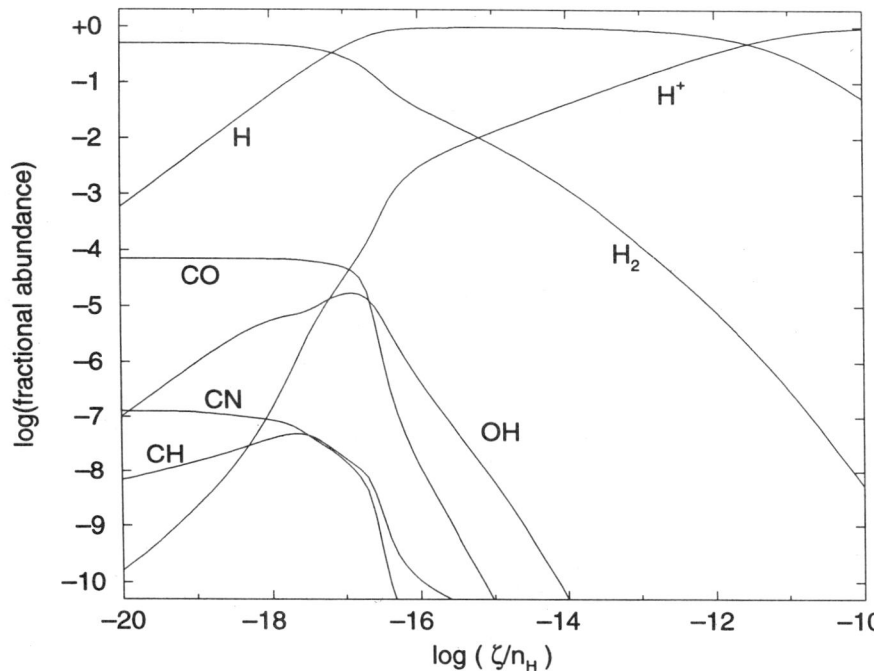

FIG. 2. Steady state molecular abundances vs ζ/n_H at $T = 500$ K and $n_H = 10^4$ cm^{-3} (adapted from Lepp and Dalgarno 1996).

The CH_3^+ then recombines to form either CH or CH_2:

$$CH_3^+ + e \rightarrow CH_2 + H, \quad (3.6)$$
$$\rightarrow CH + H + H \quad (3.7)$$

or it can radiatively attach an H_2 to form CH_5^+:

$$CH_3^+ + H_2 \rightarrow CH_5^+ + h\nu \quad (3.8)$$

which recombines to form CH_3 or CH_4 or it can proton transfer with CO to produce HCO^+ and CH_4.

The simple hydrocarbons are converted to more complex hydrocarbons by reactions with C^+

$$CH_4 + C^+ \rightarrow C_2H_3^+ + H, \quad (3.9)$$
$$\rightarrow C_2H_2^+ + H_2 \quad (3.10)$$

which then either recombine with electrons or proton transfer to produce hydrocarbons containing two carbon atoms each. Larger hydrocarbons are built by reactions involving C^+ followed by recombination or proton transfer reactions.

In the standard cosmic-ray ionized chemistry most of the ionization is of hydrogen molecules. X-rays preferentially ionize heavier elements but most of the ionization is caused by fast electrons emitted from the initial ionization. Thus, ionization of molecular hydrogen continues to dominate, though direct ionization of heavy elements is increased. As the ionization rate increases so does the H_3^+ abundance. This continues until the ionization is so high that the chemistry changes from molecular hydrogen dominated to atomic hydrogen dominated.

The increase in H_3^+ with increasing ionization leads to an increase in molecule production. As with the hydrocarbon chemistry outlined above most molecules are formed from chemical networks beginning with H_3^+. Thus increasing ionization increases molecule production until the ionization becomes so large as to begin to reduce the H_2 abundance substantially. When the H_2 is destroyed molecule production and molecular abundances drop off quickly.

The fast electrons resulting from X-ray ionization not only induce additional ionizations as outlined above; they also produce ultraviolet photons. These photons are the result of exciting molecular or atomic hydrogen or helium to electronic excited states. These states then decay emitting an ultraviolet photon. This increase in the ultraviolet flux within the cloud destroys molecules and balances the increase in entry into the chemical networks. Thus, molecular abundances may go either up or down with increasing ionization, though they will all go down when the ionization is sufficient to destroy most of the molecular hydrogen.

Figure 2 shows the steady-state abundances of molecules vs ζ/n_H for a number of species, where ζ is the ionization rate and n_H the total hydrogen nuclei density (adapted from Lepp and Dalgarno 1996). The steady-state fractional abundances depend almost exclusively on ζ/n_H because nearly all the reactions are either driven by the X-ray ionization with rates proportional to $n_H\zeta$ or are two particle reactions with rates proportional to n_H^2. Division of $n_H\zeta$ by n_H^2 leaves ζ/n_H as the only free parameter. In Fig. 2 we see that as the ionization increases the chemistry favours simple diatomic molecules such as OH and CH. The increase in abundance of these simple molecules is a result of the increased entry into the reaction networks as well as the ultraviolet flux dissociating larger molecules into smaller fragments. For ζ/n_H greater than approximately 10^{-17} cm^3 s^{-1} the gas becomes primarily atomic and all the molecular abundances fall off.

4 Infrared Emission

The infrared emission is due to the reprocessing of X-ray energy absorbed in the gas to infrared lines by atoms and molecules or infrared continuum by dust grains. The H_2 infrared emissivity of X-ray illuminated clouds is a very useful diagnostic of physical conditions. Lepp and McCray (1983) found that several per cent of the X-ray energy is redistributed as H_2 vibrational–rotational emission. H_2 is rovibrationally excited in a process similar to ultraviolet pumping in photon

Table 1 H_2 line ratios for $n_H = 10^4$ cm^{-3} and varying ionization rate ζ and temperature T.

ζ		10^{-15} s^{-1}	10^{-13} s^{-1}	10^{-11} s^{-1}
T		50 K	500 K	2000 K
$v=$ 2–1 S(1)/	$v=$ 1–0 S(1)	6.9×10^{-2}	8.5×10^{-2}	6.0×10^{-2}
1–0 S(2)/	1–0 S(0)	6.6×10^{-2}	4.1×10^{-1}	$1.7 \times 10^{+0}$
1–0 S(3)/	1–0 S(1)	3.4×10^{-2}	1.9×10^{-1}	9.8×10^{-1}
1–0 S(0)/	1–0 S(1)	$2.5 \times 10^{+0}$	2.7×10^{-1}	2.1×10^{-1}
0–0 S(13)/	1–0 O(7)	2.0×10^{-3}	1.5×10^{-1}	3.3×10^{-1}
5–3 O(3)/	1–0 S(1)	1.5×10^{-2}	1.8×10^{-2}	4.1×10^{-5}
6–4 Q(1)/	1–0 S(1)	1.3×10^{-2}	1.5×10^{-2}	6.1×10^{-6}

dominated regions (cf. Chapter 9). An important difference from ultraviolet pumping is that excitation by fast electron impact can also populate the low lying vibrational states directly. Details of the energy loss of fast electrons in molecular gas have been presented by several authors (Cravens, Victor and Dalgarno 1975; Gredel, Lepp and Dalgarno 1987; Xu and McCray 1991; Voit 1991; Gredel and Dalgarno 1995). Tiné et al. (1997) detailed the H_2 emissivity for several important lines as a function of ionization rate and density. In these models the temperature was considered a free parameter.

Table 1 presents the ratio of several observed H_2 lines from the model of Tiné et al. (1997). The line ratios are sensitive to the ionization fraction which is determined by the chemistry. For high enough temperatures, collisional excitation of the low vibrational levels will dominate non-thermal pumping by the fast primary and secondary photoelectrons. The non-thermal excitation of H_2 rovibrational levels declines with increasing fractional ionization of the gas as more of the energy deposited goes into heating the gas due to Coulomb scattering (Gredel and Dalgarno 1995). Complete tables of electron pumping entry efficiencies into the various rovibrational levels of H_2 are presented in Gredel and Dalgarno (1995) and Tiné et al. (1997).

The vibrational rate coefficients fall quickly with temperature so that at low temperatures non-thermal excitations can dominate. At $T \leq 1000$ K non-thermal electron pumping of H_2 by X-rays is potentially observable in all vibrational emission ratios including the widely used $v = 2$–1 S(1)/$v = 1$–0 S(1), though this ratio is never higher than $\simeq 0.3$ (Maloney et al. 1996; Tiné et al. 1997), while it can reach 0.5 in PDRs. At higher temperatures the vibrational rate coefficients will dominate and the ratios will converge to the pure collisional values. The convergence to LTE with increasing density, n_H, will be slower at low values of ζ/n_H because the rate coefficients in the dominant H_2–H_2 collisions are smaller than the H–H_2 rate coefficients. The critical densities may be as high as $n_H \geq 10^7$ cm^{-3} depending on temperature.

Another distinguishing feature of X-ray excited gas is that pumping of H_2 by formation is likely to be particularly important. The ratio of the rate for

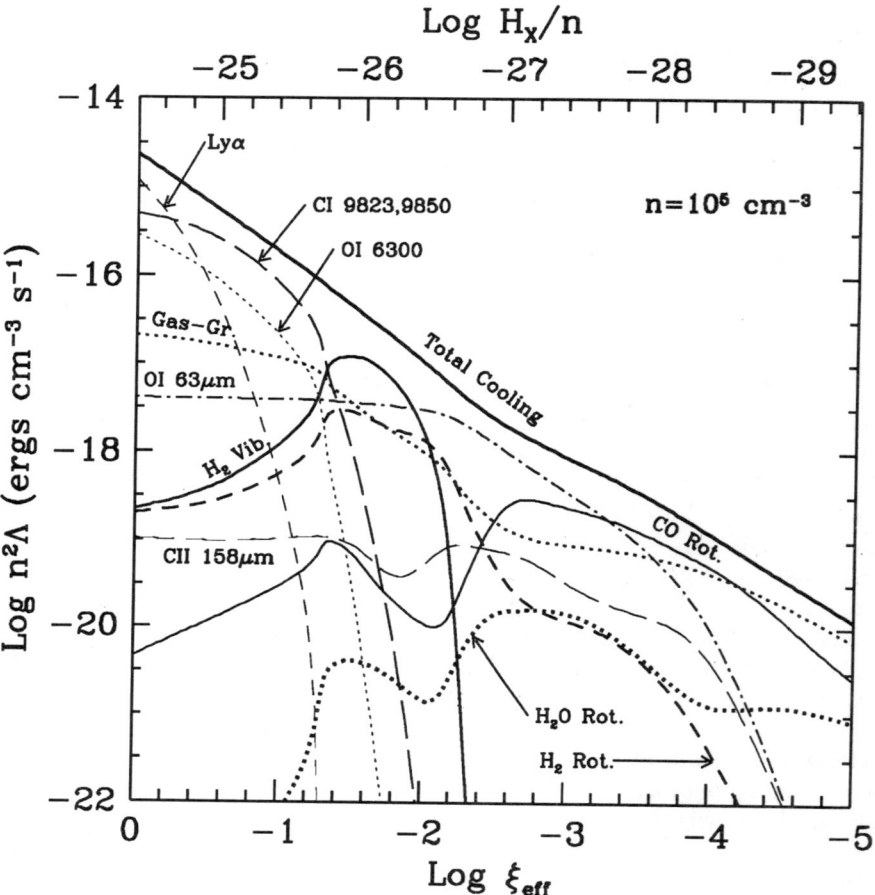

FIG. 3. Steady-state cooling rates vs ξ_{eff}. ξ_{eff} is an effective ionization parameter proportional to ζ/n_H. Also indicated along the top is the energy deposition rate divided by the density, H_x/n_H. From Maloney et al. (1996).

excitation of molecular hydrogen electronic states to the rate for dissociation and ionization of molecular hydrogen is less than 1 for an X-ray excited gas, whereas for ultraviolet excitation it is of order 10. Thus, in an XDR, the formation pumping rate may well be comparable to the rate of ultraviolet cascade from electronic excited states. Unfortunately, the state specific formation rates are still uncertain making it difficult to use this as a diagnostic. Calculations of state specific formation rates for H_2 are greatly needed.

In time dependent models of X-ray excitation of dense molecular gas ($n \geq 10^2$ cm^{-3}), Draine and Woods (1990, 1991) found that even more intense H_2 emission will be produced than in steady-state models. From their models of starburst regions, powered by high rates of supernova explosions, they predicted strong emission lines from H_3^+, OI, OH and H_2O.

Krolik and Lepp (1989) modelled the emission from Seyfert galaxies and determined that CO rotational lines are potentially observable, especially those of the rarer CO isotopes. However, their models are of high pressure ($nT \simeq 10^{10} cm^{-3} K$) isobaric regions and were specifically developed for the torus of molecular gas around an AGN, and the high emissivities of these lines are consequences of the large assumed column density ($N_H \simeq 10^{24} cm^{-2}$). More recent models of Seyfert galaxies by Neufeld, Maloney and Conger (1994) are discussed below.

Maloney et al. (1996) presented models of steady-state X-ray dominated regions. Figures 3 and 4 (from Maloney et al. 1996) show their cooling rate and infrared luminosity, respectively, for a gas at $n_H = 10^5$ cm^{-3} as functions of the X-ray energy deposition rate divided by the density, H_x/n_H. For a given X-ray illuminated cloud the X-ray flux falls off as X-rays are absorbed and the distance from a point source increases. Thus, the results in the figures can be used in conjunction with a simple calculation of the X-ray flux fall-off to obtain the depth dependence of the emissivity; in a more detailed study Maloney et al. (1996) have shown that the inclusion of line-trapping effects introduces minor differences.

Figure 3 shows rates per unit volume of the major cooling processes in the gas as functions of an ionization parameter. At low X-ray energy deposition rates ($H_x/n_H \approx 10^{-28}$ ergs cm^3 s^{-1}) the gas has a temperature of approximately 100 K and the emission is primarily from CO rotational lines. At $H_x/n_H \approx 10^{-26}$ ergs cm^3 s^{-1} the gas is at several thousand degrees and H_2 lines dominate the cooling. At $H_x/n_H > 10^{-25}$ ergs cm^3 s^{-1} the gas goes over to a warm atomic phase with Lyman alpha emission dominating the cooling.

Figure 4 shows the results for the commonly observed lines Brγ at 2.17 µm, [FeII] at 1.64 µm and the $v = 1-0$ S(1) and $v = 2-1$ S(1) lines of H_2. The line to continuum ratio is also an important diagnostic of these regions. The H_3^+ emissivity predicted in these steady-state models is low and much less than expected if transient effects are dominating (Draine and Woods 1990).

5 X-rays Sources and Observations

There are many regions in space where molecular gas is ionized by strong X-ray sources. In this section we review the observations of X-ray sources and nearby molecular gas. Though this part of the book is concerned with extragalactic sources, there are many types of Galactic X-ray sources and we briefly consider their observed effects on interstellar gas. X-ray sources in the Milky Way include X-ray binaries, Herbig–Haro objects, supernova remnants and the Galactic Centre. In extragalactic astronomy, strong X-ray luminosities are characteristic of AGNs (Mushotsky, Done and Pounds 1993) since they emit 5–40% of their radiation at these wavelengths (Ward et al. 1987).

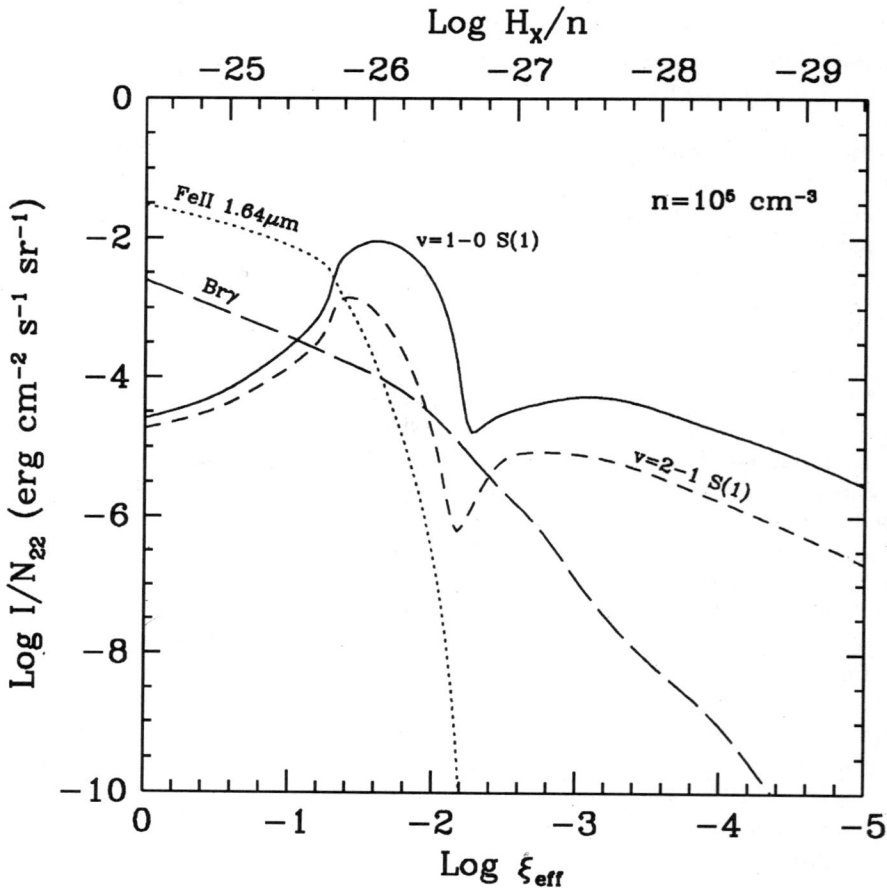

FIG. 4. Steady-state infrared luminosities of selected diagnostic lines vs ξ_{eff}. ξ_{eff} is an effective ionization parameter proportional to ζ/n_H. Also indicated along the top is the energy deposition rate divided by the density, H_x/n_H. From Maloney et al. (1996).

It is often difficult to distinguish X-ray regions from other high energy environments such as shocks or photon dominated regions. Indeed, in many sources a combination of these excitation mechanisms co-exist. Moreover, often observations of extragalactic sources do not have the resolution to separate the different high energy regions, all of which contribute to form the observed spectra.

5.1 Galactic X-ray Sources

Amongst the most interesting Galactic X-ray sources are accreting compact objects, including neutron stars and black holes. Each accretes from either the interstellar medium or a nearby companion. An example of an isolated compact object is the X-ray source 1E 1740.7−2942 which lies at ≃50′ from the Galactic Centre. A nearby molecular cloud has enhanced HCO^+ emission which then falls off in the regions of the cloud near the X-ray source, providing possible evidence that the X-ray source is closely associated with the cloud (Phillips and Lazio 1995). Yan and Dalgarno (1997) have recently modelled such an XDR on the assumption that the source is embedded in the cloud itself. Their theoretical results for the spatial distribution of the emerging HCO^+ $J = 1$–0 line matches well the interferometric observations available. They also determined intensity profiles for the as yet unobserved $J = 3$–2 and $J = 4$–3 emissions of HCO^+ which should have even higher intensities and might provide further tests of the hypothesis that 1E 1740.7−2942 is embedded in the cloud.

Several other X-ray point sources and a diffuse continuum have also been observed near the Galactic Centre. The structure of the interstellar medium in the Galactic Centre is quite complex, but radio observations of CO emission point to a molecular content in the central few hundred parsecs as high as 10^7–10^8 solar masses, depending on the CO-to-H_2 conversion factor (Güsten 1989, Sofue 1995). The molecular material is in clouds with number densities of 10^3–10^5 cm^{-3} feeding a more diffuse circumnuclear disc that extends from 1.5 pc to 5 pc from the central cavity, which contains the compact radio source Sgr A*, probably a massive or supermassive black hole. Beyond this zone the interstellar gas forms other molecular rings up to around 300 pc from the centre (see reviews by Genzel, Hollenbach and Townes 1994; Morris and Serabyn 1996). The total X-ray luminosity of the central 100 pc is approximately 10^{37} erg s^{-1}, with a possible substantial contribution at medium energies by old neutron stars accreting from the interstellar medium (Maraschi, Treves and Tarenghi 1973; Zane, Turolla and Treves 1996). The hard energy X-ray photons seem to be associated with Giant Molecular Clouds (Sunyaev, Markevitch and Pavlinsky 1993) as in the source 1E 1740.7−2942 mentioned above.

Supernova remnants are another class of Galactic X-ray sources. Observed $v = 2$–1 $S(1)/v = 1$–0 $S(1)$ H_2 intensity ratios for the IC 443 (Burton et al. 1988, 1989; Moorhouse et al. 1991) and RCW 103 (Oliva, Moorwood and Danziger 1990) remnants are in the range of 0.07 to 0.12. These ratios are consistent with thermal populations at temperatures of around 2000 K. Such high temperatures might be maintained by the shocks associated with the supernova blast waves or by the X-rays produced by the shocks. Observations of high velocities in the wings of H_2 lines, as well as of HCO^+ and CO lines, favour the shock hypothesis.

Protostellar jets interact with the interstellar medium forming shocks that heat and excite the gas. The interaction regions are called Herbig–Haro objects (cf. Chapter 10). The shocks are fast enough that X-rays and ultraviolet emission are produced. Some of these objects may then show X-ray signatures in the form of non-thermal excitation of H_2. HH 1 was identified as a potential candidate by Wolfire and Königl (1991). However, recent observations (Gredel 1996) showed no evidence of non-thermal H_2 infrared emission. All levels up to $v = 3$ are thermalized at temperatures around 2000–3000 K. Similar conclusions apply to other Herbig–Haro objects, and excitation of higher vibrational levels has not been observed.

The supernova remnant IC 443 and the Herbig–Haro object HH 7 have observed ratios of the $v = 0$–0 S(13) and $v = 1$–0 O(7) intensities at about 3 μm corresponding to excitation temperatures around 3000 K while the $v = 2$–1 S(1)/$v = 1$–0 S(1) ratios correspond to 2000 K (Burton et al. 1989). There is no clear explanation yet for this discrepancy, though due to cooling the postshock gas is expected to have a range of temperatures. It is also possible that non-thermal excitation mechanisms may produce this ratio; they include: formation pumping, which has been suggested to explain regions with enhanced $v = 2$–1 S(1)/$v = 1$–0 S(1) ratios in IC 443 (Richter, Graham and Wright 1995); UV fluorescence; and X-ray pumping. Further modelling of Herbig–Haro objects and supernova remnants is needed to make possible the identification of the contribution of each of these excitation mechanisms.

5.2 Extragalactic X-ray Sources

Starburst galaxies are galaxies in which intense star formation is occurring (cf. Chapter 20). The emission spectra are somewhat similar to those of HII regions and are typical of gas ionized and excited by the strong ultraviolet field of the newly formed O and B stars. The high supernova rates in these regions produce X-rays as well as ultraviolet radiation and shocks. Additionally, shocks are produced by cloud–cloud collisions and tidal disruptions as the starburst galaxies are often gravitationally interacting or possibly merging galaxies. A mixture of shock excitation and ultraviolet pumping has been invoked to explain the observed H_2 emission in merging systems like NGC 3256 (Doyon, Wright and Joseph 1994) and NGC 6240 (Tanaka et al. 1991; van der Werf et al. 1993). However, Draine and Woods (1990) ruled out these mechanisms for NGC 6240 on the basis of energetics and the weakness of the $v = 2$–1 S(3) line and proposed an alternative scenario in which X-rays emitted by supernova explosions are most likely producing the observed spectra (Draine and Woods 1990, 1991). They also predicted strong H_3^+ emission as a conclusive diagnostic of X-ray excitation. More recently, Mouri and Taniguchi (1995) have argued in favour of formation pumping of H_2 in NGC 6240. The same mechanism of supernova-driven X-ray excitation of H_2 was proposed for NGC 1808 (Krabbe, Sternberg and Genzel 1994) and NGC 3256 (Moorwood and Oliva 1994). However, new observations by Sugai et al. (1997) of H_2 $v = 1$ and $v = 2$ lines in NGC 6240 show that they are consistent with a thermal spectrum at 1900 K; also the $v = 2$–1 S(3) is not diminished, strongly suggesting that the emission is from non-dissociative shocks.

Seyfert galaxies and QSOs also show H_2 emission. H_2 emission in Seyferts is stronger relative to the far-infrared flux than in starburst galaxies (Kawara, Nishida and Gregory 1987), suggesting that the H_2 excitation is powered by the active nuclei. The total amount of molecular material inferred from the CO observed for these galaxies can be as high as 10^9–10^{10} M$_\odot$ (Sanders et al. 1986; Blitz 1990; Young and Scoville 1991). A Seyfert 2 galaxy is thought to have an obscuring torus or disc of molecular material surrounding the active nucleus, while Seyfert 1 galaxies are the same class of galaxies seen face on with the active nuclei exposed (Antonucci and Miller 1985; Antonucci 1993; Chapter 20). Gas closer to the X-ray source will be thermally excited to temperatures of 2000–3000 K (Lepp and McCray 1983) and H_2 emission will be thermalized. However, direct observation of gas near the active nucleus in a Seyfert 2 galaxy is obscured by the torus. Instead the observed excitation comes from X-rays which have penetrated through much of the torus and the temperature would be lower, possibly around 1000 K or less (Krolik and Lepp 1989). An example of such a galaxy could be the Seyfert 2 galaxy NGC 1068 (Rotaciuc et al. 1991). More recent observations at higher spatial resolutions (Tacconi et al. 1994) raise the possibility that the H_2 emission is due to shock excitation. Similarly, the observed strong FeII emission could also be produced in shocks from the outflows, including jets, from the nucleus (Blietz et al. 1994), though it seems more likely to be powered by X-ray excitation (Maloney et al. 1996). X-ray excitation remains the best explanation for the nuclear H_2 emission (Tacconi et al. 1994). It may also be causing the high HCN/CO ratio as Lepp and Dalgarno (1996) showed that X-ray induced chemistries increase the abundance of small carbon bearing molecules while CO remains relatively unchanged. X-ray irradiation of a molecular torus by a compact central nucleus could account for the observed H_2O megamaser in this galaxy (Claussen et al. 1984), as well as those in other Seyfert 2/LINERs (Neufeld, Maloney and Conger 1994; Chapter 23).

NGC 1275, another Seyfert/LINER galaxy, has an observed $v = 2$–1 S(1)/$v = 1$–0 S(1) ratio of 0.06 (Kawara and Taniguchi 1993) which indicates either thermal excitation (by X-rays or shocks) at around 1800 K or X-ray non thermal pumping at $T < 1000$ K (Tiné et al. 1997). However, based on the analysis of the $v = 1$–0 S(0)/$v = 1$–0 S(1) ratio observed by Kawara and Taniguchi (1993), Mouri (1994) concluded that ultraviolet pumping could contribute as much as 10% of the observed H_2 luminosity, the rest being of thermal origin at $T < 1000$ K. Indication of star formation activity in this galaxy, and the associated PDRs, have been found (Holtzmann et al. 1992). A decisive test would come from observations of high vibrational lines such as $v = 5$–3 O(3) and $v = 6$–4 Q(1), but so far only upper limits on these lines have been determined (Kawara and Taniguchi 1993).

It is possible to directly observe the X-ray excited molecular clouds near the active nucleus of a Seyfert 1 galaxy. Based on the observed H_2 spectral lines from a sample of eighteen Seyfert 1 galaxies and quasars, Kawara, Nishida and Gregory (1990) suggested that X-ray heating is the dominant excitation mechanism while winds driven by the nuclei must play this role in other AGNs. H_2 emission has been observed in NGC 3783 (Kawara, Nishida and Gregory 1989), which has a bare nucleus and in NGC 7469 (Heckman et al. 1986; Moorwood and Oliva 1988) which is a dusty Seyfert 1 galaxy. More recent observations of NGC 7469 (Genzel et al. 1995) have detected $v = 1$–0 S(1) emission in the nucleus ($\leq 10^2$ pc). They interpreted this as an indication of the presence of a face on molecular torus powered by an active nucleus.

Infrared emission from H_2 has also been proposed (Gredel and Dalgarno 1995) as a potential diagnostic of cooling flows in galaxy clusters. In these flows X-rays emitted during the radiative cooling of hot intracluster gas (see Fabian 1994; Ferland et al. 1994) should excite the cooled component, which collapses into clouds or low mass stars (Braine and Dupraz 1994; McNamara and Jaffe 1994; O'Dea et al. 1994). Very recently the first observation of the $v = 1$–0 S(1) line in such environments has been reported and ascribed to X-ray excitation (Jaffe and Bremer 1997).

6 Conclusions

The effect of X-rays on molecular clouds is just starting to be understood. For large X-ray fluxes the gas is heated to several thousand degrees making it difficult to distinguish whether shocks or enhanced ultraviolet radiation are instead responsible for heating the gas to these temperatures. Only excitation of high vibrational levels of H_2, time dependent effects and the chemistry provide signatures unique to X-ray dominated regions. The problem is compounded for extragalactic sources in which the observed molecular emission probably arises due to several excitation mechanisms. At low fluxes it is possible to observe directly the non thermal excitation caused by the secondary electrons. H_2 pumping by formation should be important in XDRs, but more knowledge about the way in which rovibrational levels are populated by formation is required before accurate line strengths can be calculated.

Many astrophysical sources produce X-rays which then influence the surrounding interstellar medium. Additional modelling is needed to fully understand how astrophysical regions respond to X-rays. The chemistry and infrared emission provide probes of X-ray dominated regions.

Alex Dalgarno was the first to suggest how an X-ray source may modify the chemistry of the surrounding gas and he has continued to contribute to the understanding of these regions. We have benefited from having him as both a colleague and a mentor and would like to express our gratitude for his continued encouragement and friendship.

Bibliography

1. Antonucci, R. (1993). *ARA&A*, **31**, 473.
2. Antonucci, R. and Miller, J. S. (1985). *Ap.J.*, **297**, 621.
3. Blietz, M., Cameron, M., Drapatz, S., Genzel, R., Krabbe, A. and van der Werf, P. (1994). *Ap.J.*, **421**, 92.
4. Blitz, L. (1990). In *Molecular Astrophysics: a volume honouring Alexander Dalgarno*, ed. T. W. Hartquist. Cambridge University Press, Cambridge, p.35.

5. Braine, J. and Dupraz, C. (1994). *A&A*, **283**, 407.
6. Burton, M. G., Geballe, T. R., Brand, P. W. J. L. and Webster, A. S. (1988). *MNRAS*, **231**, 617.
7. Burton, M. G., Brand, P. W. J. L., Geballe, T. R. and Webster, A. S. (1989). *MNRAS*, **236**, 409.
8. Claussen, M. J., Heiligman, G. M. and Lo, K.-Y. (1984). *Nature*, **310**, 298.
9. Cravens, T. E., Victor, G. A. and Dalgarno, A. (1975). *Planet. Space Sci.*, **23**, 1059.
10. Dalgarno, A., (1976). In *Atomic Processes and Applications*, eds. P. G. Burke and B. L. Moiseiwitsch. North Holland, Amsterdam, p.110.
11. Doyon, R., Wright, G. S. and Joseph, R. D. (1994). *Ap.J.*, **421**, 115.
12. Draine, B. T. and Woods, D. T. (1990). *Ap.J.*, **363**, 464.
13. Draine, B. T. and Woods, D. T. (1991). *Ap.J.*, **383**, 621.
14. Fabian, A. C. (1994). *ARA&A*, **32**, 277.
15. Ferland, G. J., Fabian, A. C. and Johnstone, R. M. (1994). *MNRAS*, **266**, 399.
16. Field, G. B., Goldsmith, D. W. and Habing, H. J. (1969). *Ap.J.*, **155**, L149.
17. Genzel, R., Hollenbach, D. and Townes, C. H. (1994). *Rep.Prog.Phys.*, **57**, 417.
18. Genzel, R., Weitzel, L., Tacconi-Garman, L. E., Blietz, M., Cameron, M., Krabbe, A. and Lutz, D. (1995). *Ap.J.*, **444**, 129.
19. Gredel, R. (1996). *A&A*, **305**, 582.
20. Gredel, R. and Dalgarno, A. (1995). *Ap.J.*, **446**, 852.
21. Gredel, R., Lepp, S. and Dalgarno, A. (1987). *Ap.J.*, **323**, L137.
22. Gredel, R., Lepp, S., Dalgarno, A. and Herbst, E. (1989). *Ap.J.*, **347**, 289.
23. Güsten, R. (1989). In *The Center of the Galaxy*, ed. M. Morris. Kluwer, Dordrecht, p.89.
24. Heckman, T. M., Beckwith, S., Blitz, L., Skrutskie, M. and Wilson, A. (1986). *Ap.J.*, **305**, 157.
25. Holtzman, J. A. *et al.* (1992). *AJ*, **103**, 691.
26. Jaffe, W. and Bremer, M. N. (1997). *MNRAS*, **284**, L1.
27. Kallman, T. R. and McCray, R. (1982). *Ap.J. Suppl. Series*, **50**, 263.
28. Kawara, K. and Taniguchi, Y. (1993). *Ap.J.*, **410**, L19.
29. Kawara, K., Nishida, M. and Gregory, B. (1987). *Ap.J.*, **321**, L35.
30. Kawara, K., Nishida, M. and Gregory, B. (1989). *Ap.J.*, **342**, L55.
31. Kawara, K., Nishida, M. and Gregory, B. (1990). *Ap.J.*, **352**, 433.
32. Krabbe, A., Sternberg, A. and Genzel, R. (1994). *Ap.J.*, **425**, 72.
33. Krolik, J. H. and Kallman, T. R. (1983). *Ap.J.*, **267**, 610.
34. Krolik, J. H. and Lepp, S. (1989). *Ap.J.*, **347**, 179.

35. Langer, W. D. (1978). *Ap.J.*, **225**, 860.
36. Lepp, S. and Dalgarno, A. (1996). *A&A*, **306**, L21.
37. Lepp, S. and McCray, R. (1983). *Ap.J.*, **269**, 560.
38. Lepp, S., McCray, R., Shull, J. M., Woods, D. T. and Kallman, T. (1985). *Ap.J.*, **288**, 58.
39. McNamara, B. R. and Jaffe, W. (1994). *A&A*, **281**, 673.
40. Maloney, P. R., Hollenbach, D. J. and Tielens, A. G. G. M. (1996). *Ap.J.*, **466**, 561.
41. Maraschi, L., Treves, A. and Tarenghi, M. (1973). *A&A*, **25**, 153.
42. Moorhouse, A., Brand, P. W. J. L., Geballe, T. R. and Burton, M. G. (1991) *MNRAS*, **253**, 662.
43. Moorwood, A. F. M. and Oliva, E. (1988). *A&A*, **203**, 278.
44. Moorwood, A. F. M. and Oliva, E. (1994). *Ap.J.*, **429**, 602.
45. Morris, M. and Serabyn, E. (1996). *Ann. Rev. Astron. Astrophys.*, **34**, 645.
46. Mouri, H. (1994). *Ap.J.*, **427**, 777.
47. Mouri, H. and Taniguchi, Y. (1995). *Ap.J.*, **449**, 134.
48. Mushotsky, R. F., Done, C. and Pounds, K. A. (1993). *ARA&A*, **31**, 717.
49. Neufeld, D. A., Maloney, P. R. and Conger, S. (1994). *Ap.J.*, **436**, L127.
50. O'Dea, C. P., Baum, S. A., Maloney, P. R., Tacconi, L. J. and Sparks, W. B. (1994). *Ap.J.*, **422**, 467.
51. Oliva, E., Moorwood, A. F. M. and Danziger, I. J. (1990). *A&A*, **240**, 453.
52. Phillips, J. A. and Lazio, T. J. W. (1995). *Ap.J.*, **442**, L37.
53. Richter, M. J., Graham, J. R. and Wright, G. S. (1995). *Ap.J.*, **454**, 277.
54. Rotaciuc, V., Krabbe, A., Cameron, M., Drapatz, S., Genzel, R., Sternberg, A. and Storey, J. W. V. (1991). *Ap.J.*, **370**, L23.
55. Sanders, D. B., Scoville, N. Z., Young, J. S., Soifer, B. T., Schloerb, F. P., Rice, W. L. and Danielson, G. E. (1986). *Ap.J.*, **305**, L45.
56. Sofue, Y. (1995). *Pub.Astron.Soc.Japan*, **47**, 527.
57. Sugai, H., Malkan, M. A., Ward, M.J, Davies, R. I. and McLean, A. S. (1997). *Ap.J.*, **481**, 186.
58. Sunyaev, R.I, Markevitch, M. and Pavlinsky, M. (1993). *Ap.J.*, **407**, 606.
59. Tacconi, L. J., Genzel, R., Blietz, M., Cameron, M., Harris, A. I. and Madden, S. (1994). *Ap.J.*, **426**, L77.
60. Tanaka, M., Hasegawa, T. and Gatley, I. (1991). *Ap.J.*, **374**, 516.
61. Tarter, C. B., Tucker, W. and Salpeter, E. E. (1969). *Ap.J.*, **156**, 943.
62. Tiné, S., Lepp, S., Gredel, R. and Dalgarno, A. (1997). *Ap.J.*, **481**, 282.
63. van der Werf, P. P., Genzel, R., Krabbe, A., Blietz, M., Lutz, D., Drapatz, S., Ward, M. J. and Forbes, D. A. (1993). *Ap.J.*, **405**, 522.
64. Voit, G. M. (1991). *Ap.J.*, **377**, 158.

65. Ward, M. J., Elvis, M., Fabbiano, G., Carleton, N. P., Willner, S. P. and Lawrence, A. (1987). *Ap.J.*, **315**, 74.
66. Wolfire, M. G. and Königl, A. (1991). *Ap.J.*, **383**, 205.
67. Wolfire, M. G., Hollenbach, D., McKee, C. F., Tielens, A. G. G. M. and Bakes, E. L. O. (1995). *Ap.J.*, **443**, 152.
68. Xu, Y. and McCray, R. (1991). *Ap.J.*, **375**, 190.
69. Yan, M. and Dalgarno, A. (1997). *Ap.J.*, **481**, 296.
70. Young, J. S. and Scoville, N. Z. (1991). *ARA&A*, **29**, 581.
71. Zane, S., Turolla, R. and Treves, A. (1996). *Ap.J.*, **471**, 248.

23
Water Molecules in the Circumnuclear Regions of Active Galaxies

David A. Neufeld
The Johns Hopkins University

1 Introduction

The circumnuclear regions of active galaxies are perhaps the most remarkable astrophysical environments in which molecules have been observed. Luminous maser emissions in the 6_{16}–5_{23} 22 GHz line of water have now been detected from at least 18 active galaxies; in every case where the size of the emission region has been determined, the water maser emission has been found to originate primarily within a parsec- or subparsec-sized region that is coincident with the active nucleus of the galaxy.

Thanks to the extraordinarily high brightness temperatures generated by the maser amplification process, observations using the technique of very long baseline interferometry (VLBI) have been possible for several sources, allowing the source structure to be probed at angular resolutions better than one milliarcsecond. Furthermore, because individual emission features are spectrally very narrow, valuable kinematic information is obtained about the line-of-sight Doppler velocities of the emitting gas. Thus, observations of water maser emission provide an extraordinarily powerful probes of the inner circumnuclear regions of active galaxies.

In this chapter, we will review the observed properties of water maser emission from active galaxies; the physical and chemical processes responsible for the generation of that emission; and the information about active galactic nuclei that has been obtained from the careful analysis of observations of water maser emission. The study of water maser emissions from active galaxies is evolving rapidly on both the observational and theoretical fronts; this chapter provides a description of the field at the time of writing, March 1997.

In Section 2, the observational properties of water masers in active galaxies are briefly reviewed. In Section 3, I discuss the physical and chemical processes responsible for the production of water and the excitation of maser emission within the circumnuclear environment in active galaxies. The use of water megamasers as probes of circumnuclear discs is considered in Section 4.

2 Observed Properties

2.1 Fluxes

By comparison with water masers in Galactic sources, water masers detected from active galactic nuclei show extremely large fluxes; they are therefore often referred to as "megamasers". Based upon the assumption that the maser radiation from these sources is radiated isotropically, apparent luminosities in the range 35 to 6000 $L_\odot$ (Koekemoer et al. 1995) have been inferred. These values probably represent overestimates of the true luminosities: since the non-linear amplification of maser radiation typically leads to beaming, the true luminosities of those sources from which maser radiation is observed are likely to be less than the "apparent" luminosities, due to selection effects. The extragalactic water megamasers are invariably associated with active galaxies, and in particular with Seyfert 2 galaxies and LINERs. A large systematic survey by Braatz (1996) has yielded a detection rate of 14% for 22 GHz water emission from nearby Seyfert 2 galaxies and LINERS, with no detections from Seyfert 1 galaxies.

2.2 Spectra and Variability

The spectra of the water megamasers are typically characterized by large numbers of narrow features of individual width one to a few $km\,s^{-1}$ that cover a total velocity extent of several hundred $km\,s^{-1}$; however, the sources NGC 1052 and TXFS 2226-184 are exceptions to this rule, showing broad water emission lines with no discernible substructure. Both the overall water line fluxes and detailed spectra of megamaser sources are observed to vary over time periods of months or less, suggesting by the usual arguments related to light travel times that the emission regions are extremely compact. Thus even in the absence of interferometric observations, extremely high brightness temperatures are inferred for the water line emission, implying that maser amplification is the only plausible emission mechanism.

2.3 Spatial Structure

The small sizes of the maser emission regions, and the correspondingly high brightness temperatures of the emergent radiation have been confirmed by interferometric observations. In 1986, Claussen and Lo (1986) reported that interferometric observations made with the very large array (VLA) placed upper limits of 1 pc and 3.5 pc respectively upon the emission regions responsible for most of the emergent water maser flux in the active galaxies NGC 4258 and NGC 1068. Furthermore, the maser emission region in each case was spatially coincident with the active nucleus, leading Claussen and Lo to suggest that the maser radiation might be generated within a dusty molecular torus of the type that had been proposed by Antonucci and Miller (1985) to explain the polarization properties of Seyfert galaxies.

More recently, observations with the very long baseline array (VLBA) have been used to probe the structures of the water maser emissions from these galaxies at angular resolutions better than one milliarcsecond. The most recent observations of NGC 4258 (Miyoshi et al. 1995) and NGC 1068 (Greenhill et al. 1996; Greenhill 1997) suggest that in each of the two sources the maser emission arises within a geometrically thin and slightly warped disc that we view nearly edge-on, rather than in a geometrically thick torus. At the time of writing, two other megamaser sources have been observed with the VLBA: megamasers in NGC 4945 appear to exhibit (Greenhill et al. 1997) a disc structure similar to that of NGC 1068 and NGC 4258, while the maser emission from NGC 1052 apparently traces a radio jet (Braatz et al. 1996). The results of VLBA observations of six additional megamaser sources are expected within the next year.

3 Physical and Chemical Processes

3.1 Water Chemistry

The strong maser amplification needed to account for the high brightness temperatures observed in megamaser sources implies that water vapour must be abundant within the emitting regions. Chemical models naturally predict that water will be an abundant constituent of molecular gas whenever the gas phase oxygen abundance exceeds the carbon abundance and the gas temperature T exceeds ~ 400 K. Under these conditions, the production of water is driven by the sequence of neutral–neutral reactions

$$O + H_2 \rightarrow OH + H \qquad (3.1)$$

$$OH + H_2 \rightarrow H_2O + H, \qquad (3.2)$$

which are rapid for $T \geq 400$ K but slow at lower temperatures (cf. Chapter 8).

X-ray heating, shock heating, and viscous heating within a differentially rotating disc are all possible heating mechanisms that could raise the gas temperature sufficiently to drive the copious production of water in megamaser sources. Models invoking X-ray heating have been discussed most extensively in the literature that has appeared to date (Neufeld et al. 1994; Neufeld and Maloney 1995) and will be emphasized here. It should be noted, however, that several groups are working actively on alternative models for megamaser emissions from gas that is heated by shocks or by viscous dissipation. Regardless of whether shock heating or viscous heating are also important in particular sources, X-ray heating would seem to be an essential ingredient in the physics because active galaxies are known to be luminous sources of X-radiation.

Figure 1 shows how the gas temperature and water abundance within an X-irradiated cloud are predicted to depend upon the incident X-ray flux. These results apply to the case of gas at pressure $p/k = p_{11} 10^{11}\, \mathrm{K\, cm^{-3}}$ that is exposed through a shielding column density of $N_H = 10^{24} N_{24}$ H nuclei per cm^2 to an X-ray source with an intrinsic (i.e. unattenuated) 1–100 keV flux of $F_X = 10^5 F_5\,\mathrm{erg\, cm^{-2}\, s^{-1}}$. The source was assumed to have a power-law spectrum of index $\alpha = 0.7$ (i.e. with the monochromatic flux at frequency ν obeying $F_\nu \propto \nu^{-\alpha}$). The results presented by Neufeld et al. (1994) are based

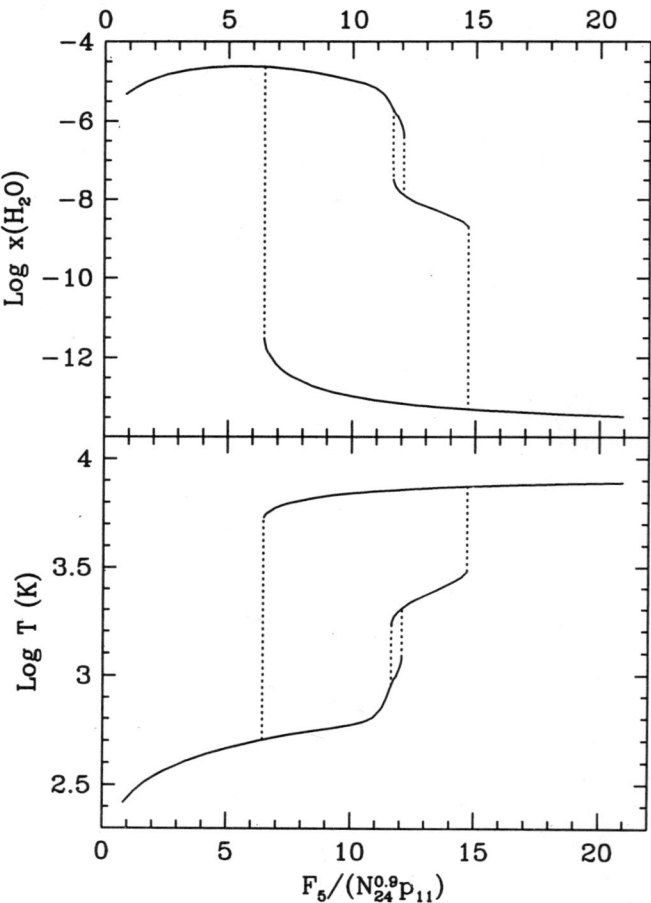

FIG. 1. Gas temperature, T, and water abundance relative to H nuclei, $x(H_2O)$, as a function of 1–100 keV X-ray flux, $F_X = 10^5 F_5 \, \mathrm{erg\, cm^{-2}\, s^{-1}}$, reproduced from the paper Neufeld et al. (1994). Results shown here were obtained for a pressure $p/k = 10^{11} \, \mathrm{K\, cm^{-3}}$ and a shielding column density $N_H = 10^{24} \, \mathrm{cm^{-2}}$. The labelling of the x-axis indicates the approximate scaling with pressure and shielding column density. These results apply in the optically thin limit where the molecular cooling rates are not diminished by radiative trapping.

upon the detailed model of Maloney et al. (1996) for the chemistry and thermal balance within an X-irradiated molecular cloud.

Figure 1 shows that there is a critical range of ionization parameters for which the water abundance is predicted to be high: predicted water abundances in excess of 10^{-5} are expected only when the quantity $F_5/(N_{24}^{0.9} p_{11})$ lies in the range ~ 1 to 10. For larger values of that quantity, the X-ray heating rate exceeds the maximum cooling rate that can be sustained by warm molecular gas; thus, the medium undergoes a transition to a weakly ionized atomic phase in which the temperature is ~8000 K and the water abundance is negligible. If, on the other, the X-ray flux lies below the critical range given above, then the X-ray heating rate is too *small* to raise the gas temperature above ~400 K; unless shock heating or viscous heating now come into play, rapid water production is no longer driven by the neutral–neutral reaction sequence described above, and the gas—although molecular—does not exhibit a high water abundance.

3.2 Maser Excitation

The collisional pumping mechanism that has been widely invoked in models of water maser emission from Galactic star-forming regions (e.g. de Jong 1973; Neufeld and Melnick 1991) and circumstellar outflows (e.g. Cooke and Elitzur 1985) appears to be a plausible mechanism for the excitation of water maser emission in extragalactic water masers as well. The level population inversion that is implied by the presence of maser amplification can arise only if the conditions in the source depart from local thermodynamic equilibrium; in a collisional pumping scheme that departure arises because the temperature of the radiation field is smaller than the kinetic temperature of the gas. In the case of the 22 GHz $6_{16} - 5_{23}$ transition of water (as well as several other higher frequency transitions that have also be observed to mase in Galactic sources), the upper and lower states are populated at roughly the same rate, as a result of inelastic collisions with H_2. However, the lower state has a shorter lifetime for spontaneous radiative decay simply because of the arrangement of the energy states (see Neufeld and Melnick 1991, Appendix A). Therefore, provided that the gas density is not so high that collisional de-excitation completely dominates spontaneous radiative decay, the upper and lower states show a population inversion.

Detailed models of X-irradiated clouds (Neufeld *et al.* 1994) have shown that the collisional pumping mechanism described above can adequately account for the large fluxes observed from extragalactic megamasers, given reasonable assumptions about the source structure. For a wide range of assumed values for the X-ray flux and gas pressure, these models predicted maser luminosities of $10^{2\pm 0.5} L_\odot$ per pc^2 of illuminated surface area, implying that the fluxes observed from typical megamaser sources can be accounted for by X-irradiated regions of area only $\sim 0.1\, pc^2$. Furthermore, these predicted surface luminosities for X-irradiated clouds are probably conservative because the calculation of Neufeld *et al.* (1994) neglected at least three effects that tend to reduce the degree of radiative trapping in non-masing far-infrared water transitions and tend to increase the effective spontaneous decay rates and thereby enhance the population inversions: (1) the absorption of non-masing infrared water line radiation by cold dust (Collison and Watson 1995); (2) bulk velocity gradients; and (3) microturbulent line broadening.

4 Water Megamasers as Probes of Circumnuclear Discs

Over the past four years, observations of water maser emission have demonstrated the enormous usefulness of spectral line interferometry as a probe of the inner circumnuclear regions of active galaxies. Not only do VLBA observations provide angular resolutions of better than one milliarcsecond; they also yield valuable kinetic information about the line-of-sight Doppler velocities of the emitting gas.

4.1 Kinematic Probes of the Gravitational Potential

A renaissance in the study of water megamasers began in 1993 with the discovery by Nakai et al (1993) of high-velocity "satellite" maser features in the Seyfert galaxy NGC 4258. These features, lying at velocity shifts of up to $\pm 1000\,\mathrm{km\,s^{-1}}$ from the systemic velocity of the galaxy, suggested that the emission might be arising in a differentially rotating disc, with the features close to the galaxy's systemic velocity being beamed radially toward the observer, and the satellite features being beamed tangentially.[1]

Spatial information about the low velocity maser features was subsequently obtained (Greenhill et al. 1995b) with the use of the technique of spectral line very long baseline interferometry (VLBI). For the low velocity maser features, the line-of-sight velocities were found to vary linearly with the projected position on the sky, exactly as expected for a rotating disc. This interpretation was further strengthened by a third important observational development: the discovery of a secular redwards shift in all the low velocity maser features (Haschick et al. 1994; Greenhill et al. 1995b). This shift could be understood quantitatively as a centripetal acceleration associated with the circular motion of masing material on the near side of the disc (Watson and Wallin 1994; Haschick et al. 1994; Greenhill et al. 1995a).

More recently, in one of the first observations carried out using the very long baseline array (VLBA), Miyoshi et al. (1995) succeeded in locating the high velocity satellite features. Their velocities and projected positions showed excellent agreement with Kepler's law, yielding an estimate of $3.6 \times 10^7\,M_\odot$ for the mass enclosed within the inner radius of the masing gas. The maser features appear to trace a geometrically thin, and slightly warped, Keplerian disc, which we view nearly edge-on at an inclination angle of $83°$. The inner radius of the masing annulus is 0.13 pc and the outer radius is 0.25 pc. By comparing the observed angular size scale of the disc with the linear size scale implied by the centripetal acceleration of the low velocity maser features, Miyoshi et al. were able to obtain a distance estimate to the source of $6.4 \pm 1\,\mathrm{Mpc}$.

The beautiful observations of NGC 4258 that have been reported in the past four years have demonstrated the extraordinary power of megamaser observations to probe the inner regions of active galaxies on subparsec scales.

---

[1]In the absence of spatial information, however, the disc interpretation was non-unique. Emission from a jet could not be ruled out (Nakai et al. 1993), and it was not even clear that the frequency shift of the satellite features was kinematic in nature (Deguchi 1994).

Such observations have yielded (1) the most compelling evidence to date for the existence of AGN supermassive black holes; (2) the first direct imaging of a circumnuclear disc in an active galaxy; (3) the first indication of a warp in such a disc; (4) the first evidence that such a disc is Keplerian; and (5) an accurate geometrical distance indicator which does not involve the usual chain of arguments needed to calibrate the extragalactic distance scale.

Since the observations of NGC 4258 were completed, further VLBA observations have revealed the existence of similar masing discs in the nuclei of NGC 1068 (Greenhill et al. 1996) and NGC 4945 (Greenhill et al. 1997) and led to mass estimates of $1 \times 10^7 M_\odot$ and $1 \times 10^6 M_\odot$ respectively for the central objects about which the masing gas orbits in those sources. In NGC 1068, the declining rotation curve that was characteristic of the satellite features in NGC 4258 is also clearly evident. However, the decrease in line-of-sight velocity appears to be shallower than Kepler's Law, suggesting perhaps that the disc itself possesses a significant mass relative to that of the central object. In all three cases, the projected positions of the maser features is suggestive of a disc that is warped.

4.2 Constraints on the Physical Conditions in the Discs

If the discs in NGC 4258, NGC 1068, and NGC 4945 are indeed warped, we may expect their surfaces to be illuminated obliquely by X-rays from a central source. Neufeld and Maloney (1995) considered the constraints imposed by the presence of water close to an X-ray source; the very existence of molecules places an upper limit upon the ionization parameter within the disc. Modelling the maser emission region in NGC 4258 as a steady accretion disc with a constant value of the dimensionless viscosity parameter α, they found that the density in the disc midplane declines roughly as r^{-3}, where r is the radial coordinate, whereas the incident X-ray flux declines as r^{-2}. The ionization parameter is therefore a *increasing* function of r, and the conditions become *less* favourable for the existence of molecules with increasing distance r from the disc centre.

Neufeld and Maloney argued that the critical ionization parameter therefore imposes a maximum radius, r_{cr}, beyond which the disc is purely atomic. Inside the critical radius, molecules may exist at least at the disc midplane where the density is highest. The structure of the disc in this picture shown in Figure 2, reproduced from the paper of Néufeld and Maloney (1995). Assuming that the observed outer radius of the masing annulus represents the critical radius beyond which molecules may not exist,[2] they estimated the midplane density in the disc as $n(H_2) = 2 \times 10^7$ cm^{-3} at the assumed critical radius of 0.25 pc. This implied a mass accretion rate of $7 \times 10^{-5} \alpha M_\odot$ yr^{-1}; α is a highly uncertain quantity but is expected in most models of accretion to be less than—or of order—unity.

---

[2]The interpretation of Neufeld and Maloney rests critically upon the assumption that the outer radius of the observed maser emission does indeed represent the critical radius at which the ionization parameter becomes too large to permit the existence of molecules. If the absence of maser emission beyond 0.25 pc has some other explanation, then the disc density and mass accretion rate derived above are lower limits.

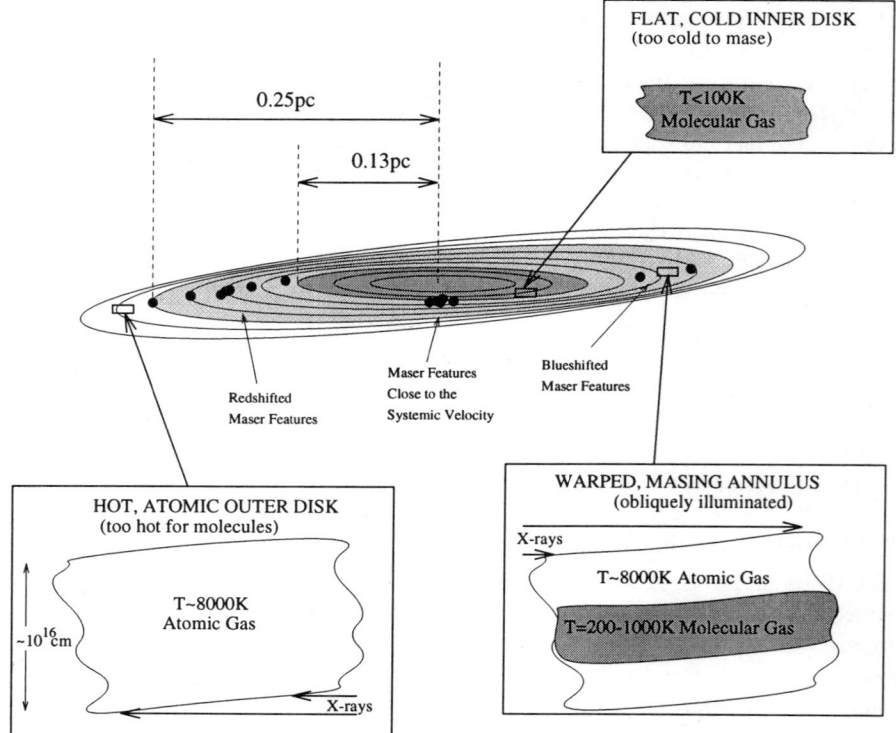

FIG. 2. Model for the warped, masing circumnuclear disc in NGC 4258, from the paper of Neufeld and Maloney (1995).

Neufeld and Maloney speculated that the *inner radius* of the masing annulus represented the point at which the warp flattens out and the disc can no longer be heated by obliquely incident X-rays. Given the assumption that the accretion rate has remained roughly constant over the several million year timescale on which material is transported from radius r_{cr} to the active nucleus, they obtained an estimate of $\sim 10^{-2}\alpha^{-1}$ for the efficiency with which the central engine converts rest mass energy into X-rays.

Bibliography

1. Antonucci, R. J., and Miller, J. S. (1985). *ApJ*, **297**, 621.
2. Braatz, J. A. (1996). PhD thesis, University of Maryland.
3. Braatz, J., Claussen, M., Diamond, P., Wilson, A., and Henkel, C. (1996). *BAAS*, **189**, #89.06
4. Claussen, M. J., and Lo, K.-Y. (1986). *ApJ*, **308**, 592.
5. Collison, A., and Watson, W. D. (1995). *ApJ*, **452**, L103.

6. Cooke, B., and Elitzur, M. (1985). *ApJ*, **295**, 175.
7. Deguchi, S. (1994). *ApJ*, **420**, 551.
8. de Jong, T. (1973). *A&A*, **26**, 297.
9. Greenhill, L. J. (1997). Private communication.
10. Greenhill, L. J., Henkel, C., Becker, R., Wilson, T. L., and Wouterloot, J. G. A. (1995a). *A&A*, **304**, 21.
11. Greenhill, L. J., Jiang, D. R., Moran, J. M., Reid, M. J., Lo, K.-Y., and Claussen M. J. (1995b). *ApJ*, **440**, 619.
12. Greenhill, L. J., Gwinn, G. R., Antonucci, R., and Barvainis, R. (1996). *ApJ*, **472**, L21.
13. Greenhill, L. J., Moran, J. M., and Herrnstein J. (1997). *ApJ*, **481**, L23.
14. Haschick, A. D., Baan, W. A., and Peng, E. W. (1994). *ApJ*, **437**, L35
15. Koekemoer, A. M., Henkel, C., Greenhill, L. J., Dey, A., van Breugel, W., Codella, C., and Antonucci, R. (1995). *Nature*, **378**, 697.
16. Maloney, P. R., Hollenbach, D. J., and Tielens, A. G. G. M. (1996). *ApJ*, **466**, 561.
17. Miyoshi, M., Moran, J., Herrnstein, J., Greenhill, L., Nakai, N., Diamond, P., and Inoue, M. (1995). *Nature* **373**, 127.
18. Moran, J. M. (1997). Private communication.
19. Nakai, N., Inoue, M., and Miyoshi, M. (1993). *Nature*, **361**, 45.
20. Neufeld, D. A., and Maloney, P. R. (1995). *ApJ*, **477**, L17.
21. Neufeld, D. A., and Melnick, G. J. (1991). *ApJ*, **368**, 215.
22. Neufeld, D. A., Maloney, P. R., and Conger, S. (1994). *ApJ*, **436**, L127
23. Watson, W. D., and Wallin, B. K. (1994). *ApJ*, **432**, L35.

24
The Suppression of Dust Formation in Evolved Stars Near Active Galactic Nuclei

T. W. Hartquist and F. Bertoldi
Max-Planck-Institut für extraterrestrische Physik

R. H. Durisen
Department of Astronomy, Indiana University

J. E. Dyson and R. J. R. Williams
Department of Physics and Astronomy, University of Leeds

and

J. M.C. Rawlings and D. A. Williams
Department of Physics and Astronomy, University College London

1 Introduction

As stated in Chapter 17, dust formation probably does not occur in environments in which photodissociation and photoionization processes are sufficiently rapid to maintain most carbon in C^+. Chapter 15 describes the influence of the external interstellar background radiation field on the chemistries of the outflowing envelopes of evolved stars; the typical interstellar background radiation field is too weak to be of importance at the depths in the envelopes at which dust formation occurs. However, near enough to an active galactic nucleus (AGN) the radiation field impinging on the outside of an evolved star's envelope is sufficiently strong to cause carbon to be primarily in the form of C^+ at the positions where dust formation would occur in the absence of an AGN (Hartquist et al. 1995).

In Section 2 we give estimates of the external radiation field strength necessary to result in the carbon being primarily in C^+ where the dust would otherwise form. Section 3 is a short comment on the effect of the suppression of dust formation on the mass loss rate from an evolved star, while Section 4 concerns possible consequences of altered stellar mass loss rates for stellar evolution and the metallicities of AGN line forming regions.

2 The Ionization of Carbon

To calculate the depth in an outflowing envelope to which carbon is ionized, one requires knowledge of the incident external ultraviolet radiation field. The absolute magnitude, M_{1550}, of the continuum at 1550 Å, in the AGN's rest frame is a suitable measure of the ultraviolet radiation field of an AGN. It has been given for a number of flat spectrum quasars by Wampler et al. (1984), who found M_{1550} to lie in the range of -13.5 to -27 for assumed values of the Hubble constant, $H_0 = 50$ km s^{-1}Mpc^{-1}, and the deceleration parameter, $q_0 = 1$, at the current epoch. Comparing with Habing's (1968) estimate for the average Galactic interstellar radiation field at 1000 Å, $\nu U_\nu = 4 \times 10^{-14}$ erg cm^{-3} (where $U_\nu d\nu$ is the energy density of photons having frequencies between ν and $\nu + d\nu$), one finds that the mean intensity of the ultraviolet radiation field of an AGN at a distance R_Q from its centre exceeds the average Galactic background by a factor

$$\chi \equiv \frac{(\nu U_\nu)_{1000}}{4 \times 10^{-14} \text{ erg cm}^{-3}} = 5.9 \times 10^{10} \times 10^{-(10+0.4M_{1550})} \left(\frac{R_Q}{1\text{pc}}\right)^{-2} \quad (2.1)$$

Here we assumed that the AGN spectrum behaves as $U_\nu \propto \nu^{-1}$ in the ultraviolet. Hence, for flat spectrum quasars, χ ranges from about 10^6 to 10^{12} at a distance of 1 pc from the central radiation source. The flux of photons with energies between the carbon and hydrogen ionization potentials at 11.2 and 13.6 eV for such a spectrum is (cf. Draine and Bertoldi 1996)

$$F_i = 1.2 \times 10^7 \chi \text{cm}^{-2}\text{s}^{-1}. \quad (2.2)$$

(This result differs slightly from a similar one given in Chapter 9 because we have assumed a somewhat different approximation to the average Galactic ultraviolet background.) The radius from the centre of the star to which the radiation field of the AGN can ionize carbon is r_+, which in the absence of photoabsorption by H$_2$ is given approximately by the solution of

$$\alpha_{\text{rec}} \left[3 \times 10^{-4} n_H(r_+)\right]^2 \lambda(r_+)/2 \approx F_i \quad (2.3)$$

where $n_H(r_+)$ and $\lambda(r_+)$ are the number density of hydrogen nuclei and the density scale height at r_+, and where an elemental abundance ratio of carbon to hydrogen of 3×10^{-4} has been assumed with carbon the dominant source of electrons; $\alpha_{\text{rec}} \simeq 2.2 \times 10^{-12}(T/10^3\text{K})^{-\frac{1}{2}}$ cm^{3}s^{-1} fits the Nahar and Pradhan (1994) rate coefficient for C$^+$ recombination with electrons in the temperature range 10^2–10^4K. Equation (2.3) is analogous to the standard equation for the Strömgren radius of an H II region (cf. Dyson and Williams 1997) and is based on the assumption that the transition region between predominantly ionized and neutral carbon is sharp compared with the density scaleheight, or equivalently, that all photons that can ionize carbon are absorbed where almost all carbon is in C$^+$.

Interior to the sonic point in the outflow, the scale height is of order 10^{12} cm (cf. Cherchneff, Barker and Tielens 1992), and we find from (2.3) that

$$n_H(r_+) \approx 3.5 \times 10^{12} \text{cm}^{-3} \left(\frac{T}{10^3 \text{K}}\right)^{1/4} \left(\frac{\chi}{10^{11}}\right)^{1/2} \left(\frac{\lambda(r_+)}{10^{12} \text{cm}}\right)^{-1/2} \quad (2.4)$$

According to Cherchneff, Barker and Tielens (1992) grain formation occurs where $n_H \approx 10^{10}$ cm^{-3} in stars that are not exposed to an AGN radiation field (also see Chapter 12). Hence (2.4) implies that grain formation will not occur in the envelope of a star in the vicinity of an AGN if

$$\chi \geq 8 \times 10^5 (T/1000\,\text{K})^{-1/2} \quad (2.5)$$

Equations (2.3), (2.4) and (2.5) do not provide good estimates if H$_2$ absorbs a significant fraction of the radiation that could ionize carbon. We therefore performed detailed radiation transfer calculations with the code of Bertoldi and Draine (1997) to determine the reduction of carbon ionization due to H$_2$ shielding as a function of the H$_2$ column density. We define an effective shielding coefficient, $f_{sh,C}$, through

$$f_{sh,C}(N_{H_2}, T_{H_2}) \int a_{\nu C} d\nu \equiv \int \exp[-\tau_\nu(N_{H_2}, T_{H_2})] a_{\nu C} d\nu \quad (2.6)$$

where the integration is performed over the spectral range 912–1100 Å, $a_{\nu C}$ is the carbon ionization cross section which we adopted from Osterbrock (1989), and τ_ν is the optical depth at frequency ν due to H$_2$ absorption lines. The latter depends on the H$_2$ level population, and the H$_2$ column density towards the AGN. We assumed a thermal rovibrational level population for H$_2$ with a fixed excitation temperature and a Doppler broadening parameter $b = 3$ km s^{-1}. We did not weight the shielding factor with the spectral energy distribution which, given that (1100–912)/1100 is a small number, would cause only a small change. Thus, at a given H$_2$ column density, $10^{22} N_{22}$ cm^{-2}, between the radiation source and the point in the stellar envelope under consideration, the carbon ionization rate is reduced by about $f_{sh,C}$, which we can fit by the expression

$$f_{sh,C}(N_{H_2}, T_{H_2}) \approx \exp(-T_{H_2}^{0.26} N_{22}^{0.43}) \quad (2.7)$$

which is accurate within 20% for 10^{18} cm$^{-2} < N_{H_2} < 10^{22}$ cm^{-2} and 200 K $< T_{H_2} < 2000$ K. For a stellar envelope temperature of more than 10^3 K, the H$_2$ shielding is negligible only when

$$N_{H_2} < 10^{20} \text{ cm}^{-2} \quad (2.8)$$

Hence we now consider conditions under which (2.8) holds between the stellar radius at which $n_H = 10^{10}$ cm^{-3} and the AGN.

We first neglect chemical processes and examine only the question of under what circumstances can the external radiation field destroy H$_2$ that is formed deeper in the star or stellar envelope and is flowing outwards. We assume for simplicity that the thickness of the H_2 dissociation front is smaller than the density scale height, that about 10% of the ultraviolet photons absorbed by H_2 induce dissociation (e.g. van Dishoeck 1987) and that a fraction q (less than but of order unity) of the photons with energies between the ionization potentials of C and of H are absorbed in the dissociation front. These ionization potentials are appropriate limits because the photodissociation threshold energy of H_2 is nearly equal to the ionization potential of C and atomic hydrogen absorbs a large fraction of the photons entering the stellar envelope with energies above its ionization potential. We then find the value, n_D, of n_H in the dissociation front by equating the outward moving flux of H_2 molecules to $0.1q$ times the flux of photons in the energy regime. That is,

$$(n(H_2)/n_H)_{in} n_D v_D = 0.1 q F_i \tag{2.9}$$

$(n(H_2)/n_H)_{in}$ is the ratio of the H_2 and hydrogen nuclei number densities deeper into the envelope than the dissociation front and v_D is the outflow velocity of the envelope at the location of the dissociation front. From (2.2) and (2.9) it follows that for n_D to exceed 10^{10} cm^{-3}

$$\chi \geq 3 \times 10^8 q^{-1} (n(H_2)/n_H)_{in} (v_D/1 \text{km s}^{-1}) \tag{2.10}$$

We now consider whether in the absence of dust H_2 can be produced in the envelope sufficiently rapidly by gas phase chemistry for $N_{22} \geq 0.01$ between the radius at which $n_H = 10^{10}$ cm^{-3} and the AGN. The most rapid means of forming H_2 in the envelope is

$$H + e \rightarrow H^- + h\nu \tag{2.11}$$

$$H^- + H \rightarrow H_2 + e \tag{2.12}$$

In principle, the radiation near 1 μm emitted by the star can remove H^- by

$$H^- + h\nu \rightarrow H + e \tag{2.13}$$

but we will neglect this process in part because the H^- in the stellar photosphere may absorb enough of that radiation to cause (2.13) to be slow. Thus, we will overestimate the rate at which H_2 is formed by (2.11) and (2.12). We take the electron number density to be 3×10^{-4} (corresponding to the electrons arising from carbon being almost completely in the form of C^+), the rate coefficient of (2.11) to be 1×10^{-15} cm^3 s^{-1} ($T/2000$ K) as according to Ruden, Glassgold and Shu (1990), and the photodissociation rate of H_2 to be 5×10^{-17} s$^{-1} \chi N_{22}^{-3/4}$ as according to Bertoldi and Draine (1996). Then, it follows that chemical production of H_2 (in the absence of significant outflow of H_2 from deeper within the envelope) gives $N_{22} \leq 0.01$ for a scale height of 10^{12} cm and $n_H = 10^{10}$ cm^{-3} if

$$\chi \geq 2 \times 10^7 \tag{2.14}$$

(2.5), (2.10) and (2.14) are three conditions that must be met in order for carbon to be kept primarily in the form of C^+ at the radius where $n_H \approx 10^{10}$ cm^{-3}. Condition (2.10) is clearly the most restrictive of the three, and, henceforth, we shall assume that when it is met dust does not form in the stellar outflow except, perhaps, on the dark side of the star where the outflow is shadowed from the direct AGN radiation.

3 The Effect on the Stellar Mass Loss Rate

Cherchneff, Barker and Tielens (1992) have presented an approximate analytical model, based on the considerations and numerical results of Willson and Bowen (1984, 1986) and of Bowen (1988), for the wind flow from an evolved red giant (also see Chapter 12). Stellar pulsations are assumed to give rise to shock formation at a radius r_0. Kinematic and thermal contributions to the pressure then influence the scale height of the gas density distribution at radii greater than r_0 but less than r_p, the radius at which the outward flow becomes supersonic. The mass loss rate is determined by the density at r_p (which in turn depends on how many scale heights r_p is above r_0) and by the sound speed at r_p.

The models of Willson and Bowen (1984, 1986) and of Bowen (1988) on which the Cherchneff, Barker and Tielens (1992) description is based are for low-mass M stars (i.e. stars with oxygen to carbon atmospheric abundance ratios of greater than unity and masses lower than about $4M_\odot$); Cherchneff, Barker and Tielens applied the model to low-mass carbon-rich stars. In the next section we will consider more massive stars, including primarily intermediate-mass stars (those having masses of roughly 4 to 8 $M_\odot$; in normal Galactic environments only stars more massive than about 8 $M_\odot$ evolve to become supernovae; see Chapter 11). While the models in the literature are not for intermediate-mass stars, we will assume that such stars share many important properties with lower mass stars and draw general conclusions about mass loss from the more focused studies reported in the literature. We do so even though we know that the Cherchneff, Barker and Tielens (1992) picture is not definitive and that the mass loss driving mechanisms may differ or at least operate very differently in stars of different type, luminosity, composition and mass; in the absence of a complete theory of stellar mass loss we have no perfect choice.

In the above-mentioned models of mass loss from stars (see Chapters 11, 12 and 13) radiation pressure on dust grains and the coupling of dust grains to the gas result in an outwardly directed force which causes the value of r_p to be less if dust is present than it would if dust were not present. Thus, if dust is present the density at the sonic point is higher than it is if dust is not present. Use of the Cherchneff, Barker and Tielens (1992) formulae for a low-mass star described by parameters given in their Table 1 and in which the value of the mean mass per particle in the envelope is 1.7 amu gives a mass loss rate of 10^{-5} $M_\odot$ y^{-1} when dust grains are assumed to form and a mass loss rate of only about 4×10^{-8} $M_\odot$ y^{-1} when dust grains are assumed to be absent. Thus, the suppression of dust formation may reduce mass loss rates from evolved stars by a couple of orders of magnitude.

The effect of the alteration by H_2 photodissociation of the mean mass per particle in the radius regime between r_0 and r_p on the mass loss rate is less clear. As long as dissociation does not occur down to r_0 (where according to Cherchneff, Barker and Tielens (1992) $n_H \approx 10^{13}$ cm^{-3}) the development and propagation of shocks will not be altered. From the considerations that led to (2.9) and (2.10), we see that $\chi \geq 3 \times 10^{11}$ is required to cause dissociation so far into the envelope; this is almost certainly a substantial underestimate of the value of χ required to affect H_2 so deep in the stellar envelope since three-body formation of H_2, even if (2.11) and (2.12) did not form H_2, is important at $n_H \approx 10^{13}$ cm^{-3} (Hartquist et al. 1995). Above r_0 a good deal of the pressure is kinematic (i.e. associated with the inertia of the material accelerated by the shock) rather than thermal if the Cherchneff, Barker and Tielens (1992) picture is correct, and in that case the mean mass per particle has little effect on the scale height; if so, the dissociation of H_2 is unlikely to affect the mass loss rate of the star through its alteration of the mean mass per particle anywhere at stellar radii greater than r_0.

Thus, we will adopt the assumption that when $\chi \geq 3 \times 10^8$ the mass loss rate of an evolved giant or supergiant star decreases by a couple of orders of magnitude relative to that that occurs when $\chi \ll 3 \times 10^8$.

4 Implications for Stellar Evolution and AGN Line Forming Regions

Low-mass stars evolve on timescales that are too long for them to turn into giants during the lifetimes of AGNs, but intermediate-mass stars would have had sufficient time to evolve in a much greater fraction of AGNs. Although it seems clear that evolved intermediate-mass stars should contribute a significant fraction of the gas returned to the Milky Way's interstellar medium, one of the great puzzles in stellar astronomy is the failure to identify any intermediate mass stars on the asymptotic giant branch (AGB) in the Milky Way or the Magellanic Clouds (Lambert 1992). Theoretical results (cf. Chapter 11) indicate that such stars should be luminous carbon-rich stars, because dredge-up from an He-burning shell should result in a C-rich atmosphere (Iben and Renzini 1983); however, no luminous carbon stars have been identified. Possibly the intermediate-mass stars lose their envelopes due to rapid mass loss at the base of the AGB, or, as they ascend, they process the carbon to nitrogen through development of a hot-bottom convective envelope (HBCE) and so are able to masquerade as O-rich M stars or S stars (Lambert 1992); see also Chapter 11.

Despite the uncertainties in the evolution of intermediate-mass stars in the Galaxy and the Magellanic Clouds, we can identify two possible consequences for their stellar evolution when their mass loss rates were lowered (as we suggest) in regions with $\chi \geq 3 \times 10^8$. When the mass loss rate is lowered, an intermediate mass star can stay in the AGB phase for a longer time.

Firstly, prolonged evolution on the AGB would provide more opportunity for dredge-up of carbon and s-process elements due to He-burning shell flashes. When the envelopes are eventually returned to the interstellar medium there should be more enrichment of carbon and nitrogen than occurs in the Galaxy and, if an HCBE develops, possibly most of the dredged-up carbon will be converted to nitrogen on the AGB. Secondly, if the reduction of the mass loss rate is extreme, enhanced supernova rates could result because then the degenerate carbon/oxygen core of an intermediate-mass star could evolve all the way to carbon ignition before loss of the stellar envelope truncated the growth of the mass of the core. The resulting carbon detonation/deflagration supernovae would probably result in unusual abundance distributions among the heavy elements.

AGN emission line regions tend to fall into two categories (e.g. Dyson, Williams and Perry 1996) named broad emission line regions (BELRs), which have linewidths ≥ 1000 km s^{-1}, and narrow line regions (NLRs). Photoionization models of the regions (e.g. Ferland and Persson 1989; Shields and Ferland 1993) have been developed in an attempt to use the emission features to diagnose the physical properties. Ordinarily, these models are specified by n_H and an ionization parameter often defined as

$$U_i \equiv F(H)/n_H c \qquad (4.1)$$

where $F(H)$ is the total incident flux of Lyman continuum photons and c is the speed of light. For a $U_\nu \propto \nu^{-1}$ spectrum, the relationship between χ, U_i, and n_H is

$$U_i \approx 1 \times 10^{-2}(\chi/6 \times 10^{10})(n_H/10^{10} \text{cm}^{-3})^{-1} \qquad (4.2)$$

The BELR models suggest $n_H \approx 10^{10}$ cm^{-3} and $U_i \approx 0.1$ implying that $\chi \approx 6 \times 10^{11}$, and the NLR models suggest $n_H \approx 10^3$ cm^{-3} and $U_i \approx 0.01$ implying that $\chi \approx 6 \times 10^4$. These values of χ obtain at distances of the order of 0.3 pc and 10^3 pc, respectively, from a powerful AGN (cf. eqn (2.1)).

χ is too low in the NLRs for the AGNs to affect the evolution of stars near them. However, χ is clearly high enough in the BELRs (and at distances of up to roughly thirty to fifty times greater from the AGNs than the BELRs are) to alter the mass loss rates and evolutionary tracks of AGB stars. From the considerations presented in the second paragraph of this section, we see that the suppression of dust formation in AGB stars by the AGN radiation fields may have led to the anomalously high abundance of nitrogen, particularly for high-redshift quasars, suggested by observations of some BELRs (Shields 1976; Osman 1980; Uomata 1984; Hamann and Ferland 1992). Enhanced supernova rates caused by the suppression of dust formation in intermediate-mass AGB stars in BELRs may be necessary for the production of supernova obstacles in an AGN global wind around which the wind–obstacle interactions lead to the formation of BELRs (Dyson, Williams and Perry 1996).

Bibliography

1. Bertoldi, F. and Draine, B. T. (1996). *ApJ*, **458**, 222.
2. Bertoldi, F. and Draine, B. T. (1997). *ApJ*, in preparation.
3. Bowen, G. H. (1988). *ApJ*, **329**, 299.
4. Cherchneff, I., Barker, I. R. and Tielens, A. G. G. M. (1992). *ApJ*, **401**, 269.
5. Draine, B. T. and Bertoldi, F. (1996). *ApJ*, **468**, 269.
6. Dyson, J. E. and Williams, D. A. (1997). *The Physics of the interstellar medium*. Institute of Physics, Bristol.
7. Dyson, J. E., Williams, R. J. R. and Perry, J. J. (1996). *ApSS*, **237**, 187.
8. Ferland, G. J. and Persson, S. E. (1989). *ApJ*, **347**, 656.
9. Habing, H. J. (1968). *Bull Astron Inst Netherlands*, **19**, 421.
10. Hamann, F. and Ferland, G. J. (1992). *ApJ*, **391**, L53.
11. Hartquist, T. W., Durisen, R. H., Dyson, J. E., Rawlings, J. M. C., Williams, D. A. and Williams, R. J. R. (1995). *ApJ*, **453**, 77.
12. Iben, I. and Renzini, A. (1983). *ARA&A*, **21**, 271.
13. Lambert, D. L. (1992) In *Elements and the Cosmos*, eds. M. G. Edmunds and R. J. Terlevich. Cambridge University Press, Cambridge, p.92.
14. Millar, T. J., Rawlings, J. M.C., Bennett, A., Brown, P. D. and Charnley, S. B. (1991). *A&AS*, **87**, 585.
15. Nahar, S. N. and Pradhan, A. K. (1994). *Phys Rev A*, **49**, 1816.
16. Osmer, P. S. (1980). *ApJ*, **237**, 666.
17. Osterbrock, D. (1989). *Astrophysics of gaseous nebulae and active galactic Nuclei*. University Science Books, Mill Valley, California.
18. Ruden, S. P., Glassgold, A. E. and Shu, F. H. (1990). *ApJ*, **361**, 546.
19. Shields, G. A. (1976). *ApJ*, **204**, 330.
20. Shields, G. A. and Ferland, G. J. (1993). *ApJ*, **402**, 425.
21. Uomoto, A. (1984). *ApJ*, **284**, 497.
22. van Dishoeck, E. F. (1987). In *IAU Symposium no. 120—astrochemistry*, eds. M. S. Vardya and S. P. Tarafdar. Reidel, Dordrecht, p.51.
23. Wampler, E. J., Gaskell, C. M., Burke, W. L. and Baldwin, J. A. (1984). *ApJ*, **276**, 403.
24. Willson, L. A. and Bowen, G. W. (1984). In *The relationship between chromospheric-coronal heating and mass loss in stars*, eds. R. Stalio and J. B. Zirken. Osservatorio Astronomico, Trieste, p.127.
25. Willson, L. A. and Bowen, G. W. (1986). In *Lecture notes in physics, 254, cool stars, stellar systems and the sun*, eds. M. Zeilik and D. M. Gibson. Springer, Berlin, p.385.

Index

Ablation of clumps by winds 102, 109–112
Absorption against background quasars due to translucent clouds 88
Absorption line studies
  of diffuse clouds 53, 73, 76, 77, 79, 80, 89
  of translucent clouds 74, 80, 81, 88
Accelerated Lamda Iteration method 321, 322, 328
Accretion discs around YSOs 141–161, 165, 169–171, 173–175
Active galactic nuclei, see AGNs
Adiabatic approximation, see Born–Oppenheimer approximation
AGB stars 265–281, 285–310, 240, 243–256, 260, 313–328, 331–344, 347–354, 357, 360–367, 372, 381, 517–523
AGN optical line forming regions, metallicities of 523
AGN radiation fields and spectra 474–478, 482, 483, 509, 518
AGN winds 523
AGNs xiii, 201, 437–460, 469–485, 497, 500–502, 507–514, 517–523
^{26}Al 174, 332
AlCl in AGB outflows 333
AlF in dust formation in M stars 294
AlO in dust formation in M stars 294, 301
Alfvén surface in a wind 144, 145, 152, 153
Alfvén waves 104, 144, 145, 151, 189, 191, see also wave support of interstellar clumps
Alfvénic Mach number 189, 191
Alpha Orionis 268
Ambipolar diffusion 107–108, 115, 130, 216, see also heating by ambipolar diffusion in YSO winds
Amorphous carbon 253, 267–269, 281, 374, 375–397, 416, see also HAC
Angular momenta coupling in molecules 4, 5

Annealing 303–305, 410
Aquilae 1982 397
Arp 220 437, 444, 445, 455, 456
Arp 299 445, 446, 448
ASCA X-ray observatory 179
Associative detachment 12, 25–28, 55, 56, 423, 426, 429, see also H_2 formation by associative detachment
Associative ionization 12, 25–27, 166
Asymmetric top molecules 9
Asymptotic giant branch stars, see AGB stars
Auger processes 419, 491, 492

B335 126–128
B5 83, 102, 109, 110
Bending modes of molecular vibration, description of 6
Bar structures in galaxies 438, 443, 460
Barnard 5, see B5
BD+303639 205, 208
Be stars 371, 374
Benzene in dust nucleation 273–275, 278, 279
Berkeley Illinois Maryland Interferometer 442
Big Bang, see Early Universe
Bistability of interstellar cloud chemistry 72, 73
BL Lac objects 440, 441, 469
Black dwarfs 238
Black holes xiii, 238, 457, 435, 440, 498, 499, 513
B0.5V star's far ultraviolet radiation field 206
Born–Oppenheimer approximation 1–3
Born–Meyer potential 295, 296
Boundary layers between clumps and stellar winds 86, 118–125, 162, 186, 230, 231
BR 1202-0725 452, 453
Brightness temperature, definition of 314
Brown dwarfs 244
Burnham's Nebula 223

CI emission from TMC-1 112
C in interstellar clouds 69–72, 111, 112
C star
    definition 241, 242, 266, 286
    evolution from M star 249
    evolution to S star 249
C^- formation by radiative attachment in supernovae 423
C^+ in interstellar clouds 69–71
C-type shock 191–193, 229
C_2 as a density diagnostic in diffuse and translucent clouds 89
C_2 formation
    by associative detachment 427
    by radiative association 24, 32, 423
C_2 in AGB stars 242, 339, 342
C_2 in basic interstellar gas phase schemes 65
C_2 in diffuse clouds 78, 89
C_2 in novae 406, 411
C_2 in PNe and PPNe 356
C_2 in R Cr B stars 375, 394
C_2 in supernovae 423
C_2 in translucent clouds 80
C_2 rotational populations in diffuse and translucent clouds 73, 77
C_2^+ formation by radiative association 32
C_2H abundance depth dependence in an interstellar cloud 69
C_2H in AGB stars 272, 273, 334, 340, 342
C_2H in basic interstellar gas phase schemes 62
C_2H in boundary layers 123, 124
C_2H in dense cores 113, 115, 116, 130
C_2H in PNe and PPNe 365
C_2H in translucent interstellar clouds 71, 80–82
C_2H_2 in AGB stars 242, 270, 272–275, 280, 333–337, 339, 340, 356, 404
C_2H_2 in dark clouds 71
C_2S basic structure and energy levels 6
C_2S in AGB stars 333, 342
C_2S in dark clouds and dense cores 84, 112, 127, 128
C_2H diffuse clouds 78
C_3H_2 in translucent clouds 81, 82
C_3H_2 in dense cores 127, 128
C_9 380
C_{60}, see fullerenes
Carbon chain linear molecules 380, 381

Carbon chain molecules, large interstellar 86, 87
Carbon chemistry
    in PDRs 210–212, 216
    in shocks 177, 180, 185, 188
    in the interstellar medium, basic gas phase 61, 62
Carbon flash 239
Carbon monocyclic rings 380, 381
Carbon nuclear burning 238, 239, 240, 243, 244, 523
Carbon star, definition, see C star, definition
Carbonaceous grains 87, see also amorphorous carbon, HAC, graphite grains, diamonds and PAHs
Cataclysmic variables 395
Centaurus A 437, 451, 453, 454
Central helium flash, see helium flashes
Cepheus A 84
Cepheus A West 222
CH abundance depth dependance in an interstellar cloud 68, 69
CH basic structure and energy levels 4, 5
CH formation by radiative association 24
CH in basic interstellar gas phase schemes 62, 64, 65
CH in dense cores 129, 130
CH in diffuse interstellar clouds 53, 78, 86, 87
CH in GMC translucent clumps 106
CH in PDRs 212, 213
CH in PNe and PPNe 362
CH in shocks 179, 180
CH in translucent cloud 76, 80, 82
CH in XDRs 493
CH in YSO winds 163
CH masers in AGNs 456
CH photodissociation threshold 57
CH^+ in basic interstellar gas phase schemes 61, 62
CH^+ in diffuse interstellar clouds 53, 79, 85, 86, 90
CH^+ in PDRs 210–212
CH^+ in PNe and PPNe 359, 362
CH^+ in shocks 85, 187
CH^+ in XDRs 451
CH_4 in AGB stars 333, 340, 342, 343
CH_4 in dark interstellar clouds 71
CH_2CO in basic interstellar gas phase schemes 62

CH_3CN in basic interstellar gas phase schemes 63
CH_3OH as a product of surface reactions 66
CH_3OH in AGB stars 339, 342
CH_3OH masers in AGNs 457
CH_3OH in hot cores 133
CH_4 as a product of surface reactions 66
Charge transfer rate coefficients 419, 420, 424, 427–429
Chinese astronomers and emperors 415
Chlorine chemistry in the interstellar medium, basic gas phase 65
Chondritic meteoretic material 174
Circinus galaxy 453, 455, 470
Cirrus 79, 82, 83
Clemens–Barvairis objects 82, 83
Close coupling approximation 26
Clumpy structure of clouds 71, 77, 79, 101–112, 201, 213
Clusters of stars 242
CN abundance depth dependence in an interstellar cloud 68, 69
CN as a diagnostic of electron density in diffuse and translucent clouds 85
CN formation by radiative association 24
CN in AGB stars 242, 334, 336–338, 340, 341
CN in basic interstellar gas phase schemes 64
CN in dense cores 130
CN in diffuse interstellar clouds 53, 78
CN in novae 400, 406, 411
CN in PDRs 214–216
CN in PNe and PPNe 356, 365
CN in R Cr B stars 375, 394
CN in translucent clouds 71, 80–82
CN in XDRs 493
CN photodissociation threshold 56, 57
CO abundance depth dependence in an interstellar cloud 69–71
CO antenna temperature to H_2 column density ratio 75, 76, 441, 445, 449, 450
CO basic structure and energy levels 4, 473
CO cooling
 in PDRs 210
 in shocked gas 183–185
 in supernova ejecta 420, 421

CO emission
 from dark clouds and dense cores 74, 125–132
 maps of dense cores 109, 116
 maps of GMCs 102, 103, 108
 studies of translucent clouds 74, 79, 80, 90, see also CO emission maps of GMCs
CO emission variability in Miras 269, 285
CO formation
 by associative detachment 31, 423
 by radiative association 13, 20, 22, 24, 31, 403, 424
CO in AGB stars 241, 242, 269, 270, 272, 273, 281, 285, 286, 333, 338, 339, 360, 361
CO in AGN tori 470, 474, 475, 477–481, 484
CO in basic interstellar gas phase schemes 62, 63
CO in diffuse clouds 73, 77–79, 90
CO in novae 399–401, 403–405, 407, 408, 411
CO in PDRs 213
CO in PNe and PPNe 351, 352, 356, 357, 360–364
CO in QSOs 450–453
CO in radio galaxies 450, 451, 453
CO in Seyfert galaxies 448–450, 453–455, 500, 501
CO in shocks 185, 186, 188
CO in starburst galaxies 441–443, 447–450
CO in supernovae 24, 30–32, 416–425, 427, 428, 430–433
CO in the Galactic Centre 499
CO in translucent clouds 79, 81, see also CO emission studies of translucent clouds
CO in ultraluminous infrared galaxies 443–448, 451, 452, 454, 455
CO in YSO winds 163
CO millimetre line and infrared continuum emission correlation in starburst galaxies 441, 460
CO in XDRs 493, 497
CO photodissociation, self-shielding and shielding by H_2 63, 70, 79, 81, 333, 351, 352, 363, 366, 401, 403
CO photodissociation threshold 56, 57

CO to H_2 column density ratio in GMC clumps 104
CO^+ formation by radiative association 24, 25, 31, 423
CO^+ in PDRs 211, 212, 216
CO^+ in supernovae 30, 31
CO^+, important role in YSO wind chemistry 167
^{56}Co in supernovae 419
^{57}Co in supernovae 419
Cold dark matter in cosmology 43–45
Collapse of dense cores to form stars 125–132
Collimation of young stellar object outflows in X-wind model 156–160, 164, 171
Collisional excitation and deexcitation of H_2, see H_2 collisional...
Collision rates, see reaction and collision rates
Collisionally induced dissociation of H_2 57, 164, 165, 185, 186, 188, 193, 222, 227, 228, 276, 277, 403
Collisional pumping of masers, see maser pumping mechanisms
Cometary tails 364
Compact symmetric objects (CSOs) 441, 456
Complex potentials 27, 28
Continuum radio emission from starburst galaxies 438, 441, 445
Convection in high-mass stars 259, 260
Convection in stars, see dredge-up in stars
Cooling by neutrinos, see neutrino cooling
Cooling flows in galaxy clusters 502
Cooling of PDRs 210
Cooling of shocked gas 183, 184, 186, 188, 187, 192
Cooling of white dwarfs, see neutrino cooling
Cooling of XDRs 496
Copernicus satellite 77
Cores in interstellar clouds, see dense cores and hot cores
Cosmic ray exclusion by YSO winds 174
Cosmic microwave background 38, 41, 46, 47, 477, see also radiative association, stimulated
Cosmic ray acceleration in shocks 373
Cosmic ray acceleration near YSOs 162

Cosmic ray induced ionization 57, 60, 62, 67, 78, 89, 90, 106, 130, 215–216, 492
  in YSO winds 167, 175, 177
Cosmic ray induced production of radioactive nuclei 174
Cosmic ray induced ultraviolet radiation field 57, 106, 112, 333
Cosmic ray induced ionization in AGB stars 333
Cosmology, see early Universe
Coupling of angular momenta in molecules, see angular momentum coupling in molecules
CRL 2688, see Egg Nebula
CRL 618 355–357, 361, 365, 367
CS and bistability 73
CS formation
  by associative detachment 429
  by radiative association 22, 32, 428
CS in AGB stars 338, 341, 352
CS in basic interstellar gas phase schemes
CS in collapsing dense cores 112, 125–128, 130
CS in dark clouds 74
CS in diffuse clouds 78
CS in PDRs 214
CS in Seyfert galaxies 448
CS in starburst galaxies 443, 447, 448
CS in supernovae 32, 428–431
CS in translucent clouds 80–82
CS^+ formation by radiative association 25, 32, 424
Cyanapolyynes in basic interstellar gas phase schemes 64
Cyanapolyynes in AGB stars 333–338, 356, 361
Cyanapolyynes in interstellar clouds 84, 112, see also HC_3N
Cyanapolyynes in PNe and PPNe 356, 357, 361, 367
Cygnus A 446, 475–478, 480, 481

Dalgarno, A. vii–ix, 12, 38, 55, 175, 217, 415, 502
Dark cores, see dense cores
Dark interstellar clouds, see interstellar clouds
Degenerate matter 238–240, 243, 244
Dense cores 80, 90, 102, 108–133, 161, 422, see also hot cores
Dense cores, line profiles of emission from collapsing 125–132

Depletion on to grains 68, 84, 105, 106, 114–118, 128–131
Depth dependent models of interstellar chemistry 67, 69–72, 128–130
Desorption from grains 68, 106, 117
Deuterated species, see fractionation
Deuterated species in hot cores 133
Deuterium, see HD
Deuterium fractionation, see fractionation
Deuterium to hydrogen isotopic ratio 50, 61, 74
Diamonds 267, 268
Diatomic molecules, structures and energy levels 3–6
Diffuse interstellar bands 53
Diffuse interstellar clouds, see interstellar clouds
Discs in AGNs, see tori in AGNs
Dissipation in shocks 181, 183, 191–193
Dissociation induced by collisions, see collisionally induced dissociation of H_2
Dissociative recombination as a basic process and in gas phase schemes 26, 57, 62, 63, 78, 86
DQ Her 393, 400
DR 21 89
Dredge-up in stars 240, 241, 243, 244, 248–250, 252, 254, 260, 286, 316, 522, 523
Dust, see grains and visual extinction
Dust destruction and sputtering 197, 367, 408
Dust formation
  in AGB stars xii, xiii, 242, 249, 252–254, 256, 265–281, 285–310, 317, 347, 353, 355, 517–523
  in novae 393–412
  in supernovae 417
  near hot stars 371–384
Dust-free environments x, 11–32, 37–51
Dust in AGNs, see tori in AGNs
Dust infrared spectral features 267, 286, 305, 374, 375, 381, 393, 397, 409, 410
Dust opacity in PDRs 202, 203, 215
Dust-to-gas mass ratio in stellar envelopes 280
DY Cen 376

"Early-time" chemistry, see time-dependent models
Early Universe 11, 18, 25, 28–30, 37–51, 165
Eddington limit 396
Effelsberg 100m telescope 132
Egg Nebula 355–357, 367
Einstein X-ray observatory 172, 173
Electron energy deposition in supernova ejecta 420, 421, 424
Electronic states of molecules 2, 3
Elemental abundances, solar and cosmic values of 59, 77, 287
Enstatite 301, 302, 307
ERE (extended red emission) 410
Escape probability in large velocity gradient approximation 321, 323
ESO 97-G13, see Circinus galaxy
ESO New Technology Telescope 443
Evolution of stars, see stellar evolution

Fayalite in dust formation in M stars 301–303, 305
Fe clusters in dust formation in M stars 299–302
Fe in dust formation in WR outflows 374
FeC in dust formation in WR outflows 394
FeO in dust formation in M stars 308–310
FeS in dust formation in M stars 307–309
Fine structure excitation and cooling in supernova ejecta 420
Fine structure excitation in interstellar clouds 73, 77, 89
Fine structure excitation
  cooling and emission in PDRs 210, 213
  cooling and emission in shocks 184, 188
  cooling and emission in starburst galaxies 448
Forsterite in dust formation in M stars 301–302
Fractionation 61, 65, 89, 90
Freeze-out, see depletion
Fullerenes 87, 375, 381
FUSE 91

G10.6 − 0.4 131
G34.3 + 0.2 132
G45.47 + 0.05 132
Galactic Centre 208, 442, 497, 499

Galaxy clusters 502
Galaxy formation xi, 41–51
Gamma rays in supernovae 419
GGD37 222
Globular clusters 42, 43, 47–49
GMCs (giant molecular clouds) 79, 101–108, 447, 497
Grain inertia in shocks 194
Grain surface reactions and chemistry 53, 55, 56, 60, 66–68, 84, 87, 88, 90, 102, 105, 129, 133, 197, 265
Grain-neutral friction in shocks 193, 195–198
Grains, see dust and H_2 formation on grains
Grains, charge carried by in shocked gas 175, 196
Grains, charge on in PDRs 209
Graphite grains 267, 268, 397
Gravitational lenses 452, 453
Gravitational stability, see Jeans mass
Gruenerite in dust formation in M stars 301
Gunn–Peterson test 42, 48, 49

H^-, see radiative attachment and H_2 formation by associative detachment
HI absorption in AGNs 455, 456, 475, 481
H_2 basic structure and energy levels of 3, 4, 471–473
H_2 collisional deexcitation and excitation of vibrational levels 206–208, 226, 228
H_2 collisional deexcitation as a heating mechanism 209, 210
H_2 collisionally induced dissociation of, see collisionally induced dissociation of H_2
H_2 cooling
  in PNe and PPNe 366
  in shocks 183, 184, 195
  in the early Universe 43–45
H_2 electron impact excitation of 226, 227
H_2 excitation by energetic electrons 495
H_2 formation
  by associative detachment 25, 27, 29, 40, 48, 59, 162–165, 187, 358, 402, 472, 520
  on grains 60, 189, 202, 222, 224, 358, 363
  in shocks 187, 189
  in YSOs 162–165, 167
H_2 in AGB stars 333, 360, 519, 520
H_2 in AGN tori 470, 472, 480–485, 501, 502
H_2 in cooling flows in clusters of galaxies 50
H_2 in novae 400–403, 405, 407, 408
H_2 in PNe and PPNe 351, 352, 356, 358, 360–366
H_2 in Seyfert galaxies 494–501
H_2 in supernova remnants 499, 500
H_2 in supernovae 32
H_2 in the early Universe 25, 29, 38–41, 43–45, 48–50
H_2 in XDRs 493–501
H_2 infrared and ultraviolet emission from Herbig–Haro objects 221–231, 499, 500
H_2 infrared emission
  from AGNs 482, 500, 501
  from PDRs 201, 204–209, see also H_2 infrared and ultraviolet emission from Herbig–Haro objects
  from Seyfert galaxies 454
  from shocks 125, 499, 500, see also H_2 infrared and ultraviolet emission from Herbig–Haro objects and H_2 cooling in shocks
  from starburst galaxies 500
  from XDRs 484, 494, 498
  general considerations 470–474, 480
H_2 intervening absorption towards QSOs 48 482
H_2 Lyman bands and Werner bands, see Lyman bands and Werner bands
H_2 photodissociation and self-shielding 60, 61, 202, 203, 227, 333, 351, 352, 363, 366, 401, 403, 519, 520, see also PDRs and XDRs
H_2 photodissociation rate from typical ISM radiating field 202
H_2 photodissociation threshold 56, 57
H_2 pumping by formation in XDRs 495, 496, 500
H_2 quadrupole transition rates 203, 204
H_2 rotational level populations in diffuse interstellar clouds 73, 77, 86, 91

H_2 shielding of C against photoionization 519, 520
H_2 ultraviolet emission from PDRs 203, 204
H_2 ultraviolet pumping and fluorescence 73, 78, 203, 222–226, 228, 230, 471, 472, see also H_2 photodissocation and self-shielding
H_2 vibrational excitation diagnostics of diffuse interstellar clouds 77
H_2^+ formation
 by associative ionization 26, 27, 29, 31, 38–40, 48, 166, 402, 403
 by radiative association 17, 20, 29, 59, 162–165, 167, 358, 402, 405
H_2^+ important role of in YSO wind chemistry 167
H_2CN in AGB stars 242, 339, 356
H_2CO as product of grain surface chemistry 66
H_2CO basic structure and energy levels of 9
H_2CO in AGB stars 338, 339
H_2CO in basic interstellar gas phase schemes 62
H_2CO in dark clouds and dense cores 90, 126–128, 130, 133
H_2CO in translucent interstellar clouds 81, 83, 88, 90
H_2CO masers 456
H_2CS in AGB stars 339
H_2CS in translucent clouds 82
H_2O as product of grain surface chemistry 66
H_2O basic structure and energy levels of 7, 9
H_2O cooling
 in PDRs 210
 in shocked gas 183, 184, 195
H_2O in AGB stars xii, xiii, 241, 338, 339, 341
H_2O in AGN tori 456, 459, 460, 470, 478–480, 483, 485, 507–514
H_2O in basic interstellar gas phase schemes 63
H_2O in dark clouds and dense cores 88, 129, 130
H_2O in dense core–stellar wind interfaces 121, 122
H_2O in diffuse interstellar clouds 75
H_2O in PDRs 211, 216

H_2O in shocks 179, 180, 183, 184, 187–189, 195
H_2O in XDRs 497
H_2O maser pumping in AGN tori 511
H_2O masers xii, xiii, 181, 285, 313–323, 325–328, 456, 459, 460, 470, 478, 485, 507–514
H_2S in AGB stars 338, 341, 342
H_2S in basic interstellar gas phase schemes 65
H_2S in dark clouds 90
H_2S in hot cores 133
H_2S in translucent interstellar clouds 81–83, 88, 90
H_3^+ detection in infrared absorption 60
H_3^+ emission from XDRs 500
H_3^+ dissociative recombination 57, 89, 90
H_3^+ in supernovae 31
H_3^+ key role in interstellar chemistry 60, 62–64, 72, 89, 90, 167, 492, 494
H_3O^+ in dark clouds 89, 90
HAC (hydrogenated amorphous carbon), 410
Hat Creek Interferometer 132
Hayashi limit 240, 241, 260
HC_3N in AGB stars, see cyanapolyynes in AGB stars
HC_3N in dark clouds and dense cores 74, 79, 84, 112–118
HC_3N in dense core–stellar wind interfaces 123, 124
HCl in AGB stars 343
HCl in basic interstellar gas phase schemes 65
HCl in diffuse interstellar clouds 79
HCN abundance depth dependence in interstellar clouds 65
HCN basic structure and energy levels of 6
HCN in AGB stars 272, 333, 337, 338, 340, 341, 356
HCN in AGN tori 478–480
HCN in basic interstellar gas phase schemes 62
HCN in dark clouds and dense cores 71, 90, 130, 133
HCN in PDRs 214–216
HCN in PNe and PPNe 357, 365
HCN in Seyfert galaxies 448, 453, 454, 501
HCN in starburst galaxies 442, 447, 448
HCN in translucent interstellar clouds 81, 82, 86

HCN in ultraluminous infrared galaxies 444–447
HCO in dense cores 130
HCO^+ as a tracer of H_2 75
HCO^+ basic structure and energy levels of 6
HCO^+ in AGB stars 333, 338
HCO^+ in basic interstellar gas phase schemes 63
HCO^+ in collapsing gas in regions of high mass star formation 132
HCO^+ in dark clouds and dense cores 74, 89, 90, 130, 132
HCO^+ in diffuse interstellar clouds 78–80
HCO^+ in PDRs 211, 212, 216
HCO^+ in PNe and PPNe 357, 364, 365
HCO^+ in Seyfert galaxies 448
HCO^+ in starburst galaxies 442, 448
HCO^+ in supernova remnants 499
HCO^+ in translucent interstellar clouds 81–83, 86
HCO^+ in ultraluminous infrared galaxies 444
HCO^+ in XDRs 499
HCS^+ in translucent clouds 82
HD 161 796 256
HD 56126 347, 348, 356
HD formation by radiative association 17, 30
HD in interstellar clouds 60, 61, 79
HD in the early Universe 30, 38, 41, 44
$^3He^+$ absorption in AGNs 481
He^+ importance for the removal of CO in dark interstellar clouds 62
He^+ two photon continuum in shocks 223, 224
He_2^+ formation by radiative association 17, 29, 38
Heating by ambipolar diffusion in YSO winds 165, 169, 170
Heating by H_2 formation 165, 189
Heating of PDRs 209, 216
HeH^+ formation by radiative association in early Universe 29, 31, 38, 39
HeH^+ in PNe and PPNe 358, 359, 365
HeH^+ in supernovae 32
Helium burning 237, 239, 240, 242, 243, 245–254, 258, 286, 522, 523
Helium core in evolved stars 237, 238, 240

Helium flashes 239, 240, 245–247, 250–254, 258, 522, 523
Helix Nebula 364, 365
Herbig–Haro objects 160, 161, 221–231, 497, 499, 500
Herzsprung–Russell diagram 47, 240, 241 245, 246, 258, 260, 348, 350
HH 1 223, 499, 500
HH 1/2 222, 223
HH 2 223–225, 228, 229
HH 2A' 223, 229
HH 2H 223, 228, 229
HH 7 223, 230
HH 7–11 222
HH 32 222
HH 40 222
HH 43 223, 230
HH 47A 223–225
HH 90 222
HH 110 222, 231
HH 111 222
HH 111/121 222
HH 417 225, 226
HNO in dense cores 130
Hopkins Ultraviolet Telescope 223–225
Hot bottom burning 245, 249, 522
Hot cores 102, 133
HS in dense cores 130
Hubble 12 205, 209
Hubble Space Telescope 75, 77, 79, 90, 91, 222, 223, 394, 470
Hydrocarbon chain radicals in AGB stars 3: 336–338
Hydrocarbon formation in novae 406, 407
Hydrocarbons, large ones in PDRs and XDRs 213, 493
Hydrocarbons, production of complex ones in basic interstellar gas phase schemes 62
Hydrodynamics, basic equations of one-dimensional 180, 181
Hyperfine splitting 4

IC 63 204
IC 342 213
IC 443 499, 500
IC 694 445, 446, 448
Ices 68, 74–76, 84, 88, 90, 117, 133, 293
IK Tau 339, 341, 342
Infrared absorption in dark clouds 74
Infrared bright active galaxies, see ultraluminous infrared galaxies

Infrared emission, see H_2 infrared emission
   from novae 393, 397, 399, 400, 403, 409–411
   from PDRs 201
   from starburst galaxies 437, 438, 441, 442
   from Wolf–Rayets 372–375, 381, 384
   from XDRs 494–501
Infrared observations of molecules in SN1987A 30, 31
Infrared Space Observatory (ISO) 74–76, 91, 194, 222, 343, 394, 437, 440, 451, 470
Infrared spectra of R Cr B stars 375, 376
Infrared spectra of supernovae 416, 417, 418, 419, 431
Initial mass function of stars 444
Initial–final mass relation in stellar evolution 242, 243
Inside-out collapse 126
Instabilities in boundary layers 230, 231
Instabilities in shocked gas 182, 194, 195, 197, 221, 229
Intercore gas (in a cluster of dense cores) 110, 111
Interferometric observations of dense core collapse 126–128, 131, 132
Intergalactic medium 42, 44, 46, 49, 50
Intermediate-mass stars
   definition 243
   evolution 245–258, 522, 523
   lifetimes relative to Galaxy's age 286
International Ultraviolet Explorer (IUE) 223, 225
Interstellar clouds 53–91, 101–133
   definition of dark 55
   definition of diffuse 55
   definition of translucent 55
   diagnosis and modelling of dark 75, 76, 83, 84, 88, 91, see also dense cores
   diagnosis and models of diffuse 73, 76–79, 85–87, 89
   diagnosis and models of translucent 74, 79–83, 88, 89, 104–108, 201
   ionization induced in by YSO X-rays 173, 174
Interstellar molecules, list of detected 54
Interstellar ultraviolet background 202, 206, 518

Inverse predissociation, see radiative association, indirect mechanism
Inversion properties and transitions of symmetric top molecules 7–9
Ionization balance in interstellar clouds 66, 69–71, 73, 89, 90, 101, 104–108, 173, 174, 193
Ionization fronts 203
Ionization in supernova ejecta 419, 420, 422
Ionization structure
   in C-type shocks 193, 196
   in XDRs 491, 492
   in YSO accretion discs 173, 174
   in YSO winds 163–165, 168–170
   of nova ejecta 398, 399
   of PDRs 212–216
   of Wolf–Rayet and R Cr B winds 378, 379
IRAM 30m telescope 444, 451
IRAM interferometer 452
IRAS 255, 267, 373, 375
IRAS 100 micron emission observations 73, 74
IRAS 10214+4724 451
IRAS 17436+5003 256
IRAS 20100−4156 458
IRAS two-colour relation for dusty AGB stars 252, 253
IRC+10216 266–268, 277, 278, 333, 336, 337, 347, 355–357, 394
Isothermal sphere, collapse of 126, 128, 129
Isotopic fractonation, see fractionation

J-type shocks, definition 191–192, 228
Jeans mass 42, 43
Jets in AGNs 440, 456
Jets of YSOs 156–160, 161, 171, 179, 221, 231, 499
Jets, origin in X-wind model 156–160
Jump conditions for hydrodynamic shocks 182
Jump conditions for perpendicular MHD shocks 150

KCl in AGB stars 333
Keck telescopes 50, 440
Kleinmann–Low nebula 133

L 134N 75, 84
L 483 126–128
L 1498 125, 127, 128

L 1527 126–128
L 1544 127
Lamda doublet transitions in OH 314
Lamda doublets 4
Large velocity gradient (LVG) approximation 321–323, 328
LiH formation by associative detachment 30
LiH formation by radiative association 17, 20, 30, 41
LiH in the early Universe 44, 47
LiH^+ formation by radiative association 18, 19, 30
$LiHe^+$ formation by radiative association 29
Linear polyatomic molecules 6
LINERs 439, 501, 507
Long-period variable stars 249
Low-mass stars
    definition 243
    evolution 245–258
Lyman alpha clouds and forest 44, 47–50, 481
Lyman bands of H_2 202–204, 227, 401, 471

M17 84, 211, 213
M star to C star conversion 249
M stars, definition of 241, 285, 286
M33 459
M81 442
M82 213, 442–444, 455, 458, 460
Mach number, definition 182
Magellanic cloud AGB stars 249
Magellanic clouds 459
Magnetic fields
    in shocked gas 187–197
    in star formation 101, 103, 107, 108, 125, 130
    in winds, bipolar flows, and jets of YSOs 141–160
Magnetic pressure in boundary layers 121
Magnetohydrodynamics (MHD), basic equations of 148, 149
Magnetosonic waves 189, 191, 229, 230
Main sequence stars 141, 237, 240, 265
Maser pumping mechanisms 319–322, 511
Masers xii, xiii, 132, 184, 285, 326, 313–328, 443, 444, 456–460, 470, 478, 485, 507–514
Mass loss, see winds and jets
Massive stars, evolution of 244, 259, 260

MBM12 70
MBM16 75
Merged-beam apparatus 26
Mergers and tidal interactions of galaxies 438, 442–448, 450, 459, 500
MERLIN 314, 316, 328, 459
Metals in interstellar clouds and their effects on ionization structure 6, 89, 90, 105, 107, 108, 113, 129
Metasilicates 301, 302, 307
Meteorites 174, 280, 281
Mg and dust condensation 300
MgO clusters and dust condensation 295–298, 302
Microwave background, see cosmic microwave background
Milky Way, mass of molecular gas in 443
Millimetre absorption against radio sources in the study of translucent clouds 80–82
Miras 249, 269, 313–328, 339
Mixing in supernova ejecta xiii, 424
Mk3 460
Mk231 447, 454, 455
Mk876 450
Mk1014 450
Multifluid models of shocks 190–198, 229, 230
MV Sgr 376, 377

N_2 formation by radiative association 24
N_2 in AGB stars 333
N_2 in basic interstellar gas phase schemes 64
N_2 in novae 408, 409
N_2^+ formation by radiative association 32
N_2H^+ in dark clouds and dense cores 90, 130
N_2H^+ in translucent clouds 82, 83
NaCl in AGB stars 333
Neutrino cooling 238, 239, 243, 244, 246, 257
Neutron stars 238, 489, 498, 499
New Technology Telescope, see ESO New Technology Telescope
NGC253 213, 442, 443, 455, 457, 459, 460
NGC1052 508
NGC1068 437, 439, 440, 448, 453–455, 460, 482, 483, 501, 508, 509, 513
NGC1275, see Perseus A
NGC1808 500

NGC2023 205
NGC3077 442
NGC3256 500
NGC3690 445, 446
NGC3783 501
NGC4151 455, 460
NGC4258 459, 470, 483, 508, 509, 512, 513
NGC4945 478, 455, 509, 513
NGC5793 455
NGC6240 500
NGC7023 205
NGC7027 211, 357, 363, 365
NGC7293, see Helix nebula
NGC7469 438, 483, 484
NGC7469 501
NH in basic interstellar gas phase schemes 64
NH in diffuse clouds as produce of surface reactions 87, 90
NH_3 in AGB stars 333, 356
NH_3 in basic interstellar gas phase schemes 64
NH_3 in dark clouds and dense cores 74, 84, 90, 112–117, 125, 128–130
NH_3 in hot cores 131, 132, 133
NH_3 in PDRs 216
NH_3 in PNe and PPNe 356, 357, 362, 367
NH_3 in starburst galaxies 447
NH_3 in translucent clouds 83, 88, 90
NH_3 in ultraluminous infrared galaxies 447
NH_3 produced in surface reactions 66
NH_3 structure and spectrum 7, 8
Nitrogen chemistry
    in interstellar clouds 63, 64
    in PDRs 210, 211, 216
NML Cyg 315
NML Tau 339
NO formation by radiative association 24
NO in basic interstellar gas phase schemes 64
Nobeyama Millimeter Array 127
Novae xiii, 11, 371, 376, 393–412
NQ Vul 399, 400
NRAO 12m telescope 444
Nucleation theory of dust formation 270, 271, 288–294, 404
Nucleosynthesis in the Big Bang 38, 41

O as dominant gas phase oxygen bearing species 88, 89
O VI stars 258
O, importance of for the removal of species in dark clouds 71
O^- formation by radiative attachment 428
O_2 formation by associative detachment 31, 423
O_2 formation by radiative association 22–24
O_2 in AGB stars 341
O_2 in dark clouds 88, 90
O_2 in interstellar gas phase schemes 63
O_2 in PDRs 211
O_2 in supernova ejecta 31, 32
O_2 formation by radiative association 25, 423, 424
OB stars 206, 208
OCS in AGB stars 338, 342
OCS in translucent clouds 82
OH 231.8 + 4.2 341, 342
OH absorption in active galaxies 455, 456
OH cooling
    in PDRs 210
    in shocked gas 183, 184
OH formation
    by associative detachment 31
    by radiative association 24
OH in AGB stars xii, 338, 339, 341, 343
OH in AGN tori 470, 474, 475, 478–481
OH in diffuse clouds 84
OH in GMC translucent clumps 106
OH in interstellar gas phase schemes 63, 64
OH in PDRs 210, 211, 216
OH in PNe and PPNe 359, 363, 364
OH in shocks 179, 180, 184–189
OH in supernova ejecta 31, 32
OH in translucent clouds 81
OH in wind–dense core boundary layers 121–124
OH in XDRs 493, 497
OH variability in Miras 269
OH in YSO winds 163
OH masers xii, 285, 313–315, 321, 328, 443, 444, 456–459
OH megamaser–IR intensity correlation 458, 459
OH photodissociation threshold 57
OH production by surface chemistry 66
OH structure and spectrum 5, 473

OH$^+$ role in YSO wind chemistry 167, 168
OH/IR stars 255, 285
Olivines 304–307, 397
Omicron Ceti 316
ON1 ultracompact HII region 131
Opacity in YSO winds 166, 172
Ophiuchus cloud 84
Optical absorption observations of interstellar molecules 53, 77, 79, 80
Optical emission lines
  from AGNs 438, 439
  from Herbig–Haro objects 221, 223, 228, 229
  from supernovae 419, 422
Optical light curve of a nova 396–398
Optical line emission from PNe and PNNe 364
Optically violent variables (OVVs) 441
Orion GMC 84
Orion star forming region 125, 133, 205, 207, 208
Orthosilicates in dust formation 301–307
Ortho states of H$_2$ 3, 4, 471
Overshoot of convection in stars 244, 254
Owens Valley Radio Observatory 132, 446, 447
Oxygen chemistry
  in interstellar clouds 62, 63
  in PDRs 210, 211, 216
  in shocks 179, 180, 184, 185, 187, 188
Oxygen rich stars, definition 266

P1159 stars 258
PAHs, see polycyclic aromatic hydrocarbons
Para states of H$_2$ 3, 4, 471
Parkes Tidbinbilla interferometer 460
Parsamyan 18 205
PDRs, see photon dominated regions
Perseus A 451, 484
Perseus cloud complex 83
Pg 0838 + 77 450
Photodissociation 56, 57, 66
Photodissociation in basic interstellar gas phase schemes 63
Photodissociation regions, see photon dominated regions

Photoelectric emission from grains and large molecules as a heating mechanism 209
Photoevaporation 203
Photoionization and ionization structure of interstellar clouds 89, 106
Photoionization and photodissociation in YSO winds 166, 167, 172
Photoionization in PDRs 210–213, 215
Photon dominated regions (PDRs) xii, 66, 67, 85, 88, 165, 180, 201–216, 255, 332, 357, 365, 366, 470, 471, 481, 497
Photospheres of AGB stars, typical conditions in 270
PKS 0528-250 480
PKS 1409-65, see Circinus galaxy
Planetary nebula central stars 208, 254, 258
Planetary nebulae (PNe) and protoplanetary nebulae (PPNe) xii, 205, 208, 211, 242, 255, 257, 336, 344, 347–367
Polycyclic aromatic carbon clusters 380, 381
Polycyclic aromatic hydrocarbons (PAHs) 6, 86, 87, 89, 90, 209, 272–275, 278–281, 352, 367, 375, 400, 408–410, 483
Population I stars 43, 47
Population II stars 46, 47
Population III stars 41–44, 47
Post-AGB stellar evolution 254–260
Potential curves 2
Predissociation 406
Proton exchange reaction 2
Pulsations of evolved stars 268, 269, 271, 278, 313, 317, 318, 322, 323, 327, 332, 254, 376, see also Miras and long-period variable stars

Q 0013-004 480
QSOs 15, 16, 42, 47, 49, 50, 437, 439–441, 443, 450–453, 469, 480, 481, 500, 501
  molecular absorption seen against 50
Quantum defect theory 26
Quantum theory of reactions 15, 16
Quasars 440, 441, 451–453, 469, 470, 481
QV Vul 409

R Coronae Borealis stars (R Cr B stars)
    371, 375–379, 381–383, 394
R Cr B 375, 376
R Crt 327
R Leo 316
Radiative association x, 11–32, 39, 41,
    48, 55, 56, 62, 162, 358, 403,
    423–425, 428, 429
  indirect mechanism 12, 21–25
  stimulated 11, 17, 18, 20, 30, 41
Radiative attachment 25, 28, 40, 48,
    162, 187, 358, 402, 423, 426,
    429, 472, 520
Radiative charge transfer 19
Radiative transfer in masers 320–323
Radio continuum emission from starburst galaxies 438, 441, 445
Radio galaxies 437, 440, 441, 451, 452,
    469, 470, 481
Rate coefficient, definition 55
RCW 103 499
Reaction and collision rates, experimental and theoretical studies of
    x, xi, 11–32, 57, 58, 62, 70,
    71, 78, 184, 207, 279, 419,
    420, 424, 426–430, 480
Rearrangement collisions 57, 58
Recombination era in early Universe 29,
    38–41
Red giant stars 239, 240, 288
Reflection nebulae 204, 205
Reionization in the early Universe 42,
    44, 46, 49
Ring nebulae 372
ROSAT 173
Rosette Molecular Cloud (RMC) 102–108
Rotation of molecules 3–9
Rotation of stars, effect on evolution 244
Rotational excitation in shocks 181–183,
    186, 195, see also H_2 infrared
    emission from shocks
RV Tauri stars 353
RX Boo 327
RY Sag 375, 376

S 140 206
S Persei 315, 316, 326
S Scuti 253
S star definition 241
S stars evolving from C stars 249
S^- formation by radiative attachment
    429, 430

S-process nucleosynthesis 252, 254, 523
Saturation of masers 322
Selection rules for radiative transitions 6,
    9
Semiclassical theory of reactions 14, 15
SEST 444
Seyfert galaxies 438–441, 448–450, 453–456, 459, 460, 469, 479, 482,
    483, 489, 500, 501, 507, 512
Sgr A* 499
Sgr B2 87, 89
Shell burning in evolved stars 237–240,
    245–254, 258, 286, 522
Shocks xii, 48, 49, 51, 63, 85, 109, 110,
    118, 119, 129, 133, 160, 162,
    172, 179–198, 221–231, 489,
    497, 499, 500, 509
  in collisions of winds of binary stars 373,
    382, 383
  in mass loss from AGB stars 269,
    271, 276–280, 315, 318, 332,
    355, 521
  in nova ejecta 408
  in PNe and PPNe 348, 349, 351,
    356, 357, 361, 366
  in Wolf–Rayet winds 373, 382–384
  ionization upstream and downstream
    by radiation emitted in cooling 187, 188, 190, 228, 229
  radiation from 194, 195, 221–230,
    see also individual species
Si^- formation by radiative attachment 426
Si_2 formation by radiative association 32
SiC and SiC grains in AGB stars 267,
    269, 275, 280, 281, 286
SiC in novae 397, 400
SiC_2 and dust 393
SiC_2 in AGB stars 338
SiH_4 in AGB stars 333
Silicon chemistry in PDRs 213, 214, 216
SiO condensation to form dust 291–294,
    301
SiO cooling in supernova ejecta 122
SiO emission from shocked regions near
    YSOs 197
SiO formation
  by associative detachment 426
  by radiative association 22, 31, 425
SiO in AGB stars xii, 333, 338, 354,
    see also SiO masers
SiO in collapsing regions of high mass
    star formation 132
SiO in novae 397, 400, 411

SiO in supernovae 24, 30, 31, 416–418, 422, 425–428, 430–432
SiO masers xii, 285, 313, 315–325, 332, 338
SiO^+ formation by radiative association 25, 31, 425
SiO_2 and dust 393, 357, 411
SiS in AGB stars 333, 338, 341
SN1987A 24, 30–32, 415–433
SO formation
    by associative detachment 429
    by radiative association 429
SO in AGB stars 338, 341
SO in dark clouds and bistability 73
SO in supernovae 429, 430
SO in translucent clouds 81–83
SO^+ in translucent clouds 82
SO_2 in AGB stars 338, 341, 342
SO_2 in dark clouds 90
SO_2 in translucent clouds 82, 83
Sobolev optical depth 321
Solar wind and cosmic rays 174
Sonic point or surface in a flow 144, 150
Spin 3
Spin–orbit interactions 4, 5
Star formation xi, 41–47, 84, 91, 101–133, 161, 173
Star formation efficiency 444
Starburst galaxies 179, 201, 213, 437–460, 469, 470, 483, 496, 497, 500, 501
Stars, central temperatures of 240
Stellar evolution 237–260, 285, 286, 522, 523
Storage rings 26
Stretching mode, definition 6
Sulfur chemistry
    in interstellar clouds 64, 65
    in PDRs 211–214, 216
    in shocks 179, 180, 185
Sunspots 223
Supergiant stars 260, 353
Supernova remnants 497, 499, 500
    in starburst galaxies 438, 442, 445, 500
    in ultraluminous infrared galaxies 445
Supernova types, defined 415
Supernovae xiii, 11, 24, 30–32, 46, 101, 165, 179, 238, 239, 266, 415–433, 485, 523

SVS 13 163, 165
Symmetric top molecules 7–9

T-Tauri stars 131, 145, 163, 169, 170–172, 221, see also Young Stellar Objects
Tauras cloud complex 83, 84
Tauras–Auriga complex 108, 109
Temperature, definition of brightness 314
Thermal balance in an XDR 490, 491, 509–511
Thermal pulses in stellar evolution 240, 145–251, 253, 254, 258, 286
Thermonuclear runaway in novae 395, 396
TiC in dust formation 280, 281
Tidal interactions of galaxies, see mergers and tidal interactions of galaxies
Time-dependent interstellar cloud chemistry models 66, 67, 71, 72, 84, 106–108, 110–118, 130, 131
Time-dependent and position-dependent interstellar cloud chemistry models 128–130
Timescales, chemical 66, 67, 112, 113
^{44}Ti in supernovae 419
TiO in AGB stars 241
TiO in dust formation 294, 301
TMC-1 75, 83, 84, 90, 102–119, 123, 129, 333
Tops, see asymmetric top molecules and symmetric top molecules
Tori in AGNs 439–442, 453–456, 459, 460, 469–485, 501, 502, 507–514
Translucent interstellar clouds, see interstellar clouds, translucent
Troilite, see FeS in dust formation
Turbulent mixing, transport and heating 63, 72, 86, 111, 112, 118–125, 221, 230, 231, see also boundary layers
Turbulent support of clouds 101–108
TX Cam 324, 339, 340–342
TXFS 2226-184 508

U Her 327
UKIRT 267
Ultracompact HII regions (UCHIIs) 131–133
Ultraluminous infrared galaxies 437, 439, 443–447

Ultraviolet background, *see* interstellar ultraviolet background
Ultraviolet radiation induced by cosmic rays, *see* cosmic ray induced ultraviolet radiation field
Ultraviolet spectra of novae 393, 396, 399

V 348 Sgr 376, 377
V 705 Cas 399, 408–410
V 838 Her 409
V 842 Cen 397, 398, 405, 409
V 1500 Cys 400
V 3795 Sgr 377
V 4169 Sgr 398
V Cra 377
Very low-mass stars, definition 244
Vibrational excitation in shocks 186, 195, *see also* H_2 infrared emission from shocks
Vibrations of molecules 2, 6
Visual extinction
    definition 55, 104
    relationship to column density 68, 75, 76, 104
VLA 442, 455, 460, 508
VLBA 324, 325, 456
VLBI 314, 315, 328, 437, 441, 445, 455, 458, 507, 512
VO in AGB stars 241
Voids in the early Universe 48, 49
VX Sgr 319
VZ Sgr 377

W3 (OH) 132
W51 IRS 2 132
W52 e2 132
Wardle instability 194, 195, 197, 229
Waves, role in supporting clouds 101–108
Werner bands of H_2 202–204, 227, 401, 471
White dwarfs 238, 240, 242, 245, 257, 258, 350, 395–397
Wind driving mechanisms xii, 141–164, 242, 243, 249, 251–253, 257, 268, 269, 271, 276–278, 317, 318

Winds
    effects on stellar evolution 242–244, 249, 251, 252, 260
    of AGB stars xii, 11, 242, 243, 249, 251–253, 265–281, 285–310
    of AGNs, *see* AGN winds
    of YSOs xi, 11, 101, 102, 109–112, 115, 118–121, 126, 127, 161–175, 179, 197, *see also* X-wind model
Wolf–Rayet stars and their nebulae 242, 258, 260, 371–384, 394
WR 104 373, 381, 384
WR 112 381
WR 121 384
WR 140 373

XDRs, *see* X-ray dominated regions
X-ray binaries 497, 495
X-ray dominated regions (XDRs) xiii, 173, 174, 469–472, 480–484, 485, 489–502, 509–511
X-ray emission from Wolf-Rayet stars 373
X-ray shadowing by clouds 76
X-rays from AGNs 439, 454, 483, *see also* AGN radiation fields and spectra
X-rays from YSOs 162, 172–175
X-rays in supernovae 419
X-wind model 141–161, 164, 170–172, 174, 175

YO 241
Young Stellar Objects (YSOs) 109–112, 115, 118–121, 141–175, *see also* X-wind model and winds of YSOs

Zeta Oph cloud 70, 75, 77, 78, 86
Zeta Per cloud 70, 75
ZrO 241

1E 1740.7-2942 499
1Zw 1 450
3C 48 451
3C 84, *see* Perseus A
3C 405, *see* Cygnus A
III Zw 35 459
4C 12.50 451
4C 31.40 456, 457